D1387731

0521 258 448 0 058 2E

THE CAMBRIDGE
ILLUSTRATED HISTORY OF THE
WORLD'S SCIENCE

THE CAMBRIDGE ILLUSTRATED HISTORY OF THE WORLD'S SCIENCE

Colin A. Ronan

CAMBRIDGE UNIVERSITY PRESS

NEWNES BOOKS

Published by
The Press Syndicate of the University of Cambridge
The Pitt Building, Trumpington Street, Cambridge CB2 1RP
32 East 57th Street, New York, NY 10022, USA
296 Beaconsfield Parade, Middle Park, Melbourne 3206, Australia
and
Newnes Books, 84-88 The Centre, Feltham, Middlesex, TW13 4BH
and distributed for them by The Hamlyn Publishing Group Limited
Rushden, Northants, England

First published 1983

ISBN 0 600 38423 3 Newnes Books
ISBN 0 521 25844 8 Cambridge University Press

Filmset in Bembo by Input Typesetting Limited
Durnsford Road, London SW19 8DR

Printed in Great Britain by The Pitman Press, Bath

Contents

Introduction

Wars we have sung. The blind, blood-boltered kings
Move with an epic music to their thrones.
Have you no song, then, of that nobler war?
Of those who strove for light, but could not dream
Even of this victory that they helped to win,
Silent discoverers, lonely pioneers,
Prisoners and exiles, martyrs of the truth
Who handed on the fire from age to age. . .

Alfred Noyes, *The Torchbearers*, Prologue:
The Observatory.

At the end of 1917, during the last twelve months of what has been called 'the bloodiest war in history', the poet Alfred Noyes and two American astronomers drove up the newly laid road to the peak of Mount Hamilton in California. Their aim was to test a giant telescope which had just been completed. With a power to penetrate space far further than ever before, this vast instrument was the practical result of the imagination and perseverance of George Ellery Hale, one of Noyes's two companions. The telescope had cost a fortune, and that night was to be the moment of truth: would it fulfil its theoretical promise and open out great vistas of the universe hitherto unknown to man, or would it prove no more than an expensive disappointment? As things turned out it was a resounding success. All the sweat and toil of five years spent in grinding the giant 100-inch mirror, all the inventiveness and effort put into designing and building the huge but delicate engineering structure to carry the mirror, had proved their worth. Here was a tool to help mankind forge ahead into a new age of scientific discovery.

Noyes was very moved by the occasion. The quiet dedication of the scientists, their cherished hopes and fears, fired his own imagination. And for a moment he glimpsed the wider aspects of their work; he saw what they were striving to achieve set against the vaster canvas of man's intellectual struggle to understand the universe in which he finds himself. This, he realized, was the really great conflict; the war against ignorance that could only be won by an increase of knowledge of the natural world. And, no doubt, as he mused on this, as he thrilled to the sight of distant suns, he also remembered that at that very moment his own countrymen were being slaughtered in the mud and filth of Flanders, sent to suffer and die for no other purpose than to thwart a king's power-drunk ambition for conquest. Little wonder he should ask his question; the tragedy is that it is still as relevant today as it was more than half a century ago.

Certainly the world has learned the true horrors of war, and has begun to try to avoid the ultimate catastrophe, in spite of the aspirations of political ideologists and nationalists who will still turn to war if this is the only way they can achieve their ambitions. Unhappily this is still the mainstay of history. Man's success in subjugating other men is the substance of history books, not men's achievements in thrusting off the cloak of ignorance and superstition. In English there exists no simple history of man's achievements in science, though there are plenty about his struggles to dominate others.

Noyes's plea is echoed here because it is essential to realize that there is a vast gap in general historical literature even today. The history of science is nowhere to be found. This is not to suggest that the history of power politics should be erased from the record, but to emphasize that it is but one aspect of the historical drama and perhaps, in the long run, the least memorable aspect. That it has received more than its share of attention seems only too evident, and today, when science is playing an increasing part in the life of every man, woman and child, it is imperative to know something of its origins and development throughout the civilized world. Only in this way can we see this glorious adventure of the human intellect in anything like its true perspective. For science is something which has appeared in every culture on Earth; it has grown naturally out of man's innate curiosity about the world in which he finds himself. Fear of science is based on ignorance and misunderstanding about its nature and its genesis.

In a book of this size it is plainly impossible to cover in detail all the scientific and technical knowledge of every age and every civilization of world history – and indeed to attempt such a task would require the inclusion of a great deal of information that would be impossible for a reader who has not been trained in science to understand. For this reason a line had to be drawn on what to include and what to leave out. The first, and vital, decision was to cover what may generally be called 'pure science' rather than technology; that is, to include those scientific ideas that have been developed in the interests of 'knowledge' rather than those conceived with some practical application specifically in mind. Obviously it is as impossible to draw a hard and fast line between the two when discussing the science of ancient Egypt as it is when discussing modern chemistry; but, in general, the policy has been to concentrate on those developments in science that have brought about a conceptual change, rather than a technical one. Sometimes, too, the question of whether or not a field comes within this definition of 'science' differs from age to age. Medicine is a case in point here; and it has been included in many early civilizations since it provides us with a valuable clue to the state of biological knowledge of that time, whereas in the twentieth century it has developed into a field of its own more akin to a technology than a conceptual science. The third decision on planning the present volume was to attempt to give equal coverage to those sciences superseded by the Scientific Revolution as to the development of 'modern' science. This decision partly reflects the relatively familiar nature of much of the history of science since the Renaissance, while it also seems desirable to give the earlier civilizations credit for much creative, bold and imaginative work, the value of much of which seems only recently to have been recognized.

Such then is the aim of this book; to take an overview of the development of science and scientific thought the world over from early times until now. It will prove an exciting adventure, partly because the thrill of discovery is exciting and partly because the story will carry us into stimulating times remote from our own when the whole outlook was different, often based on assumptions unfamiliar to modern man and frequently part of a culture disparate from our own. We shall have to divest ourselves of accepted ideas and use our imagination to project ourselves back into another age. Sometimes this may not be easy, but it is always worthwhile for its surprises as well as for the fresh understanding it brings.

Chapter One

The Origins of Science

Science has proved a vast intellectual adventure. To engage in it requires a vivid creative imagination, tempered by firm discipline based on a hard core of observational evidence; and science has attracted some of the best minds in every civilization that has developed to a stage when it can tackle the challenge of Nature in this way. For science is not the mere collecting of facts – though this is necessary; it is a system of logical correlations of those facts cementing together a hypothesis or body of theory. This theory is itself tempered by the general outlook of the times in which it is formulated. The theory must be sound enough to attract minds trained in logical thought, and at the same time be open-ended enough to leave room for development and adjustment in the light of later evidence. Such a theory, sometimes known as a paradigm, will change from time to time for a whole host of reasons, as we shall see. To the extent that these changes are occasioned by ever more complex experience, science is a growing and expanding body of knowledge; but when they are brought about by religious, philosophical, social or economic reasons, the history of science engages with all the fluctuations of more general history.

It is impossible to discuss the history or theory of science without coming face to face with magic. This was a way of looking at the world which was a complex amalgam of spiritism and arcane knowledge. To anyone not seduced into thinking that modern science is merely miracle-working writ large, the very mention of magic in this context may seem strange or even unacceptable. Yet what seem on the surface to be utterly disparate approaches to Nature do in fact contain many common factors. The magical view was a legitimate way of expressing a synthesis of the natural world and man's relationship to it. When, in a primitive society, the magician, shaman or witch-doctor performs a rain-making act, he expresses his realization of a link between rain and the growth of crops, between one aspect of Nature and another, and his appreciation that man's survival depends on the behaviour of the natural world. He has an understanding of some connection between man and the world around him, some primitive comprehension that, given the right procedure, man can control the powers of Nature and set them to work for his own good.

What were the essential beliefs of magic, as it was found among the most ancient of peoples and as it persists in some more primitive cultures today? Magic gave expression to what was, by and large, an animistic view of Nature. The world was populated with and controlled by spirits and hidden spirit forces, residing perhaps in animals, or trees, or in the sea and wind, and the magician's task was to bend these forces to his purpose, to make the spirits co-operate. He made invocations, cast spells and prepared potions, because he saw a world of affinities and sympathies. This outlook might lead to sympathetic or imitative magic, where men might eat the flesh of an animal in order to absorb some of its qualities, or dress up like the animals and enact their capture and death so that their hunt might be blessed with success. Drawing or painting pictures of animals or making figurines of them extracted power from them, so weakening them and helping in their capture. The magical world was a world of relationships rather than independent objects, and was based on man's own interrelations with the life and conditions he found around him in a world where forces were personified and everything had a specific influence.

Illustrations page 25

The magician might have a very subtle insight into the general relationships of nature, and his acts of manipulation, however mistaken they sometimes were, led, as they were bound to do, to a certain empirical knowledge of various substances. The ingredients of potions, for instance, may have been chosen originally for their magical associations, but gradually, their success or failure would show which were genuinely efficacious and which were not. Slowly, a body of practical knowledge would be collected, used and expounded in the light of experience such that by degrees the magician came to be the first in the lineage of experimental investigators, and the remote ancestor of the modern scientist. And when, in due course, man turned to taking more down-to-earth steps for his wellbeing by, say, building irrigation schemes, he began, consciously or subconsciously, the process of relegating the powers of the spirit world to a role more of co-operation than of direct intervention. For thousands of years the two approaches co-existed side by side in a state of more or less easy truce; and as man's techniques of controlling Nature grew more powerful, the spirit world was forced to redefine its role.

When it was thought that the world was one of affinities, controlled by spirits and animistic forces, the magical outlook was an appropriate way of correlating the phenomena of the natural world. But as society developed in the ancient Middle East, an interest in the details of natural phenomena gave rise to a more 'solid' form of knowledge. Meanwhile, magic slowly became degraded: its mystical qualities were abused either for private ends, giving rise to witchcraft, or for public ends, creating a powerful priestly caste and humouring the ignorant and credulous. Such degradation in its turn led the philosophers of ancient Greece to adopt a completely non-magical approach. In doing so they formed the attitude of mind which has remained central to Western scientific culture.

There are those who would deny that there was any genuine science in prehistoric times. To them primitive medicine, prehistoric surgery and technology were all purely practical, with no abstraction of underlying principles. Yet from what has been said of magic it is clear that there was indeed an underlying doctrine and set of basic principles. These stated that the world was peopled not only by a visible collection of human beings, animals, plants and minerals, but also by an invisible world of spirits and spirit forces. These forces could sometimes be seen in action by everyone, as in the case of thunder and lightning, or when there was an earth tremor or a flood. Disease and pestilence among men and animals were taken as evidence of acts by evil spirits. Thus natural phenomena of the physical world were correlated with the spirit world, and procedures were developed for dealing with both worlds. Certainly these basic principles would not be considered scientific today, but in primitive times to postulate such interventions was an act of rationality; they offered an acceptable paradigm to explain the diverse phenomena that man experienced.

While the concept of a divine world operating on the world of Nature was the current view, the priest or the priest-magician had a scientific aspect to his knowledge; he had the knowledge of Nature on the one hand, and the access to the gods on the other. There was no conflict between science and religion; both were interlinking aspects of the real world. In prehistoric and early civilizations science was an amalgam of natural and spiritual explanations. It is described here as science both because it was a rational way of correlating what was observed, and because it contained some nuggets of truth, some observations or explanations that were gradually to be bonded together and one day provide a non-magical view.

Priests, as in ancient Egypt, often acquired power through their role as guardians of scientific knowledge. In most lands, as we shall see, scientific knowledge was closely associated with the calendar and the agricultural year; such knowledge, therefore, meant power over the people by regulations and controls, so some aspects of science – astronomy, for instance – were sometimes closely guarded state secrets. The possession of such knowledge, secret or otherwise, was a mark of high social standing. In some later societies, such as the Greek, this led to great emphasis being placed on the intellectual side of science compared with its more practical (manual) and experimental aspects.

What was the essence of the 'new approach', inklings of which were found, for example, in late Babylonian times? How did it differ from the esoteric and manipulative knowledge that it supplanted? The new synthesis was a rational correlation of experience, a scheme of explaining natural phenomena without recourse to any occult or supernatural elements. It eschewed the intervention of divine beings: thunder was not the divine anger of Marduk made manifest, instead it was the result of some 'blind force', operating without any supernatural implications. There were gods – the new view did not necessarily espouse atheism, although its practitioners were sometimes accused of it – but the deity or deities were kept in their place. As

Galileo was fond of quoting, millennia later, 'the Bible shows the way to go to Heaven, not the way the heavens go'. Natural events were taken to be the result of natural causes. General inexorable patterns of behaviour were sought, true for the past, present and future, subject not to the whims of capricious spirits but due only to the way the world is constructed. In itself this scientific view is not necessarily more logical than the magical viewpoint; it is just a different way of looking at Nature, and is based on different premises. Yet the scientific outlook has provided a far more powerful means of understanding, predicting and controlling the world than any path trodden by the magician.

The struggle to comprehend the strange world in which we live is a noble one. It is a continuing struggle. Our present scientific synthesis is another step along the road to a more comprehensive picture, but it is not the final one. Our current paradigms will one day give place to new and improved bodies of theory, just as those we now accept replaced the paradigms before them. For example, whereas it was once universally agreed by Western scientific philosophers that the stars and planets were fixed to crystal spheres centred on the Earth – a belief which was pregnant enough with puzzles in celestial motion to tax the most brilliant minds – this was replaced by a concept of motion in empty space, which again presented new problems to challenge the intellect. Now we have moved on to motion guided by universal gravitation in a relativistic space-time universe; this represents the present pinnacle of modern cosmological thought. It is greatly superior in many ways to the doctrine of the spheres, but it is not the last word. A new and more widely embracing paradigm will doubtless come to replace it.

This new paradigm will not contain magic in its old form because magic is now discredited. But of course there are some who would seek a new paradigm with magical overtones, of association, of interrelationship, even of spiritism. They believe there are aspects of the world which lie beyond or outside the purview of modern science, yet they try to give them scientific explanations. They talk of forces or influences which cannot be defined, either because the idea of them has not been explored properly or because their existence is a matter of faith rather than a question of reason. Such suggestions are generally rejected by science today, partly because present paradigms are largely unassailed and still productive, but more importantly because no alternative theory so far proposed is broad enough, nor has it generated new ideas that can be tested independently and proved to be either true or false by the touchstone of experiment. Modern quasi-magical theories have so far failed, not because they do not fit in with modern theories but because they are not the result of sufficiently disciplined investigation, intellectual or experimental. For science today is nothing if not a severe practical and mental discipline in which unprovable hypotheses survive only if they prove extremely fertile.

Early Science

The flame of science, as we have described it, first glimmered some ten thousand or more years ago in the Middle East. It began when man started to gather knowledge, mainly but not only for day-to-day living. Details of plants were compiled, including some that were neither of use medicinally nor for food, but were described purely because of their intrinsic interest. Animals were captured and catalogued, those which could not be domesticated as well as those which could. And as time progressed the requirements of living, too, brought extra knowledge: means were found for lifting heavy loads; rollers, pulleys and the wheel were devised, agricultural techniques developed, hides were tanned, weaving invented, pottery manufactured and some materials were smelted. Sometimes great ingenuity was shown; the first use in Central America of the cassava plant is a prime example. The plant is cultivated for its tuberous roots which are used for making flour, bread, tapioca, a laundry starch and an alcoholic drink, yet in the natural state the tubers are poisonous. The poison – a form of cyanide – is removed by a combination of grating, squeezing and heating the tubers. But how did the Indians of Central America discover this? Recognizing that the tubers were poisonous might not be difficult, but the removal of the poison and the recognition that what remained was not only edible but could be used as a staple diet shows an investigative logic at work. This investigative logic was built first into patterns of straightforward material relationships, and then, later, into more general ideas or theories.

Illustration page 26

In prehistoric times man discovered the use of herbal drugs and sometimes added other materials to his primitive pharmacopoeias, while shepherds and farmers who kept livestock – animals were first domesticated in about 7000 BC – must have learned much about the way animals reproduce and something of their diseases, of curing ailments, and by such techniques as setting a broken leg. The ministrations of the midwife would have been one of the earliest medical services and a medical calling one of the earliest professions, linked though it undoubtedly was with religious observance.

The primitive doctor would apply herbal and animal remedies, but his ministrations would not end there. He would also use charms for driving away the evil spirits visiting his patient, and practise divination. This last could take a number of forms, but it might involve dosing an animal with a powerful concoction to see whether or not it would survive; this would give an indication whether the good spirits were with him and whether his magic – and his herbs – were going to be successful or should be changed. He might try to drive out the illness into some other living creature – an early example of the scapegoat principle. But whatever means he used, he would be bound to gather experience and compile a body of treatment.

One of the most surprising aspects of early health treatment was the surgical practice of trepanning, that is drilling a hole in the skull. Why it was done we can only guess; perhaps to relieve pressure caused by concussion or, possibly, to allow the escape of evil spirits. But whatever the reason, this operation was carried out on living people, and one can only assume that the patient must have been given either some kind of herbal anaesthetic or a hefty dose of alcohol, for drilling bone with a stone drill must have been a protracted business. Trepanning, coupled with other simple surgery, as well as sewing up wounds, would lead to some knowledge about the inside of the body.

Man's knowledge of what we would now call the biological sciences was nevertheless slow to form into a science. For a long time he could only gather disconnected facts, and here and there amass detailed evidence, but fitting all this together into a coherent scheme of knowledge was a different matter. There were so many variations, even in animals or plants of the same kind, that it was difficult to know how to place things together. It was hard to say for certain what depended on what. But things were different when it came to the world of physics. Here it was much easier to observe cause and effect, and possible to find an underlying idea that could be applied in a wide variety of cases. The idea of number is an example.

That number can be applied to a wide variety of things – to everything – must have been realized very early. Man is an individual, a one; he has one mouth, one nose, one head, one body. He also has two eyes, two ears, two arms, two legs. There are two sexes, they were a duality. So were qualities such as hot and cold, wet and dry, dark and light. A family – a man, his wife and a child – made a trinity, a threesome. A three-legged stool also displayed threeness. The hand, with its thumb and four fingers was a unity, a one (the hand, the thumb), but the fingers were a four, a quartet of ones. All together, thumb and fingers made a five – a four and a one. So the foundations of arithmetic arrived.

First, then, there was the idea of counting: an abstract idea that could be thought about without any material objects present. You could think of oneness, or twoness, or whatever number you liked. And what is more, such 'numbers' seemed to have properties of their own. The number one entered into them all; it was universal. Two also entered into many numbers, into the whole class of 'even' numbers. But there were also other 'odd' numbers, some of which were not divisible at all, except by one. These seemed to be special numbers, with an individuality that was unique, a significance apparently mysterious and potent, and it was not long before there grew up a kind of number magic, a mystic numerology.

The useful, and powerful, technique of arithmetic developed side by side with numerology, and numbers soon rose well above those which could be counted on fingers and toes. Before the days of writing this presented certain difficulties; it was no great problem to cut as many notches in a piece of wood as were needed, but it was tedious to have to count right through them each time one wanted to know the total.

The solution was to use groups. Groups of five could readily be recognized; sets of five notches followed by a space before the next notch was cut were easy to recognize and did away with the need to count right through. And, of course, five was only one possible grouping; there were others. As we shall see later, the Mayas chose twenty, but the most frequently used was the grouping of ten, based on fingers or toes.

Once the grouping or 'base' for a number system had been devised, it allowed the easier development of the four main divisions of arithmetic – addition, subtraction, multiplication and division. In particular, it helped with the whole concept of subtraction, for once there was a base, it might be more convenient to express the fact that one had, say, a number which was less than twenty, or thirty, or some other group multiple, by a small amount, rather than specify the number by counting up from one. Thus 29 is (30 − 1), 47 is (50 − 3), and so on. With groups of notches, or twigs in bundles, this subtractive method of counting could be very convenient. And, of course, counting the complete groups would soon lead from addition to multiplication, which is essentially an extension of addition, with the advantages of being easier and quicker when the numbers are large.

Illustration page 26

Astronomy may have been the first separate study to incorporate application of mathematics. To use the heavens as a clock or a calendar required numbers. And to measure the distances between the Moon or the stars and the horizon also involved numbers. But there was an extra dimension to this problem. If a person wanted to know how far above the horizon the Moon might be, he had to measure a distance he could not reach. He would have been forced to find this distance in a new way, either by stretching out his arm and seeing how many fingers' breadth he could fit in between Moon and horizon, or by holding a thread between his outstretched hands and determining the distance that way. He had to stretch his arms out to the full to do this; if not he would get a different answer. The measurement was therefore one which was different from an ordinary length: and was the first step towards measuring an angle, a type of measurement that was later to prove of great importance.

Illustration page 48

This may all sound very fine, but is to a great extent conjecture. We do not know when man measured angles though angles were certainly being measured in early Mesopotamia, and were clearly known when Stonehenge was built in the second millennium BC. The positions of the Moon and stars were very important to prehistoric man, and determining these positions meant measuring angles. Today, so many people live in an urban environment and seldom if ever look up at the night sky; the Moon and stars make no impact at all. But in the country, away from artificial light, the situation is quite different, and in the Middle East the starry sky is a particularly remarkable and unmistakable feature. There can be no doubt about prehistoric man looking at the night sky; he must have done so with awe as well as curiosity.

The changing pageant of the heavens was, then, something that was bound to captivate the mind and imagination of early man. The slow majestic turning of the sky throughout the night, carrying the stars across from one side of the horizon to the other, was a sight well worth watching. So, too, was the movement of the Moon, which not only rose and set like the stars, but also changed its shape, waxing from a thin crescent at the beginning of the month until it was a great globe in the sky, and then waning again. It was also a nearly ideal timekeeper, taking no more than 29½ days to complete its cycle of phases. All early calendars were based on the Moon.

The stars themselves moved as a whole across the sky as if the whole dome of heaven turned, and such patterns as could be recognized remained the same night after night, year in and year out. They were patterns that one community might identify in a different way from another, but the underlying principle behind the choice was the same; to gather the stars into groups depicting animals, or heroes and heroines, or the gods themselves. There were also some wandering stars which appeared from time to time in the sky – what we call planets (a word derived from the Greek for 'wanderer'). Their apparently erratic behaviour must have been a source of wonder to the prehistoric astronomer and their motions were to act as a powerful stimulant to scientific research.

The heavens, then, presented a continually varying display, an amalgam of regularity and apparent surprise, for not only the planets seemed capricious but so too did other phenomena in the skies – the appearance of stars which seemed to hurtle to Earth ('shooting stars'), the unheralded arrival of bright flaming stars (comets), as well as rainbows and haloes round Sun and Moon. No man could ignore them or fail to experience their fascination. Indeed the heavens have always appealed to the imagination, so that man's changing beliefs about them, the growth of his ideas of the nature of the skies, form a thread guiding us through the labyrinth of cultural differences in various civilizations. Furthermore, ideas about the heavens act like a mirror, reflecting man's growing scientific attitudes, and will be particularly useful to us as our story unfolds.

Egypt

Between 4000 and 3000 BC Neolithic or New Stone Age culture became well established in Mesopotamia – now the west and south-west of Iraq – and in Egypt. Here the first organized cities and states were established, but the two areas gave birth to somewhat different civilizations.

Egypt was a country centred on the Nile, with a hostile environment to its south and on its eastern and western borders. Indeed, it has been likened to an island, limited in the north by the sea and on its other borders by desert, and in many ways the Egyptian civilization displayed a certain insularity. It was conservative and inward-looking; by and large it was not interested in expansion or in the conquest of other lands. To an ancient Egyptian, Egypt was a self-contained universe: it had its independent gods and its unique way of life. The Egyptian language and hieroglyphic writing went hand-in-hand; the very system of hieroglyphics was insular, unsuitable for expressing any other language, and in diplomatic correspondence with other countries a different system of writing had to be used. Effectively, the ancient Egyptians lived in a cultural isolation.

But if isolation was the keynote of ancient Egypt, its civilization was nevertheless magnificent; it was looked on with envy by those outside its borders and only the surrounding deserts prevented it becoming the victim of jealous neighbours. Some nomads did indeed come in and settle in the sparsely populated area of the Delta, but they did not disturb the basically peaceable nature of the country which was essentially a land of farmers and scribes.

The annual inundation of the Nile, which usually occurred in July, was the foundation of Egyptian life. There was a well-ordered system of irrigation at all times, and particular care was taken of the waters available at the annual flooding. Good crops were the rule – often three a year – and there were fine herds of cattle, mostly pastured in the Delta area. There was no question of the Egyptians scraping a living from a hostile or barren land, though methods of agriculture were primitive and conservative. There was also a high standard of gardening resulting in a good deal of garden produce.

Illustration page 28

Egypt became a united kingdom in the fourth millennium BC and except for two periods of instability it remained united for well over two thousand years. The main periods of unified rule are known as the Old Kingdom, Middle Kingdom and New Kingdom; with an early dynastic period and the gaps of instability, they ran as follows:

		Approximate dates
Early dynastic period	Dynasties I and II	3100 to 2686 BC
Old Kingdom	Dynasties III–VI	2686 to 2160 BC
Period of instability	Dynasties VII–X	2160 to 2040 BC
Middle Kingdom	Dynasties XI and XII	2040 to 1786 BC
Period of instability	Dynasties XIII–XVII	1786 to 1567 BC
New Kingdom	Dynasties XVIII–XX	1567 to 1085 BC
Decay and foreign domination	Dynasties XXI–XXXI	1085 to 332 BC

The rulers were the pharaohs, whose despotism was tempered by ideals of responsibility towards the ordinary people and, considering the early times in which they lived, they did see to it that their subjects had reasonably comfortable and happy lives, governed by the rule of law, which seems to have been generally regarded as just.

Egypt had a large and efficient administration. Much of this seems to have been centred on the great temple foundations, though from time to time the pharoahs themselves proved great administrators, most notably at Thebes during the XIIth dynasty with Amenemhet and his successors. The administration standardized weights and measures, while its employees, the scribes, wrote either in hieroglyphics or in the more cursive hieratic script, (so called because *Hieraticos* means 'priestly' in Greek and the scribes were, by and large, clerics). The Egyptians wrote on papyrus, which was made in Egypt at a very early stage; it seems to have been in use before 3500 BC, in Pre-Dynastic times. Made from the pith of the stem of a tall sedge (*Cyperus papyrus*) which was found in abundance along the marshes around the Delta, its manufacture into sheets was simple. It was an ideal material for use in the dry conditions of the Middle East and was later used extensively in Rome. In the damper climate of Europe, papyrus was less stable, but in Egypt it was superior to any of the other writing materials and remained in use there until the ninth century AD. (Incidentally, it is worth noting that the Greeks, who believed the Egyptians to be a nation of immense wisdom, called a strip of papyrus *biblion*, from which our own word 'bible' is derived; our own word 'paper' is derived from papyrus although, in fact, paper is quite different material and was invented by the Chinese not the Egyptians.)

For the spread of paper around the world, see page 271

The Greek overestimate of Egyptian wisdom may have been due, in part at least, to the impression gained by those who visited Egypt and were overawed by the magnificent buildings they found there. However, to be impressed by the grandeur of Egyptian monumental building is understandable enough, for building was one of the Egyptians' greatest forms of expression. The Nile valley is itself a vast quarry and although they had to import all the timber they required either from Libya or Syria, they soon learned to excel in the arts allied to the indigenous materials – they were expert stone-cutters, superb sculptors, they painted well and were master craftsmen in metals, especially gold.

Egyptian monumental building began early, well before the Old Kingdom, although it was not until the Third Dynasty that Imhotep, the principal officer of King Djoser, designed and had built the first really great stone tomb, the Stepped Pyramid at Saqqarah. This and the slightly later Great Pyramid of Khufu, which contains upwards of 2,300,000 limestone blocks each weighing 2.5 tonnes, is an eloquent example not only of Egyptian imagination and building technology but also of their administrative ability in organizing a vast army of builders – 100,000 according to the Greek historian Herodotos. It was certainly a feat keeping so large a workforce free of dysentery, cholera and typhoid in what must have been crowded camps without modern methods of hygiene. Recent research has shown that it appears that diet had much to do with it, for it contained radishes, garlic and onions, which are natural inhibitors of the bacteria involved. This is not, of course, a claim that the Egyptians knew of bacteria but does show that their practical medical knowledge was developed enough for them to be aware of the use of certain plant foods in a situation of this kind.

Illustration page 26

The Egyptian ability to construct vast buildings and statues is not itself science: what we should now call the principles of mechanics were involved, but it seems that there was no basic body of scientific knowledge or theory to which the builders could refer. Their constructional prowess was based on solid practical experience and a flair for structural engineering. Indeed the Egyptians seem essentially to have been a very practical people, more concerned with effective results than with philosophizing about the basic principles involved. In the short term such an attitude brings successes but in the long run it does not encourage either speculation or new ideas. It meant, for instance, that when Akhenaten built his great temple at Karnak in the 1370s BC the techniques used were not substantially different from those used by Khufu some thirteen centuries earlier.

Egyptian Astronomy

The Egyptian lack of interest in philosophical speculation and bias towards the practical can be seen even in astronomy. To them astronomy was the necessary utilitarian basis for their time-keeping, for the Egyptians were more preoccupied than any other early people with the reckoning of time. Perhaps this was a consequence of a large administration which would be bound to busy itself with events occurring on set dates, with taxes due at specific times. But whatever the reason, Egyptian astronomy was not concerned with theories about the Sun and Moon, nor did it contain any specific ideas about the motion of the planets, although the planets were known to wander among the fixed stars. Certainly they had their own specific groupings of stars, but their cataloguing of them was so imprecise that it is not now possible to be certain, except in a couple of instances, about which of their constellations coincide with those of later traditions. Of the northern group of stars, for instance, the only constellation which we can now identify is the Plough or Dipper.

An example of the Egyptian approach can be seen in the picture of the cosmos as drawn in the Greenfield papyrus. In this funerary papyrus of Princess Nesitanebtashu, priestess of Amen-Ra at Thebes in about 970 BC, which is therefore a late post-New Kingdom document, we see a purely symbolic representation of the universe. It is peopled by a host of gods and goddesses. The sky is the body of the goddess Nut, the Earth is represented below by the god Qeb, lying on his side, while the god of the air, Shu, stands in the middle, having helped up Nut to her somewhat uncomfortable position. In some other drawings of this scheme two small boats are shown moving over Nut's body, one carrying the Sun, the other the Moon. Such drawings as these had religious significance, not surprisingly, for the Egyptian astronomers were also priests. But the drawings here bear little relation to the physical appearance of the heavens. They do show that the Egyptians, whose country was long and narrow, tended to think of the heavens above them as a long narrow box – indeed, another drawing of Nut (in the Nineteenth Dynasty cenotaph of Seti I at Abydos) gives Nut just such a very elongated body – but there seems to be no other item of belief about the material

Illustration page 27

universe that we can draw from such pictures.

The reason for this apparent lack of concern with the nature of the physical universe was that the main attention of the priest-astronomers was centred elsewhere; their interest lay in the afterlife. This alone gives point to the drawings of Nut, Shu and Qeb in the Greenfield papyrus. What we have here is a religious cosmogony, a mythological picture of the beginning of the universe. Admittedly, the earliest Egyptians had a rather more physical picture of the beginning of the cosmos: that there was a primeval flood from which a hill emerged, bearing the first living things. But this simple projection back into time of the annual inundation of the Nile gradually gave place to a more elaborate religious description. Known now as the Heliopolitan cosmogony, this explained that the primeval god Atum (originally a local deity located at the ancient city of Onu, known to the Greeks as Heliopolis) spat out the first pair of gods Shu and Tefnut (air and moisture), and it was Shu who separated Heaven (Nut) from the Earth (Qeb). To these five gods were added those of the Osiris myth: the fertility god of the underworld, Osiris himself, his wife Isis, and his brother and murderer Seth, whose wife Nephthys is the 'Hostess of the House'. There is also the falcon god, Horus, the son of Osiris and Isis, whose incarnation was the ruling pharoah. The Osiris myth, the myth of the dead god who begets a son out of the underworld, was originally connected with the cult of the first kings of Abydos, where the death of Osiris was re-enacted annually. Coupled with the Atum (one of the manifestations of the Sun-Creator god) and his four secondary created gods, the Osiris family became the 'Children of Nut' and so the significance of the cosmological picture was otherwordly and of much greater moment to the priest-astronomer than any purely physical description of the universe could be. The priest-astronomers had no urge to enquire into the detailed positions of stars and were, apparently, not moved to make any speculations about their nature; they concentrated on the spiritual, not the physical world. And they were powerful – the priesthood of Amon saw to it that virtually all state records and every stone memorial or sculptured relief of the 'heretic' pharoah Akhenaten (who had abolished the old gods) were eradicated and his memory defiled. The old custodians of learning and their established view of the universe prevailed. To the ordinary Egyptian the whole family tree from Atum to the pharoah, the 'living Horus', was pregnant with the hope of life after death; to the later Greeks it was all part of the wonderful and mysterious knowledge ascribed to the ancient Egyptians.

To the priest-astronomers of ancient Egypt the heavens did, however, have the practical use of time-determination. The constellations were primarily used for finding out the motion of the Sun across the heavens as the year passed. And they devised an eminently satisfactory calendar. It was not astronomically sophisticated but it represents the most advanced civil calendar in ancient times.

Illustration page 27

It was noticed early on that the annual inundation of the Nile coincided with a striking astronomical event. This was the pre-dawn appearance on the eastern horizon of Sirius (known to the Egyptians as Sothis), the brightest star in the sky; it appeared at this time after a long period of invisibility and in the clear skies of Egypt its arrival must have been striking. This pre-dawn or heliacal rising of Sirius became called 'The Opener of the Year' and the civil calendar was pegged to it. Thus the first calendar was made up of twelve months, each of 29 or 30 days, and being thus tied to the Moon's 29½-day cycle of phases, it gave a total of only 354 days. To this an additional or intercalary month was added therefore every three or, sometimes, two years. The calendar therefore marched in step – or nearly in step – with the arrival of the Opener of the Year and the festivals associated with it. All this was devised in the Pre-Dynastic period, and once the country had a more rigid administrative system, there was a demand for a more precise seasonal calendar that did not have months of varying length, nor the addition of an extra month at irregular intervals.

As a result, the seasonal year, the period from one summer solstice to the next was measured and a value of 365 days obtained. The measurement itself was possibly done by using a vertical rod in the ground and observing the changing length of the shadow at noon each day. As the seasons pass so the Sun climbs higher in the sky and the noonday shadow shortens until at midsummer it is at its minimum, then as the Sun sinks lower again, the noonday shadow lengthens until it reaches its greatest length at midwinter. This phenomenon was well-known in every settled civilization, and the fact that some of the monumental buildings were aligned with the heavens, for example with the rising of Sirius at the Egyptian new year or with midsummer sunrise, makes it clear that the Egyptians were aware of such things. On the other hand a day count may have been taken between one pre-dawn or heliacal rising of Sirius and the next. After a few years the 365-day period would have been readily found.

In drawing up the new calendar the administrators retained the three Egyptian seasons which custom had long ago established – the Inundation, the Emergence and the Harvest – but the four months in each of them were each given a length of 30 days. The calendar also operated a series of ten-day weeks. This gave a total of 360 days to which was added a small intercalary period of five days. It was practical and administratively sound, and was the first calendar based on the seasons to be produced by civilized man.

As with other civilizations, the earliest settlers in Egypt had used a lunar calendar, and religious festivals were still determined on this basis even when the 365-day civil calendar was adopted, (just as we do today when we determine Easter). However, although the Egyptian civil calendar was an administrative triumph, and was put into use, probably between 2937 and 2821 BC, at a time when the lunar and civil (solar) calendars coincided, it was not exactly correct. The real year contains 365¼ days, and gradually the new civil calendar became out of step with the seasons, the error amounting to no less than 50 days after a couple of centuries. The easiest way to rectify this would have been to add 50 days after the 200-year period but this was not done – perhaps, it has been suggested, because the scribes did not favour it – so a new lunar year was devised to run alongside the civil calendar and in step with it.

From about 2500 BC Egypt therefore had three calendars, the civil one, of 365 days, its lunar counterpart and the original lunar year, kept in its proper place by the heliacal risings of Sirius. The civil calendar came back in step with Sirius every 1460 years (because it is one whole day out in every four years and $4 \times 365 = 1460$): this was known as the Sothic cycle. Erring calendars will always come back into step after a long period, of course, and in what seems to have been the fifth century BC a calendar cycle incorporating civil and lunar calendars was devised, but it was a late development, long after the classical period of Egyptian civilization.

The Egyptian day ran from sunrise to sunrise and was divided into two periods of twelve hours. The Egyptians were not only the first people to make such a division but also the first to make these hours equal in length. Twelve hours were chosen because these fitted in to the movement of the stars across the sky as they rose and set during the night. Indeed, it appears that some learned scribe went through the civil calendar and picked a star, or perhaps a group of stars, that rose just before dawn on the first day of each of the ten-day weeks of the year. Once he had done this he called the time between the heliacal rising of one such star or star group (known as a 'decan') and the heliacal rising of the next group, an 'hour'. Sometime around 2150 BC the number of these hours was fixed at twelve, for although the scheme would give 18 decans passing per 'night' some of these would be lost in the twilight after sunset and before dawn. We know this from the examination of painted 'star clocks' on the roofs of tombs of the period. A day of twelve hours was presumably adopted to balance the twelve hours of the night.

The hours by day were measured by sundials, or more correctly, shadow-clocks. These might be quite simple, as one from the time of Thutmose III (1490 to 1436 BC) bears evidence. It was a straight piece of wood with five divisions and a raised horizontal arm at one end. At noon it was turned round to measure the Sun's shadow because after noon it fell in a different direction; although it indicated only ten hours in all, the first and last would have been lost in twilight. An Egyptian water-clock or clepsydra for determining the hours from the reign of Amenhotep III (1397 to 1360 BC) has been preserved, but as it depicts the calendar situation of 1540 BC it may well be that such clocks were first designed in the Eighteenth Dynasty. At all events, its invention heralded a new type of painted star-clock constructed by the more precise methods which the clepsydra could provide. Such are the star-clocks on the tombs of Rameses

VII (*c.* 1142 BC) and other Ramesside rulers.

On the ceiling of the East Osiris Chapel in the temple of Hathor at Dendera in Upper Egypt was a circular Zodiac (the band of constellations through which the Sun, Moon and planets seem to move). It is quite well-known but it is only dated 30 BC and so is a late importation into Egyptian astronomical lore. This and other Zodiacs seem to have come only after the Greek conquest of Egypt in the third century BC, for the Egyptians were never concerned enough with planetary motions to devise anything like this themselves. Another comparatively late import into Egyptian astronomy arrived with the Persian invasion of the country in the sixth century BC: the belief in the casting of personal horoscopes. Though the planets had been noticed in early times, the ancient Egyptians had not been much concerned with precise movements, only with their spiritual significance. They were all considered to be aspects of Horus; Jupiter was 'Horus who illuminates the Two Lands' or 'Horus who opens the Mystery', Saturn was 'Horus the Bull' or 'Horus, Bull of the Sky', Mars 'Horus the Red' or 'Horus of the Horizon', to name only some of the titles, so they must have been thought to have some kind of impact on national life; indeed, Mercury was considered malevolent when an evening star but seems to have had a different aspect when a morning star. But this was all quite different from the procedure of casting a personal horoscope based on the aspects of the stars and planets at the time of birth; that practice was something originally devised in Mesopotamia and was based on a quite different cosmology to the one the Egyptians knew.

For the origins of astrology, see pages 43–44.

Egyptian Mathematics

If ancient Egyptian astronomy was concerned primarily with the practical art of timekeeping, ancient Egyptian mathematics was confined to arithmetic, and practical arithmetic at that. Mathematics was not itself considered to be a form of knowledge irrespective of its application, as was to be the case in Greece. Thus research into mathematical principles was negligible. There was no underlying theory of mathematics, no theoretical system of geometry: mathematics only concentrated on counting, addition, subtraction, multiplication and division, but was well-adapted to the kinds of problems which scribes in an administration would meet.

Egyptian arithmetic was based in essence on the two-times table, as well as an ability to find two-thirds of any number, whether a whole number or a fraction. The scribes mostly wrote their calculations in hieratic script although some numbers are found on carved reliefs written in hieroglyphic form. The symbols for different numbers varied widely, and the need to write numbers in carefully aligned rows was less necessary with the Egyptians than it is with us, but like us they counted in tens. Subtraction was done by asking the question '5, plus how many to make 9?' rather than 'subtract 5 from 9'. In this way tables of addition could be used for the process of subtraction.

Illustration page 45

Above Sorcerer wearing stags' horns, an owl mask, wolf's ears, the forelegs of a bear and a horse's tail – drawing on the qualities of all these animals by imitative magic. A reconstruction after Henri Breuil from the Late Stone Age cave in France known as Les Trois Frères.

Left Tibetan spirit trap of wood, straw, string and an animal skull, fastened on a rooftop to guard against demons approaching from the sky. A typical example of early prophylaxis against the dangers of the universe. De Otto Samson collection.

Below Late Stone Age painting of a bison wounded by arrows from a cave at the archaeological site of Niaux in Ariège in the south of France. This painting dates from about 10,000 years ago; the impulse to draw such hunting scenes probably stemmed from an attempt to gain magical strength in the hunt, while the accuracy of the drawing bears witness to the complex anatomical knowledge which prehistoric man derived from practical experience.

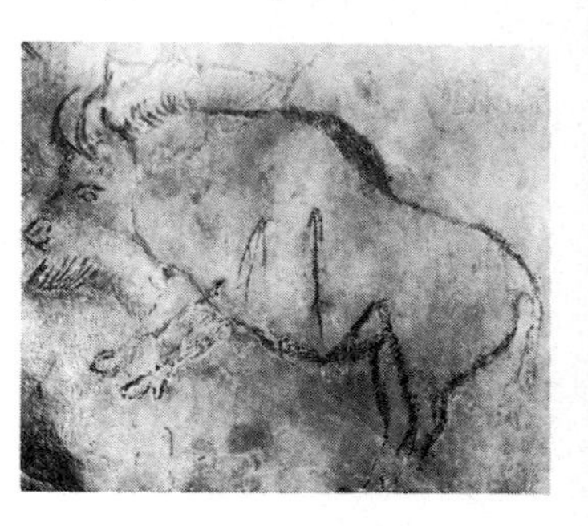

Above Late Stone Age flint arrowheads and harpoons. Long experience gave an ability to break stone accurately into complex shapes. Nevertheless, this knowledge was practical and technological and owed nothing to a study of the nature of the stone itself. Musée des Antiquités Nationale, Saint-Germain-en-Laye.

Above right Tribesmen in Borneo measuring the length of the midday shadow; such measurements were done by many primitive peoples and led to a determination of the movement of the heavens and the length of the year. After a photograph taken in about 1910 by Kose and McDougal.

Right The pyramids of Khufu, Khafre and Menkaure at Giza in Egypt. Symbolizing the primeval hill from which the world originated, the pyramids display great mathematical accuracy, and face the cardinal points of the compass.

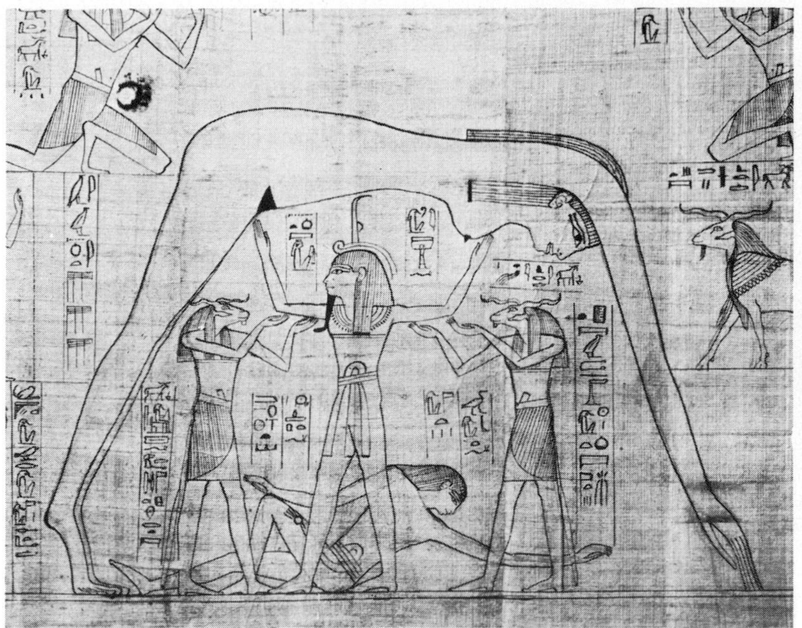

Above The pharaoh Akhenaten and his wife Nefertiti, sacrificing to Aten, the Sun god, who in his turn dispenses his gifts to mankind. Akhenaten abolished the old gods to concentrate on Aten-worship; but he failed to establish a permanent new world-picture. Egyptian Museum, Cairo.

Above left The Egyptian sky-goddess Nut supported by Shu, god of the air, with Qeb, a god of the Earth, lying on his side. Like many other primitive cosmogonies, the Egyptian's explanation for the origin of the universe involved the separation of sky and earth from their original embrace. From the Greenfield Papyrus, British Museum, London.

Left Some Egyptian constellations depicted on the roof of the tomb of Seti I (c. 1318-1304 BC).

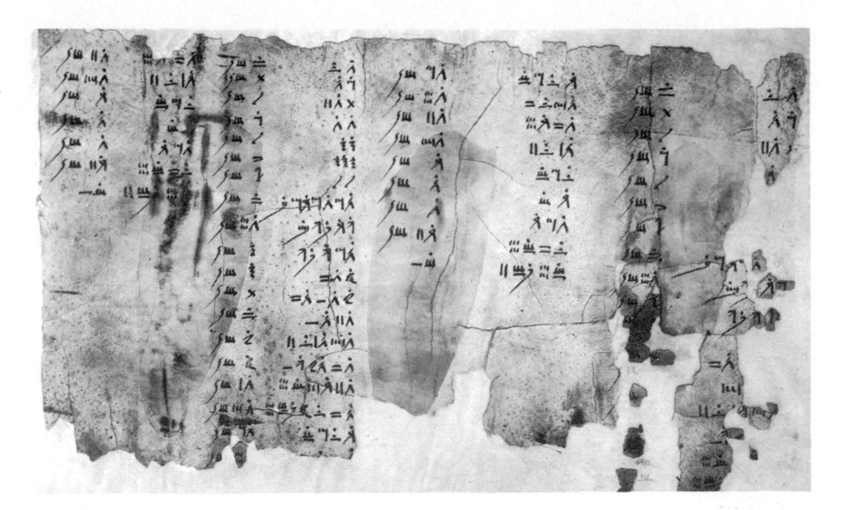

Right Egyptian leather scroll displaying a series of sums including the addition of fractions, c. 1700 BC. British Museum, London.

Right Scenes of agriculture and river life in ancient Egypt, demonstrating the early use of the plough, an essential invention for agriculture and thus for all settled civilization. British Museum, London.

Multiplication was just a case of using the two-times table. Thus to multiply 8 by 7 the scribe would write:

1	8
2	16
4	32
7	56

If the multiplier were a number like 21, he would therefore have to write out a longer two-times series and then pick out the appropriate terms (shown here by a ★); thus 21 by 8 would be written:

★1	8
2	16
★4	32
8	64
★16	128
21	168

Multiplication was also done for ten times and for one-tenth. There were also rules for multiplying reciprocals (i.e. multiplying by one divided by a number thus, $^1/_{21}$ or $^1/_{17}$, or whatever) using their unit fractions in a similar way to the whole numbers used in multiplication. As far as the addition of fractions was concerned they had tables of these, though they could be worked out; the Egyptians always used only unit fractions (i.e. ½ $^1/_3$ ¼ $^1/_5$ etc.) except for the fraction of $^2/_3$. Division was done in a way similar to subtraction but using the two-times table to achieve the result. Tables of useful multiplications were available.

As far as geometry was concerned, the only results sought were purely practical ones. When surveyors wanted to parcel out land and measure out right-angles they used the '3–4–5 rule', setting out a right angle by using a triangle of rope with sides of 3, 4 and 5 units. This is a particular case of Pythagoras's theorem; the more general theorem was not known in Egypt, however. It has been suggested that in order to construct some of their buildings they would need to know the value of π (pi), that is the ratio of the circumference of a circle to its diameter. But the value could have been obtained for all practical purposes by rolling a disc along the ground and measuring the length marked out, and there is no evidence whatsoever that they ever worked out the mathematical ratio.

Egyptian Science in General

Ancient Egyptian astronomy has left its record in tombs and buildings, Egyptian their mathematics in temple records and palace papyri, but what of the rest of Egyptian science? The Egyptians were masters of many arts and crafts, and each master craftsman must have amassed a wealth of detailed information. For instance those dealing with mining and metallurgy will have noticed geological strata and veins of metallic ore, they would know about the melting temperatures of metals and how to generate sufficient heat in furnaces, and they would have knowledge about alloys and of making moulds and casting. In other words they would presumably have some basic facts about geology and the physics and chemistry of materials. Glass-making, in which the Egyptians excelled, requires similar knowledge.

In medicine the Egyptians were competent surgeons, and their dentists in particular were adept at draining abscesses (a condition brought on possibly by the high incidence of arteriosclerosis), while gold stoppings for teeth were also widely used, at least by the well-to-do, and all this by the Fourth Dynasty. Two notable medical papyri have survived, and their saneness, soundness and freedom from superstition are remarkable; the men who compiled them were not only experienced but wise. In one medical papyrus a description is given of a man 'having a gaping wound in the head', with details which make it clear that the physician had observed the membranes surrounding the brain, the convolutions of the brain itself and the fluid around it, while it is evident that the author also thought that the brain was the control centre of the body.

Considerable knowledge of human anatomy would have come from the Egyptian practice of embalming. Preserving all parts of the dead was a serious matter for a people who believed in a material afterlife. The general idea was that just as Osiris was killed and dismembered by Seth and rose again when his body was reassembled, so would an individual be resurrected when the various components of the living person – the soul, shadow, name, heart and body – were brought together again. On the physical side these parts not only had to be carefully preserved but the work had to be done tastefully enough to lure back the spiritual components. The most elaborate methods of all were reserved, originally, for royal personages and involved a certain amount of surgery. The brain, intestines and other vital organs were removed and, after washing in wine, were placed with herbs in canopic jars. The body cavities were filled with perfumes and sweet-smelling resins and the body sewn up. It was next immersed in nitre for 70 days, then it was washed and wrapped in bandages which had been dipped in some gummy (resinous?) material. Finally the body was placed in its sarcophagus and sealed. A far less elaborate method was to inject the corpse with oil of cedar, immerse it in nitre for 70 days, then after removal from the solution, withdraw the oil and the fleshy parts, leaving only the skin and bones. For the poor, the intestines were simply purged and the corpse covered with nitre for the 70-day period. But in the first two methods of treatment the embalmers come to know the human body and its parts extremely well, and with their surgical experience, amassed a valuable amount of anatomical knowledge. But it seems to have stimulated no research into the way the body actually functioned.

The social standing of the medical man in ancient Egypt is uncertain. The first important physician of whom we have knowledge is Imhotep, the grand vizier of King Djoser and builder of the Step

Pyramid, so Egyptian physicians had a noble forebear; one, indeed, who became the god of medicine. As to the position of his successors we can, however, only hazard a guess; since they seem to have been only straightforward practitioners with no formally constituted medical college to license them, they may have been outside court or priestly circles, and have been considered no more than skilled artisans, as was the case in nearby Mesopotamia. As for the nature of animals and plants, relief sculptures make it clear that the Egyptians were careful and competent observers of Nature, stylised though their pictures often were. The relationship between the Sun and the growth of crops was also expressed in carvings, perhaps nowhere so well as in the bas-reliefs commissioned by the heretic pharaoh Akhenaten, whose monotheistic worship of the Sun led him to wish to be depicted with his family offering symbolic sacrifices to the Sun, whose many radiating arms brought all manner of blessings to mankind.

Illustration page 27

The Egyptians were also great gardeners; they loved formal gardens, sometimes with a large pool stocked with fish, and here they cultivated moisture-loving crops. They also domesticated animals, especially the cat. They must certainly have gained a vast amount of basic botanical and zoological knowledge. Again, with their use of cosmetics, paints and dyes, and with their pickling and curing of fish, their preserving of food in general, as well as in their embalming, they assembled a treasury of chemical knowledge. None of this was science in the sense in which Greek knowledge became science, but it was an essential beginning. Facts had to be collected, collated, and separated from legends, myths and fables.

For the similar Chinese interest in gardening, see page 183.

This the Egyptians did, and there were some enquiring minds like Imhotep, a veritable polymath, and the scribe Ahmose who, in the reign of Auserre Apopi I (1607 to 1566 BC), wrote a magnificent mathematical treatise (the Rhind Papyrus). This last was not only a set of mathematical tables but also a course in arithmetic, from addition to subtraction to fractions and reciprocals, as well as some geometrical problems concerned with the measurement of areas and volumes, though once again the results are all practical ones; there is no theoretical reasoning. But though the spirit of enquiry became stultified after the sixteenth century BC, the ancient Egyptians certainly amassed many facts which would at least contribute to the development of science in its infant state.

Illustration page 45

Mesopotamia

Mesopotamia, the land 'Between the Rivers', occupies the flat alluvial area between the Tigris and Euphrates in what is now Iraq. Together with the area north and west, which stretches in a curve from the Tigris round to the coasts of Syria, Lebanon and northern Israel, it forms what has been called the 'Fertile Crescent'. Between present-day Baghdad and the Persian Gulf the land slopes only gently, to give a total difference in height of no more than ten metres; in consequence the rivers flow slowly, depositing great quantities of silt, overflowing their banks, and slightly changing course from time to time. In the extreme south there are marshes and reed swamps. The water supply is irregular, and rainfall is low, so cultivation must be close to the rivers or be supported by irrigation. North of the crescent the soil in the plains is hard, and unsuitable for crop rearing for two thirds of the year.

Although not such an easy area to farm as Egypt, it had a plentiful supply of raw materials – agricultural products included animals, fish and date palms, and very early on a reed industry appeared, providing plant-fibre products as well as reeds themselves. In addition there are bitumen springs and limestone some 55 km (35 miles) to the west, but there is no wood except for the coarse type obtainable from the date palm and suitable only for rough beams, while there is no hard stone and but little metal. All through its history Mesopotamia needed to trade, and the southern part of the country particularly became a vast market place and therefore a centre for the exchange and dissemination of ideas.

Between 10,000 and 5000 BC large permanent settlements grew up. To begin with these were merely the places where hunters and cattle breeders began to practise agriculture and change to a more sedentary way of life, but in due course permanent houses were built, and even the gods settled into temples made for them. Clay vessels were made and fired at ever higher temperatures to give greater hardness, specialist crafts developed and metal was discovered and worked; the Neolithic Age gave way to the Chalcolithic or Bronze Age. By 4000 BC definite cultures were emerging on the Tigris at Samarra, in the surrounding Jazira area, in the south at Ur (near present day Nasiriya) and at Hajj Muhammad (some 240 km or 150 miles southwest of Baghdad).

The new cultures had their cities, for there had been cities for some time in southern Mesopotamia, (although the earliest city of all for which there are records is Jericho, a long way to the west and dating back to around 8000 BC). None of these was yet walled, the earliest-known walled city being Uruk in the south, close to the Euphrates. The area between the two rivers was peopled by the Sumerians, who in 3000 BC were recording that they considered themselves superior to the nomads and the mountain dwellers of the north 'who do not know houses and who do not cultivate wheat'. Certainly the Sumerians were very different from the Semitic people in the north whom they criticized: well before 3000 BC they had drained the marshes in the south, had irrigated the land and domesticated oxen, donkeys, cattle, sheep and goats, as well as using wheeled chariots and living in mud-brick houses. They seem to have been the heirs of a culture which had its beginnings at least a thousand years earlier.

The Invention of Writing

The mention of language and the Sumerians brings us to the invention of writing – vitally important for the growth of abstract science and its transmission. For it seems that it was the Sumerians who first devised this development of language by inventing special signs specifically adapted to a writing material, in their case clay tablets. The very earliest such recording appears to have been begun by priests of the early Sumerian (pre-3000 BC) civilization, who had to prepare records of the surplus grain and other products held for the state at the temples. These would undoubtedly have been in some form of picture-symbols – wheat in store were recorded by an ear of wheat, oxen by an ox head, and so on. One tablet containing these temple accounts probably represents a receipted bill for material supplied. The earliest Egyptians hieroglyphs were another form of writing with picture-symbols or pictographs, as was the first form of written Chinese.

Picture-symbols stood for objects as did sounds in the spoken language; it was the genius of the Sumerians that made an identification between picture-symbol and sound. Then, when the original picture-symbols were found to be too limited in application, they were modified to extend the written vocabulary. To achieve this the Sumerians added extra lines; for example, lines under the chin in a picture-symbol of a head meant that the mouth only was concerned (they thus changed the symbol for the sound 'sag' into the sound 'ka'). In due course there were about 2000 signs representing syllables, whereas the simpler Egyptian language had only about 732 signs, even as late as the Middle Kingdom. Gradually, as writing became more widely used, the symbols began to represent sounds only, not objects. However one symbol could now represent a number of words having similar sounds (i.e. homonyms – words like sew, sow, so), Sumerian examples of which go back to 3000 BC; the range of expression of the symbols increased, but the price was some ambiguity, especially in a monosyllabic language. To overcome this problem, other symbols were placed before or after the original symbol to help indicate a particular meaning; for instance, a symbol for 'wood' might be placed in front of the symbol for the sound, to specify what sort of object was being referred to.

Although the marshy land above the Persian Gulf had reed-beds, the fibre was not suitable for making papyrus; the Sumerians therefore needed some other readily available material on which to write. They chose their inexhaustible supply of clay, forming the clay into tablets or cylinders or even prisms, and writing on it with a sharpened piece of reed. Metal or stone were only used as writing surfaces on monuments, not for everyday work. To begin with a reed with a sharp blade was used, but drawing curves in the mud was not easy, and gradually the scribes cut a more wedge-shaped end to the reed, and their symbols became increasingly stylised and composed only of straight lines. Thus a syllable such as 'sag', which had begun as the pictogram of a head, became more and more a collection of wedge-shaped marks, which tilted over because of the way in which the scribe held the clay tablet. In Egypt the scribes used a reed pen, because the smooth surface of the papyrus made it possible to develop a more cursive script but, like their hieroglyphics, it could not be tailored to other languages. Cuneiform, on the other hand, was used in the Syrian Hittite states and soon became the diplomatic language of the Middle East.

Illustration page 48

Neither hieroglyphic nor cuneiform scripts were alphabetic; they were syllabic, each sign representing a syllable. Alphabetic writing seems to have originated with the peoples on the coasts of what are now Syria, Lebanon and Israel, and the scripts are known as Ugaritic and Phoenician. Ugaritic gets its name because it was the writing found when the ancient city of Ugarit (now Ras Shamra) was excavated in 1929. Here the people spoke a dialect resembling Hebrew and Phoenician, and in the temple there the excavators discovered both cuneiform tablets and a cuneiform alphabetic script in which the signs corresponded in order to Hebrew and Phoenician. However, the Ugaritic script contained more signs than either and corresponded in many ways to the later Arabic. The city reached its peak of development between about 1800 and 1200 BC. The Phoenicians lived a little further south, occupying the coastal regions of present-day Lebanon as well as a little of what are now Syria and Israel, and their earliest inscriptions date from 1700 to 1500 BC. These are truly alphabetic, containing 22 letters, – each of which represents a single sound and which can be combined to form syllables with the words written from right to left. This Phoenician alphabet spread throughout Syria, Arabia and Palestine and even further afield, for the Phoenicians were great traders and established many colonies and trading posts on the Mediterranean coast. Inscriptions in the Phoenician alphabet which have been found in Cyprus, Malta, Sardinia, at Carthage and at Marseilles, have been dated to the fifth century B.C.

Mesopotamian Culture

The discussion of the origins of writing has taken us to comparatively late times. We must return to the Fertile Crescent in the third millennium BC, when the southern area between the Euphrates and Tigris rivers was under the control of the Sumerians. Their sovereignty lasted until about 2350 BC when the Akkadians conquered the country. Originally a nomadic people, the Akkadians took over power in Sumer but nevertheless were 'conquered' culturally by their subjects, having brought no new enlightenment with them and being content to promote that which they found, proudly styling themselves kings of Sumer and Akkad.

Mesopotamia and the Fertile Crescent were not as peaceable as Egypt, and there were a number of power struggles, incursions and conquests, of which the most notable was the Assyrian in the twelfth century BC. Yet the general enlightenment which the Sumerians first showed continued and seems only to have waned in the sixth century BC with invasion by Persia.

Originally Sumeria had a number of important cities each of which exercised control over the surrounding areas: Ur is perhaps the most famous, but another was the ancient city of Lagash which had been founded in the sixth millennium BC and reached its greatest fame and eminence around 2125 BC under Gudea, who seems to have been a governor rather than a king. Ur, the city supposed to have been the home of Abraham, was the Sumerian capital from 2800 to 2300 BC. Founded about the fourth millennium BC by people from northern Mesoptamia, Ur I seems to have been destroyed by a flood (the Biblical Flood?), the Euphrates then running closer to it than it does now. It was the place where kings were buried along with a whole retinue of court and other officials, women and servants chosen to serve the monarch in the life after death. It was adopted as capital by the Akkadians and it was here that the almost legendary Akkadian king, Sargon I, reigned. Ur was also the site of many great monumental structures of which the chief was the ziggurat, a vast, stepped pyramidal structure built in successive stages with a shrine at the top approached from the ground by way of huge broad staircases. Most notable was the ziggurat to Ur's patron deity, the Moon goddess Nanna or Sin, which covered an area 64 by 46 metres (210 by 150 feet) and had a height of 12 metres (40 feet). Three of its sides had sheer walls and three great staircases, each with one hundred steps, ran up the fourth side. The building shows clearly that, by the third millennium, Sumerians were familiar with all the basic forms of architecture, the column, arch, dome and vault. What is more, each giant wall was curved slightly outwards both from base to top and from side to side, thus giving an impression of great strength, whereas a straight wall might have seemed to sag slightly under the great superstructure above it. Technically this is the principle of 'entasis', or bulging, which was to be rediscovered 15 centuries later by the builders of the Parthenon at Athens. Entasis was primarily an artistic discovery, rather than a scientific one, but it underlines the sophistication of Sumerian knowledge. By about 2000 BC Ur had declined, and suffered some destruction during uprisings, though its geographical position helped it later to regain some of its commercial prowess as a centre of foreign trade. Something of its previous religious eminence was also recovered.

Traces of a prehistoric settlement also exist for Babylon but its development as a great city came later. For a time it was a provincial satellite of Ur, but after Ur had fallen it became in 1894 BC the centre of a small kingdom, one of many set up in the area by a large tribal federation from Arabia – the Amorites or Canaanites – who, like the Akkadians before them, seem to have become imbued with Sumerian culture. Later kings of Babylon were also of Canaanite stock, including the justly famous Hammurapi (1792–1750 BC) who conquered surrounding city states and established Babylon as the centre of a kingdom, Babylonia, which included all southern Mesopotamia and part of Assyria (northern Iraq). Babylon grew and prospered to become the envy of foreign princes, and about 1570 BC control of the city passed to the Kassites, a northern people possibly from Iran, who

Illustration page 45

founded a dynasty that lasted more than four hundred years. Then other power changes took place – there was Assyrian domination, a return to Kassite power, another sacking – then, in about 1124 BC, it came under control of the Amorite king Nebuchadrezzar who founded his own dynasty. After this, power over the city seems to have alternated between the Assyrians and the Chaldeans, a people from the southern area bordering the Persian Gulf. With the fall of the Assyrians in the seventh century BC, a Chaldean dynasty was at last established under Nabopolassar, and with his son, Nebuchadrezzar II, Babylon became famous once more, and 'Babylonia' synonymous with 'Chaldea'. It was at this time that Babylon became the largest city in the world, covering an area of 10,000 hectares (25,000 acres).

We can study the culture of Sumer, Akkad and Babylonia from the host of cuneiform clay tablets that have survived. Although these could not be rolled like papyrus, nor bound into books, they were nevertheless collected into libraries and stored as archives. In fact one historian of science has gone so far as to say that 'the Egyptians invented books, while the Sumerians invented record offices!' At any event we are in their debt for their librarianship, since the vast collection of 25,000 tablets at Nineveh, and the immense collections at Tell el-Amarna and at Nippur, allow us to reconstruct something of Sumerian and later thinking about man and his environment. Besides these there are two additional sources of information, that provided by palace reliefs carved on walls and monuments, and the information which can be derived from seals. Every man of standing had his own seal, the imprint of which he would put on tablets written by him, or for him by a scribe, and these seals were often elaborately detailed with writing or pictures, or both. Plenty have survived and some thousands have found their way into museum collections.

Mesopotamian Medicine

As in Egypt a glance at medicine will help give some insight into the biological knowledge of the Mesopotamians. It will also involve magic and divination, because magical as well as scientific means were used in treatment of a disease or to cure an illness, while in applying his medical remedies a physician would seek help from divination in foretelling the likely success of his ministrations.

In company with every early civilization, the Mesopotamians made extensive use of herbal drugs. Roots, stems, fruits and leaves were used, but, unfortunately, it is difficult to make certain identification in every case of the drug prescribed for a particular illness. All the same, it is clear that they recognized dropsy, fever, hernia, itch and leprosy, as well as various skin complaints and conditions affecting the hair, throat, lungs and stomach, and had drugs for them. But medication was not confined to herbal remedies; minerals such as alum, crushed stones and salt were prescribed, as well as medicines containing parts of animals. And as in some western European medicine in medieval times and later, treatment was affected by a preference for magic numbers. Of these three and seven (and their multiples) were the favoured ones, and prescriptions might be made up with three sesame seeds (say), or seven, or 21 or whatever such combination seemed appropriate; non-magical quantities would be avoided. Again, a dosage might be given seven times, a cauterization applied three times, and so on. Always the numbers would be applied because of their hidden power which would give greater effect to the medicine. Then there was also the idea, to be found in many early civilizations, that herbal drugs should be made from material gathered at an auspicious time – at Full Moon, for example, or seven days before or after this. Such considerations were sometimes made also when collecting animals or animal parts for incorporation in a formulation. And when a potion was being compounded it was sometimes deemed necessary to have a special person present – a child or a virgin, perhaps, whose purity or innocence might influence the very mixing of the ingredients.

While there is evidence that medical texts and medical handbooks were available, much seems also to have depended on the physician's experience. Compared with Egypt little or no major operative surgery was practised, and specialists seem to have been rare. The status of the medical man is interesting: he was classed with diviners, innkeepers and bakers! This classification is, however, not quite what it seems, for it really only indicates that he was not an official of the court, but one who lived on what he could earn from the fees he charged. In the time of Hammurapi (about 1750 BC) these fees followed a specified scale, as do the penalties for failure. In the collection of laws which he codified, we find that if, for instance, a physician set a nobleman's broken bone, he was to be paid five shekels of silver, but if his patient was only a commoner, the fee was three shekels, and two if a nobleman's slave. The fee for a surgical operation was larger, and also graded, as were the punishments for failure, ranging from the replacement of a slave if his treatment had resulted in death, to the loss of a hand for the death of a member of the nobility. Surgery has always been fraught with dangers for the patient, but in Babylonia its practitioners too seem to have had good reason to be chary of it, at least where members of the aristocracy were concerned.

Veterinary surgery was also a feature of Mesopotamian medicine. To help the vet, and the human physician as well, there were various incantations and curses which could be used since, like everything else, diseases were thought to be the creation of the gods. Drugs would act as palliatives but the medical man would only get to the seat of the trouble by appeasing the god who caused the disease. As in more primitive times, the medical practitioner was something of a priest.

The belief in the divine origin of diseases meant, of course, that divination – a form of communication between gods and men – was of singular importance to the doctor. By divining the will of the gods, he could gain useful insight into the reasons for the visitation of the disease in the first place and, by looking into the future (which was known to the gods), he could obtain guidance about the effects of courses of treatment. Various forms of divination were used: the flights of birds, the appearance of monstrous or peculiar births of animals, the appearance of the stars were all examined, while dreams had a special place of their own. Because of the vividness and sense of reality which dreams can bring as well as their sense of inconsequentiality, they were considered direct messages from the gods, though to understand their language often required professional interpretation by priest or physician. The famous biblical story of the pharoah's dream of the seven fat and seven thin kine and its interpretation by Joseph was a typical case in point, for in this the Hebrews and Egyptians were following beliefs similar to those held by the Babylonians and other early peoples. But where the Babylonians differed from their neighbours was in their adoption of the technique of hepatoscopy. This was a means of divination carried out by the examination of animal livers, usually those of sheep or goats. The diviners noted the five lobes of the liver, and their predictions were based on the state of these, as we know both from many texts and from clay models which were made of divinatory livers and carried explanatory instructions. These have been found not only in Babylonia but also far away from it with the inscriptions in various languages, so it is obvious that they were used for transmitting information about the divinatory technique. They provide interesting evidence of one way ideas could be spread from one centre of culture to another.

Illustration page 46

The idea of using animal livers gives some information about Babylonian knowledge of the workings of the bodies of man and animals. Clearly they must have been aware of the importance of blood: blood was life. If a man loses blood he faints; if he loses too much he dies. And the bloodiest of all the organs in the body is the liver, so it was obviously of signal importance, and in due course they concluded that it was the seat of the emotions as well as of life itself. The heart they classed as the seat of intellect, a view echoed by the Hebrews. The Babylonians also appear to have realized that diseases could be transmitted from one person to another (although the evidence for this is not conclusive). Possibly, then, the biblical injunction that those with contagious diseases should be isolated, together with their possessions, was derived from Babylonia.

Mesopotamian Biological and Geographical Knowledge

The Mesopotamians seem to have gathered a considerable amount of information about animal species, and by Babylonian times, if not before, they had made an attempt to bring order out of the chaos of so many different kinds of animals by attempting a systematic classification. As an example of the everyday recognition of different species, in 1910 BC at the fish market in Larsa on the Euphrates, some thirty varieties were on sale. However, their scientific knowledge was broader than a classification of species required for food. Cuneiform

Illustration page 45

tablets have been found which list hundreds of different animals giving both Sumerian and Akkadian names, while others detail more than 250 varieties of plants. And some of these tablets contain not just lists but also a primitive classification: animals are separated into fish and other creatures living in the water – shellfish such as mussels and oysters were given their own separate classification – then there is a differentiation between serpents, birds and four-legged animals. The larger groups were broken down further – there were dogs, hyenas and lions in one sub-group, and camels, horses and asses in another. Plants, too, were divided into trees, cereals, herbs, spices and drugs, and plants which bore fruits looking rather similar, such as apples and figs, were classed together.

It is likely that at least as early as the Babylonians it was known that one plant, the date palm, reproduced sexually. At least, they knew that the palm could not reproduce by itself and required two kinds of sexes, one with offspring (fruit) and one which appeared sterile, if a crop was to appear. How did they know? Presumably by experience, for if the apparently sterile male plants had been dug up, it would in due time have become clear that the hitherto fruitful plants were fruitful no longer. Later the idea of plant sexuality was extended (by the Assyrians) to cypresses and mandrakes.

Illustration page 46

Like other peoples in other areas of the world the Mesopotamians knew of the existence of other countries and other climes; this is shown by tablets enumerating, for example, King Sargon's conquests, or the geographical lists which the Sumerians prepared for their scribes, while lists have also been found of places with which they traded. That this resulted in some map-making may not, then, be surprising, and two maps have come down to us. One is a map of the city of Nippur – a map which was precise enough to help archaeologists when they excavated it towards the end of the last century. The other is a map of the world with a description in cuneiform. Babylonia, Syria and other territories are represented as a circular area surrounded by the Persian Gulf; Babylon lies at the centre. This was the first of such 'wheel-maps' of the world and was to be copied later by the Arabs and in medieval Europe. It enshrines the viewpoint, so often to be repeated in the maps of other and later civilizations, that the map-makers lived in the centre of the world. Other countries were seen as surrounding areas (which they were) and distances were only approximately correct, based on journey time rather than on geometric measurement. Such maps gave a picture of one's country and its immediate surroundings, which was their aim. They can also be seen to have a more general cosmological meaning than this, since they showed a vision of a flattish Earth with oceans at its extremities, and so depicted man's place under the dome of heaven – the general concept of the time.

Illustrations pages 46, 47

Trade and Weights and Measures

In Mesopotamia, trading was carried on and business conducted without any formal currency system although pieces of precious metal were used for barter and in usury, which carried a high rate of interest. But if they had no standard coinage – this did not arrive until the seventh century in either Assyria, or, more probably, in Lydia (now western Turkey) – the Sumerians very early on developed a first-class system of weights and measures. As with all early measurement standard lengths were based on parts of the body. There were hands or palms, digits (fingers) and feet. The Egyptians had two kinds of cubit, the royal cubit of seven palms and the short cubit of six palms, but in Sumeria there was only one cubit which had a length of 495 mm (19½ inches) and lay between the two Egyptian standards.

Weighing was first used for measuring amounts of gold-dust, and not for commercial usage, although the Sumerians (and also the people of the Indus Valley civilization in northern India) made use of weights for trading sometime about 2500 BC, more than a millennium before it was the practice in Egypt. The basic unit of weight was the shekel; in Sumeria this was 8.36 gm (129 grains) with a larger unit, the mina, 502 gm or 60 times the shekel. In Sumeria the unit of capacity was the log of 541 cc (33 cubic inches) which, with variations, was adopted in Phoenicia, and by the Israelites and Judaeans as well, while about 1400 BC it was taken up by the Egyptians.

The Sumerians used a standard factor of reduction based on 60 and its multiples. Thus the mina was 60 times larger than the shekel, and the homer 720 (12 x 60) times greater than the log. Even the Sumerian foot was exactly two-thirds (forty-sixtieths) of the cubit. Thus they were the first people to have a completely self-consistent standard series of measures, and this not later than 2000 BC, possibly even earlier. The standard factor of 60 was useful because it was one into which many numbers will divide exactly: 2, 3, 4, 5, 6, 10, 12, 15, 20 and 30, and it still has many uses today, as in our measurement of angles and of time. As they had only two cuneiform signs for writing all numbers, there was some ambiguity. The sign 𒌋 alone could mean 10, or 600, or 36,000 or even larger multiples of 10 x 60, while 𒁹 could be 1, 60, 3,600 and other multiples of 60, because there was no symbol for zero. This was made even more ambiguous when fractions were considered, since all fractions were expressed as multiples of 1/60. In consequence 1½ would be written as 1,30 (i.e. 1 and 30/60), or 𒁹𒌋𒌋𒌋, which could equally well be 60 + 30 or 90. The correct number depended on the context, which makes it difficult to decipher such tablets today.

The usual arithmetical operations of addition, subtraction, multiplication and division were carried out using tables which gave values in unit steps up to 20, and then in 10s up to 60. Repeated applications were then required which may have been tedious but at least had the merit of preventing the cuneiform tablets on which they were cut being too cumbersome.

The Old Babylonians (so-called to distinguish them from the Neo-Babylonian civilization of the seventh and sixth century BC) developed what we should now call algebra, and were therefore able to solve mathematical equations. These equations, which were needed for solutions of problems in building, in land surveying and in commerce, were written out in words and their solutions carried out step by step according to certain well-tried rules. They could solve not only simple equations but also quadratics and even cubic equations.

Geometry was another mathematical discipline which the Old Babylonians, if not the Sumerians, studied. They did not lay down a logically formal system as Euclid was to do about a thousand years later, but they were able to calculate the areas of plane figures and the volumes of many solid ones including pyramids, cylinders and cones. They knew of isosceles triangles and were aware of the general relationship between the sides of a right-angled triangle which, as we have seen, was not known to the Egyptians. On the other hand their value for π was inferior to the Egyptian one.

Unfortunately the Sumerian and Babylonian mathematical achievements were not followed up in later times. Their algebra was forgotten, and, except for a period when, as we shall see, some Greeks took up the subject for a brief spell, it lay dormant until Islamic mathematicians revived it in the ninth century AD. The way the Mesopotamians extended the scale of numbers to fractions or submultiples of 10 and 6 was also lost, and had to wait until the sixteenth century AD for revival, while their adoption of a division by 10s as a basis for all official measurements was not in fact revived in the Western world until the advent of the metric system towards the close of the eighteenth century. Lastly, their use of the position of number symbols to indicate their value was also lost and not regained in the West until the arrival in the tenth century of the Hindu-Arabic numerals which we still use today. That the Sumerians and Babylonians made great advances in mathematics may be an understatement; they seem to have laid the very foundations of the entire subject.

For Hindu-Arabic numerals, see page 192.

Mesopotamian Astronomy

The caches of cuneiform tablets which have shown us the glories of Old Babylonian mathematics have also brought about a knowledge of Sumerian/Babylonian astronomy and of the Chaldean developments. They make it clear, for instance, that these people created the art of scientific astronomical observation. And even in so speculative a subject as cosmology – the nature of the universe – they managed to draw a picture which was not wholly an exercise in mythical imagination.

The Sumerian universe was peopled with gods and goddesses, the offspring of Apsu and Tiamat, whose union also begat mankind and the animal world. But their scheme of the universe was more materialistic and descriptive of Nature than was the Egyptian. The Earth was somewhat in the shape of an upturned gufa (a coracle-shaped boat), its shores leading down to a salty sea. Overhead spread the sky, a vast dome for ever unreachable. What happened at the point where they met? To the Egyptians it would have seemed an irrelevant question, but to the Sumerian/Babylonian mind it was indeed legitimate; to begin with they thought of some structure such as a bank or ridge at the rim of the salty sea and imagined that the dome rested on this. Later, by Old Babylonian times, the ridge had become a range of mountains which not only supported the sky but also gave access to heaven. In both cases their explanation was a physical not a metaphysical one.

The Sun moved across the sky by day and back under the Earth at night; so did the Moon. When they came to discuss the phases of the Moon, they were careful to tie these in with the position of the Sun and so, by implication, realized that the Moon shone only by reflecting sunlight. The stars themselves were thought of as fixed to the dome of heaven, but, characteristically, the Mesopotamians did not leave it at that. They not only organized the stars into constellations but, like the Egyptians, assigned to them seasonal appearances and, this time unlike the Egyptians, studied the motions of the planets in detail, observing that their paths lay close to the ecliptic (the Sun's apparent seasonal path in the sky). This planetary work was to lead to important results later.

The dome of heaven had its own colour – blue – because it was made of blue gemstone, for in the *Epic of Gilgamesh* the whole of heaven was imagined as having three layers, each one made of a precious stone, and echoing the three-part construction of the Earth itself – the mountains inhabited by gods, the flat lands by man, and the underworld by the dead. The idea that the heavens were made of precious stones is, of course, echoed later in the Bible (in the books of Exodus and Revelation, for instance), as too is the idea that rain was stored in heaven and then released when holes in the sky were unplugged. There was, however, another view which claimed that rain came down from the clouds. Considering the general state of knowledge in Old Babylonian times and earlier, their picture gave a very rational description of the cosmos.

Observing was certainly carried out. Heliacal risings and settings were observed, and a particular interest was taken in the planets, whose positions were noted not only at times of rising and setting, but also at other specific times such as opposition (at which time a planet is at its highest point in the night sky at about midnight). Great interest was also taken in the movement of Jupiter, which is the brightest planet in the night sky, and as early as King Ammisaduga's time (about 1921 to 1901 BC) in Venus, which appears in evening or early morning skies and is then even brighter than Jupiter. Although we do not know precisely what observing instruments were used, it is clear that these included sundials and waterclocks and, by the time of King Tukulti-Ninurta I (1260–1232 BC), some type of 'transit instrument' for determining the moment when a celestial body is due south. Other celestial phenomena such as shooting stars and comets were observed and recorded, as well as eclipses of the Sun and the Moon. Observations were often made from the top of a ziggurat.

Illustration page 48

The stars and planets, Sun and Moon, were all looked on as placed there by the gods for the benefit of mankind. Their purpose was to shower down their influences, to give an indication of the nation's fortunes, and to provide a basic guide for a calendar by which the people could regulate their agriculture and arrange for celebration at the proper time of religious festivals to Marduk and their other deities. To begin with, the Sumerians thought, as the Egyptians did, that the length of the year was 360 days – a glorious confirmation of their sexagesimal system – and they divided each day into six watches, three for the day and three for the night. These were unequal periods because they varied with the seasons which, of course, brought different lengths of day and night. They proved inconvenient for astronomical use, and finally the whole day was divided into 12 equal periods of 30 'gesh' each: this was another division which gave a total of 360 units. The division into 360 was applied also to sectors of the sky, giving us the 360-degree circle.

The 360-day year therefore had mathematical overtones as far as the Sumerians were concerned, but it was also tied to a lunar calendar, having a regular series of months of 30 days and 29 days to fit in with the lunar cycle of phases. There were 12 months giving a total of 354 days, and then an extra (intercalary) month was added when necessary to keep the year in step with the seasons. The intercalary month must have been introduced very early on, for in Ur during the period 2294 to 2187 BC it was already recognized that the introduction of intercalary months followed an eight-year cycle. This Babylonian calendar formed the basis of the Hebrew one and of the first Greek and Roman calendars as well.

We come now to the last period of Mesopotamian astronomy. It is called Chaldean because it was a Chaldean dynasty which ruled the Neo-Babylonian kingdom in the seventh and sixth centuries BC, and they were Chaldean priests in the temples who observed and carried on their scientific work both then and later, after the Persian conquest (sixth to fourth centuries BC) and the conquest by Alexander the Great, which occurred in 332-323 BC. In post-Alexandrian times the Chaldeans made a great stride forward by applying mathematical analysis to astronomy in a way quite different from their contemporaries, the Greeks.

The Chaldeans inherited the zodiac from their predecessors, the Old Babylonians, and they made a long series of observations, especially of the Moon and the planets. Unlike the Greeks, they formulated no planetary theory, but instead they drew up detailed tables of past planetary motions so that they could 'predict' or anticipate future motions in fresh tables. This they achieved by using arithmetic to express the changing speeds with which the planets appear to move across the sky. Their methods can be most easily understood by drawing a diagram which shows how they plotted the way in which the movement would vary with time. Their first attempts at predicting the movements of the Sun, for example, worked on the

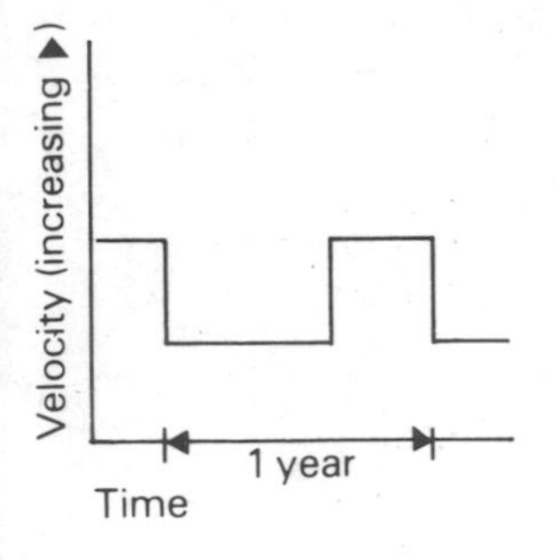

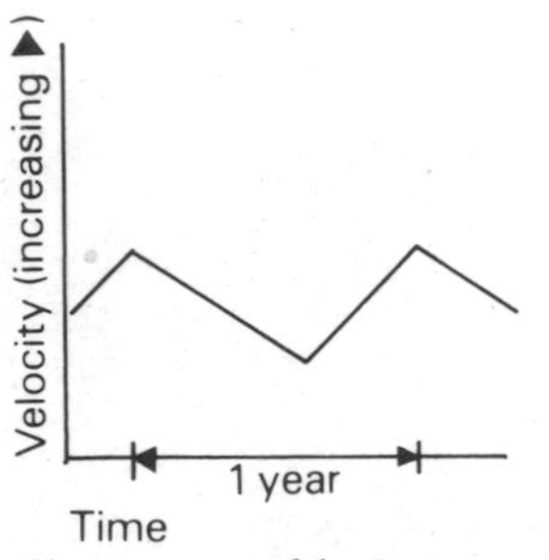

The movement of the Sun across the sky: above, the earliest view proposed by the Chaldeans, and, below, the view as later modified to accord with observational experience.

For the astrology of ancient Egypt, see page 24; for Plato's contribution to astrology, see pages 97–98; and for the Islamic development of the subject, see pages 208–212.

assumption that it had two speeds – faster in winter and slower in summer. Later, between 181 and 49 BC, with more observational experience to go on, they realized that they had over-simplified the situation. In fact the Sun only gradually changed its speed across the sky; this is shown in the lower figure. Here a zig-zag line represents the velocity changes and it was such 'zig-zag functions' that they worked out mathematically, using their arithmetic, for the Sun, Moon and planets. This was something new and truly scientific, which was not to be followed up in this kind of way until very much later in Western Europe.

It was also the Chaldeans who helped introduce astrology in a form in which it is still known today. This was the degenerate form in which the arrangement of the heavens at birth was thought to indicate not only personal characteristics but also the future fate of an individual. In the days when there was a general belief in two worlds – the world of governing gods and spirits and the physical world – the idea that the heavens should depict on the large scale the attitudes of the gods to the well-being of a people was acceptable enough. Applying such major indications to every person was an extension of this idea, but its further extrapolation to laying down the course of that person's future life was of doubtful validity. Indeed, one can term it degeneration because it led to much baseless superstition. Nevertheless by 410 BC horoscopes setting out the positions of the heavens at the time of an individual's birth were in existence, and when later the Chaldeans were able to calculate future planetary positions the whole apparatus of predictive astrology was there. But although the Bible and Greek sources look on the Chaldeans as magicians and astrologers, diviners and soothsayers, much of the development of astrology using horoscopes was not in fact carried out until the third century BC and later, and was done in Egypt, often by Greek-Egyptians. It is a distortion of the situation to lay on the Chaldeans the blame for all the superstitious practices that arose out of the extension of astrology to the personal level. It is more of a distortion still to ignore the very real scientific contribution to astronomy that the Chaldeans made, yet this is what happened. Their true achievement was forgotten until the astronomical and mathematical cuneiform tablets were discovered and analysed in our own times.

Astronomy in Western Europe

It has long been thought that the first developments of scientific astronomy took place in Babylonia, and this seems generally to be true. But there were developments elsewhere, perhaps helped along by information which had filtered through, either along trade routes or by way of reports by travellers and explorers. A known recipient was China, where there developed a great civilization interested in science, while there were also some astronomical achievements in some parts of Europe. Perhaps these early Europeans achieved something in other branches of science as well, but there seems to be no surviving evidence for it.

Left Mathematical papyrus written in Egyptian hieratic script from c. 1575 BC though a copy of a papyrus of three centuries earlier. It deals with the measurement of the area of the triangle and of the slopes of pyramids. British Museum, London.

Below Reconstruction of the ziggurat and temple precinct at Babylon. These buildings were at the heart of the Mesopotamian cities yet were open only to the priests – symbolizing the arcane, priestly nature of much early knowledge. Vorderasiatisches Museum, Berlin.

Left Assyrian bas-relief of a lioness. The sophistication of Assyrian anatomical knowledge is shown in the paralysis of the hind legs caused by the arrow through the spine, 9th century BC. British Museum, London.

Above Babylonian model of a liver for use in hepatoscopy. Each part of the liver (normally a sheep's) was thought to denote a specific event; for instance a long cystic duct denoted a long reign for the king. Study of animal entrails for prediction purposes continued until at least Roman times, and served as the basis of much work in animal anatomy. British Museum, London.

Right Assyrian deity shown fertilizing a date-palm. Bas-relief from the temple of Assurbanipal at Nimrod. Musée du Louvre, Paris.

Right Babylonian 'wheel-map' of the world, where Babylonia, Assyria and surrounding areas are represented on a round plain encircled by the Persian Gulf. Babylon is in the centre. British Museum, London.

Opposite Statue of Gudea of Lagash (2250 BC) showing a plan of the city on his lap, the world's oldest surviving map. Musée du Louvre, Paris.

The best known example of this Megalithic period, so-called because huge stone blocks were used in construction, is Stonehenge. Though built in stages over the centuries, the basic construction was made as early as 2800 BC, in Neolithic times. The last stages of its development were not finished until 1100 BC, in the Early Bronze Age. It is clear that the building techniques, using carefully dressed stones up to 9 metres (30 feet) long and weighing 50 tonnes, as well as the practice of entasis, were derived independently, but in recent years it has also become clear that Stonehenge was far from unique; it was only one of a host of megalithic structures, spread over Great Britain (with a preponderance in the West of Scotland), and Brittany. The old idea that Stonehenge was a Druid temple has long been shown to be without foundation, but archaeologists are still of the opinion that it was a religious building. A hypothesis less often accepted is that it was also an astronomical observatory designed to give calendar information and probably to warn of forthcoming eclipses as well. Yet the evidence is impressive.

Careful surveys of Stonehenge and other stone circles, or stone rings, and detailed statistical analysis of the results leave little doubt that they were indeed astronomical observatories, designed and built on the basis of experience and the need to watch for risings and settings of the Sun and Moon for the purpose of determining a seasonal calendar. They were needed because in the kind of latitudes in which they are to be found heliacal risings and settings cannot be observed as satisfactorily as they can in Egypt and Mesopotamia. In the higher latitudes neither the Sun nor the Moon sink so rapidly to the horizon, nor do they rise so swiftly. Their apparent paths as they set and rise are more oblique, with the result that turbulence of the air and other factors make it more difficult to detect the moment when they disappear or reappear – a difficulty increased by the longer twilight in these latitudes. Twilight does not, of course, affect observations of the risings and settings of the Sun and Moon, but it does make horizon observations of stars and planets virtually impossible.

The stone ring overcomes the problem by using Sun and Moon alone as calendar indicators and observing the most and least easterly points of rising, and the equivalent westerly points for their settings. In the summer the days lengthen and the Sun rises at more northerly points along the eastern horizon, and sets at more northerly positions along the western horizon. At midsummer, on the longest day of the year, the Sun has reached its most northerly rising and setting points. In wintertime the opposite happens, and at midwinter these rising and setting points are at their most southerly. These extreme points can be observed by using markers fixed vertically in the ground, so that the position of the most northerly or southerly points of rising or setting can be determined by the positions of the two markers. A natural feature on the horizon – a dip between two crags, for instance – was sometimes used as one marker, so that only one stone marker had to be set up. Plenty of such cases are known. Stones were, of

Opposite above Aerial view of Stonehenge, southern England. This megalithic complex has alignments for observing risings and settings at extreme western and eastern points on the horizon.

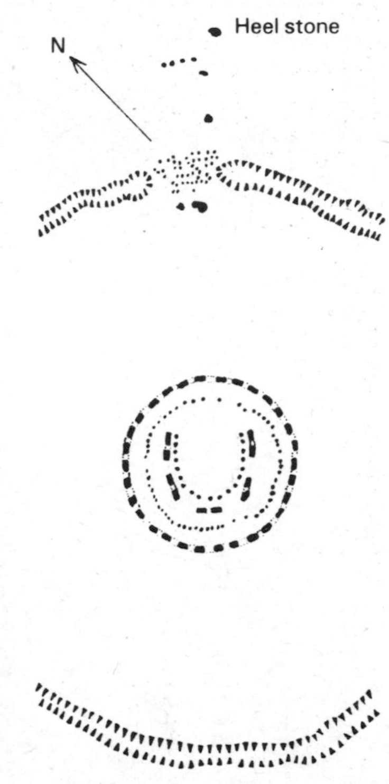

Stonehenge: For an observer at the head of the inner horseshoe of stones, the heelstone marks the midsummer sunrise, while the relative positions of many other stones could be used to determine the rising and setting of both Sun and Moon at various times in the year.

Opposite below left Babylonian cuneiform tablet giving details of the appearances of the planet Venus, dating from the second millennium BC. British Museum, London.

Opposite below right Alignment of standing stones at Carnac in Brittany. This complex is the most extensive of the many megalithic arrangements of western Europe, and astronomical observations could be made there very accurately.

course, mainly used because they would survive the damp climate better than other natural materials.

There can be no doubt that long experience lay behind the idea of setting up stone-ring observatories, and it was certainly long experience which led to the establishment of special markers for the Moon, since it would become evident after centuries of watching that eclipses occurred only when the Moon and Sun were in certain positions with respect to one another. The complex arrangement of one of the stone rings at Stonehenge makes it a distinct possibility that this was, among other things, an 'eclipse computer', though built by a people with virtually no written language. We must, however, be careful about making statements of too definite a kind about how little – or, indeed, how much – the stone-ring builders knew. Some undeciphered signs have been found on some stones and there is evidence that they may have had a 365-day seasonal calendar divided into 16 solar 'months'. Certainly the stone rings, tombs and pottery which they have left behind make it clear that this was a civilization that would repay further study.

Illustration page 48

It has also been claimed that in constructing all the stone circles over what is quite a large area they used a standard measurement – the megalithic yard. There is at present some argument about whether there ever was such a standard, which is claimed to be 0.829 metres (2 feet 8 inches) but statistical and other arguments make it appear increasingly likely that it did exist. Thus it may well be that we have here a civilization which had at least reached the stage of having an organized system of standards for measurement.

Ancient Meso-American Culture

During a part of the period when the Egyptian and Babylonian cultures were flourishing, other civilizations were developing in Meso-America – that is the area of what is now Mexico and some adjacent districts of Central America. The origins of the peoples who formed these civilizations are obscure; some groups of Mongolian hunters seems to have come across the Bering Strait from north-eastern Siberia, perhaps using a land bridge which may once have existed. Precisely when this happened is unknown, but there is evidence that by 11,000 BC they had occupied most of the New World south of the ice cap which covered the north of North America. Yet migrations may have occurred much earlier than this: a blade fashioned from the obsidian (a volcanic glass) found near Pueblo in Mexico has given a radio-carbon dating of 21,800 BC. The climate grew warmer round about 7000 BC, and by 6500 BC an incipient agriculture was practised in the Tehuacan valley in Meso-America; maize, beans and chilli peppers were grown as well as squash (a plant of the gourd family which could be used both as a vegetable and as an animal feed). Between 5000 and 3500 BC mutant forms of maize which had husks were cultivated side by side with the wild forms. Pottery from 2300 BC has been found in this area, possibly diffused north-west from Colombia and Ecuador. Domestication of animals occurred comparatively late; by 1500 BC the dog was the only domesticated creature. Progress was slow, and the years 1500 to 900 BC are called the 'Formative Period' of the region.

The Mexican highlands were dry, but the lowlands were more hospitable, especially the area near the shores of the Pacific near Chiapas and Guatemala. Here there was a village culture which made highly developed pottery figures of women showing evidence which makes it seem likely that by then they were able to carry out Caesarean births. But our main interest must lie with the Olmec civilization which existed at the same time as the village culture, and extended on into the 'Middle Formative Period', 900 to 300 BC.

The Olmecs occupied the humid lowlands on the other side of the Meso-American isthmus facing the Gulf of Mexico, close to what is now Veracruz and extending westwards as far as Tabasco. Around the area of San Lorenzo, on a compact plateau, a dominant land-owning community took control over surrounding land where there was abundant rainfall and which had a fine alluvial soil, an area that has been termed the 'ecological equivalent of the Fertile Crescent of Mesopotamia'. Here they built a city with dwelling houses and a large ceremonial centre. But the aspect of this civilization which has most captured popular imagination is their sculpture, especially the vast heads that they carved. These heads weigh 44 tonnes and are of basalt, probably found among the lava flows of the Cerro Cintepec mountains, some 80 km (50 miles) to the north-west.

The Olmec heads are not only remarkable for their size, but are unique for another reason; they all depict men wearing what looks like an American football helmet. And the surprising thing is that this is just what they may be – heads of games-players with a defensive covering – for the Olmecs had a fast national ball game in which protective clothing was a necessity. It was played using a large rubber ball, the rubber for which came from the *Hevea* tree. Later this rubber was also made into garments, and adopted medicinally as an ideal material for plasters. In the seventeenth century AD Spaniards reported that they saw the material being moulded round earthenware formers for fabricating footwear and bottles, but how early such developments occurred is unknown. Whether it was the Olmecs who extended its use is a matter for conjecture.

Between 800 and 400 BC the most important Olmec centre was at La Venta in Tabasco (southwest of present-day Tonala). By this time they were carrying out some elaborate burials; in the centre of La Venta itself there is a large artificial mound made of earth and clay some 30 metres (100 feet) high which may house a tomb, while north of this area some actual tombs have been excavated, among them one containing the remains of two children surrounded by magnificent jade ornaments. By this time, too, pottery was more elaborate, and a number of trading stations were set up for importing jade, iron ore, cinnabar (the chief ore of mercury), the mineral serpentine, and other commodities, but details of all the uses to which they put this material are not clear. Nor is there any firm evidence about what theoretical science, if any, existed under the Olmecs. Like other early cultures they had a pantheon of gods, though theirs were part human and part jaguar in appearance. Some were concerned with those phenomena of the natural world which were important in daily life; there were gods of fire and of rain, and one whose concern was with the growth of maize, the staple crop. But this is, at most, an extension of primitive nature worship. There is, however, one interesting fact which archaeological excavation has brought to light, and this is the presence of a type of necklace made of small concave iron mirrors each with a hole pierced in the centre. The very existence of such mirrors must mean that something was known of reflectivity and, probably, of burning glasses as well, but to what other uses might they have been put?

In tracing the development of civilization and possible science in Meso-America it must be remembered that it was an area in which development happened piecemeal; in some regions civilization flourished while others were still primitive. During the time of the Olmec culture there were still plenty of villages untouched by any high culture at all. But this should not surprise us: a similar situation arose in the Old World. When the Mediterranean civilizations were basking in the Bronze Age, most of Europe was still bordering on Neolithic barbarism. And just as, in the West, Europe owed so much to early Mediterranean culture, so in Meso-America all later cultures were indebted to the Olmecs. Such was the Izapan, which made many artefacts of rare beauty and where religious rites were heightened by the use of hallucinogenic mushrooms, as well as the two civilizations with which we are more concerned for their scientific interest, the Zapotec and the Mayan.

The centre of the Zapotec civilization, which overlapped with the Olmec in the post-800s BC, was at Monte Alban, a series of connected hills near Oaxaca, south of the area which the Olmecs occupied. The Zapotecs are of interest because of the hieroglyphs (or more properly 'glyphs') and reliefs they carved on sandstone slabs surrounding a large courtyard on the main plaza. The reliefs are known as *danzantes* as they appear to represent human figures in dance postures and show evidence of a study of human movement, while the glyphs are evidence of writing. The writing itself concerns various Zapotec places and also describes a 52-year cyclic calendar, with days and months expressed by a system of bar-and-dot numerals. This 52-year cycle or 'Calendar Round' was adopted because after 52 repeats of a 365-day calendar, a specific day occurs again in the same position in the year. Once again, then, we encounter what appears to be a sophisticated calendar system which would have been administratively admirable, like the Egyptian, and shows clear knowledge that the 365-day period was not exact. It was the first written calendar in Meso-America, and was adopted by the Mayans.

Mayan Civilization

The first of the Mayan civilization dates from about 300 BC, in the southern part of their area, near Seibal and Altar de Sacrificios; possibly they were originally Olmecs who had migrated after the collapse of San Lorenzo about 900 BC. Later the Mayas spread over the entire Yucatan peninsula, and created the greatest New World civilization, which lasted from about 100 BC until its collapse towards the end of the ninth century AD.

Early on in their development certain traits showed themselves and, by the time of the 'early classic' Mayan period from AD 100 onwards, they were much in evidence. Thus their architecture, which soon reached an advanced level, had its own specific characteristics: they built vast temple platforms using a core of cemented rubble faced with thick layers of plaster, and when finished the whole buildings formed huge stepped pyramids. Like the Mesopotamian ziggurats, these had a temple at the top, and also a large stairway, flanked on either side by masks representing their gods. The largest Mayan city – and the largest city in Meso-America until the time of the Spanish conquest – was Teotihuacán which was planned from the start in four areas, separated by two great avenues. At the peak of its development (about the close of the sixth century AD) its population was more than 150,000 and the temple complex covered some 20 square kilometres (8 square miles), crowned by a stepped pyramid no less than 60 metres in height.

Another notable Mayan city from 'late classic' times (AD 600 to 900) was Palenque; indeed it is said to have been the most beautiful of all the Mayan sites, with graceful pyramid-shaped temples and palatial buildings with mansard-style roofs, embellished with carvings of rulers, gods and ceremonies. Its principal building, the 'Palace', contained not only a vast honeycomb of galleries but also a large four-storey tower which may have been both a look-out post and an observatory.

The Mayan picture of the universe was both conservative and primitive. The world contained four 'directions', each associated with a colour and a tree with a bird perched on top of it. The directions emanated from 'the great tree of abundance' at the world's centre, and were reached by roads, perhaps the counterparts of the limestone causeways found at some Mayan sites. The heavens were built in thirteen layers, and the underworld, associated with the jaguar, in nine. The Earth itself was thought of as the back of a giant lizard or crocodile lying in a vast pond with water-lilies and fish – a picturesque though hardly scientific description of the isthmus where they lived. The Mayan ruling class, who were all noticeably taller than the peasants who served them, were believed to have been created separately.

Ancestor worship was practised and much attention was lavished, as in Egypt, on deceased royalty who were thought of as descended from the gods, and it is possible that there was a hereditary priesthood. The Mayan religion, however, in spite of claims to the contrary, seems without doubt to have had a dark and gruesome side. They practised self-mutilation and human sacrifice, sacrifice they carried out either by torture followed by decapitation, or by taking the heart from a living victim.

Mayan Numbers and the Mayan Calendars

Illustration page 57

The Moon, the Sun and Venus were carefully observed for astrological reasons. Nevertheless the Mayas had elaborate calendars, and from AD 300 were able to record astronomical and historical events using a system of glyphs, which they soon developed into a tool of considerable flexibility for expressing words and sentences.

Mayan counting was far more advanced than that in any other Meso-American territory, and seems to have had its origins in 300 or 200 BC; it was well developed by the 'classic' period of Mayan civilization, in the third century AD. They used a count of 20 – a vigesimal system. Thus the Mayas would speak of the number 41 as 'one in the third score' because it was two-score (i.e. 40) plus one. Similarly 379 would have been 'nineteen in the nineteenth score' (i.e. 19 + (18 × 20) or 19 + 360). However they did have an alternative way of looking at things, even closer to our own system; here 41 would be 'two score and one' instead of 'one in the third score', and 51 would be 'two score and eleven' rather than 'eleven in the third score'. Just as we have names for the digits 1 to 10, so the Mayas named the first twenty numbers. For writing comparatively low numbers they used symbols based on the tally, or on counting on the fingers, but special symbols were introduced for higher numbers.

The Mayas have left us no mathematical treatises, nor do they seem to have had any astronomical theories; we do not even have any note of the methods they used in astronomy. So we are probably right in saying that they developed no scientific astronomy nor any formal mathematics, though of course they must have used their number system in trade and administration. There are, however, two practical astronomical contexts in which Mayan numeration has come down to us; these are the drawing up of calendars and their concern with eclipses.

The Mayan calendars were the foundation for all later date reckoning in Meso-America, but were themselves based on the 'Calendar Round' of the Olmecs, namely a 365-day year coupled with a 52-year cycle. However, the Mayas also had a 'Count of Days', sometimes now known as the 'tzolkin'; this was a 260-day period based on the meshing together of two smaller cycles, one of 13 days, the 'trecena' and the other of 20 days, the 'veintena'. This 260-day calendar was the one used primarily for divination, since all the 20 days of the veintena, each of which had a specific name, were thought to be linked with fate in one way or another.

The 365-day calendar was broken down into eighteen months of 20 days each, the remaining five days being termed 'days of evil omen'. Such a calendar year was not, of course, exact, since it did not contain the extra quarter of a day to make it keep in step with the seasons, and the Mayas allowed it to drift; a complete circuit to bring calendar and seasons back in step would take 29 'Calendar Rounds', that is a total of 1,508 calendar years (29 × 52 years).

The Mayas also had a lunar calendar, which ran concurrently with the tzolkin and the calendar year. Days on this lunar calendar were counted according to the age of the Moon (that is, its phase), and lunar half years were counted as well as full years. The Mayas also had a special lunar cycle concerned with the prediction of eclipses of the Sun, a cycle which must have been based solely on experience, since they possessed no theoretical astronomy to account for these spectacular events. Yet experience would teach them that after certain periods of time (measured in lunar months), coupled with a cycle based on the lunar phases and linked in with the tzolkin, solar eclipses might be expected to occur. They thus arrived at a cycle of 405 lunar months (11,960 days) or 46 tzolkins; this was divided into intervals of five and six lunar months.

Two other interesting calendar calculations were made by the Mayas; these were the Venus cycles and a method they devised for expressing long intervals of time. Venus is so bright an object in the evening and pre-dawn skies that it is not surprising it should be picked out for special notice, particularly by a people concerned more with astrology than with astronomy proper. What they did was to observe Venus and record the regular cycles of its appearance; in modern terms they measured its 'synodic' period, that is the time between one appearance at a certain point in the sky (its first pre-dawn rising, say) and its next reappearance at this point. They recognized this period as 584 days (modern value 583.92 days), and they then worked out a cycle of 2,920 days – a figure which is equal to 8 × 365 days and 5 × 584 days – and so linked together their year and the planet's synodic period. They had a still longer period, a 'great cycle' of 37,960 days, which incorporated all the other cycles, the tzolkin, the 365-day calendar and the synodic period of Venus.

Illustration page 58

It should be clear by now that the Mayas were interested in long time cycles, a characteristic of many calendar makers. This meant, of course, that they had to handle long periods of time, and having an interest in history and such historical events as were known to them, they realized that they needed a better way of reckoning time than using years and divisions of years which, with so many calendar reckonings in use, could lead to ambiguities and even confusion. They therefore devised a long-term day count which was thought of as beginning some time in the third millennium BC, and counted the days consecutively ever after. It is an excellent system for the dating of past astronomical and historical events, and it is much to the Mayas' credit that they had the insight to think of it. A similar system, the so-called 'Julian day count', was devised in Western Europe by Joseph Scaliger in 1583, but this was not until at least six centuries after the Mayan.

For the Chinese calendar, with its exceptionally long cycles, see pages 155–156.

Later Meso-American Cultures

By the close of the ninth century AD the Mayan civilization atrophied, but others rose to take its place, all indebted to it to a greater or lesser extent. One such civilization was that of the Toltecs which probably flourished between AD 900 and 1519. It was late Neolithic, for such metal as was used was confined to ornaments and, interestingly enough, it made no use of draught animals; all power requirements were based on manpower alone. Nevertheless they had a highly productive agriculture, with maize as the main source of their protein, and a wide range of secondary crops which included tomatoes, chilli peppers, cotton, tobacco, cacao, pineapples, peanuts, avocados and cassava. In the lowlands they carried out intensive cultivation without difficulty, but in the highlands they had to undertake extensive terracing, irrigation and reclamation of swamps, a course rendered necessary to feed their very dense population.

Yet if the Toltecs had no more than a stone-based technology, they could nevertheless write – on bark from the paper-fig tree – for there are records of astronomical observations (made for astrological reasons). They too had a 260-day calendar and a 365-day one, both derived, presumably, from the Mayan; their multilevel universe was doubtless Mayan as well. They seem to have been a deeply religious people, at least judging them from the art and architecture of the region. But historians are still doubtful if there was genuinely a separate Toltec culture at this time. Such history as there is of the Toltecs is so full of magic and myth that it may have been a later fabrication. However, archaeology does provide evidence for an empire of some kind in the area at this time.

Toltecs apart, the great post-Mayan civilization in Meso-America was the Aztec. Believing themselves the chosen people of the Sun god Huitzilopochtli or Tonatiuh, this poor nomadic tribe of the northern areas was led by a magician-priest to the Basin of Mexico and settled on a series of islands in the Lake of Texoco. From here, some time in the fourteenth century AD, they spread outwards until they covered a large area of Mexico.

Much of the Aztec land was sloping and subject to erosion, while a quarter of the flat areas they occupied were chains of lakes with waterlogged shores. In spite of these disadvantages they farmed intensively, applying plant and animal fertilizers and using the techniques of terracing and swamp reclamation, in the latter case making it possible for some lakes to be colonized as well as turned into highly productive land. They used knives of obsidian – the volcanic glass for these coming presumably from the area of the volcano Popocatepetl – and millstones of basalt, and they obtained salt from Lake Texoco itself. Yet in spite of their own intensive agriculture, there were many crops they could not grow and these, such as tobacco, rubber, cotton, paper, tropical roots and fruits, cacao and honey, as well as precious commodities such as jade and turquoise, they obtained by conquest.

Left The Caracol or observatory at Chichen Itza Mexico. Passages cut in the walls are aligned for observing Sun, Moon and stars. The structure dates from Mayan times.

Below The Aztec 'sunstone' used for calendar regulation. 3.7 m in diameter, the carving represents the cyclic motions of celestial bodies. Museo Nacional de Antropologia, Mexico City

Right Part of the 'Venus calendar' from the Mayan manuscript known as the Dresden codex. Its hieroglyphs and illustrations depict astrological predictions related to the heliacal risings of Venus. Sächsische Landesbibliothek, Dresden.

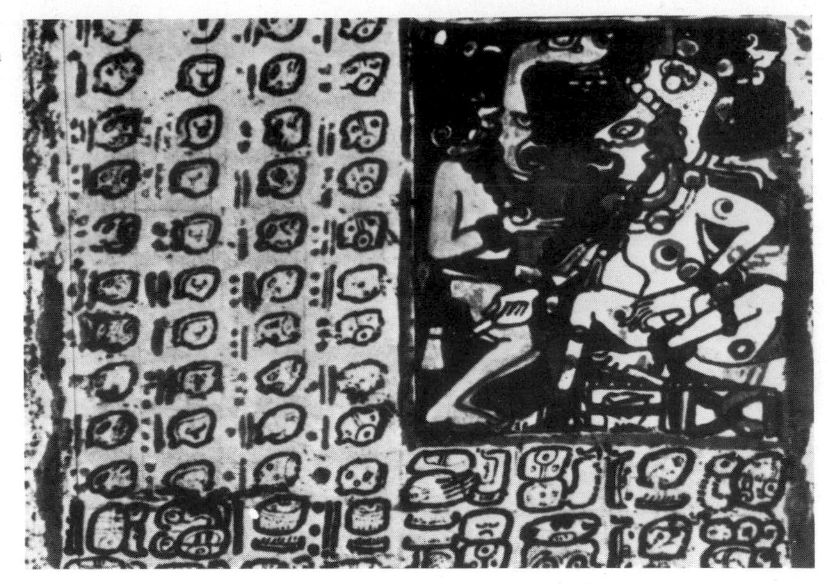

Right A quipu from Inca Peru. As well as being used for numerical purposes, these knotted cords served mnemonics for history, legend and all sorts of records. American Museum of Natural History, New York.

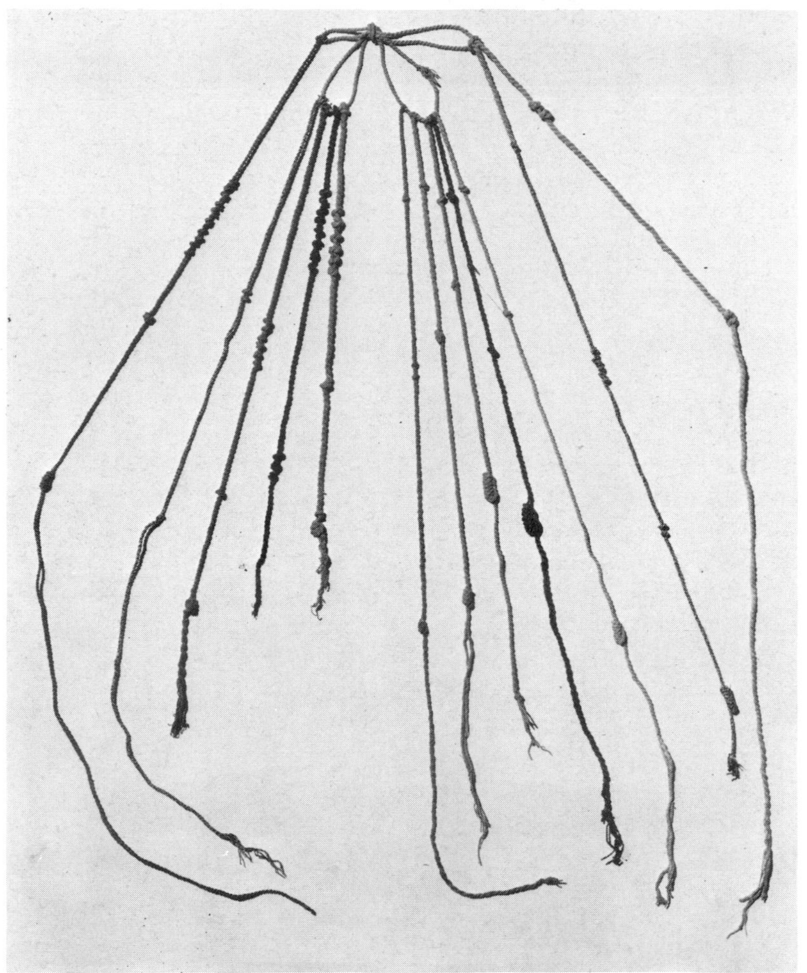

The Aztec religion was grim; it was centred on the belief that the Sun must be nourished, otherwise it would vanish from the sky. This demanded human sacrifice of blood and a living heart, and such sacrifice formed the most important part of their ritual. Man's place on Earth was to fight and die for the gods and for the preservation of the world order, and life depended on men's fulfilment of their role. Witchcraft dominated everyday life, even though sorcery was forbidden and punishable by death.

The Aztec civilization does not sound very attractive, but it brought a political unity to Mexico, and its priests made an attempt to rationalize a vast collection of gods and a wide variety of cults. In their study of Nature, though, they appear to have made no new contributions. They perpetuated the Mayan calendars, placing special emphasis on the 260-day tzolkin, and set great store by their medical profession which subscribed to the old Mesopotamian belief, of three thousand years before, that diseases were caused by the gods; medical treatment was therefore a palliative until the appropriate gods could be persuaded to effect a cure themselves. The glimmerings of science which had appeared in the culture of the Olmecs and the Mayas had disappeared.

Illustration page 57

South American Civilization

While primitive settlements practising a form of agriculture were beginning in Meso-America, South American Indians were settling in the Andes area of what is now Chile and Peru. The earliest coastal settlements were founded around 3500 BC, the people fishing in the sea with nets and by hook and line; a few crops – beans, squash, cotton and chilli peppers – were cultivated. At some time between 2100 and 1800 BC, pottery was being made and weaving done and later, dating between 1400 and 1000 BC, the beginnings of a civilization – the Chavin civilization – are to be found at Chavin de Huantar. Temples with stone platforms and masonry walls built in courses have been found; they were honeycombed with galleries at different levels and ventilated by shafts. These were ceremonial centres and show a great flair for sculpture; in one a shaft of natural light illuminates a white, granite, low-relief carving some 4.6 metres (15 feet) high, and this 'Smiling God' in the form of a human figure with snakes for hair and great fangs in the upper jaw, is still an awe-inspiring sight.

The Chavin civilization had its successors – the Huari and the Tiahuanaco – though these were much later and take us into the period 200 BC to AD 1000. However, before they ended, the Incan civilization, the greatest in pre-Columbian South America, was already in being. Even in its earliest stages certain notable characteristics appeared: a feline face motif used widely in decoration, and a wide range of metalwork, including the welding and soldering of thin hammered gold, copper ornaments and spear-heads, the fashioning of silver and the preparation of alloys. Lost-wax casting was also known. By AD 600 mummification was practised, and two centuries later the Incas could be described as a civilization with a considerable technology. They practised irrigation and water management on a larger scale than their predecessors, invented a method for building in mortarless masonry, used the crowbar and promoted the use of other metal tools, and had an arm balance for weighing. Yet in spite of their mechanical prowess they never considered the basic principles involved; each invention was the solution to a particular problem, not an invitation to formulate a science of mechanics.

The Incas had standard weights and measures, based, as in the case of all early measurements, on the parts of the human body; thus there was the handspan and the fathom (arms outstretched), sticks of which were used for land measurements. For distance there was the pace and the 'topo' which was 6,000 paces (7.2 km or 4 miles); this last was much used in their extensive road system which, by the fifteenth century, consisted of two main routes, one coastal, about 3,600 km (2,200 miles) long and the other of a similar length inland along the Andes. These were for foot travellers and pack animals only, since the Incas had no wheeled vehicles.

Like other civilizations in Meso-America the Incas practised animal and human sacrifice, the latter particularly at times of national disaster such as famine or pestilence, but (it appears) without the accompanying cruelty of the Mayas and Aztecs; it was important to them that a human sacrifice, when made, should be without blemish. The Sun was worshipped and so were other celestial bodies. Silver was spoken of as the 'tears of the Moon' while the constellation of stars we call Lyra (the Lyre) they saw as a llama, and they invoked it on all occasions for protection. Closely connected with their worship and their agriculture was the Incan calendar, about which much still remains to be discovered. Nevertheless the present view is that they had both a lunar calendar and one based on the seasons. What is not clear is how the two were related, but in one Inca area at least every third seasonal year was made up with a 13-month lunar calendar, a scheme which would bring the lunar calendar almost in step with the seasonal one. The seasonal year was one of 12 months, each month having three weeks of ten days each, to which were added five days for the most important Inca religious ceremonies, bringing the total to 365 days. The year began at the summer solstice, which fell in December since the Andes lie in the southern hemisphere. Observations to determine this were made from a raised platform in the middle of the great square in Cuzco using prearranged markers.

Illustration page 58

One Inca invention still in use in some parts of the Andes is the *quipu*, a device used for counting. It consists of a main cord on which are tied a number of hanging cords (usually 48) of different colours. Knots tied in the hanging cords record numbers, a knot furthest away from the main cord designating units, one nearer tens, and so on; absence of a knot means zero. It was a simple device, useful for counting but cumbersome for calculating.

Conclusion

The earliest civilizations have been seen to be remarkably inventive in devising systems – often incorporating mathematical constructions far in excess of any practical value – to forge a link between the natural phenomena they observed and their cosmological perceptions of the universe. Even among the Egyptians, whose science seems more down-to-earth than that of many other ancient peoples, the formal grandeur of the pyramids appears to reflect a perception that mathematics offers a key to the relations between man and the universe.

Yet for the most part the activities of these ancient astronomers and mathematicians eventually proved to be misdirected. Their work in observing and calculating the movements of the heavens was often astonishing, but much of it proved of little importance to the ultimate development of science. It was their failure to inquire more carefully into the nature of the heavenly bodies, and into the nature of the mechanisms that drove them across the sky, that led to this dead end. Nevertheless, the subtlety of their work is a fine tribute to the inspiration of animistic religion, and of the belief in the kind of magic that ascribes more importance to the relationships between phenomena than to the nature of the objects themselves.

Chapter Two

Greek Science

We come now to a civilization whose culture has profoundly affected our own, whose attempts to make sense of the natural world in which they found themselves have had more influence than any other on the way we think. The outlook, literature and science of the Greeks underpin the very foundations of our own view of the world.

Among all the peoples of western antiquity it was the Greeks who not only collected and collated facts but also welded them into a grand scheme; who rationalized the entire universe without recourse to magic or superstition. They were the first philosophers of Nature who formed ideas and devised interpretations which could stand on their own, invoking no gods to bolster weakness or obscurities in explanation. But, of course, all this did not appear overnight. The Greeks themselves did not spring, like Athena, fully armed from the forehead of Zeus; they were the heirs of earlier cultures of the eastern Mediterranean, and only gradually developed their scientific attitudes of mind.

The Aegean Civilization

Greek culture owed something to the Egyptians, to the Phoenicians and, later, to the Mesopotamians but above all it was the product of two earlier cultures, the Minoan and Mycenaean. These last two were centred on the Aegean, bounded in the west by Greece and in the east by Turkey. Here, between 3000 and 2200 BC – a time contemporary with the Sumerians and the Egyptian Old Kingdom – the Bronze Age arrived for the neolithic peoples who occupied Crete and the Cyclades, the numerous small islands southeast of the Greek mainland. Flourishing metalwork crafts existed in Crete by 2500 BC, in the islands of the Cyclades and on the southern part of the mainland. Houses with several rooms and roofed with tiles were built at this time, but there were some differences between the two main cultures, most notably in the way burials were made, communal tombs being used in Crete and small graves, never taking more than six, and usually less, in the Cyclades. There was trade between the two areas: for instance, fine white marble figures went from the Cyclades to Crete for gold and silver ware. Some of the Cretan jewellery of this period shows Mesopotamian inspiration, underlining what was surely an important link between the two cultures.

Between 2300 and 2000 BC the general unity of the Cycladean and Cretan way of life was broken by an influx of migratory invaders. Into the Cyclades came a people who probably originated in Anatolia (Western Turkey), while mainland Greece was invaded some time after 2300 BC by invaders who seem to have been the Kurgans, originating in the Balkans and South Russia. The Kurgans were a pastoral, horse-breeding, Indo-European people. Like so many migratory groups, they took over the culture they found, though they also brought in new ideas such as the use of horses and new burial customs. At this time, however, mainland culture lagged behind that in the Cyclades and particularly in Crete. Indeed Crete seems to have been little affected by the upheavals in the north-west, continuing to develop unmolested until it reached its peak about 1500 BC.

Crete is not a large island; it is no more than 55 km (35 miles) broad at its widest point and only 245 km (150 miles) in length. Yet here there grew up a powerful civilization with centres in Phaestos in the south, Mallia on the northern coast and, above all, Knossos, which was itself only a few kilometres inland and 25 km from Mallia. The Cretans were the first maritime power in the Mediterranean basin but they are often remembered for their powerful ruler King Minos. Minos is, however, a legendary figure, the son of Zeus and Europa, who obtained the Cretan throne with the help of the sea god Poseidon, a somewhat picturesque way of underlining the fact that the Cretans were so formidable a maritime people.

Wall paintings and paintings on pottery show that the Minoans were fine observers of nature – they depicted with great realism a variety of plants and a host of animals, including flying fish, octopuses, and wild duck.

The peak period of Minoan culture was followed by a series of disasters. First, in 1500 BC, there was a vast volcanic eruption on the island of Thera some 110 km (70 miles) north of Crete, burying the island's settlements under many metres of ash and causing great tidal waves to beat on the northern shores of Crete itself and on islands of the Cyclades; indeed a city on the island of Melos seems to have been completely swamped. The Thera eruption is possibly the source of Plato's story of Atlantis which, if he did not invent it, may be an echo of some Egyptian record of the disaster. The nub of Cretan civilization appears to have suffered no permanent ill effects from the eruption but, rather more than a generation later, by the middle of the fifteenth century BC, most of the important places in central and southern Crete were suddenly destroyed by fire.

The Cyclades were also overrun by people from the mainland soon after the destruction of Crete, and again settlements were destroyed by fire. Then another wave of destruction came to Crete, sometime between 1400 and 1150 BC, and this time the palace of Knossos was destroyed, never to be rebuilt. The centre of power in the Aegean gravitated to Mycenae on the Peloponnesian mainland, though Tiryns on the Gulf of Argolis and Athens were also important centres.

However, towards the end of the thirteenth century BC the chief places on the Peloponnese were burned down in a 'barbarian' invasion from the north, an invasion which heralded a collapse of the whole Aegean civilization.

The Coming of the Greeks

The classical Greek language falls into two dialects, eastern and western, and this may reflect two movements of Greek-speaking peoples into the area we now call Greece. The origins of the eastern dialect are somewhat in dispute; of one thing, however, we can be reasonably certain, and this is that between the tenth and eleventh centuries BC the Peloponnese was entered by Dorian Greeks from the north-west and north-central mainland, who spoke the western dialect. They came down southwards in a series of waves, and settled not only in the Peloponnese, but also in the south Aegean islands, as well as in the islands of the Dodecanese (the area off the south-west coast of what is now Turkey) and on the south-western (Turkish) mainland itself. Except in Attica and Argolis, the coming of the Dorians was followed by some depopulation, but how much of this was due to the invasions and how much to a (probable) change of climate bringing with it conditions of drought and famine, we do not know.

The Dorians brought with them new attitudes of mind, and when these impinged on the Mycenaean culture they led to what is sometimes called the 'Proto-Geometric' culture. The name is due to the fact that pottery and other art forms developed in Athens now showed a particular sensitivity for shape and proportion; old forms and patterns were reworked with a precision never achieved by the Mycenaeans in their adaptation of Minoan art. This was, indeed, the beginning of what we think of as Greek classical art, and is evidence of an outlook which was soon to have the most important effects on Greek science and philosophy.

The World of Homer and Hesiod

Homer and Hesiod were poets in whose work we find the first flowering of this new Greek culture, although their achievements still had a Mycenaean and Minoan flavour, for they were the heirs of the older Mediterranean civilizations as well as the progenitors of a new age. As inheritors of these previous cultures they followed the age-old tradition of the minstrel and the teller of tales. It is a tradition that is as old as man, because people have always felt the need to explain origins and to commemorate great events, to dwell on past glories and extol historic times.

Nothing is known of Homer's life, or even if he was one man or many, but it is likely that he lived in Ionia, on the west coast of Turkey, the region from which the first Greek philosophers came.

Ostensibly the *Iliad* is a record of events of the legendary Trojan War, in which forces from Troy, on the western Turkish mainland, vied with mainland Greeks. The date of the war is uncertain, it could have been any time between 1280 and 1180 BC, and the epic was composed probably some three or four hundred years later, in the middle of the ninth century BC. The *Odyssey* was most likely composed a century later which means, of course, that the author of the *Iliad* could not have completed it; scholars often refer to the *Odyssey* as the work of Homer II. This would explain its different tone, for whereas the *Iliad* is a story of war and strife, the *Odyssey* is about peace, about colonists, travellers, merchants and domestic life. It is at once more romantic and more moral.

Illustration page 75

These two epics were of great importance in the early days of Greek civilization, and are mentioned here because of what they taught. It is clear that they helped to standardize the Greek language as well as to record something of the outlook and behaviour of Mycenaean times. They enshrine the geographical knowledge gained by Aegean and Phoenician sailors, who had reached the Atlantic and had brought home with them the idea that the disc-shaped Earth was surrounded by a vast circular ocean. We can gain some impression of beliefs about the natural world, and even of their medical knowledge, for wars bring wounds and wounds need medical treatment. We learn, too, something of agriculture and husbandry, of arts and crafts, of what is considered good and bad, of what is chivalrous and what is not. In brief we gain an insight into the moral attitudes and world view of Homer and his times.

Hesiod was a master of instructive poetry; he lived in Ascra in Boeotia in the central region of the mainland and is remembered now for two poems, the *Theogony* and *Works and Days*. The first can be translated as the 'genealogy of the gods' and as a poem about the myths of the Greek pantheon of gods and goddesses need not concern us here. *Works and Days* primarily concerns itself with rules for husbandry and navigation, though it also provides a calendar of lucky and unlucky days and offers a moral homily as well. Living in the aftermath of war, a time of moral decline, Hesiod is concerned with problems of good and evil, justice and injustice, and he looks back to a golden age, extols a love of tradition and a sense of good conduct. Nevertheless, the rules for navigation and husbandry occupy more than a third of the whole, and are composed of simple instructions. Thus 'when the Pleiades, daughters of Atlas, are rising, begin your harvest, and your ploughing when they are going to set. Forty nights and days they are hidden and appear again as the year moves round, when first you sharpen your sickle. This is the law of the plains, and of those who live near the sea.' Indeed, this section is in essence a farmer's almanac, but it was not the first. A thousand years before, the Sumerians had just such an almanac, and Hesiod's follows an old and sound tradition that shows us just what the peasant's daily round was like. Like the calendar, his navigational rules are simple but none the less effective for that. The final section on lucky and unlucky days is pure superstition and not in keeping with either the rationalism or the poetic power of the earlier material; possibly it is a later addition.

Early Ionian Science

One thing which may have struck the reader so far is that here for the first time we are able to name our authors. No longer do we have to rely on an anonymous text; now we can come to grips with the man who originated it – at least we can as far as Hesiod is concerned. Homer and Hesiod were the harbingers of a new culture, of an age which even now is wonderful to contemplate, although it was still a literary age rather than a scientific one. But what we come to in the seventh century BC is something different – a scientific age. Though Greek religion was at least as animistic as other ancient religions, relying on sacrifices to the gods and divine intervention in human affairs, Greek science represented the remarkable achievement of separating the investigation of the laws of Nature from any religious questions of the relationship between man and the gods. At last we are face to face with the beginnings of the astonishing phenomenon of Greek science.

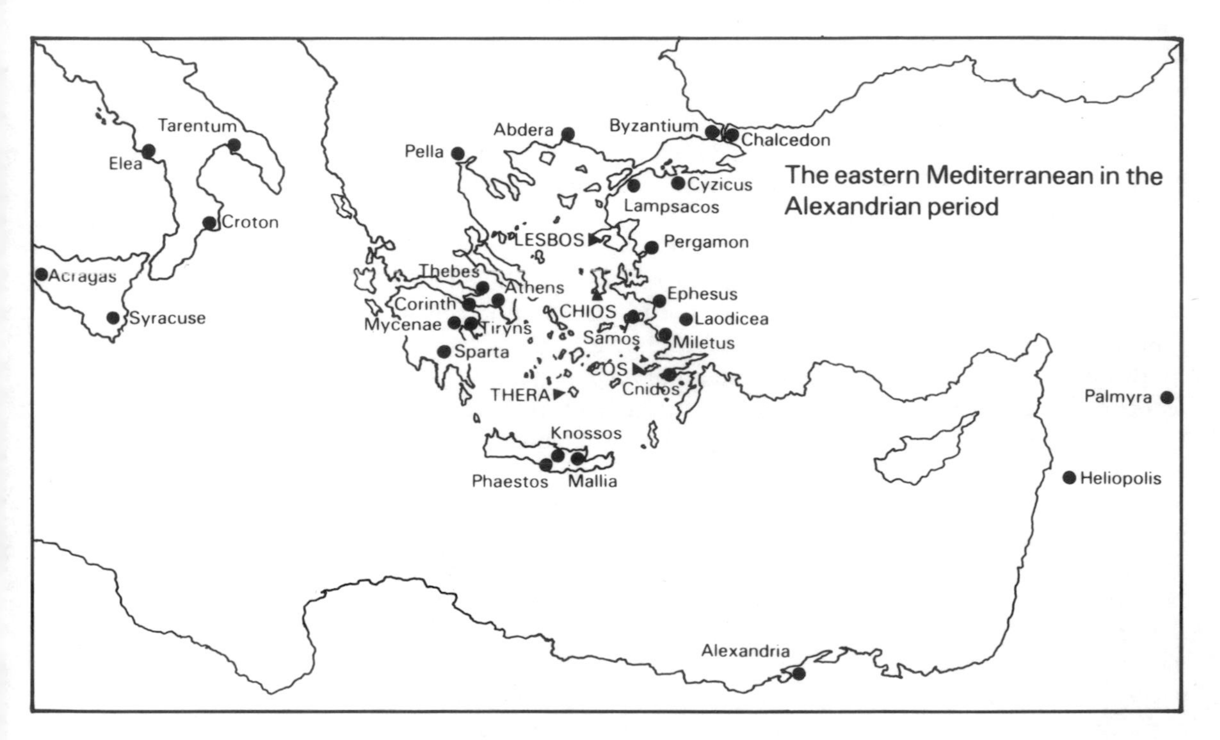

The eastern Mediterranean in the Alexandrian period

Why science should have suddenly begun to blossom here, on the eastern side of the Mediterranean, is a question to which there is no categorical answer. There would seem to be no geographical or racial reason why this should have been so; all we can say is that here were colonists living in a new political environment, entirely of their own devising and not imposed from without, and in an area which was also new to them. They were bound to ask questions and seek answers, which they might not have done had they remained settled in a traditional way of life. They could see that frequently there was more than one solution to a particular problem, and that it was not always desirable to do things as they had always been done. A new look at things might bring about all kinds of improvements. What is more, Ionia was a trading area, a focus for merchants from the east and south-east, from the Fertile Crescent or from further afield, from Iran and from India and even from China. The Ionians lived, then, in a stimulating environment, and nowhere was this more true than at Miletos, its main harbour and richest market.

Thales of Miletos

Thales was born about 624 BC of parents who may have been of Phoenician origin, although he was usually considered a true Milesian by descent, and was a statesman, mathematician, astronomer and business tycoon, as well as being one of the traditional 'Seven Wise Men' of ancient Greece. His statesmanship was exercised in persuading the Ionian cities to unite against the Persians, whose growing power was a continual worry to him. Yet in the end it proved impossible for them to withstand the Persian onslaught. As to business acumen, Aristotle tells that when Thales was criticized for being impractical and spending too much time on philosophy instead of making money, he decided to confound his critics. Predicting a glut of olives during the next summer, he paid deposits on all the olive presses in Miletos and nearby Chios, hiring them at a low price because no one bid against him. When the olive harvest came the presses were all needed at the same time and Thales let them out at any rate he pleased. 'Thus', said Aristotle, 'he showed the world that philosophers can be rich if they like, but that their ambition is of another sort.' However, there is another story which tells of Thales falling into a well while star-gazing, and being mocked by a pretty Thracian servant girl for trying to discover what was going on in the heavens while unable to see what was at his feet. So we have two opposite traditions, one showing how practical a philosopher could be, the other how impractical. But Thales's philosophy had a practical side – he worked out how to determine distance at sea, for instance – and in the century after his death in about 547 BC he was extolled as a model of practical ingenuity. Perhaps there is more to be said for Aristotle's story than for the other.

Thales's fame rests primarily on his astronomy, and on a feat he could not possibly have done! He is said to have predicted the total eclipse of the Sun which occurred on 28 May 525 BC and led to the cessation of six years' hostilities between Lydia and the Medes. Thales is supposed to have achieved this by using a cycle of eclipses – the 'saros' – familiar to the Babylonians, which Thales had learned about during travels in Egypt. But in fact the Babylonians knew of no such cycle; all they could do was to see if an eclipse was likely by observing the Moon well after its last quarter, when it was close to New Moon. Did Thales know this and if he did, could he have worked it out? Most historians of science are satisfied that he could not have done so, at least not as the fifth-century Greek historian Herodotus described, for Herodotus states that Thales was able to tell 'within the year' when it happened. We can certainly rule out a lucky chance and must conclude, however regretfully, that this was something fathered on to him posthumously. It was just the kind of thing which occurred in the past, and happened not only to Thales; a century later the philosopher Anaxagoras was supposed to have predicted the fall of a large meteorite at Aegospotami – a chance event quite impossible to forecast.

The Greeks claimed to have received their mathematics from Egypt by way of Thales; Herodotus, Aristotle and his pupil Eudemos, who wrote a history of mathematics, all claim that Thales 'after a visit to Egypt, brought this study to Greece'. Indeed, Eudemos goes so far as to specify what Thales brought, namely a number of propositions in theoretical geometry. Yet our present knowledge of Egyptian mathematics gives us no grounds for supposing that the Egyptians actually possessed any geometrical theory; theirs was a practical rule-of-thumb geometry. But if Thales did not bring such geometry from Egypt did he, perhaps, devise it himself? Certainly geometry was that branch of mathematics at which the Greeks were to excel: their art shows their love of symmetry and elegant shapes but, though they were later to display geometrical genius, there is no firm evidence that Thales began the process. Certainly he did some geometry, but it would all seem to have been of a practical kind – and that was just the kind of thing he could have brought from Egypt. Yet, of course, it was from such practical geometry that the whole theoretical structure was later to be developed.

Thales may not have invented geometrical theory, nor predicted a total solar eclipse, but it is evident that he had a penetrating mind. He considered the nature of the world and concluded that water was the basic constituent of all things. The Earth, he believed, was a flat disc floating on water. Today this may seem simple and naive but to one who had travelled in Egypt and seen its barren land brought to life by the Nile floods, it might well appear a logical, rational thing to suggest. Yet it was more than this. For Thales was using no gods to account for the fertility of the land; he was attempting to find a natural physical explanation. In short, he was taking as scientific a view as he could. And it was this mental approach he adopted for explaining earthquakes, using his floating-Earth idea as a basis, since he postulated they began as eruptions of hot water in the surrounding oceans that shook the Earth. At all events Thales clearly seems to have been the first to display those qualities which were to characterize Greek science: to give natural, not supernatural, explanations of the world, and to try to derive underlying theories from the facts of observation and experience.

Anaximandros, Anaximenes, Hecataeos and Heraclitos

Anaximandros (often known as Anaximander) was a younger contemporary of Thales, being born about 610 BC; he died some time after 547 BC. As with Thales we have only the unreliable reports of later Greek philosophers to help us. Thus Anaximandros is often credited with determining the equinoxes and the obliquity of the ecliptic, i.e. the fact that the Sun's apparent path in the sky is inclined at an angle to the celestial equator (that imaginary line across the sky drawn at 90° from the celestial pole, the point around which the heavens appear to rotate). Yet this seems unlikely, both because they were already known in Babylonia, and also because such work would run quite contrary to other ideas he certainly appears to have entertained.

Anaximandros did, though, draw a map of the inhabited world and write a book to explain the Earth and its inhabitants: he believed that there were an infinite number of worlds, all having been divided from an infinite universe to which, one day, they will return and be reabsorbed. He then went on to explain that the material from which the Earth and its surroundings were formed was separated off and given a rotating motion; as a result the heavier materials fell to the centre, forming the Earth itself, while fire and air were left at the edges to form the celestial bodies. The Sun and Moon were, he thought, rings of fire surrounded by air. This air had passages shaped like pipes through which the light from the fire escaped. Thus the phases of the Moon were explained by the Moon-pipe opening being seen from various angles. He explained eclipses in the same way.

The Earth itself, said Anaximandros, was like a short cylinder, man living on the surface at one end. Water, too, played an important part in his scheme of things, for all animals were made from substances in the sea when acted upon by sunlight. Man himself originated from fish. So here, about the same time as Thales, we have a coherent scheme of the whole of creation, all explained from one basic premise – the generation of the Earth and the heavens from some primeval material. It may not be a scheme we could subscribe to today but it enshrined an approach (the devising of hypotheses and the study of their corollaries) which is not so very far from that still used at the present time, more accurate though our modern theories may be.

Anaximenes may have been a pupil of Anaximandros, and his view of the cosmos was very similar. He too subscribed to the belief that the infinite was the source of the Earth and the heavens, but it was his opinion that air was the basic substance from which everything came: it stretched outwards to infinity. Anaximenes was led to this view of air because he had noticed the processes of rarefaction and condensation: he said that when air is distributed all around us it is invisible, but when condensed it turns into water; when air is heated it turns in due course into fire. To support these contentions, he quoted the observation – or can we call it 'experiment'? – that when we blow out air from our mouths it is cold, whereas when we open our mouths wider and so do not compress the air, it is hot. To Anaximenes this primal substance was essentially composed of tiny particles. He used the name 'air' also for the primal substance itself because, like the real air, it was everywhere, permeating everything.

Breath was identified with the soul – a view which seems to have been widespread in many early civilizations – and it was the soul, the breath, which kept man's physical body together. Moreover, Anaximenes believed the whole of creation, the whole cosmos, breathed, and thus was the soul of all things.

As far as the disc-like Earth itself was concerned Anaximenes, like Anaximandros, thought that it and the whole material world was formed by condensation from a rotating mass of air. The Earth floated on air. The Sun and Moon were discs made of fire and turned around the Earth; they became invisible when far away and behind the high northern parts of the Earth. And if we look at a map, we shall see that there are mountains to the far north of Miletos – the Rhodape Mountains of Bulgaria. In his approach, then, Anaximenes was following in the footsteps of his mentor Anaximandros, and once again attempting to build a world-picture from hard scientific facts.

Hecataeos, about whose life very little is known, is remembered now in some quarters as one of the earliest writers of Greek prose and in others for writing the earliest Greek work on geography. Unfortunately his original geographical text no longer exists, but from later descriptions it appears to have been accompanied by a map and to have consisted of two sections, one on 'Europe' and the other on 'Asia and Africa'. It was, it seems, concerned with details of people and places, especially in coastal regions. Hecataeos still thought of the inhabited world as disc-shaped and surrounded by the waters of Oceanus, but within these limitations his map seems to have provided a general layout of the Mediterranean area, as well as showing the presence of Libya, Egypt, Mesopotamia and part of India, and even indicating something of Europe; the lands of the Celts and of the Scythians (a nomadic people in southern Russia and the Crimea) were marked. As we might expect in a geography based mainly on hearsay and reports from sailors and merchants, the fabulous and the genuine are to be found side by side; crocodile hunting, the hippopotamus and the phoenix are all described with equal credence. Yet as a whole the book was sound, even if Hecataeos set too much store by the ocean ring Oceanus, using it, for example, to explain the annual inundation of the Nile rather than taking the more realistic view of Anaxagoras that it was caused by water from melting snows in the Libyan mountains.

Heraclitos, who seems to have been a contemporary of Hecataeos, was a native of Ephesus, a port some 50 km (30 miles) north of Miletos. A stern critic of other philosophers, he is said to have been an arrogant misanthrope; his work was both abstruse and peppered with cutting remarks about lack of intelligence in his readers, but in spite of these failings it contained some useful ideas. He maintained that the universe is balanced between opposites and in perpetual tension, just like a string in a musical instrument, an attitude of mind which colours all his explanations, especially those on human behaviour. It may be that in all this he was trying to put over some message about the human soul but if this is so, he was unsuccessful. His fame rests now, as it did in his own day, on what he taught about the natural world. He saw everything there in an unstable state of flux, so that what we perceive with the senses is something transitory, not true knowledge – a view which was later to receive wide currency and militate against setting too much store by practical observation.

In this universe of Heraclitos fire held a primary place as the agent of change; fire consumes things, changing them until they themselves become fire, while without it substances can condense or solidify. The heavenly bodies are cups containing fire, and the phases of the Moon are due to the way the mouth of its cup is turned towards us; in a very clear sky and at a high altitude above the horizon, the Moon's disc can indeed look as if the surface is the inside of a container, so his idea was not as wild as it may seem.

Pythagoras

Of all the figures of Ionian science none had more influence on later generations than Pythagoras. Born about 560 BC on the island of Samos, some 50 km (30 miles) north-west of Miletos, he was a religious leader as well as a prototypical scientist, and there are many legends about him. What is not in doubt is that he took part in a religious revival which occurred all over Greece during the sixth century BC, and in due course became the leader of a brotherhood which conceived a new kind of holiness. This, the Pythagorean Order, expected its members to perform ascetic exercises, abstain from certain actions and avoid specific foods – it seems they had a vegetarian and alcohol-free diet, and had to avoid wearing animal products such as wool. Women as well as men were members of the Order and all wore a distinctive dress, went barefoot, and lived a simple life of poverty.

The Pythagoreans believed that the soul can leave the body, either temporarily or permanently, and move into the body of another human being. Possibly such a view had an oriental origin, but the Pythagorean movement itself was probably a reaction against the extravagances of the cult of the wine god Dionysus. In due course political ideas became welded to the religious, as might be expected of an Order which kept itself separate from the general community, and Pythagoras was eventually forced to leave Samos. He moved to Croton (now Crotona) in southern Italy where he founded an ethical-political-philosophical academy which advocated aristocratic rule. It was at first welcomed by those who opposed the growth of democracy, but later its doctrines were found unacceptable and Pythagoras was forced to leave the city. He travelled northeast to Metapontion on the Gulf of Tarento, where he died about 500 BC. When, some fifty years later, a violent democratic revolution occurred throughout the Greek cities along the southern Italian coast, the Pythagorean Order was attacked and its meeting houses destroyed. However, many of the members managed to escape, either northeast to Tarentum or to Philiasos (or Phileius) on the Greek mainland.

At Tarentum they continued as a political power until about 350 BC.

Historians of science are now in some disagreement about precisely what it was that Pythagoras taught, and to whom he imparted his teaching; quite probably he instructed only an initiated inner circle of disciples. Moreover it is perhaps impossible to decide which Pythagorean doctrines are due to Pythagoras himself and which to his close followers. All the same, by the fourth century BC other Greek philosophers believed, erroneously perhaps, that certain ideas were due to Pythagoras, an assumption there is little point in disputing; after all, it is the influence rather than the precise origin of these ideas with which we are concerned.

Pythagoras visited Egypt and Babylonia when a young man, and perhaps it was this visit which gave him the urge to study mathematics and declare that 'all things are numbers'. Certainly it seems that Pythagoras and his followers were much concerned with numbers, and there is no doubt that they developed a whole number theory. This theory appears to have been based on three kinds of observation. In the first place they noted that there was a mathematical relationship between the notes of the musical scale and the lengths of either a vibrating string or a vibrating column of air, as in a 'panpipe' or a flute. An air column or a string of specific length would give one note, and then reduced to half its length would give a note one octave above. A ratio of lengths of 2 to 3 gave the musical interval known as the fifth, and a ratio of 3 to 4 gave a fourth. Thus, if one takes a vibrating string which is 12 units (inches, millimetres, etc.) in length, and reduces its length to 8 units, it will sound a fifth above the original note; reduce it to 6 units and it will sound the octave. Therefore, as the octave and fifth were considered harmonious sounds, Pythagoras said that the numbers 12, 8, and 6 were in a 'harmonic progression' and he considered this so important that he extended the idea to geometry, with the result, for instance, that he claimed that a cube was in geometrical harmony because it has 6 faces, 8 corners, and 12 edges.

Illustration page 75

For comparison with the Chinese study of sound, see pages 171–172.

The second observation was concerned with right-angled triangles. From Egypt Pythagoras would have learned of the 3, 4, 5 rule about the lengths of the sides, but modern research has shown that in Babylonia he would have come across what we call the Pythagorean relationship. The Babylonians had, in fact, realized that the numbers could be 3, 4, 5 or 6, 8, 10 or any such combination where the largest number squared is equal to the squares of the other two numbers added together. This was a distinct step forward and the Pythagoreans were to make good use of it. Their third observation was that there were definite numerical relationships between the times taken by the various celestial bodies to orbit the Earth.

The Pythagorians drew the reasonable conclusion from their studies that 'everthing is numbers'. The modern mathematician, particularly with the recent development of the computer sciences, might reach an overtly similar conclusion; but there is a vital difference. The

Pythagorians' idea was fundamentally a mystical one and ascribed an absolute, even divine, status to the numbers and their relationships. Today, even though some philosophers claim that mathematics represents a more pure form of knowledge than any other, the scientist uses numbers not as a divine principles but as an extremely powerful and flexible tool for describing and predicting all manner of natural phenomena.

At this time counting was often done using pebbles, and since the Pythagoreans, like other Greeks, had an inherent feeling for shape, it is not to be wondered at that figurate numbers captured their interest. Figurate numbers are just what their name implies, numbers made from counting in patterns. Start, for instance, with one pebble and then count in triangular shapes; this will give the figurate numbers • or 1, ∴ or 3, ⛬ or 6, ⛬ or 10, and so on. The numbers in this series were considered to have special significance for they gave the 'perfect' number 10, since 1 + 2 + 3 + 4 = 10, which, because it had four dots on each side, was called the 'tetractys'; it was considered holy and the Pythagoreans literally swore by it. There was a second group of numbers which were called 'perfect'; these were numbers which were equal to their separate factors added together. Such were 6 (because 6 = 1 + 2 + 3), 28 (1 + 2 + 4 + 7 + 14), and so on. The next such number is 496, then 8,128 and 2,096,128, . . .; clearly it was not easy to work these out at the time, although a century later a general formula for calculating them was devised by Euclid – an instance of the close connections between geometry and arithmetic.

A similar search by the Pythagoreans led to 'amicable' numbers, that is two numbers each of which is equal to the sum of the factors of the other. Thus the pair 220 and 284 are amicable (because the factors of 284 are 1, 2, 4, 71 and 142 and these add up to 220, while 220 has the factors 1, 2, 4, 5, 10, 11, 20, 22, 44, 55 and 110, which total 284). The numbers 220 and 284 are supposed to have been discovered by Pythagoras himself, and certainly they were the only pair of amicable numbers known in antiquity.

Figurate numbers were very important in Pythagorean arithmetic and of course there were many kinds besides the triangular numbers just mentioned. There were square numbers, • ∷ ⁙ etc. (1, 4, 9 . . .), pentagonal numbers, numbers formed from rectangles with unequal sides ('heteromeke' numbers), numbers formed from pyramids with square bases and pyramids with triangular bases, cubic numbers and even 'altar' numbers (numbers formed by pyramids whose bases were rectangles with unequal sides).

Another aspect of numbers which intrigued the Pythagoreans were 'means'. To begin with they were concerned with the 'arithmetic mean' (i.e. the middle number of three in an 'arithmetic' progression: thus in the progression 4, 5, 6 the arithmetic mean is 5, in the progression 4, 8, 12 it is 8, etc.) and it seems probable that Pythagoras learned of this when he visited Babylonia. Later they moved on to the 'geometric mean' (i.e. the middle number of three in a 'geometric

Opposite above A Greek amphora depicting Ulysses and the Sirens in a scene from the Odyssey. Many scholars have tried to interpret poetical images such as the Sirens with natural or geographical phenomena, but without conspicuous success. British Museum, London.

series' such as 2, 4, 8 where 4 is the mean, or 3, 9, 27 where it is 9), and to the harmonic mean (in the harmonic series 6, 8, 12 mentioned above, the harmonic mean is 8). But what did all this signify? In the first place, on the purely practical side, it meant that those of the Pythagorean Order who were interested in numbers were able to develop arithmetic and the techniques of handling numerical quantities. Secondly, as far as the religious side was concerned, it all enshrined a great deal of mysticism, with plenty of occult relationships between numbers themselves, a kind of magical numerology which had no scientific significance whatever.

Pythagoras is still popularly remembered because of the theorem about right-angled triangles, yet it is not clear whether the general theorem was proved during his lifetime or later, and indeed it is not at all certain what contributions to geometry he or his followers made. We do know, however, that the five-sided figure, the pentagon, was of great mystical significance, as well it might be, since when its sides are extended to make a five-pointed star, the sides and diagonal lines across the figure intersect in proportions which give rise to the 'golden section'. (The golden section is a proportion which classical antiquity thought to be very pleasing to the eye and it was frequently used in architecture; it is the division of a length so that the relation of the smaller section to the greater is the same as the relation between the greater section and the whole.) The Pythagoreans used the pentagon as a sign of recognition among themselves but they did not invent it; it was known in Babylonia.

Of all the mathematical knowledge attributed to the Pythagoreans the most important was that which comes out of the Pythagorean theorem, the fact that not every quantity can be expressed in whole numbers. For although the long side or hypotenuse of a right-angled triangle may have its length expressed in whole numbers, more often it does not: whether or not it does depends upon the lengths of the other sides. Thus if the shorter sides are 3 and 4 then the hypotenuse will be the whole number 5 (because $3^2 + 4^2 = 25$ and the square root of 25 is 5), whereas if the shorter sides are 4 and 5 the length of the hypotenuse is not a whole number but 6.4031242 . . . The early Greek philosophers were unhappy about this; it worried the Pythagoreans and it worried later mathematicians since it threatened the idea that geometry was the foundation of mathematics, but it led to much careful work, and in that way acted as a stimulant.

For a later Greek approach to this problem, see pages 99–100; for the Chinese attitude to irrational numbers, see page 150.

The Pythagoreans were probably able to construct three of the five 'regular' solids (solid figures like the cube in which all faces are equal as well as the angles between them). They knew of the cube, the pyramid and the twelve-faced dodecahedron, and doubtless the symmetry of these delighted their mystical souls. The belief in the fundamental importance of number also coloured their attitude to music. We have already seen how numbers were used for expressing musical intervals, and this work was extended so that entire musical scales covering many octaves could be worked out. Later still, in the middle

Opposite below The temple of Aesculepios at Cos, an attractive complex where healing was done through ritual and divine intervention. Hippocrates is said to have trained there, before undertaking his travels throughout Greece.

Above A late medieval woodcut illustrating Pythagoras' appreciation of the mathematical relationship between the notes of a musical scale and the lengths or sizes of vibrating strings, bells or columns of air in wind instruments.

Right Roman mosaic from Pompeii probably depicting philosophers in the Academy. Plato, the central figure sitting under the tree, is teaching geometry using a stick for drawing figures in the sand. Museo Archeologico Nationale, Naples.

Right The Tower of the Winds in Athens, dating from the first century BC. At the top were sundials and inside was an elaborate clepsydra or water-clock which showed the time on a dial. Though a rare example of Greek mechanical contrivance, its ingenuity suggests that other similar devices may have also been constructed.

of the fourth century BC, Archytas of Tarentum was to develop this into a complex musical theory. The Pythagoreans also believed there was a musical aspect to the heavens – the so-called 'music of the spheres' – a view based partly on their knowledge of the mathematics of sounds and partly on their studies of the orbital times of the planets.

For this idea in the work of Kepler, see page 339

Pythagorean astronomy obviously owed much to the Babylonians; like them it conceived the celestial bodies as divine while it also accepted the idea that the planets were at different distances from the Earth and closer than the stars. The Pythagoreans' love of numbers led them to study this by determining the periodic motions of the planets, and in the end they adopted an order of distance for them based on the speed at which they moved in their apparent orbits around the Earth. This gave them the order Earth, Moon, Mercury, Venus, Sun, Mars, Jupiter and Saturn, although they later refined it by placing Mercury and Venus after the Sun, having observed no transits of either planet across the Sun's disc. We now know there are such transits, although they are rare and can only be seen through a telescope, but as the modified sequence of distances was what we now know to be correct (as measured from the Sun not the Earth as starting point, and omitting the Moon), we may be grateful that the Pythagoreans were not aware of such transits.

The Pythagoreans' love of beauty and symmetry, as well as their preoccupation with numbers, led them to some important views about the universe. The first of these was that the planets must all move regularly round the Earth in the simplest of curves, that is to say, in circles; it was a view, as we shall see, which was to have the deepest effect on Greek and medieval European astronomy. The second was that the heavens and the Earth were spherical. Why this view was adopted is uncertain, but it may well have been partly for symmetry; a spherical heaven is much more elegant than the hemispherical one Homer had described. And as for the Earth itself, quite possibly Pythagoras had observed a ship sinking over the horizon and come to the conclusion that the Earth was curved, not flat, but even so, accepting a sphere rather than a mound would have been an act of faith, faith in the essential beauty of creation.

However, the most surprising of all the Pythagorean views of the universe was their suggestion that the Earth was a planet, in orbit like the other planets. This view is usually credited to Philolaos, one of Pythagoras's disciples, who was born in Croton and worked in the second half of the fifth century BC, and it may well be that the idea is his. The basis of this theory of a moving Earth lay in the significance of the number ten, the tetractys; it was believed that it must express the number of moving bodies in the universe. To get a number as large as this meant that some rearrangement was necessary, and this was achieved by placing, not the Earth, but a 'central fire' in the middle of the universe, putting everything else in orbit around it. Thus there was the central fire (stationary), about which

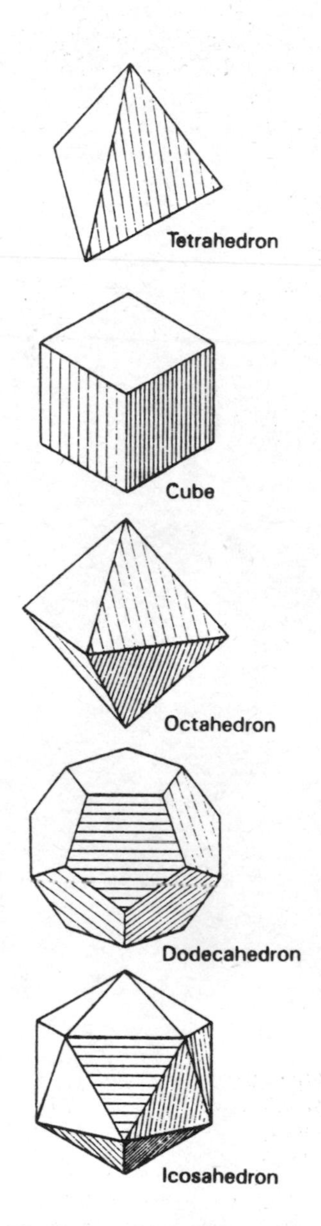

The five regular solids beloved of the Pythagoreans for their unity of shape and angle.

orbited the Earth, the Moon, the Sun, the five planets and the slowly moving sphere of the stars. Yet these give us only nine moving bodies, not ten. To overcome this problem Philolaos, if Philolaos it was, proposed the existence of an 'antikhthon' or 'counter-Earth' which also orbited the central fire, and did so at precisely the same speed as the Earth itself. In consequence, the counter-Earth always lay between the Earth and the central fire (the light of which was reflected by the Sun), so that when the inhabited part of the Earth turned away from the Sun to give us night – for Philolaos believed the Earth to rotate – the central fire still could not be seen directly.

This whole idea was indeed ingenious; it solved the aesthetic and mystical problem which demanded ten orbiting bodies, and at the same time it accounted not only for observations of planetary motions but also explained why the central fire was invisible. Although untrue, in 450 BC this was a bold attempt to formulate a world-picture and, with its orbiting and rotating Earth, it was to exert no little influence on some later philosophers.

Science on the Greek Mainland

Anaxagoras, Parmenides, Zeno and Empedocles

Greek scientific thinking began in Ionia but was not confined there for long. As we have seen, Pythagoras left Samos and moved to southern Italy, and so spread the scientific approach to a different part of the Mediterranean coast. Another who did the same was Anaxagoras, who was born some sixty years after Pythagoras in Clazomenae in Lydia, about 65 km (40 miles) northeast of Samos and not much further from Miletos, but he left there at the age of twenty to visit Athens, where he stayed for the next thirty years. Here Anaxagoras became friendly with the statesman Pericles, through whose efforts Athenian democracy reached its height, becoming for a time the greatest power among the city-states. It was a friendship which had its drawbacks, however, for political opponents who could make little impact on Pericles attacked Anaxagoras instead, charging the 'atheistic philosopher' with impiety, and had him banished.

Anaxagoras wrote only one book, which he completed some time after 467 BC. In this he maintained that in the beginning the universe was a motionless uniform mixture. 'Mind' then entered and caused the whole system to whirl around; in the ensuing vortex, dark, dense, cold matter fell to the centre, forming a disc-like Earth, while all hot, dry, rarified matter was thrown to the outside. The Sun, Moon, stars and planets were, he thought, torn out of the Earth, and were heated by friction as they were then carried round in the maelstrom of matter. He believed the Sun to be no more than a red-hot stone. It was all a very rational and logical view, even if it lacked the inspiration of the ideas of the Pythagoreans, and it too played a part in influencing some later thinkers.

Parmenides and Zeno were natives of Elea, a seaport on the western coast of Italy that had earlier been a refuge of the Ionian philosopher

Xenophanes, who had fled when the Persians overran Phocaea, the northernmost city of Ionia. Xenophanes had been a believer in a kind of monotheism, a supporter of one god above all other gods and men, a god who was the cause of all motion. He also taught that land and sea had once been mixed together, and said that shells found in mountainous areas supported this contention; indeed he seems to have recognized fossils for what they were – true animal remains – and it may well be that Parmenides had once been his pupil. Parmenides is important to us not so much because he followed the monotheistic idea but because he tried to see the ultimate 'reality' behind natural phenomena, the essentials which lie behind what we observe. He claimed that the essence of everything is 'Being'; Being fills all space, so the universe must be one and unlimited. Of course it is possible to think of non-Being, that is of utterly empty space, of a vacuum, but this is not reality. Being is changeless, eternal and motionless; change, transitoriness and motion are non-Being and are unreal and illusory.

Obviously Parmenides reached conclusions quite the opposite to those of Heraclitos, to whom flux and change were the true reality, but Parmenides's real importance lies in his approach to the problem of trying to explain the essential continuity of the universe. For a time the view of Parmenides exerted a considerable influence.

Zeno was a friend and follower of Parmenides, with whom he may have visited Athens about 450 BC. According to legend, he met an unpleasant end some twenty-five years later when he was tortured and put to death by the ruler of Elea or of Syracuse, against whom he had plotted. Zeno is now remembered not for his politics but for his famous paradoxes, devised to prove the truth of the Parmenidean philosophy that the universe was a continuous and unchanging entity. The best known of his paradoxes is that of Achilles and the tortoise, in which Zeno 'proves' that Achilles can never overtake the tortoise, even though he can run a hundred times faster. Zeno's argument is that when Achilles has reached the tortoise's starting point, the tortoise will have moved ahead by one hundredth of the distance; and when Achilles has covered this one hundredth part, the tortoise will have travelled another one hundredth of that distance, with the result that it will still be ahead, and so on *ad infinitum*. Then there is the famous arrow paradox. Here Zeno claims that at any moment an arrow can only occupy a place equal to its size. It cannot occupy a larger space, nor can it be in two different places at the same time. Since between one instant and the next instant there can be nothing, no interval at all, it follows that the arrow cannot move. Aristotle made attempts to refute these arguments and later philosophers and mathematicians have followed suit. It is now realized that these two paradoxes, far and away the most profound and most difficult of all which Zeno invented, are due to confusing two quite different aspects of a particular problem. Either we look at the continuous flow of motion or we think of the objects – the arrow, the tortoise and

For the use of paradoxes in Chinese thought, see page 141.

Achilles – occupying various positions along its course. In the latter case we are bound to fix the object at a particular point, so putting it at rest for an instant. The two ways of looking at the problem are mutually exclusive and so they should not be confused.

Zeno has an important place in the development of Greek mathematics because his paradoxes underline that strange phenomenon uncovered by the Pythagoreans in their theorem about right-angled triangles, the phenomenon of incommensurability – the fact that some values exist which cannot be expressed in whole numbers. Zeno wanted a continuum, a continuity between one unit of time or of distance and the next. The idea of coming to terms with incommensurability by using an infinite number of very small units, as the Pythagoreans had done, was one that he rejected because it did not suit the outlook of Parmenides, and the paradoxes were designed to refute the Pythagorean approach. The paradoxes do, of course, question some basic geometrical ideas as well as numerical ones, particularly that which considers a straight line made up of a number of consecutive points: Zeno attacked this too with a paradox.

Parmenides and Zeno were philosophers rather than scientists, and some of their ideas may seem a little strange, but stranger still was Empedocles of Acragas (modern Agrigento). Acragas, which lies on the southern coast of Sicily, was a very beautiful city and a centre of Greek culture until destroyed by the Carthaginians near the end of the fifth century BC. Empedocles, who was born there about 492 BC, lived and worked in the city until his death some sixty years later, and is the subject of many legends. He is mentioned as a physician by medical writers and is supposed to have written a medical treatise, while he is also claimed to have been the author of books on other subjects, but we have no details of them and it may be that they never existed. However, Galen, the great Greek surgeon of Roman gladiators in the second century AD, referred to Empedocles as the founder of the Sicilian medical school, and Aristotle, only a century after Empedocles was supposed to have lived, credited him with the invention of rhetoric.

Empedocles is also said to have performed various wonders, including stopping an epidemic by diverting two rivers so that their waters were mixed, to have improved local climate by setting up a windbreak across a gorge and, rather more in the realms of fantasy, to have revived a woman who had no pulse and had not breathed for thirty days, and to have calmed the summer winds of the Mediterranean by trapping them in sacks. His death is variously reported: some say he threw himself into one of the craters of Etna – or slipped into it while making observations? – another claims that he ascended into heaven. Certainly his poetry – by which he communicated his ideas – shows vivid imagination, great eloquence and liveliness, and a touch of theatricality, but was there really anything for which later Greek scientists remained in his debt? Or was Empedocles a legend and little more?

In fact, Empedocles made some real contributions to Greek science, the chief of which was his doctrine of the four elements. What he seems to have done is to have modified the extreme views of Parmenides and arrived at the idea of four unchanging substances or elements or, as he called them, 'roots of all things', and two basic forces. The elements were named earth, air, fire and water and the two forces, more poetically, as love and hate, i.e. attraction and repulsion. The elements are not to be considered identical with the ordinary substances that go by those names, but rather with their essential and permanent characteristics. Every material substance is, however, composed of them: thus a piece of wood contains the earthy element (that is why it is heavy and solid), the watery element (which is why, on heating, it will first exude moisture), as well as air (it smokes) and fire (it emits flames when it burns). The relative proportions of these elements determine the kind of wood it is. As we shall soon see, this four-element theory was to prove of capital importance.

Empedocles appears also to have made some experimental investigations. Thus he used a clepsydra (water clock) to prove that the substance air really had a material existence, noticing the way bubbles were emitted when air was trapped below the water surface – an important observation if he was to continue to hold the Parmenidean view that emptiness did not mean a vacuum. He discussed the question of light and vision, making observations which led him to the view that luminous bodies emit something which meets rays issuing from the eyes; this was not correct but it was a step forward from the Pythagorean belief that vision was caused solely by something emitted from the eye. Empedocles also taught that light takes time to travel through space, though he must have arrived at this result by reason alone, for the first observational proof was not obtained until more than two thousand years later.

For Römer's demonstration of this, see page 376.

As far as the universe itself was concerned, Empedocles believed there were four stages in its development. First there was a complete mixture of the four elements inside the spherical universe; after this the elements were increasingly separated by repulsion (by the force 'hate'). The third stage was a period of total separation of the elements, followed by the fourth, a partial and increasing mixing once more due to attraction ('love'). Whether Empedocles considered this to be a cyclic process or a process which occurred only once is not clear and is still a matter of debate. What is not in doubt is that he believed not only that the universe was spherical and full of material, but also that it was contained in a surrounding transparent globe. At this time the only truly transparent material known was rock crystal (a variety of quartz), hence the belief that the universe was a crystal sphere, an idea later generations elaborated as a series of spheres nestling inside one another. The stars were fixed to the crystal sphere and, like the planets, were lumps of fire, although the Moon shone, he realized, because it reflected sunlight, a view held by Parmenides.

Empedocles then extended this explanation also to the Sun, which he thought to be an image of the entire daylight sky reflected by the Earth's surface, for unlike the Pythagoreans, he does not appear to have thought of the Earth as a sphere. He did, however, give a correct explanation of solar eclipses as being due to the Sun's light being obscured by the Moon.

Another interesting idea of Empedocles's was one concerned with the fourth stage of evolution of the universe; the creation of animals. He believed that in the beginning various animal parts – various limbs and organs – were formed and that these later came together. In the earliest stages this resulted in the creation of monsters – hence the legendary stories about them – but they were not well adapted to their environment and did not survive or reproduce. Eventually, though, satisfactory forms in harmony with their surroundings did arrive; they succeeded in producing offspring and so survived. This was, in a sense, a doctrine of evolution and natural selection and was, indeed, quoted by Darwin in his preface to the famous *Origin of Species*. But with Empedocles the selection stops once suitable creatures have arrived and propagated themselves, yet it is just here that Darwin's theory of evolution begins, calling on the effects of heredity.

For the development of the theory of evolution in the 19th century, see pages 420–430.

Lastly we must note that it was Empedocles who put forward the idea that the blood in the body ebbed and flowed, taking his example from the way water behaves in a clepsydra. This view was generally accepted to be true until 350 years ago.

The Greek Atomists

Greek atomic theory was born in Abdera, a flourishing seaport on the northern shores of the Aegean established by refugees from the Persian invasion of Lydia. Here Leucippos, the first Greek atomist, settled after he had moved from Miletos some time around 478 BC, yet although it is evident that Leucippos put forward his theory in Abdera, we do not know precisely what he taught, nor the source of his ideas. Nothing – no book, no written fragments – of his teaching survives and the theory as we know it comes from his pupil Democritos, the most able advocate of the new teaching. But from what Democritos tells us it seems that Leucippos displayed many traits of Ionian philosophy, especially about the formation of the universe.

There are two traditions about Democritos: one that he was born in Abdera about 500 BC and died as a very old man about 404 BC; the second and more likely account is that he was not born until about 460 BC. This would make him a contemporary of Socrates whom, he says, he went to Athens to see, though it seems he was too shy to introduce himself. At all events he inherited riches, and determined to do some research abroad. This was not unusual; philosophically minded Greeks often travelled about the Mediterranean in search of knowledge, but some commentators have claimed that Democritos went eastwards either to Persia or, even further, to India. These stories seem to have arisen to try to show that the origins of the atomic theory came from the East, but recent research makes this appear unlikely. There was, admittedly, an 'atomic theory' in India but it was more like an equivalent of the Greek four-element theory than the kind of thing Leucippos and Democritos were to produce. Indeed, there would now seem to be every reason to suppose that Greek atomism was indigenous and a natural development of previous ideas; a growth of a reaction against certain aspects of the teaching of Parmenides and Zeno. This, certainly, is what Aristotle later supposed and he may well have been right, for by the middle of the fifth century BC it seemed to a number of thinkers that Parmenides had shown that attempts to explain the beginning of the universe were doomed to failure.

For atomic theory in India, see page 194.

Parmenides had shown to his own satisfaction and without much doubt to the satisfaction of many others, that nothing could have come into existence from non-Being, i.e. from nothing. But this view was coupled with his teaching that Being, the essence of things, cannot alter (if it did alter it would become non-Being). Thus any explanation of the origin of the cosmos using primary substances was bound to come to grief because it would involve change. This was an unsatisfactory state of affairs, and Anaxagoras, and no doubt others, had sought a way out of the impasse without success. An atomic theory appeared to be the one means of solving the dilemma.

The basis of the Leucippan-Democritan theory is that two things exist, atoms and the void: the world is thus composed of lumps of material in a sea of total emptiness. The atoms are solid substance, infinite in number and shape, and most, if not all, too small to be seen; indeed in a later form of the theory held in the next century at Athens by the Ionian Epicuros the atoms were certainly invisibly small. An atom cannot be cut up or divided up in any way at all and is completely solid. All atoms are perpetually in motion in the void.

When a collection of atoms becomes separated a vortex occurs and like atoms tend to gravitate together, forming a kind of skin as they hook on to one another. This is a spherically shaped covering and it is this which contains our whole universe. Yet this spherical bubble is not alone. As there is no limit to the extent of the void nor, as we have seen, to the number of atoms, so there can be many other spherical bubbles, other universes, besides our own. These all vary in size and in what they contain – one will have no Sun, another no animals, and so on – and now and then such universes will collide and be destroyed.

Everything we see about us is composed of atoms and void: this must be so because nothing else exists. Substances differ from each other because their atoms differ either in shape, or in the way they are arranged. Moreover, the atoms may be close enough to touch, giving a dense rigid material, or they may be some distance apart, in which case we have a soft pliable material. Strangely enough it seems that Democritos did not postulate that atoms had any weight of their own; it was left to Epicuros to suggest this. On the other hand Democritos believed that atoms did not cease their individual motions

Opposite above left A 17th-century illustration of the universe according to Plato and his followers. The Earth and its four elements are at the centre, the sphere of fixed stars at the outside, and the planets, Moon and Sun in between. From Giambattista Riccioli *Almagestum Novum*, 1651.

Opposite above centre A 16th-century illustration of the homocentric or concentric spheres suggested by Eudoxos. Below are diagrams of the solar (left) and lunar (right) eclipses. Biblioteca Estense, Modina.

Opposite above right Aristotle's proof of the curvature of the Earth from a 16th-century manuscript of *De Sphaera*, combining his observation of objects on the horizon with his understanding of lunar eclipses. Biblioteca Estense, Modena.

once they were part of a body; this merely restricted how much freedom of movement they possessed. Such movement was due to 'necessity', which he took to be some inherent cause unaffected by outside conditions.

The range of everyday experience which the atomic theory could explain was certainly extensive. Taste, small, touch, sight and hearing were all results of atomic behaviour: taste was caused by direct contact between atoms of the substance and those of the mouth, sounds were generated by atoms imprinting themselves on the intervening atoms of air, which carried the imprint to the ear; sight and smell were similar air-borne imprints, while touch was a contact mechanism like taste. But this was not all; the atomic theory could explain a wider range of phenomena too. Fire and the human soul were both atomic, each composed of very fast-moving spherical atoms that were unable to link together. At death the atoms of the soul departed from the body, but they did this only slowly; that is the reason the hair and nails of a corpse continue to grow for a while. The atoms of the soul have the task of generating warmth in a body and making it able to move: they are in fact a vital force, the very essence of life. But once they have departed there is death and, what is more, nothing is left. The atoms of the soul disperse and all is over. There is no after-life because there is no soul to experience it.

It was not only a novel doctrine but a materialistic one. To Democritos everything was predetermined in the sense that it was the result of plain cause and effect among the atoms. Chance certainly played its part, but this did not alter the predetermination of events, only our inability to predict them. It was an ingenious, logical theory and explained many phenomena, but it was essentially an exercise in scientific speculation. It differed markedly from modern atomic theory because our theory, the foundations of which were laid in the nineteenth century, is based on careful measurement and precise chemical analysis; the Greek atomic theory should not be confused with it for, intellectually brilliant though it was, it was the result of no experimental techniques.

For the elaboration of the atomic theory by Dalton in the 19th century, see pages 436–437.

In spite of their many obvious attractions, the atoms of Democritos never had a permanent place in the mainstream of Greek science. They were accepted by Epicuros, certainly, and were part of his materialistic philosophy; they were also an integral part of the long expository poem *De rerum natura* (On the nature of things) by the Roman Lucretius (Titus Lucretius Carus) written in the first century BC. This was not so much a scientific poem as one commending Epicurean philosophy to the Romans, yet it was in the philosophic doctrines of Epicuros that the atomic theory was to remain embedded and, in the end, buried.

Opposite below left A section of a late manuscript of Aristotle's *Analytics*, a text on logic; it indicates the close association in the Greek mind between logic and geometry. Universitäts Bibliothek, Basel.

Opposite below right A page of Euclid's geometry containing four separate theorems. From *Euclidis megaresis philosophi platomy*, 1505.

Socrates and After

We must pause now and turn from the growth of Greek science to look for a moment at Socrates (470–399 BC), perhaps the most famous of all Greek philosophers, 'the best and wisest and most righteous curio Martem, huic Iouem, Ioui Saturnum, Saturno Fixas ſuperponit. Auctor tamen de Mundo ad Alexandrū, *Porphyrius*, *Apuleius*, *Marſilius Ficinus*, & quidam Platonici Mercurium ſupra Venerem collocarunt, reliquis retentis: ideòq. duas ſeries in ſequenti diagrammate vides. Sed de hoc ſyſtemate plura lib.9. ſect.3. cap.3.& 5.

II. Syſtema Platonis, aut Platonicorum.

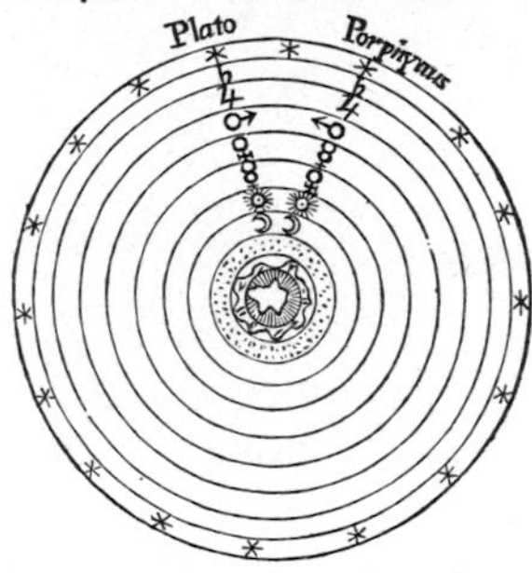

Non deſunt tamē rationes & authoritates, quibus probabile fieri poſſet Eudoxum, Calippum & Ariſtotelem ſecutos fuiſſe ſyſtema Pythagoræ, de quibus ſuo loco. Interim falſum eſt, quod ait *Clauius* in ſphæra, ſolum eſſe Authorem libelli de Mundo ad Alexādrum, qui Mercurium ſub Ioue ac ſupra Venerem ponat. Demum Plato concedit in Timæo, Terram verti circa ſuum centrum, quod negat Ariſtoteles.

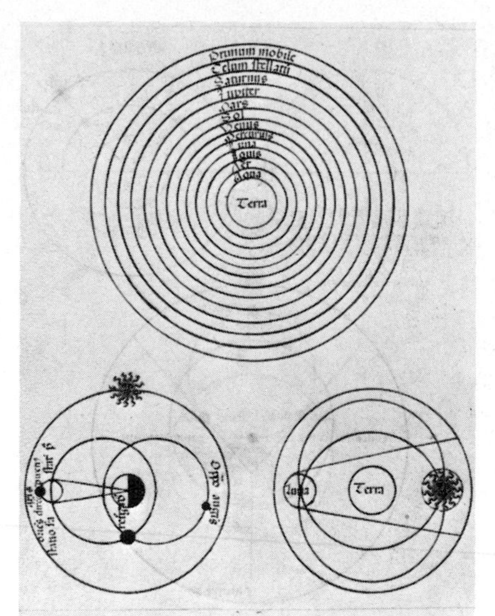

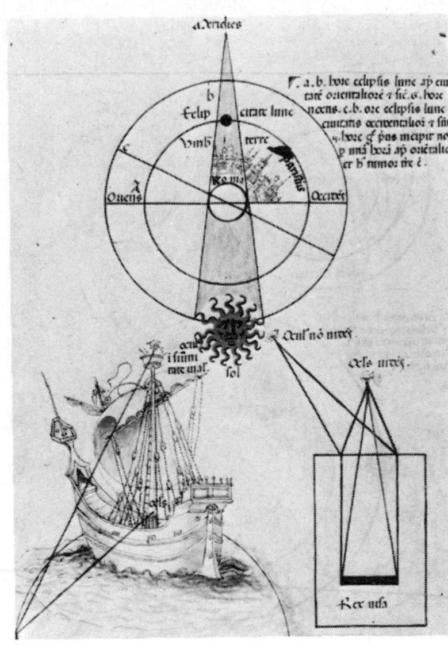

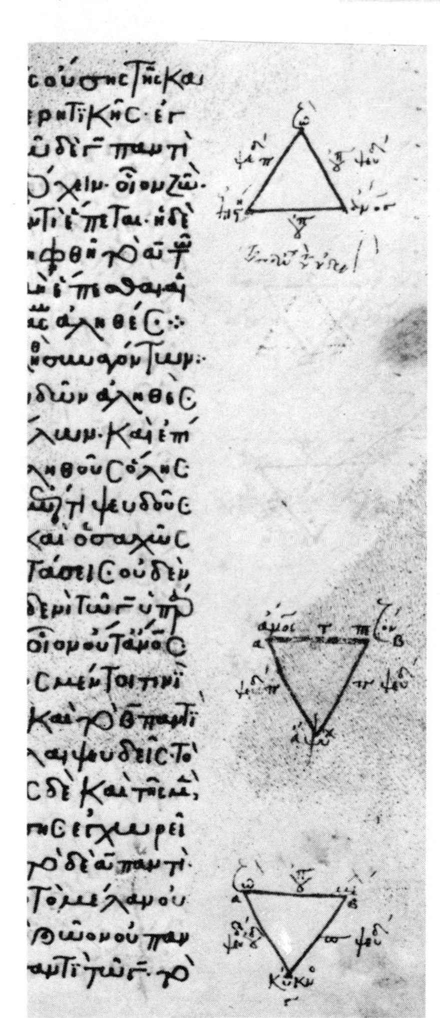

Specularia

Sit rurſus craſſitudo.ae.ſpeculū uero conuexū ſit.adc.oculus aūt ſit.b.uiſus aūt refracti in.eh.ſint.bce.bdh.reliqua uero ſicut & in planis.

Theorema nonum.

Obliq̄ lōgitudines a planis ſpeculis ſicut ſe hn̄t: ſic videt̄.
Sit oculus.b.longitudo aūt obliqua ſit.de.ſpeculum uero ſit.ac.igit̄ refractis uiſibus uidet̄ quidem.d.in.a.&.e.ſup.c.ſicq̄ ſe hēt in phātaſia ſicut uero ſe hēt: propius propius: & remotius remotius.

Theorema decimum.

Obliquae longitudies a conuexis specul' ſicut ſūt vere: ſic ſpectāt̄. Sit longitudo.ed.oculus aūt.b.ſpeculū uero cōuexū.ac.aſpect⁹ porro refracti in.ed.ſint.ba.bc.reliq̄ uero eadem.

Theorema vndecimum.

Elſitudines & craſſitudies a cauis ſpeculis quaecūq̄ ſunt it.a coincidentiam viſuum conuerſa videntur: quēadmodū in planis & conuexis ſpeculis: quaecunq̄ autem extra coincidentiam ſicut ſunt: ſic & ſpectantur.

Sit cauum ſpeculū.ac.oculus aut ſit.b.uiſus uero refracti ſint.ba.bc.eorū coīcidētia porro ſit.f.celſitudo ſit.de.&.kn.&.kn.qdē itra.f.coicidētiā ſit at.de.ſit extra coincidentiā igit̄.productis uiſib⁹ ſicut in planis & cōuexis ſpeculis apparet.k.ſup.m.&.n.ſup.l.quare conuerſæ uident̄: rurſus ſuper exteriorem coincidentiam celſitudinis apparet quidem.d.ſuper.g.&.e.ſuper.h.ſicut ſe habet ſic ſpectatur.

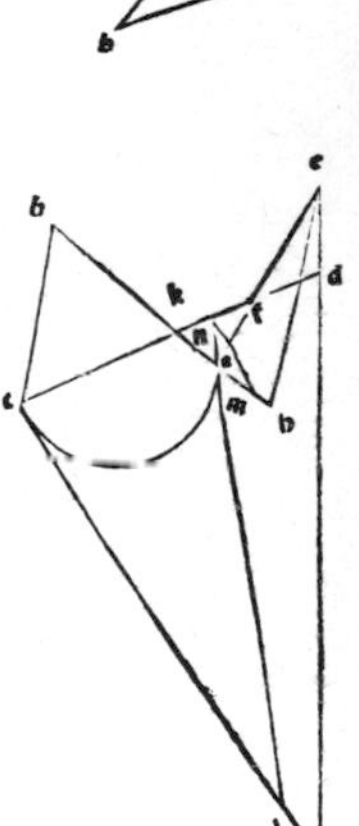

In craſſitudinibus.

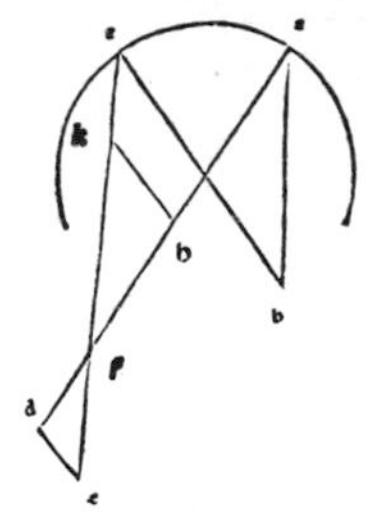

Rurſus craſſitudo qdē ſit.de.&.kh.cauū autē ſpeculū ſit.ac.oculus uero ſit.b.uiſus aūt refracti ſint & cōcurrentes in.f.ba.bc.igitur productis uiſibus ſimiliter.kh.conuerſæ apparet.k.quidem per.c.&.h.per.a.Sicut eſt in planis & conuexis ſpeculis ad.de.ſicut ipſum quidē.e.infra per.a.&.d.ſup.c.

Theorema.xij.

Obliquae longitudines a cauis ſpeculis quaecunq̄ intra coincidentiā viſuū iacent: vt vt ſunt ſic ſpectantur: quaecunq̄ vero extra: conuerſae.

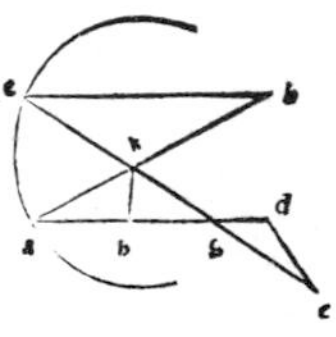

Sintq̄ inq̄ longitudines obliquę.ed.hk.cauum uero ſpeculū ſit.ac.oculus aūt ſit.b.uiſus refracti & cōcurrētes in.g.ſint.bad.bce.& ipa quidem.hk.obliqua longitudo ſit intra coincidētiam.g.&.de.ſit extra.Igitur.hk.iuxta naturam apparet: ſicut & in planis & conuexis ſpeculis.Sed.ed.conuerſa: nā ipm

Right The great lighthouse or Pharos at Alexandria, depicted on a vase found at Begram in Afghanistan. It symbolized the permanence and range of Alexandrian culture. Kabul Museum.

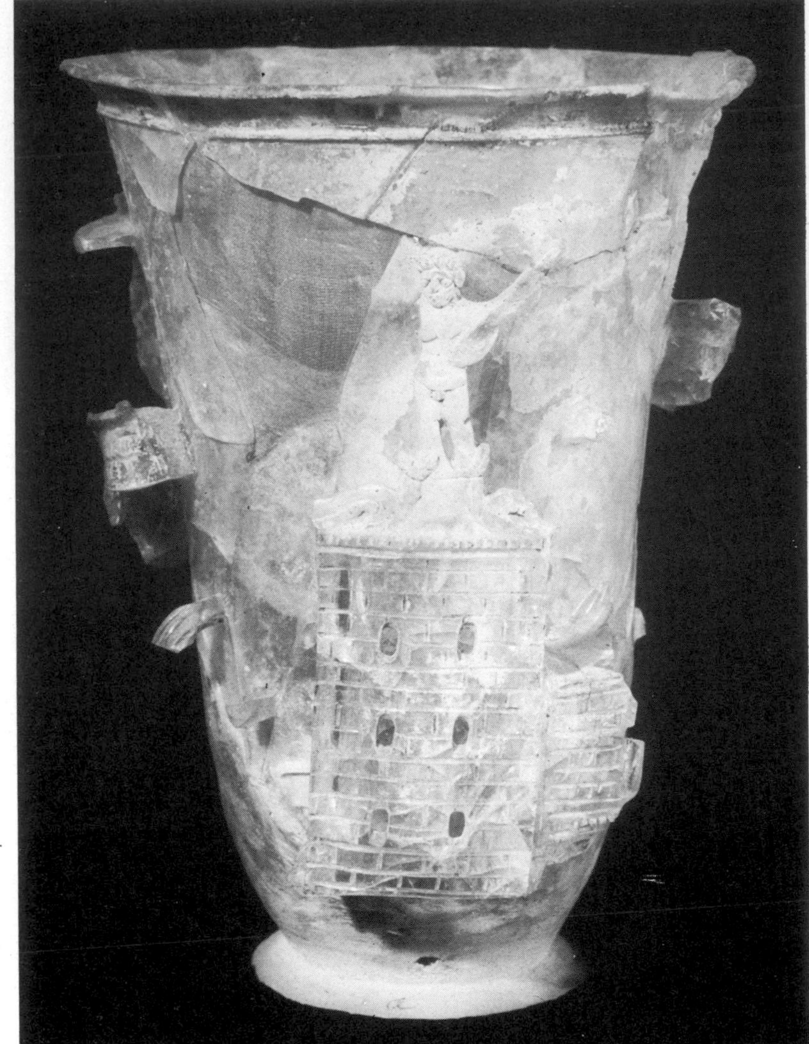

Below right A 16th-century illustration of Archimedes in his bath considering the problem of Hieron's crown. From Gaultherius Rivins, *Architecktur . . . Mathematischen . . . Kunst*, 1547.

Below A 17th-century illustration of the path of a cannonball, according to Aristotle's laws of motion. These taught that a body could undergo only one motion at a time; hence the two separate motions shown in straight lines. From Daniel Satbech, *Problematum Astronomicorum*. Basel, 1561.

Above Ptolemy's 'ruler' for measuring the zenith distance of celestial bodies. From William Cuningham *The Cosmographical Glasse*, 1559.

Above left A 16th-century illustration of Ptolemy (wearing a crown because he was wrongly identified wth the Ptolemy royal family), guided by the muse Astronomy and using a quadrant. An armillary sphere is shown, lower left; from *Margarita Philosophica*, Gregor Reisch, 1508.

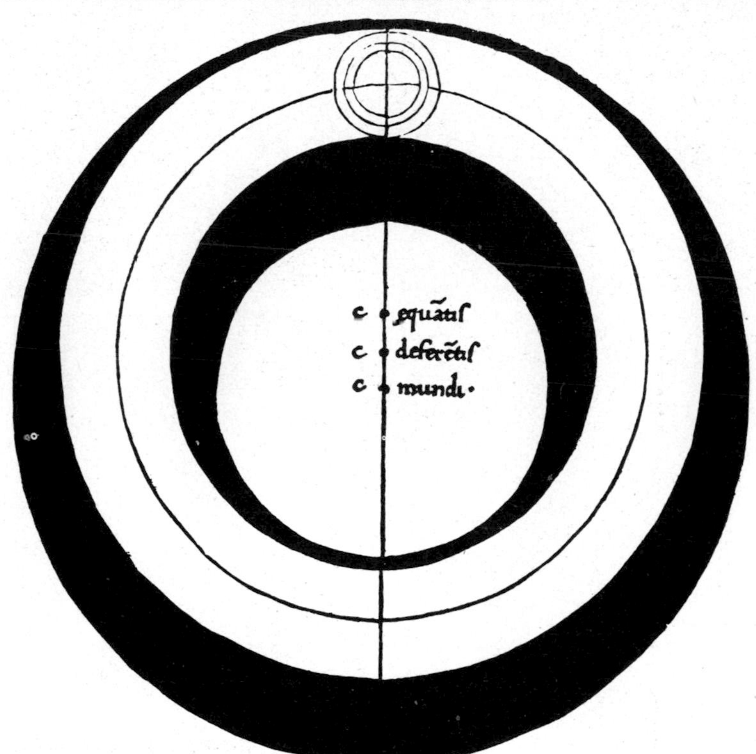

Left Diagram of Ptolemy's theory of the motion of the outer planets (Mars, Jupiter and Saturn) and of Venus. From Georg Peuerbach, *Theoricae Novae Planetarum*, 1472.

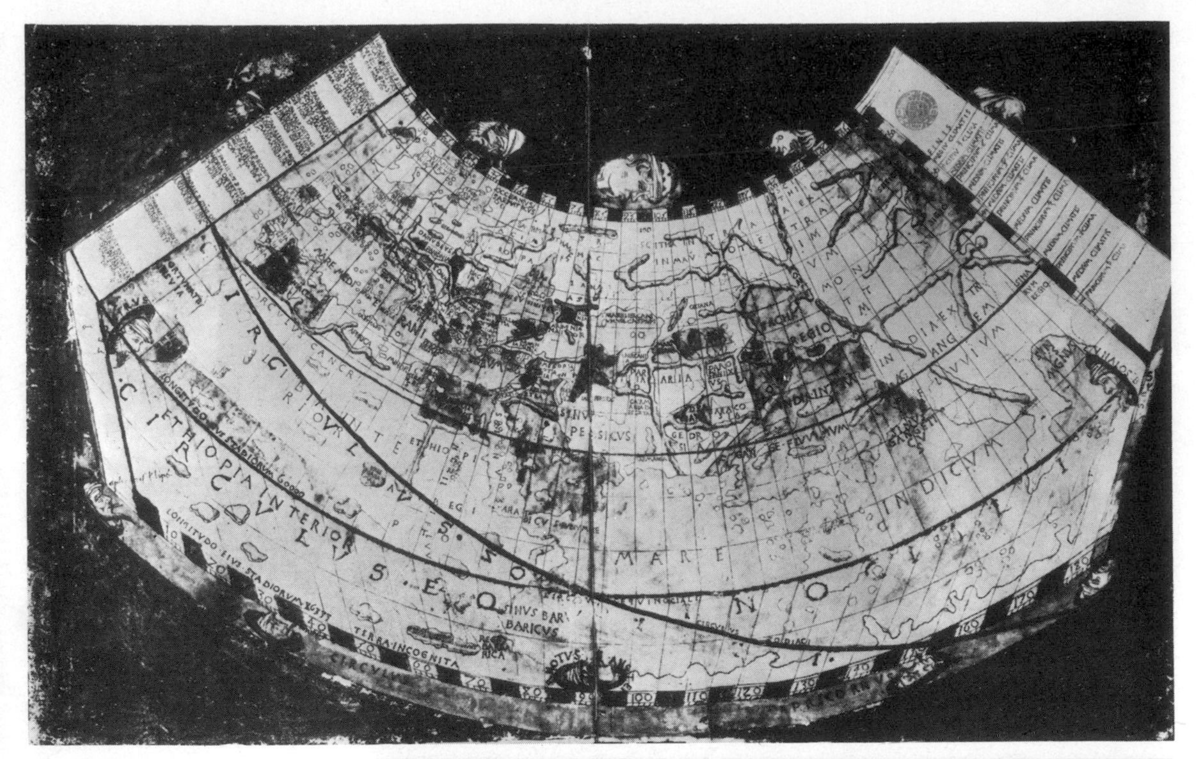

Above A reconstruction of Ptolemy's map of the world, reproduced on a conical projection with latitude and longitude, 1486. British Library, London.

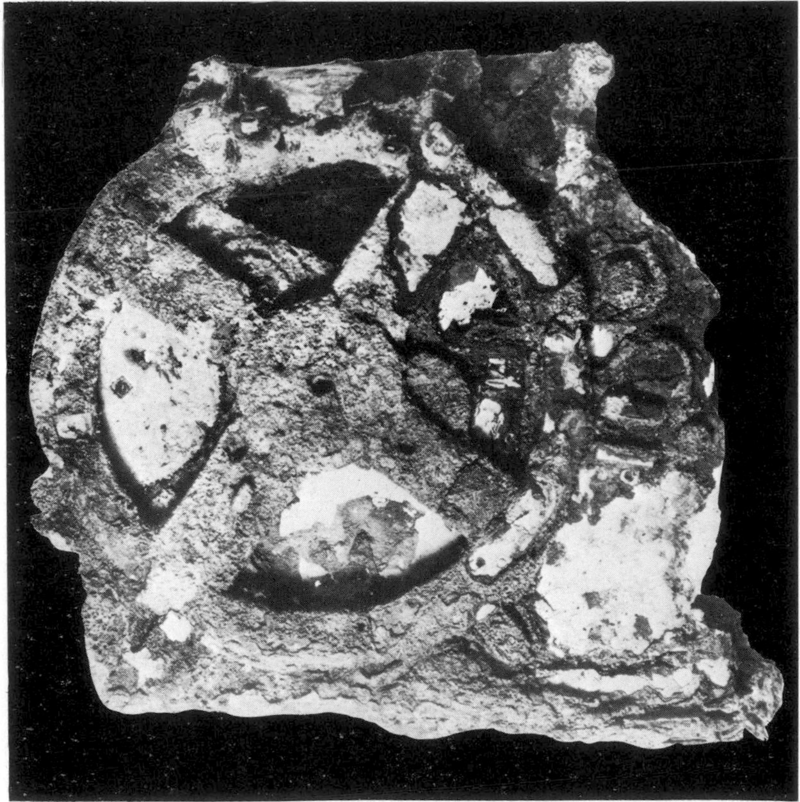

Right The antikythera, a Greek geared calculator for showing the positions of the Sun, Moon and possibly the planets. It dates from c. 80 BC.

man', as his pupil Plato called him. He is, indeed, so central a figure and marks such a watershed in Greek thought that Greek philosophers are usually referred to as pre- or post-Socratic: no wonder Democritos wanted an introduction. A man of legendary ugliness with a wife, Xanthippe, who was the archetypal shrew, Socrates seems to have inspired generations of Athenian thinkers with high moral standards and a love of truth. He was strongly opposed to the Sophists, the Greek professional teachers and writers, whose high fees saw to it that their clientele were the sons of the well-to-do, ambitious for success in public life. Socrates's opposition to them was not, however, for their exorbitant fees, though he found these despicable, but rather for their levity and scepticism and their failure, in his eyes, to inculcate absolute moral standards. Integrity and temperance were the virtues Socrates taught by his 'Socratic method', a dialectical technique in which he led his students by questions which demanded answers that could be arrived at from previous knowledge. But Socrates was often cynical and always fearless in his criticisms, and he made plenty of enemies in spite of his benevolence; it is not perhaps surprising that he should finally have been arraigned on a number of counts, including 'corrupting' Athenian youth, carrying the death penalty.

The circumstances of the death of Socrates, and in particular his dignity and lack of bitterness, had not a little to do with his subsequent fame. Plato and Xenophon transmitted his thoughts to posterity and Socrates's martyrdom consecrated his teaching. But what was the effect of this teaching on science and on its pursuit by the post-Socratic philosophers? Socrates spoke against the astronomers and meteorologists; he had no time for those who would try to fathom the nature of the physical world; they would, he believed, be better employed studying ethics and the relationship of man to man, so that men could learn how to live contentedly and peacefully as good citizens. On the face of it, his influence was catastrophic though in practice it was not as bad as that.

Socrates has always been considered to form a watershed in the history of Greek – and even in world – philosophy. Effectively, he discovered the dialectic, a method of argument whereby one begins with an incontrovertible statement based on simple experience and builds up a complex argument upon it by means of some clear and logical rules. The result was that enquiry into the natural world according to Socratic precepts – which were developed by Plato – tended to take the form of abstract and theoretical arguments from various human requirements, rather than of the careful investigation of the phenomena themselves and the derivation of hypotheses to explain them. It was in the propagation of this method of logic – which is invaluable to mathematics – that the work of Socrates and Plato is often said to have been harmful to the development of natural science. Indeed, it was only in the sixteenth century that western science freed itself from this method, and an alternative method of observation, hypothesis, prediction and experiment was developed.

In the description of the scientific views of the Ionian philosophers and those who followed them we have detected the beginnings of a scientific description of the natural world, but it is science of a very undisciplined kind. Speculation is almost unbridled, although it must be admitted that at such an early stage in the development of scientific ideas a good deal of speculative thinking is inevitable; without it little progress would be made. But it is possible to argue that the time had come for a halt to speculation alone; things had reached a stage when some caution was necessary. Speculation required to be modified by careful observation, and possibly one of the results of Socrates's attitude was to make subsequent philosophers more observant and rather more critical. But as will become evident, it did not cause a halt to Greek science.

Hippocrates of Chios

To most readers the name Hippocrates conjures up some vague picture of Greek medicine, probably connected with the ethics of medical practice, with the 'Hippocratic oath'. Yet in the fifth century BC there were two Greek scientist named Hippocrates: Hippocrates the mathematician from the Ionian island of Chios, and another Ionian, Hippocrates the physician, from the island of Cos, some 220 km (140 miles) to the south.

Hippocrates of Chios, who was a contemporary of Socrates, was the founder of a school of mathematics in Athens, and under his influence it soon became the leading mathematical centre of the Greek world, a position which it held until the rise of Alexandria some two centuries later. Originally Hippocrates was a merchant, but he lost much of his money, either from being captured by pirates or, so Aristotle tells us, because he was a fool and was defrauded by customs officers in Byzantium (modern Istanbul). But victim or fool, there is no doubt about his abilities as a geometer.

At the time Hippocrates was first in Athens, there were three problems facing mathematicians: duplication of the cube, squaring the circle and trisecting an angle. He solved the first of these and went some way to solving the second. The duplication of the cube was another name for the problem of finding the length of the sides of a second cube which is twice the volume of a given cube. It is unnecessary to go into not trouble the reader with the details of how Hippocrates solved this problem except to say that he used a mathematical method known as geometrical reduction; that is, he reduced the problem to a simpler one, solved that, and then used the result to solve the more difficult problem with which he began. Whether Hippocrates invented this method or had it from the Pythagoreans we do not know.

The second problem, squaring the circle, is that of finding the area of a circle. Hippocrates decided the most satisfactory way to tackle it was to find the area of a lune, ⌓, so called because of its similarity to the Moon at first or last quarter. Again, the details of how Hippocrates dealt with this are superfluous although his success was the main reason for his fame. It is worth noting, though, that the method he adopted was to find the area of a figure bounded by straight lines and then prove geometrically that this area was equal to that of the lune. It was not easy, and Hippocrates had to develop his method using a number of closely argued stages, but he achieved his aim in the end; however, the claim that his results enabled him to square the circle is false. A partial solution of the problem, but one which was spoken of disparagingly by Aristotle, was devised in Athens by a contemporary of Hippocrates, the Sophist Antiphon, who obtained the area by measuring the areas of polygons drawn inside a circle. The first mathematician to achieve a truly geometrical result was Hippias some 50 years later, while another solution was devised by Archimedes.

For a Chinese method of calculating the area of the circle, see page 151; for the measurement of π by Archimedes, see page 112.

The fame of Hippocrates does not rest only on his prowess in dealing with problems facing the mathematicians of his time, but also on the fact that he drew together the Greek geometrical knowledge of his day and systematized it. His was the first attempt to bring order into a multiplicity of theorems and proofs which had grown up by the latter decades of the fifth century BC, although no complete formal scheme of geometry was to be drawn up until the time of Euclid, over a century later.

Hippocrates of Cos and Greek Medicine

The traditional founder of Greek medicine was Aesculepios who is referred to by Homer in the *Iliad,* but whereas in Homer's time he was a blameless physician he later became deified as a son of Apollo, who was taught the art of healing by the centaur Chiron and then killed by a thunderbolt from Zeus in case he rendered all men immortal with his medical art. Aesculepios is usually depicted holding a staff with a serpent entwined but, curiously, the more modern medical symbol of a winged staff with two intertwined serpents has no relationship to this. Indeed it has no medical significance; it represents the magic wand of Hermes or Mercury, the patron of trade and the messenger of the gods.

Whether in fact Aesculepios existed at all is unknown, but his cult flourished: it was celebrated in special temples and involved a ritual type of treatment considered suitable for many forms of sickness: a purifying bath followed by 'incubation', a period of rest accompanied by dreams which the Aesculepian priests would interpret. Those who were cured made gifts to the temple. It seems that few drugs were used at the temple – they were prescribed by physicians elsewhere – and there was no surgery: the treatment was primarily psychological. Such temple treatment was not, however, a Greek invention but had been practised in Egypt and may, indeed, have derived from there. Nevertheless we should not lose sight of the fact that Greek medicine always paid attention to the psychological side.

Illustration page 76

The Greek physician used herbal drugs which for centuries had been obtained by the *rhizotomoi* or root-gatherers. They collected their plants and roots for sorcery as well as medicine, and over the ages had amassed a wealth of knowledge about their efficacy. They believed, too, that gathering must be done at appropriate times – at night, or at a particular phase of the Moon – sometimes accompanied by spells, and the task was considered to have its dangers. As one historian has aptly put it, plucking herbs or digging roots from Mother Earth was considered 'analogous to pulling hairs from the back of a sleeping tiger', safe only if the proper precautions were taken. It was, of course, the task of physicians to take over this knowledge and determine proper dosage and correct application.

The Greek mind was not, of course, satisfied with medical practice alone; there had also to be some medical theory. As we have already seen, different Greek schools of thought tended to look at the world in their own particular way, so it is not surprising to find that this was so in medicine too, and that in the early stages there were four main medical schools. One was Pythagorean, whose leader, Alcmaion of Croton, taught that health was due to a balance of forces within the body and, unusually for this date, considered the brain as the centre of sensations. Philolaos, the astronomer, who also took an interest in medical theory, was another member of the school and has the distinction of being the first to separate sensory, animal and 'vegetative' functions.

The second Greek school of medicine was the Sicilian, apparently founded by Empedocles of Acragas whom we have already met in connection with the theory of the four elements. He had two followers, Acron and Philistion, who both stressed the importance of air, inside and outside the body, while Acron is supposed to have written a regimen of principles for maintaining health. A third school was the Ionian, where some anatomic dissections were done, while the fourth was centred at Abdera. Here great stress was laid on the medicinal use of gymnastics and diet, while one of the leaders, the atomist Democritos who, incidentally, may well have known Hippocrates of Cos, took a great interest in what we should today call psychosomatic medicine, as well as busying himself with many other aspects of medical science. Another member of this school was Herodicos, who is said to have been Hippocrates' teacher.

The four schools of medical thought were all early, and by the time of Hippocrates towards the end of the fifth century and during the first few decades of the fourth BC, they gave way to two primary centres of medical study, one at Cnidos and the other at Cos, lying within a few kilometres of each other and only separated by the Gulf of Kerme. The medical fraternity at Cnidos concentrated on certain diseases, and were specialists when it came to obstetrics and gynaecology. The school of Cos was more general in its approach and dealt with a wider spectrum of medicine, and it seems fair to say that it was the leading centre of classical medicine. It was here that Hippocrates was born about 460 BC.

His teaching, and that of his colleagues and disciples at Cos, is enshrined in a collection of some sixty important texts known as the

Hippocratic Corpus, but it is now difficult to be certain which parts were written by Hippocrates and which by other hands. They date from the last decades of the fifth century BC and seem finally to have been collected together by Greek scholars at Alexandria. Some of the treatises seem to have come from the Cnidos school rather than originating in Cos, while the famous one *Nature of Man* was almost certainly by Polybios, Hippocrates's son-in-law, yet it does seem that quite a number were by Hippocrates himself.

This is not a history of medicine but of science in general, so it is not the place to describe the contents of each book or to go into detail about them. But because of the immense prestige of Hippocrates and the lasting effect of his teaching – it extended right down to medieval times – something must be said about what attitudes it shows to have existed at the medical school at Cos. And first and foremost it must be realized that anatomical knowledge was rudimentary. The bones were known, but the physicians of Cos were somewhat vague about the internal organs. However, they had to have some general approach to the way the body worked if they were to treat patients systematically and they therefore formulated the doctrine of the 'humours', or liquids, an idea which was not new but which they put on a rational basis. Originally it was doubtless the result of observing that human and animal bodies contained various fluids like the blood and biles which were obviously of importance. Indeed some conditions are accompanied by the excretion of liquids – a runny nose is a symptom of a cold in the head, vomiting or diarrhoea are evidence of other conditions – and these observations, coupled with the Pythagorean concept of health as the effect of balance in the body, led to the doctrine. The four elements of Empedocles also played their part in the Hippocratic version, and they were accompanied by the 'four qualities': dryness, dampness, heat and cold. As a result of this the body was thought to contain four humours: blood, black bile, yellow bile and phlegm. With them the four qualities were associated, and in a healthy person they were all in balance; an excess of one or two would lead to some bodily disorder or another. Later on, in the second century AD, the physician Galen extended the doctrine to include the four temperaments, a form of classification of people into sanguine (warm and pleasant), phlegmatic (slow-moving, apathetic), melancholic (sad, depressed) and choleric (hot tempered, quick to react). It is a classification which with the Hippocratic four humours and qualities survived in medicine until the seventeenth century.

For the career and teachings of Galen, see pages 247–249

Diseases or fevers, then, were put down to imbalance of humours and qualities, but great care was taken to recognize the various kinds, especially chest troubles and the various effects of malaria. Malaria itself causes difficulties for the physician since an attack of it can mask other diseases or at the least colour their symptoms, and in the Mediterranean area it was rife, so it is clear that the Hippocratic physicians were often up against difficulties. Yet although they were careful in their examination of patients, it is surprising to learn that

Illustration page 244

they seem not to have noticed the change of pulse rate with fever; indeed it appears that taking the pulse was a stage of examination which hardly occurred to them, perhaps because they were less concerned with diagnosing a fever than with forecasting its course (prognosis). After all, the Hippocratic physician's job was to make use of the healing power of nature, and treatment was made with this in mind. Thus, although the physician might use purgatives and emetics, starvation diets and even blood-letting, he would also prescribe baths and massage, barley water, wine and infusions of honey in attempts to assuage pain and bring relaxation and comfort so that natural healing could take over; the patient's mental well-being was also a matter for the physician's concern.

Hippocrates wrote the first treatise on medical climatology, *Airs, Waters, Places,* which describes the effects of climate and environment on medical conditions and especially the spread of epidemics, concerning itself with the nature of local water and food and even the nature of the people themselves. It broke totally new ground. But of all the books in the Hippocratic corpus, the most popular have certainly been the collections of aphorisms. Even today, well over 2300 years since they were compiled, most people have heard the one which begins: 'Life is short, Art long, opportunity fleeting, experience treacherous, judgement difficult'. The second sentence, though, is less well known: 'The physician must be ready, not only to do his duty himself, but also to secure the co-operation of the patient, of the attendants and of externals'. It is reminiscent of the so-called Hippocratic oath which has been adopted as a guide to conduct by medical men throughout the ages and which emphasizes the duty of a physician to work for the benefit of his patients, and the sanctity of confidence between them.

The Hippocratic school clearly expected high standards, although critics of Hippocrates have sometimes accused him of being more interested in general knowledge than in individual cures. Yet it is true that in Hippocrates and his successors we find the first evidence in the West of medicine as a science. Hippocrates inculcated a scientific point of view and used scientific methods in an area of activity played by magic and credulity. His judgements were careful and moderate, and he rejected all irrelevant philosophizing and rhetoric as well as a host of prevalent superstitions. Moreover, Hippocrates kept detailed case histories of his treatments, recording his failures as well as his successes in true scientific manner. Indeed, he was the originator in the West of medical records, but unfortunately his example was not followed until the ninth century AD in Islam and not until the sixteenth century in Europe.

Plato

We move now to the time of the post-Socratic philosophers, of whom the first is Plato, a pupil and intimate of Socrates and a man who exerted considerable effect on those who came later, both within Greece and elsewhere. Plato was born in 427 BC, probably in Athens, and died there in 348 or 347 BC, so most of his life was spent in troubled times. From 431 to 404 BC Athens and Sparta waged the Peloponnesian War, which ended in the total defeat of Athens. By 403 BC, a year after their defeat, the Athenians had again become a self-ruling democracy, although their social structure had somewhat changed. The devastation of the surrounding countryside meant that the great landowners were no longer the ruling aristocracy; a merchant class grew up which was both powerful and wealthy, and for the next century Athens was to enjoy a period of considerable material prosperity. Oratory reached its peak under Demosthenes, and creative thought a climax with Plato and Aristotle.

Plato was an aristocrat, always mindful of his nobility, yet he decided to take no part in public affairs, perhaps because he distrusted those in power, possibly because he set too much store by moral principles and human goodness, like his master Socrates before him, or even just because he was too interested in learning and philosophical investigation. At all events, when in his thirties he decided to travel. He first moved westwards to visit Italy and the Pythagoreans, although there is a story (from Cicero) that he began by visiting Egypt. It also seems clear that Plato visited Syracuse and may have become embroiled in Sicilian politics. It was while in Sicily that Plato met Archytas, a native of Tarentum who, besides being an able politician, made important contributions to the theory of means and proportions in mathematics, and the theory of music, including the observation that the higher the pitch of a sound, the higher the frequency of the vibrating string or column of air giving rise to it. It was Archytas, too, who became much concerned with the foundations of the various sciences and concluded that the art of calculation is the basis of all science, even more fundamental than geometry. Yet in spite of this concern with calculations and numbers Archytas did carry out some geometrical work and became famous for his solution to the problem of duplicating the cube. All this doubtless coloured his later insistence that a study of mathematics, through the insights it offers into the connection between reason and knowledge, is necessary for every man who aspires to govern.

For duplicating, or doubling, the cube, see page 90.

Plato returned to Athens in 388 BC, and by then he seems to have been fully launched on his philosophical career and confirmed in his desire to teach. About 387 BC we find him buying a plot of land on the banks of the Cephissos outside the city's western gate; originally owned or set aside for the legendary hero Academos, who is reputed to have helped Castor and Pollux take Helen of Sparta back home, the site was known as the Academy. It probably contained some buildings, perhaps a 'museum' (i.e. temple to the Muses), an assembly hall, refectory and possibly some other rooms as well; there was also an olive grove, and teaching would probably have taken place there or in the shade of the porch of one of the buildings. The existence of such a school was nothing new; there had been other schools in Greece, Egypt and Mesopotamia, but the uniqueness of the Academy

Illustration page 76

lay in the kind of 'postgraduate' training it furnished, not only in Plato's time but also long after. Indeed the Academy itself lasted for some 900 years, only closing in AD 529 by order of the Byzantine emperor Justinian.

Plato's theory of ideas underpinned his entire philosophical outlook and dominated all his scientific speculations. Basically it supposes that everything we see, everything we observe with the senses, is no more than appearance. Though there is a basic reality, this is something we cannot see; true reality is an essential Form or Idea and it is permanent and unchanging. What we observe has no such permanence; it is always an inadequate imitation of the real essence, of the Form, of the Idea. Thus when we see a cat, what we are observing is an imperfect example of the essential cat: our cat will grow old and die, but the essential Idea of cat will always be there. And it is this essential Idea that has true and permanent reality; the observed world is only its shadow. Indeed, Plato himself used the simile of a man in a cave, chained in such a way that he can only look in to the wall: he sees the world pass by as shadows on that wall. So do we when we observe Nature; the true reality escapes our senses. True reality is something we can never observe; it can be contemplated only by the mind. And this, according to Plato, is the true aim of science, to investigate and understand Ideas.

Plato's theory of ideas, sometimes rather confusingly known as 'Realism', had a far more profound effect on the history of science in the long term than any of his more specific scientific theories. For his argument that the natural world does not provide an adequate guide for learning about 'true' or 'perfect' reality, which could only be discovered through contemplation or revelation, became, through the teachings of St Paul, a cornerstone of Christian thought. To Plato even more than to Socrates, experiment and observation were not only irrelevant but positively misleading in the search for knowledge; and theories of the universe were to be evaluated not by their power to explain or predict nature but by their adequacy in expressing divine perfection. Even though Platonism was to become submerged by the teachings of Aristotle in the medieval church, only to be revived in the Renaissance, Plato's assumptions about true knowledge haunted the medieval debates about the relative roles of faith and reason. Likewise, he contributed to the continuing fascination that geometry would hold over the medieval mind as it had done over the Greek; for geometry was the great example of the 'deductive' method of reasoning that could deduce many disparate conclusions from a few original propositions, as opposed to the more experimental 'inductive' method of observing the many and varied phenomena of nature and attempting to use this knowledge to build a single unified explanation to link them together.

Plato's political theories appear in three books, the *Republic*, the *Statesman* and the *Laws*. In these he advocated an elitist society, with a group of one-fifth of the total population, the 'rulers and guardians', controlling the rest. The rulers were a self-perpetuating caste who held all things in common including wives and children. The mass of the people, he thought, had no ideals but only desires; they, the merchants and traders, artisans and manual labourers, were fit only to be ruled. Plato, of course, was distrustful of all desires and passions; he abhorred the love of money, or even the love of family, and turned his back on sexual love. Yet, astonishingly, he seems completely to have ignored the overiding passion of the politician – the love of power – although since his ruling elite was to be specially educated perhaps he thought that such training, coupled with his insistence on their communism of property and family, would render the power inoperative. Incidentally, it is interesting to see that Plato insisted that mathematics should be an integral part of this education, not only for its own sake but also because of its use in training the mind. But if mathematics was good for his state, Plato was equally certain there were some actions which were not – freedom of thought, the acceptance of new religions, criticism of the political establishment and allowing the young to consider new ideas. These were all to be capital crimes. In short, Plato was an aristocratic reactionary who advocated an idealized totalitarian state.

This digression into Plato's politics has been necessary because politics occupied so central a place in his interests and coloured all his philosophy, even affecting his approach to the world of nature and his ideas of the cosmos, as can be seen in his greatest scientific work, the *Timaios* (latinized as *Timaeus*).

The book is essentially a dialogue in three parts; the first is an introduction which contains an account of the Atlantis myth. This is followed by the 'soul of the world' which contains the four-element theory, the theory of matter itself and of objects observed by the senses; this is by far the longest section of the book. Lastly there is a section containing some physiology and a discussion on man's soul as well as his body. As to Plato's universe, this is an ordered, reasonable place, as the regular motions of the heavenly bodies declare. The soul of the universe is comparable with the soul of man. Planets and stars are the most sublime representations of essential reality; they are examples of Plato's 'Ideas' and may even be gods. Mathematics expresses the divine motions of the stars, which emit a heavenly music as they move, and when men die it is to their native stars that their souls return.

Illustration page 85

Underlying this is the doctrine of the microcosm and the macrocosm, the small world of man echoing the vast world of the universe. It was a theme Democritos had used and it has occurred to philosophers the world over, featuring prominently in the thought of medieval Europe; with Plato it was developed in such a way that it allowed him to indulge in astrology of a serene and spiritual kind, derived most probably from Babylonian sources. Its presence, however, was unfortunate, for those who read the *Timaios* in later times did not understand the Platonic outlook and used the microcosm-

For Chaldean astrology, see page 44; for its development in Islam, partly under the influence of Plato's ideas, see pages 208–212.

macrocosm doctrine as an excuse for the personal predictive astrology of the horoscope.

Not all Plato's astronomical ideas are, however, in the *Timaios*; some are to be found in other works, even in the *Republic* and in the *Laws*, for he thought astronomy, as a description of the universe, was part of the necessary education for his ruling elite; indeed he thought it should even be taught to all. What this teaching should be is not always clear, for he often used highly fanciful visionary pictures, as when he wrote of planetary motions that these occurred because the planetary bodies were fixed to 'whorls' each of which carried a siren sitting on the rim, all of them spinning about the central axis of the universe, an axis which was itself kept moving by the Fates (the goddesses Clotho, Atropos and Lachesis). The exact nature of the whorls is difficult to decide: in Plato's day large whirling discs were used as flywheels to keep the hand spindles for spinning thread moving at a regular rate, but one would be unwise to press this analogy, or even to worry about what precisely he did mean. When Plato wanted to be specific he was; when he used more colourful language he did so because he was trying to create an impression, not a detailed scientific description. Fortunately for the future history of scientific astronomy, Plato's exact meaning – if he had one – is not significant.

Plato's idea of the universe owed much to the Pythagoreans, though not to Philolaos, for he was firmly convinced that the spherical Earth was fixed in the centre of the universe, with the Sun, Moon and planets moving round it at different speeds. He did, however, accept the Philolaon order for the way these bodies were arranged with respect to the Earth. There was only one universe – Plato rejected the idea of the atomists that there were many universes – and he believed the different bodies were all formed out of the four elements: fire for the divine celestial bodies, air for winged creatures, water for those living in water, and earth for the inhabitants of dry land. The celestial bodies were not only divine but were also endowed with souls.

What then can we say of Plato's astronomy? His views contained no really new conception, no novel theory of the universe, and they were often muddled. But there was one theme which ran through them, namely the significance he attached to mathematics; this was to prove scientifically productive. However, there was another theme that had just the opposite effect, and this was Plato's lack of faith in observation, for he believed matters of the intellect superior to those perceived by the senses. Philosophical speculation about the universe was more enlightening than precise observation of apparent motions. The truc revolutions of the bodies in the universe are to be grasped by the mind, not by the sight. This was yet one more example of the aristocratic Greek idea that the search for scientific knowledge was a lofty, philosophical activity in which reasoning was more important than the more mundane activities of detailed recording and observation, the same attitude of mind that hindered the technical application of so many of their discoveries.

Was Plato good or bad for science? Did his work promote it or not? There is no doubt that his influence on later philosophers was immense, both through the efforts of his pupils and the power of his writings, and the interminable commentaries on them. Plato's emphasis on mathematics was benign but, on the other hand, he did not promote experimental science one iota; indeed he positively despised it. Certainly Greek science always tended more to philosophical speculation than to practical tests, but this failing was exacerbated by Plato's theory of Ideas. On balance we must conclude that Plato's influence on science was more inhibiting than inspiring.

Eudoxos of Cnidos

Eudoxos was a pupil of Plato, but only for a short time. Born in Cnidos in Ionia about 408 BC, he studied geometry with Archytas of Tarentum and from him may have inherited his interest in music and numbers. Eudoxos also studied medicine with Philistion, a distinguished anatomist and a disciple of Empedocles. Eudoxos was a young man of promise, but not of wealth – he first attended the Academy as an impecunious student and only when friends contributed funds could he travel abroad. When he did travel he went to Egypt and stayed for a time at Heliopolis (about 11 km or 7 miles north-east of present day Cairo), where he is said to have computed an eight-year calendar cycle. On leaving he returned to Ionia and set up his own school at Cyzicus in what is now north-west Turkey. Here he had great success, later taking some of his pupils to Athens (his second visit) where Plato gave a banquet in his honour. He finally returned to Cnidos where he taught theology, astronomy, meteorology and mathematics, wrote books, helped to provide the city with laws and became, not unexpectedly, a much honoured man. Certainly, as a scientist he far outshone Plato.

Today Eudoxos is most remembered for his theory of homocentric spheres, an astronomical concept which was to exert an immense effect for the next 1800 years. But before describing this idea, a word must be said about his mathematics because this was another field in which he made notable advances. First, he devised the formal way of presenting geometrical theorems and axioms, the technique we call 'Euclidean'. Second, he investigated the whole subject of mathematical proportions and put forward a new theory. Third, he developed the proportions of the golden section and, far more important if less glamorous, the method of exhaustion.

A new approach to proportions had become necessary ever since the Pythagoreans had discovered 'irrational numbers', because these numbers (such as the square root of 2) cannot be expressed as simple proportions. This meant either that one had to reject any correspondence between arithmetic and geometry, or to recognize that the irrational was a new kind of number. Eudoxos decided on the second course, but in doing so he had to convince other mathematicians that he was right and this was not simple; it involved proving rigorously that such numbers do exist and then that they could be handled in exactly the same way as other numbers, as well as demonstrating beyond all shadow of doubt that there was geometrical justification for them. It was not just a matter of accepting the idea. As to the method of exhaustion, this was a technique Eudoxos developed for calculating the volumes of solids like the cone and the sphere. His method was to use infinitesimally small sections, calculate the volume of these and then add them together. In arriving at it he was forced to define precisely what he meant when he took an infinitesimal piece, and in doing this he took an important step towards that kind of mathematics which a couple of thousand years later was to become known as the 'integral calculus' and is now associated with the names of Newton and Leibniz.

For the development of the calculus by Leibniz and Newton, see pages 370–371.

Eudoxos also studied the geometry of lines and circles on the sphere, and from this was able to arrive at his important theory of the homocentric spheres. This was a means of accounting for the observed motion of the Sun, Moon and planets using a number of spheres which were all homocentric, i.e. concentric with each other, rather after the fashion of those Chinese sets of carved ivory balls one within another. The scheme was most ingenious because it accounted for the fact that the planets do not appear to sweep out simple paths in the sky but move across the background of the stars sometimes in one direction, sometimes in another, and even have periods of standstill. Ever since Pythagoras it had been assumed that all celestial bodies moved round the Sun in circles, and the challenge Eudoxos set himself was to explain these observations ('save the phenomena' as the Greeks called it) by using circular motions or, rather, motions on the surfaces of spheres. He achieved his aim by using concentric spheres moving at different speeds and rotating about different axes. His final scheme was complex, and there is no need to explore it in detail; one example will be sufficient to show how it worked.

Illustration page 85

Consider the motion of the Moon. This moves round the Earth once every day so that it rises and sets, just like any other celestial body. It also moves across the starry background, completing a circuit every month. Thirdly it moves – or its orbit moves – in such a way that it undergoes a cycle of eclipses which runs for a little over 18 years. So Eudoxos's scheme required three spheres – one to account for the daily motion, a second to deal with the monthly motion and a third for the eclipse cycle – although, of course, the spheres need not be in that specific order. Indeed what he seems to have done is to have considered the daily motion first and to have accounted for this by having a sphere which rotated once every 24 hours: the axis of this sphere was in line with the north and south poles of Eudoxos's stationary Earth. Inside this diurnal sphere was another. This second sphere was the eclipse sphere and had its axis running through the 'poles of the ecliptic', i.e. at right-angles to the apparent path in the sky of the Sun, and it rotated only very slowly, taking 18 years to complete one revolution. The third sphere was to account for the Moon's monthly orbit, and this was achieved by making the inside sphere rotate once every month and arranging its axis at an angle of five degrees to the axis of the middle sphere (because the Moon's orbit is inclined by five degrees to the Sun's apparent orbit).

I have said that this appears to be what Eudoxos did, because no one can be certain; his actual work on this has not survived, and all we have to go on are the remarks of commentators such as Aristotle. In recent years a few scholars have raised questions about the middle sphere (the one concerned with eclipses) because they wonder how much Eudoxos really knew about the Moon's orbit. However, he need not have had any well-developed theory; he had only to know of the cycle, not its cause, yet the general consensus of opinion is that observational records did supply sufficient information. But any doubts need not worry us: what matters as far as we are concerned is that Eudoxos had devised a scheme to 'save the phenomena', a scheme flexible enough to account for planetary motion and the motions of the Sun and Moon on a thoroughly sound mathematical basis. Certainly some critics of his time complained that his homocentric spheres did not account for the changing brightness of the planets as they pursued their orbits, but it was to be some time before a theory which could account for this phenomenon could be devised.

For an example of the ingenuity applied to 'save the phenomena' according to Eudoxos's system, see page 217.

There is no doubt about how great an impression Eudoxos's astronomical and mathematical work made on his own generation and on those that followed. His homocentric spheres became the crystal spheres of western astronomers of medieval times, and his mathematics was to exert a profound effect on later mathematicians. Eudoxos lived at a time which has often been called 'the age of Plato', but one science historian at least has raised the question of whether it would not be more appropriate to call it the 'age of Eudoxos'.

Aristotle

With Aristotle we come to the most significant figure in Greek science. His birth in 384 BC heralded a new age. Aristotle was born in Stagira in Chalcidice on the Mount Athos peninsula in the north Aegean coast, originally colonized by Greeks from Chalcis some 230 km (145 miles) further south. Chalcidice was near Macedonia, and when Aristotle's father was appointed personal physician to Amyntas III of Macedonia (also sometimes known as Amyntas II), the grandfather of Alexander the Great, the family moved to the capital, Pella. The Macedonians had gradually become more friendly towards the Greeks and by this time considered themselves Hellenised: indeed, during the reign of Archelaos (413-399 BC) the capital became a centre of Greek culture, and it was here that Euripides wrote his great tragedy, the *Bacchae*. Internal power struggles followed the death of Archelaos, but in 359 Philip II restored peace, annexed Chalcidice and then, by a mixture of force and diplomacy, gained control of the whole of Greece. This Macedonian power was to have important consequences for Greek culture, as will become evident later.

Aristotle's parents died when he was a boy and when he was seventeen his guardian, Proxenos, sent him to Athens to complete his education. Here Aristotle enrolled in Plato's Academy and was soon appreciated for his precocity and his enthusiasm; Plato called him 'the reader' and 'the mind'. Officially Aristotle remained at the Academy for the next twenty years until Plato's death, but it seems that his enquiring outlook would in fact have led him to learn oratory and politics from others besides Plato. In due course he began to differ from Plato over many things, and according to the later Greek historian Diogenes Laërtios, Aristotle actually left the Academy before Plato died, causing Plato to remark that 'Aristotle spurns me as colts kick out at the mother who bore them'. At Plato's death his nephew Seusippos became director, and perhaps because of this or because of a wave of anti-Macedonian feeling in the city, Aristotle left Athens. He was accompanied by Xenocrates, another academician, and they set sail across the Aegean to the court of Atarneos, near Lesbos in Ionia. Here they were welcomed by Hermeias, the ruler and himself a one-time student of the Academy, and it was while he was there that Aristotle married his first wife, Pythias, a niece of Hermeias.

Aristotle stayed at nearby Assos for three years, visiting the neighbouring island of Lesbos where he met and became friendly with the naturalist Theophrastos and where he also made a number of fine biological observations on his own account. In 343 BC Philip of Macedon invited Aristotle to Pella to act as tutor to his son Alexander, and Aristotle accepted. He tutored the prince for three years until, in fact, the young man had to act as regent while his father was away on military expeditions. Aristotle remained in Macedonia until 335 then, once Alexander succeeded to the throne, returned to Athens to set up his own school and research centre. Aristotle established this in a grove once sacred to Apollo Lyceios and so it became known as the Lyceum – a term which it might be more appropriate to use for modern scientific institutions than 'academy'. Since Aristotle had the habit of walking about the grove while he lectured his students, they became known as the 'walkers-around' or 'peripatetic' philosophers. Aristotle lived and worked at the Lyceum for the next thirteen years, setting up not only a lively school and library but also a museum for natural objects of various kinds – a very different approach from Plato who would have spurned such aids since he believed one could visualize a picture of the universe by thought alone. Not so Aristotle, who believed one needed all the hard facts one could muster before it was possible to gain an insight into the natural world. The museum was furnished by Alexander himself, who also contributed funds to the Lyceum, but when he died in 323 the anti-Macedonian party of Athens once again became vindictive. They accused Aristotle of impiety and he took refuge in Chalcis, where he died within a few months. The Lyceum continued under the direction of Theophrastos.

Aristotle's work may conveniently be divided into two periods – that which was done while he was still at the Academy, and that carried out after he had left it. The first period is characterized by a number of books which show Platonic influence, the second by a more independent state of mind.

What did Aristotle teach? His early works show a great respect for mathematics but are primarily concerned with using dialectical, question-and-answer methods to form axioms. They show that he accepts certain Platonic ideas, the immortality of the soul and the divine nature of the celestial bodies, but even then, interestingly enough, Aristotle was looking on the latter as tangible bodies, not archetypes; he was thinking that one looked, not at an Idea, but at perfect motions of what were real stars and planets, genuine physical bodies. Nevertheless, he accepted that there was something different about them, and it was at this time that he turned over in his mind the possibility that they were composed of an incorruptible fifth essence, something different from, and in addition to, the four elements of Empedocles.

By accepting that physical matter was as significant as its Form or Idea, Aristotle defined a sphere of influence within which truly scientific work could be done. To this extent he was far more of a scientist than Plato, who saw no value in the investigation of matter. As a result Aristotle built up a great deal of knowledge on all manner of subjects, including biology, astronomy and physics. To all these observations he applied his rigorous logical method, which inquired into the various causes of the things he observed; only one of these causes, the 'prime mover' did he consider to be beyond the scope of reasoned investigation.

One field in which Aristotle's scientific method can readily be seen in operation is that of logic, the precision instrument of philosophic discourse. In a number of works, such as *Categories* and the two books *Analytics*, he began to lay down the laws of reasoning. Propositions, fallacies, correct reasoning procedure and a deductive system of formal argument (syllogism) were all specified in detail. Admittedly, what he wrote would be counted as very wordy, by the standards of modern symbolic logic which expresses conditions and relationships with special algebraic symbols, not words, but that is a comparatively recent development. What Aristotle did – and it was no mean achievement – was to lay down the firm foundations of the subject.

Illustration page 85

In mathematics Aristotle did not make many direct contributions; his one enduring work here was his discussion of the concepts of continuity and infinity. Infinity, he pointed out, exists only potentially, not actually, a view which could be remembered with advantage today by many commentators on the scientific scene. But the real fruits of Aristotle's teaching on these two subjects were soon to be found in the work of Archimedes and Apollonios and later, in the seventeenth century AD, helped those like Newton and Leibniz in devising the infinitesimal calculus.

Aristotle, not surprisingly, devoted much time to discussing questions which we should today classify as astronomy and physics. The attraction of these two scientific subjects was that they presented a number of clear-cut problems which could be isolated and to which some specific answers could be given. Thus one could try to account for the regular motions of the planets, how far away they were, how large they might be; or one might enquire into the motions of bodies on Earth, the reason water flowed downhill or flames burned upward. And, of course, this specificity was the reason why these two subjects were so much in evidence in early science. They were the ones with which one could really come to grips.

To Aristotle the universe was a sphere with the Earth fixed at the centre; it was finite in size because, he argued, if it were infinite it could have no centre. Aristotle's opinion that the Earth was fixed was not just accepted out of hand; he did consider whether or not it might move as, for instance, Philolaos had suggested. He rejected the idea because, in his opinion, the Earth did not experience the kinds of phenomena – rushing winds or unsteadiness – one would expect if it moved; admittedly these effects were only to be expected on the basis of Aristotle's own laws of motion as described below, but at the time the arguments – and the laws – seemed reasonable enough, the logical outcome of the evidence available.

He agreed that the stars and celestial bodies moved in circular paths. Aesthetically this was satisfying, and with a mechanism like the homocentric spheres to explain it, it appeared to account for the observations. However, Aristotle seems to have thought of the spheres as having a real physical existence; not for him some disembodied mathematical explanation. So the idea of a universe of clear crystal spheres gained currency. But what moved them? Why should all these spheres rotate? Again this question depended on Aristotle's laws of motion, which demanded that a moving body required a constant force to keep it in motion, and in the end he concluded that the outermost sphere – the sphere of the stars – was the 'prime mover'. He went even further when he suggested that behind it all was an 'unmoved mover' who drove the whole system, but here we are in the region of metaphysics, not physics. We have slid from science to divine intervention, from physical explanation to supra-physical motivation.

To Aristotle the heavens were incorruptible and changeless, a reasonable enough assumption considering the fact that no alterations had ever been observed. The same stars and the same planets had appeared for countless generations. On the other hand things were quite different on Earth; here change and decay were part of the everyday scene. He rationalized this by claiming that change was confined to that part of the universe contained within the sphere below the Moon, that is, to the innermost sphere of all which had the Earth at its centre. The celestial bodies were formed of a fifth essence, which was itself eternal and without blemish; change and transformation were confined to the four ordinary elements.

For the implications for astronomy when this idea was at last refuted, see page 335.

With this separation of the universe into changeless and changeable,

Aristotle could tackle questions like the nature of 'shooting stars', or meteors, which appear as flashes across the night sky lasting anything between a fraction of a second and a few seconds, and the arrival of comets with their hazy heads and long glowing tails, that stay in the sky for weeks or even months before vanishing as mysteriously as they arrived. Clearly they could not be true celestial bodies for they were ephemeral, transitory phenomena; they must exist in the upper air, in the sublunary sphere. They were both meteor-like or, as we should put it, meteorological. But, of course, it was a great stride forward to explain comets and meteors as physical bodies rather than gods or demons. Clouds, rain, and winds were also part of this sublunary meteorological scheme of things.

The Earth was spherical: Aristotle had no doubt about that. The reasons for its sphericity were partly aesthetic – a sphere was a totally symmetrical figure – and partly physical. The physical reasons were the results of observation. Firstly, there was the way a ship appears to sink over the horizon – a natural consequence of a spherical or a curved Earth but not of a flat one. Secondly, there was the observation that wherever one went, even from one side of the Aegean to the other, objects always fell vertically to the ground. Admittedly this could happen with a flat Earth as well as a spherical one, but not with a simple curved Earth. Together the evidence was conclusive. But why did things fall to the ground? Why did water always 'find its own level', or air spread out into the surrounding space? Why did flames always burn upwards? These, too, were physical questions which required an answer, and it is a measure of Aristotle's greatness that he was able to provide one.

For his and others' calculations of the circumference of the Earth, see page 118.

Illustration page 85

His solution was to say that everything had its natural place. The natural place of earthy materials was the centre of the Earth, and the more of the earth-element a body contained the more strongly it strove to get there. Thus, according to Aristotle, heavier (i.e. more earthy) things would fall to the ground faster than lighter ones. Water spread out over the ground if it were spilled because the natural place of the watery element was the surface of the Earth. The natural place of the air-element was around the Earth, covering it like a blanket, while the fire-element's natural place lay in a sphere above our heads. Flames burned upwards because they sought to return to their natural home. It was a very complete, consistent system.

As to the motion of bodies, Aristotle distinguished three sorts. First, there was 'natural' motion; this was observed, for instance, when a body fell to the ground due to its 'gravity', or rose like smoke due to its 'levity'. The second was 'forced motion'; it was caused by outside forces and interfered with natural motion, as when one picked up a load or shot an arrow. Thirdly, there was 'voluntary' motion; this was performed by the will of living creatures. A force was always needed to generate forced motion, the imposed velocity being proportional to this force, a consequence which incidentally, made it impossible to have a vacuum because then an infinitely great velocity

For the rather more satisfactory Indian idea of impetus, see page 194.

Illustration page 86

would result from a finite force. In consequence Aristotle rejected outright the views of the atomists. All this seemed logical enough, but it caused great difficulties when one came to consider the forced motion of a projectile, as medieval European scientists were to realize.

Aristotle's biology. It is only within the last century that Aristotle's biological work has really begun to be appreciated; previously it was overshadowed by his achievements in the physical sciences. And though we cannot distinguish precisely all his own observations of the natural world from those of others, no doubt remains of his greatness as a biologist. He named about 500 kinds of animals and showed some suspicions when it came to accepting travellers' tales, as for instance that of the manticore, a monster probably derived from garbled accounts of the Indian tiger. But he did describe the lion's appearance and the way it walked, as well as that of the elephant, which shows he had observed them himself; indeed in the latter case a discussion of the leg joints allowed him to explode the current belief that it had to lean against a tree in order to sleep.

There is evidence that his observations included dissections, and full descriptions are given of the chameleon and of crabs, lobsters, cephalopods (squids, octopuses, etc.), as well as several fishes and birds. His observations were always meticulous: he investigated the pairing of insects, the courtship behaviour, nest building and brood care of birds, but above all he studied marine life. He noted how a cuttlefish anchored itself to a rock in stormy weather, while his description of the mouth-parts of a sea urchin was so detailed that they are still known as 'Aristotle's lantern'; his assertion that the sea urchin's eggs are larger at full moon has only recently been confirmed for the species he observed. Again Aristotle noticed that the female catfish left her eggs once she had laid them and that it was the male who looked after them, though this was later disbelieved and even ridiculed. Not until 1856 was it discovered that this indeed is an exact description of the behaviour of the particular species he was observing. Moreover, Aristotle was not content only with passive observation and dissection, for he made tests of sense perception in scallops, razorfish and sponges.

The Greeks used honey as a sweetening agent and it is not therefore surprising to find that Aristotle studied bees with considerable care. Though he did not realize that the ruler of the hive is a female, a queen not a king, he described the birth of bees in the hive, the behaviour of the drones and of the workers and the way honey is collected, and gave details of the bee sting. All this is the more remarkable when we realize that he had no magnifying glass to help him detect the detail he described.

Another of his fields of study was embryology. He described the growth of the chick embryo and noted the heart beating, as well as its appearance ahead of the other organs – perhaps this originated or confirmed his idea that the heart is the seat of the soul or the mind. He knew, too, that most fishes bring forth their young in the 'potential' form of eggs but stated that one group gave birth to fully shaped active young. What, in fact, he described was the birth of the placental dogfish, yet it seemed so extraordinary to later zoologists that his observations were ignored: they were only confirmed in the early 1840s. He also observed hectocotylization, the process whereby male cephalopods use one arm to fertilize the eggs of the female – yet another observation which had to wait until the nineteenth century for confirmation.

Aristotle made a good attempt to classify living things, though his basis was not one that we should use today. To Aristotle 'the soul is the first grade of actuality of a natural body having life potentially in it'. All living things had, he believed, a nutritive soul which guided intake of food and material well-being, and animals had also a sentient soul so that they could feel. Some higher creatures have also an appetitive and locomotive soul, while man himself has a rational soul. Although we may now baulk at the use of 'soul' in this way, what Aristotle was groping towards was that perennial question of what it is that makes a living creature alive, that separates the animate from the inanimate. His term 'actuality' was part of his doctrine of 'becoming' whereby what is 'potential' becomes actual, a true reality; applied to animals it meant that the various parts were organized for the creatures' greatest good. With all this as a basis Aristotle formed a 'scale of Nature' which, in essence, was a scale of increasing complexity of 'soul'. It ran from inanimate matter at the lower end to plants, thence to sponges, jellyfish, molluscs and so on up the scale to end with mammals and man. Certainly it was a static scheme – Aristotle envisaged no emerging evolution – and its basis of classification was not one to be accepted by later men of science. But it broke new ground; it was a valiant attempt to make order out of chaos and this it achieved. It found much favour in medieval times, especially in the Muslim world, and it acted as the prototype for the classifications that were to come in the eighteenth century, more than twenty-one centuries later.

For the Confucian idea of the 'ladder of souls', see pages 134 and 147.

Aristotle discussed animal and human anatomy and the function of the body's organs. His studies were to a large extent comparative, as we should expect from a man whose work was orientated towards zoology; by and large he was good on animals and poor when it came to men, since he did not make human dissections. He also made an error over the functions of the brain and the heart since he believed the former to be an organ to cool the blood and the latter to be the centre of consciousness. Furthermore, in assessing bodily functions he accepted the mistaken doctrine of the humours which he combined with the four qualities. Compared with his outstanding achievements in zoology, Aristotle's anatomy and physiology were less than brilliant, nor was his botany much better. Though it is clear that botanic questions were discussed in the Lyceum, the interest in plants centred on their practical value. Aristotle certainly made some botanical observations but his real concern was with animals. His friend Theophrastos, who succeeded him at the Lyceum, was the superior botanist, indeed we may call him the 'father of botany' provided we then give Aristotle the title of 'father of zoology'.

What an astonishing polymath Aristotle was! There was hardly a field of scientific endeavour in which he did not make valuable contributions or in which he did not give a lead for others to follow. Certainly his influence on later generations of western thought and science was crucial, greater than that of any other Greek philosopher and man of science.

For his especial influence on Islamic thought, see page 217.

Theophrastos

Born in Eresos in Lesbos about 371 BC, Theophrastos was thirteen years younger than Aristotle and worked with him for more than twenty years before taking charge of the Lyceum which he ran the next thirty-five years. He taught some 2,000 students, among whom was the famous physician Erasistratos, and gained such standing in Athens that later attempts to prosecute him for impiety failed, and a law against philosophers was repealed.

Theophrastos did not accept all Aristotle's teaching out of hand; he had his own ideas on some matters and seems always to have been critical of some Aristotelian opinions. Of course, this is how it should be; no scientific progress will ever be made if one slavishly follows all the views of one's teacher, however brilliant, but it is a measure of the greatness and the success of Aristotle and his Lyceum that this could happen without bitterness or rancour. What did Theophrastos question? A number of Aristotle's ideas about the universe caused him to have doubts; for instance, if the outermost sphere of heaven were rotated by a 'prime mover', how was it that some celestial bodies moved faster than others? And why should the rotation of celestial bodies not extend to bodies in the sublunary sphere? He also questioned Aristotle's explanation of the purpose of some of the phenomena he observed: what was the purpose of tides, for example, or why, if Nature desired what was best for its creatures, did deer grow horns, which were harmful to them?

Such questions often raised interesting points for further research, but Theophrastos did not act only as a challenger of old ideas; he is remembered now primarily for his positive contributions. Unfortunately not all his writings have survived but, even so, those which we have make it clear he was a pioneer in at least three fields of study. In the first place he was an early contributor to the history of science, writing a book *Opinions of Natural Philosophers*, a source book for many later writers. Secondly, Theophrastos was an able mineralogist. He followed up research on minerals, ores and stones suggested by Aristotle, although tests led him to query the basic classification which Plato and Aristotle had adopted. He also gave a full description of a wide variety of substances, detailing the way they reacted to fire, their feel, colour and other characteristics, and so produced the first methodical mineralogical treatise in the West.

The third, and most important, field in which Theophrastos

worked was botany. His results are contained in his *Account of Plants*, in which he mentions some 550 species and varieties taken from an area stretching from the Atlantic in the west to the eastern Mediterranean shores, and with a few from as far afield as India, using information obtained by himself or brought back by travellers and others. Of course this meant that the *Account* contained some fables about plants, and it is also true that some of the observations were obviously limited by lack of equipment – Theophrastos had no magnifying glass, any more than Aristotle did – but there is one crucial difference in his botanical work compared with what had been done before, and this was the way he classified plants. He devised a method which was to be of inestimable use to later botanists. What is more, he not only assembled this data impartially, but discussed it critically and withheld judgement whenever he was hampered by insufficient facts.

Plants were classified into trees, shrubs, undershrubs and herbs, and general and specific differences between wild and cultivated varieties were noted. Theophrastos also discussed plant juices, medicinal herbs, the types of wood provided by various trees and the uses to which they could be put. But above all, he gave special technical meanings to some words – e.g. he used 'pericarpion' (our 'pericarp') for the seed case – and this was a vital step if there was to be a real science of botany. His descriptions of plants were first-class, and of permanent value have been those he gave of the pericarp, of petalled and non-petalled flowers, of the tissues present in higher plants (parenchymatous and prosenchymatous tissues), the precise way in which a floral envelope grows, and the mode of development and arrangement of flowers on a plant (inflorescence). In addition, he described and distinguished between angiosperms (plants with enclosed seeds) and gymnosperms (plants such as conifers with naked seeds) and, most notably, monocotyledons (plants like barley and wheat with one seed leaf) and dicotyledons (those with two seed leaves, such as peas and beans); indeed his descriptions of these last were the most accurate available until the seventeenth century.

Alexandria and Hellenistic Science

The independence of Greece ended in 338 BC with the Macedonian conquest. Two years later Philip of Macedon was assassinated and his son, Alexander, who had been Aristotle's pupil, succeeded him. It has been well said that Alexander's accession closed one age and opened a new one: it initiated a vast new empire which was to extend from the western shores of the Mediterranean to the Indus, and from Egypt and Babylonia up to the Caspian and beyond. This huge empire carried Greek culture as far east as India, and brought eastern influences across to the West. However, there was little inherent unity in Alexander's empire and on his death in 323 BC it broke into three regions ruled by his generals and their descendants: Greece and Macedonia eventually fell to Antigonos and his dynasty (the Antigonids),

Persia and Babylonia came under Seleucos and his dynasty (the Seleucids), while Egypt was given to Ptolemy Soter, who founded the Ptolemaic succession. It is this last which is of interest to us, because at Alexander's new seaport Alexandria, which Ptolemy Soter completed, he also established a museum and library. It was a centre for advanced study which was to form the focal point of the new Hellenistic culture, a place that was to attract men like Euclid and Archimedes and to flourish for seven centuries.

Illustration page 85

No certain remains of the museum survive, but the Greek geographer Strabon or Strabo, who lived in the first century BC, described it as an extension of the royal palaces (which were close to the harbour), with a public walk, a covered colonnade with seats and a large common refectory. Presumably, though, there were also other rooms where discourses were given and research done as well as an actual museum and the library itself. Moreover it must have possessed some astronomical instruments. It was developed mainly by his son Ptolemy II Philadelphos, both monarchs having help from two Greeks, Demetrios and Straton. Demetrios came from Athens and was a writer, statesman and former pupil of Theophastos, whose collection of books probably formed the nucleus of the library. Straton came from Lampsacos in the Dardanelles; a generation younger than Demetrios, he too had sat under Theophrastos. He was called to Alexandria about 300 BC as tutor of Philadelphos, and remained for twelve years until the death of Theophrastos, when he returned to Athens to direct the Lyceum. It was Straton who helped give the museum its scientific tone.

Euclid

One of the greatest men of science connected with the Alexandrian museum, and this at a very early stage, was Euclid, who worked at Alexandria between about 320 and 260 BC. He it was who founded the great mathematical school at the museum. Euclid's fame primarily rests on the *Elements*, a systematic synthesis of Greek geometry which, until comparatively recently, was the basis of all geometrical teaching in the West. Indeed, its effect was much greater; its methods of synthesis – its axioms, postulates, theorems and proofs – have been said to have affected the Western mind more than any other book except the Bible. Certainly it has had a profound effect on the way problems are tackled, for the logical way Euclid makes each proposition follow from those previously demonstrated is masterly. Its proofs show the intellectual power of this technique and some are ingenious to a degree. The proof of Pythagoras's theorem is a case in point, and there is a story that when he first saw the proposition, the seventeenth-century English philosopher, Thomas Hobbes exclaimed, 'By God, this is impossible', and 'fell in love with geometry' after reading the proof. But the *Elements* became a legend in its own time. Indeed, there is a story that Ptolemy Philadelophos (or Soter?), when discussing geometry with Euclid, asked whether there was not some quicker way of learning the subject than ploughing through all

Illustration page 85

the propositions. Euclid replied, 'There is no royal road to geometry'. No results would be achieved except using the step-by-step logical progress he had detailed. We must not, however, be led to suppose that the *Elements* was all that Euclid produced; he did some original geometrical research and wrote on mathematical astronomy, on the mathematical theory of music and on optics, though in the latter case he seems to have moved little from the views of Plato and the Academy.

For the development of a new geometry in the 19th century by questioning two of Euclid's basic postulates, see pages 478–479.

Apollonios

Apollonios was another famous member of the Alexandrian mathematical school. He was born in Perge, a city in the south of what is now Turkey, between 246 and 221 BC, and certainly visited Ephesus and Pergamon, where there was another large library; it is clear that he also worked at Alexandria for some time.

Like Euclid, Apollonios is now remembered as the author of one book, *On Conics*, which was a study of the curves obtained when slices are taken through a cone to produce the ellipse, parabola and hyperbola. Later the ellipse was to become a curve of vital importance to mathematicians like Kepler and Newton in working out planetary orbits, but in the third century BC the subject of these curves was in a fragmentary state. The achievement of Apollonios was that he generated all the curves from a double oblique circular cone; this was a new approach, allowing the mathematics to be of a more general kind, and therefore of wider applicability, than was possible before. In brief, he laid the foundations for a subject that was to assume great importance for the mathematicians of seventeenth-century Europe.

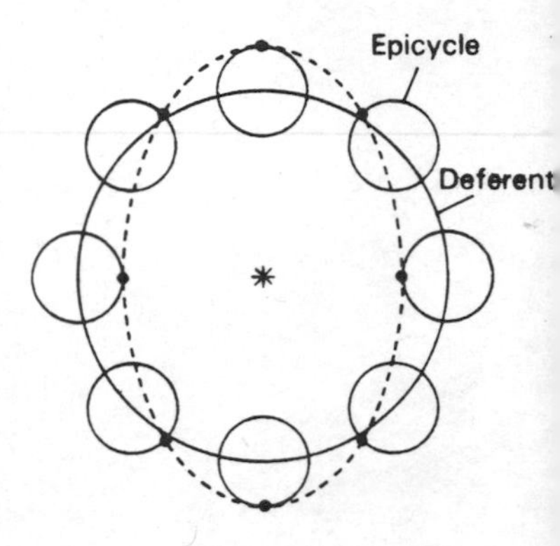

Diagram of the theory of epicycles, developed by Apollonios to show how an arrangement of circular motions (attractive to Greek astronomers because of their elegance) could yield an elliptical movement. This theory dominated ideas about the paths of moon and planets until the Scientific Revolution.

Apollonios interested himself in other branches of mathematics, especially in methods of expressing very large numbers; he worked on irrational numbers and above all on astronomy, devoting a lot of time to the Moon – indeed, he became known as Epsilon (because the shape of the Greek epsilon (ϵ) is similar to that of the crescent Moon). Apparently it is also to him that we owe the important geometrical devices of the epicycle, the deferent and the movable eccentric, which we shall come to shortly when we discuss the motions of the planets and the work of the Alexandrian astronomer Ptolemy.

Archimedes and the Alexandrian School of Mechanics

The life of Archimedes is little better documented than that of Apollonios. We know, however, that Archimedes was born in Syracuse, Sicily, about 287 BC and died there in 212 BC, killed by a Roman soldier. It seems reasonably certain that Archimedes spent some time at Alexandria. He is probably best known today for the Archimedean screw and the story of King Hieron's crown.

The Archimedean screw is an efficient and convenient method of raising water. A pipe in the form of a helix is rotated with one end in water; the water moves up the pipe as it rotates. The story of King Hieron's crown involves the question of hydrostatic weighing. King Hieron, ruler of Syracuse and friend of Archimedes, suspected that a crown (or wreath) of pure gold that he had caused to be made was really gold alloyed with silver, although it weighed the same amount as the gold given to the goldsmith. He asked Archimedes how he could determine whether or not the gold had been adulterated without damaging the crown and Archimedes is said to have hit upon the answer while having a bath. He noticed that the amount of water which overflowed from the bath tub when he entered it, was equal to the amount of his body which was immersed. He realized that if the crown were of pure gold it should displace an amount of water equal to that displaced by a lump of gold of equal weight. If, on the other hand, it was alloyed with silver, which weighs less than gold, the crown would have a greater volume and displace a greater amount of water than would be displaced by pure gold. Archimedes was so pleased with his discovery that he is supposed to have jumped out of his bath and rushed naked through the streets on his way home, shouting *heureka* 'I have found it'. Whether the story is true or not we do not know – it was first recounted by the Roman architect and engineer Marcus Vitruvius in the first century BC. But Archimedes went on to demonstrate the truth of his idea mathematically.

Illustration page 86

For the investigation of the principle of displacement by Simon Stevin in the late 16th century, see page 314.

There are other stories about Archimedes. One is that when asked by Hieron how great a weight could be moved by a small force, Archimedes demonstrated how he alone could pull a three-masted ship which had been lugged ashore by a host of men. For this demonstration Archimedes used compound pulleys, and while it seems that he did invent the compound pulley, the story, which is as late as the first century AD, has a distinct ring of fantasy about it. Not so Archimedes's supposed remark, 'Give me a place to stand on and I will move the Earth', which underlines the principle of the lever. Yet whatever the authenticity of the many stories of Archimedes, they all emphasize the fact that his reputation as a mechanical inventor was immense. Indeed, he is even credited with designing and using vast burning mirrors to set fire to ships in the Roman fleet which later besieged Syracuse – the siege which led to Roman occupation and the death of Archimedes – though the report that he designed and built celestial globes depicting the heavens and a kind of planetarium for displaying planetary movements seems to have more of a ring of truth about it, as it was referred to by Cicero (106-43 BC) who had seen the instruments.

It is for his mechanical inventions that Archimedes is remembered now, yet his more fundamental interest seems to have been in the underlying principles involved; in what we should today call the sciences of mechanics and hydrostatics. After all, he was very much a geometrician and mathematician: he worked on problems of the areas of curved figures using the powerful method of exhaustion invented by Eudoxos, he calculated the ratio of the circumference of a circle to its diameter – π (pi) – with unprecedented accuracy (his value for π was between 3.1408 and 3.1429; it is actually 3.1416). Indeed, in the sixth to tenth centuries AD Archimedes's mathematics

For Eudoxos, see pages 90–91; for more accurate Chinese and Islamic calculations of the value of π, see pages 151 and 223.

formed a subject for special study by scholars at Byzantium when that city was a major centre of Christian learning. But how would Archimedes himself have assessed his achievements. How would he have wished to be remembered? Strangely enough this is a question which can be answered. It seems Archimedes expressed the wish that his tombstone should be adorned with one of his geometrical triumphs: a cylinder inside which a sphere fitted exactly, together with an inscription giving the ratio by which the volume of the cylinder exceeded that of the sphere. In 75 BC Cicero visited his grave, and after slaves had cleared away the undergrowth he saw that this had in fact been done; the sphere and cylinder were there as well as the inscription 'with about half the lines legible, as the latter portion was worn away'.

Archimedes was both a mathematician and a practical man, and this mixture of theory and practical ingenuity seems to have been characteristic of at least some others who worked at Alexandria. The traditional Platonic outlook that the philosopher does not lower himself to deal with appearances, with the transitory mundane world of things, and instead keeps his eyes fixed on essential Ideas, was no longer the guiding force here. Indeed, so different was the intellectual atmosphere that it is no surprise to learn that Archimedes did practical experiments and was not the only mechanically minded philosopher to work there. Another was Ctesibios.

Yet again, little personal information about Ctesibios has survived: he was the son of a barber but we do not know when or where he was born or the date of his death; we do know he worked at Alexandria around 270 BC because the records tell us of a singing horn of plenty devised by him for a statue of Ptolemy Philadelphos's wife. Ctesibios's fame certainly rests on his inventions which began, it seems, with an adjustable mirror in his father's barber's shop. This could be adjusted to any position because it had a counterpoise, and the counterpoise was worked by a cord attached to a ball of lead that moved inside a tube, expelling air with a loud noise as it moved down the tube. The invention led him to investigate the whole pneumatic principle involved; a principle which is still used in many devices today, such as the commonly-used door-closing spring. Ctesibios invented an air pump with valves – presumably along the same lines – and this was connected to a series of organ pipes operated from a keyboard. He also invented a force pump for water, and designed catapults operated by bronze springs and one worked by compressed air, but in spite of this excursion into engines of war, his main work was on devices that appear to have only limited practical use. Thus his most famous inventions were clepsydrae (water clocks) having a constant flow of water which operated all kinds of jacks or automata from ringing bells and moving puppets to singing birds – the latter obviously a precursor of the cuckoo clock! These clearly exerted great fascination and were widely written about, especially by the mechanician Philo of Byzantium, who visited Alexandria about

250 BC. Fortunately, though, some description has also been preserved of another clepsydra designed by Ctesibios; this contained water-driven gearwork that operated a rotating cylinder on which the unequal hours of day and night were displayed – a less picturesque but more practical device than the singing-bird clock.

The mechanical school at Alexandria seems to have continued for many years, though the most notable figure about whom there is evidence is Heron or Hero who lived and worked in AD 62, some three centuries after Ctesibios. But the evidence we have of Heron gives us no personal details; it merely concerns his inventions and his books. His inventions were, by and large, mechanical toys: he invented a miniature automatic theatre whose figures were worked by weights counterbalanced by containers from which millet seeds spilled out, and a model in which temple doors opened when a fire was lit on a miniature altar, and closed when the fire went out, as well as trick jars, automatic sounding trumpets and the like. However, like Ctesibios, he also designed military equipment, especially a crossbow (the stomach bow or 'belopoiika') which could be drawn simply by the archer putting his weight against the stock and, again on the more practical side, an odometer for measuring distances along roads. There was also the 'dioptra', a instrument for surveyors that acted partly as a theodolite (for measuring angles) and partly as a level.

Why did a learned institution like the museum support what seems to have been a succession of toy-makers or fabricators of semi-magical and mysterious devices? Certainly some useful inventions were devised but most reports of the work of Ctesibios and manuscripts recording the work of Heron are full of mechanical contrivances designed more to entertain than to edify. What was the purpose of it all, if purpose there was? The answer is not hard to find and, indeed, a clue is given in the books which Heron himself wrote. He prepared a text on pneumatics and another on mechanics; he wrote about the dioptra, about catapults and on the optical principles of reflection. In other words, Heron wrote a series of works dealing primarily with what we should call physics, and it seems that many of the devices he and others before him made were built either to test out physical principles or to demonstrate them to a wider audience. Thus when we learn of Heron applying his knowledge of mirrors, both curved and plane, to conjuring illusions, we should not be surprised: clearly such an application would be an ideal demonstration of certain optical principles. One can ask, of course, why they were not satisfied with basic laboratory experiments rather than elaborate toys, but there was no tradition of formal physical demonstration, though there was of formal mathematical proof, and ingenious devices would have a great effect in bringing home certain principles. Instructive toys of this kind were a well-recognized way of teaching in Victorian times. Moreover it was an age unaccustomed to mechanical marvels, and perhaps physicists felt some need to dress up the principles they demonstrated, just as later in the seventeenth and eighteenth centuries scientific instrument makers felt it part of their duty to make their instruments works of art as well as devices for practical use.

However, there seems also to have been a tradition of scientific instrument making in Greece. There are no literary references to this, but an important artefact which provides a significant clue is a calendar computer recovered in this century from a wreck close by the island of Antikythera off the north-west coast of Crete. Amongst a cargo which dates from the first decade of the first century BC, one of the items on board was a mechanical calculator, which recent detailed research by Professor Derek Price has shown to be quite complex, containing an elaborate arrangement of gear wheels for counting and then displaying the results. This is an important find, since it would certainly be unlikely that this particular calendar computer was unique; it is much more probable that it was an example of a device well-known among mariners and others who could make practical use of astronomical information. Moreover, this tradition of instrument making seems to be supported by Professor Price's recent reconstruction of the Tower of Winds in Athens, which was built between 100 and 50 BC by the astronomer Andronichos, who came from Cyrrhos in the Fertile Crescent. This was an octagonal building with carved figures representing the eight principal winds, and a wind vane. There were sundials outside, and inside a water-driven astronomical clock, in which a rotating disc showed the movements of the stars and on which a ball representing the Sun recorded the seasonal motion of the Sun through the constellations. Perhaps then, mechanical contrivances for displaying scientific information were not as unusual in Hellenistic times as their absence from literary sources would lead us to suppose.

Illustration page 88

Illustration page 76

The Alexandrian School of Medicine

The museum of Alexandria boasted a notable medical school whose founder was the physician Herophilos in the early third century BC. Although his date of birth is unknown, he seems to have hailed from Chalcedon, a town on the Bosporus opposite Byzantium. We do know that he studied medicine at Cos and was invited to Alexandria by Ptolemy Soter, and that he continued to work there under Ptolemy Philadelphos. He gained a great reputation as a medical practitioner and as a teacher, and students flocked to learn from him.

It is interesting that at Alexandria the dissection of the human body did not meet with disapproval as it did in other Greek cities; this was a great bonus for medical research of which Herophilos took full advantage. He based his medical theory and practice on the four humours, but it is in anatomy that he did his greatest medical work, investigating the brain, the nervous system, the system of veins and arteries (he distinguished between the two), the genital organs and the eye. He correctly identified the brain, rather than the heart as the centre of the nervous system, and traced the nerves down from the brain through the spinal cord, though he wrote also about the *rete mirabile* (marvelous net), a complex of blood vessels and nerves which is present in animals but not in man. Some parts of the human anatomy are still named after him – thus a cavity in the heart is called the *calamus Herophili,* and the meeting point of the sinuses and the tough fibrous covering of the brain (the *dura mater*) *torcular Herophili*. He also emphasized the importance of the pulse, following here in the steps of his teacher Praxagoras, though Herophilos was perhaps the first to use a clepsydra for timing it. He traced the optic nerves from eye to brain and took a great interest in the liver and in the intestines, coining the name 'duodenum' (or rather its Greek equivalent) for part of the small intestine, and was the first to isolate the lymphatic vessels there. He made some advances, too, in gynaecology and first recognized the Fallopian tubes; a discovery later forgotten and made again independently in the seventeenth century by the Italian anatomist Gabriele Fallopio. And as if all this were not enough, Herophilos wrote on dietetics and recommended gymnastics as a healthy form of exercise.

For Gabriele Fallopio, see page 287.

Unfortunately, the followers of Herophilos at Alexandria and Laodicea, where he founded another medical centre, tended to spend their time in disputation rather than dissection and observation, although there was one notable exception, the physician Erasistratos. Erasistratos was born about 304 BC into a medical family on the Ionian island of Chios, first studied medicine in Athens, then in his mid-twenties went to the medical school in Cos. Later he moved to Alexandria and came under the influence of the palace doctor of Ptolemy Philadelphos, and when in old age Erasistratos gave up medical practice, he moved to the museum to conduct research. Here he did some valuable work.

Erasistratos was the author of a large number of books, dealing not only with anatomy but also with the abdomen, with fevers, gout, dropsy and the vomiting of blood, and with hygiene. Most of these are lost but we still know something of his anatomical work and his studies of the functions of the body, while one of his pioneering achievements at Alexandria was conducting post mortems in order to study causes of death. He was also the first to conduct experiments leading to the discovery of bodily wastage, and so laid the basis for the study of metabolism, work which was not revived until the Italian physician Santorio did so in the seventeenth century. Erasistratos's metabolic investigations led him to suggest that all bodily parts of living creatures were a tissue composed of veins, arteries and nerves, an astonishing achievement in the days before the invention of the microscope. He also disposed of the erroneous belief that digestion was either a form of cooking or fermentation and described the action of the stomach muscles; in addition he showed how the larynx is closed during swallowing and so proved that it was impossible for drink to enter the lungs as some physicians believed.

Illustration page 411

Erasistratos's studies of respiration were a considerable improvement on those previously conducted; it was he who went a step further than Herophilos and separated sensory and motor nerves. He also realized the pumping action of the heart but did not recognize the circulation; as he saw it, one side of the heart pumped pneuma, the other side blood which he thought was manufactured by the liver. He rejected the doctrine of the four humours, and supposed disease was due to an excess of blood. Strangely, though, he hardly ever practised blood-letting, as his contemporaries did, but prescribed a greatly reduced intake of food.

Like Herophilos, Erasistratos had his disciples, but they seem to have achieved nothing. Yet the two men were important, because between them they laid the foundations of anatomy and physiology in the Western world. Certainly some of their work was forgotten but in view of the disasters which befell the museum in the third and fifth centuries AD, and its utter destruction in the seventh, what is astonishing is that so much survived. For this we have to thank a Greek surgeon, Galen, who lived in Pergamon in the second century AD. who visited Alexandria and recorded many of the medical school's achievements.

For the career and teaching of Galen, see pages 247–249.

Astronomy in Alexandria

Studies of the universe are a constantly recurring theme in ancient as in modern science, and Alexandria could boast a very active and effective school of astronomy. This was due partly to the influence of Straton and to Eratosthenes, who became the second chief Librarian in about 235 BC. Eratosthenes, a geographer and mathematician, was born in Cyrene (now Shahhat in Libya) probably in 276 BC, but he spent most of his working life in Alexandria, dying there in 195 BC. He was one of the foremost scholars of his time and wrote on philosophy and on literary matters in addition to his more scientific work. As a mathematician he solved the problem of doubling the cube, giving a solution which he thought superior to all previous solutions, and in arithmetic invented what is still known as the 'sieve of Eratosthenes' – a method for finding prime numbers. His reputation among his contemporaries was immense – Archimedes dedicated a book to him – but it is for his geography that he is primarily remembered today.

For previous attempts at doubling the cube, see pages 90 and 95.

Eratosthenes's treatise *Geography* long remained a standard work, Julius Caesar was still consulting it over a century after it was written. Although people and places are described in it, it is the first book to attempt to give geography a mathematical basis, being concerned with the Earth as a globe and dividing it into zones; it also describes changes in the surface and has much to say on mapping, giving numerous distances measured along what we should call parallels of latitude or meridians. As a baseline for his mapping, Eratosthenes used a parallel running from Gibraltar at one end, through the middle of the Mediterranean to the Himalayas in the east. This was important, since one of his aims was to correct previous Greek maps, all of which were centred on Delphi and showed a circular ocean surrounding the entire land mass (an argument he justified by noting the similarities between the tides of the Indian Ocean and those of the Atlantic), whereas his new base-line bisected the then-known world. By using information from travellers he was able to insert other parallels as well as various meridians. This helped him when he later divided the globe into Frigid, Temperate and Torrid zones.

The best-known geographical achievement of Eratosthenes was his measurement of the Earth's circumference. He found, presumably from reports sent to him, that at the summer solstice the Sun shone directly down a well at Syene (present-day Aswan), casting no shadow. This meant the Sun must be directly overhead. If then the altitude of the Sun were measured at Alexandria at the same time on the same day, this angle would allow the difference in latitude between Syene and Alexandria to be determined. The difference turned out to be almost 7¼ degrees, i.e. $^1/_{50}$ of the circumference of the Earth. Eratosthenes next had the distance between Syene and Alexandria determined, probably by a *bemetatistes,* a surveyor trained to walk in equal paces. At all events he seems to have used a rounded-up figure of 5,000 stadia, and this therefore gave him 50 × 5,000 or 250,000 stadia for the Earth's circumference, although some sources say he adopted 252,000 stadia equivalent to 46,660 km (29,000 miles). This compares quite well with the modern value for the Earth's polar circumference of 39,941 km (24,819 miles), and was a considerable improvement on the over-large values of Aristotle of 400,000 stadia, and of Archimedes (300,000 stadia).

Aristarchos of Samos. Aristarchos studied under Straton and presumably did so at Alexandria, not at Athens where Straton was later to run the Lyceum, since his own work makes an earlier date more likely. Where Aristarchos carried out his astronomy we do not know; indeed we know no more about him than that he was born at Samos and lived some time between about 310 and 230 BC. But if we are ill-informed about his life, we are in no doubt about his achievements. Aristarchos spent much effort in trying to determine the sizes and distances of the Sun and Moon and was also the first astronomer to propose a thorough-going heliocentric theory, putting the Sun, not a central fire, in the centre of the universe.

Eratosthenes's observations of the curvature of the Earth from the lengths of shadows at different latitudes. By careful measurement he was able to calculate the circumference of the Earth with reasonable accuracy.

Aristarchos's treatise on the distances and sizes of the Sun and Moon was really an exercise in astronomical geometry. Working after Euclid and a generation before Archimedes, he determined the Moon's distance by an ingenious method: observing the angle of the Sun at the moment when exactly half the Moon's disc appeared illuminated. However, this was not very satisfactory in practice, both because it is difficult in this particular configuration to measure the angle between the Sun and Moon with precision, and because it is virtually impossible to be sure when exactly one-half of the Moon's disc is lit by sunlight. The observation should give the ratio of the distances of the Sun and Moon from the Earth, but due to the inherent inaccuracies Aristarchos's calculations showed that the Sun was only some 18 to 20 times further off than the Moon, whereas the true figure is nearer 400. But it was the mathematical principle which seems to have intrigued him; his observations both of the angle involved and of the apparent diameter of the Moon were grossly in error. Yet although a poor observer he developed useful geometrical methods for calculations involving very small angles, and clearly grasped how to deal with problems where answers can only be given as lying within certain limits, not in exact numbers.

As far as his heliocentric theory was concerned, this was different from the idea of Philolaos previously mentioned where the Earth and the Sun were thought of as orbiting a central fire. Aristarchos accepted the theory of the Earth's daily rotation on its axis, probably first put forward by Heraclides of Pontus – but it was his idea to set the Sun stationary at the centre of the universe. Such a theory raises an observational problem, namely that if the Earth orbits the Sun, the stars should appear to keep altering their relative positions by a small amount. No such shift was observed and Aristarchos put this down to the immense size of the celestial sphere containing the stars; indeed he even gave a ratio intended to express how vast this sphere must be. We now know that such a shift does occur but it is extremely small; to observe it requires a telescope and a very refined observing technique, so refined that it was not detected until the 1830s, more than 2,000 years later. But in spite of his advocacy and mathematical ingenuity the theory soon fell into oblivion. It seemed a most unlikely solution of the problems of planetary motion, and had to wait seventeen centuries before it could be successfully revived.

For Philolaos, see pages 77–78.

Hipparchos of Nicaea. We come now to a vitally important figure whose work was to affect the Alexandrian school but who, as far as we know, never went there. Born in Nicaea some time in the first quarter of the second century BC, he seems to have spent most if not all of his working life at Rhodes, but the details we have of his life are meagre in the extreme and have all come to us from the later Alexandrian astronomer Ptolemy. It is to Ptolemy, also, that we owe what we do know about Hipparchos's work, for the only original text which has survived is a short commentary on the astronomical work of the Stoic poet Aratos. Nevertheless, from what information we have it is clear that Hipparchos made some notable astronomical achievements in the fields of planetary motions, the behaviour of the fixed stars, the length of the year and the distances of the Sun and Moon. What is more, Hipparchos prepared a mathematical table of chords; this was a tedious job but one of great importance for computing positions of celestial bodies in the days before trigonometry.

On the subject of planetary motions, Hipparchos made no new contributions of real significance, but he did put his finger on certain weaknesses in the use by Apollonios of his epicycle system and in the way all previous systems had been applied. Thus he cleared the way for a new theory of planetary motions which would 'save the phenomena' with mathematical precision, an achievement which was to come with Ptolemy. Here, then, Hipparchos was a herald, not a conqueror, but when it comes to his observational work it is a different matter, for Hipparchos was by and large a careful and excellent observer – indeed he is sometimes said to have been the greatest observational astronomer of classical antiquity. His observing was mostly carried out with the usual instruments of his time – the armillary sphere and the ring dial – instruments made of metal rings which acted as scales, and fitted with a sighting bar which the astronomer aligned with the celestial body being observed. To these Hipparchos added the 'plane astrolabe' – a disc on which a movable chart of the skies enabled one to make calculations of the rising and setting times of celestial bodies, and an instrument which could also be used for measuring angles. In addition he used the 'dioptre', a beam of wood along which a wooden prism could be moved: it was useful for measuring the apparent size of the discs of the Sun and Moon.

Illustration page 87

With these instruments Hipparchos made a catalogue of star positions – the first in the Western world – setting out the positions as angles measured along the Sun's apparent path (the ecliptic) and north and south of it. The starting point of the angular measurements along the ecliptic was the point where this path crossed the celestial equator (the circle on the celestial sphere aligned with the Earth's equator); it is the point which marks an equinox because when the Sun reaches it on its annual circuit of the heavens the hours of night and day are equal. In due course this led Hipparchos to discover that this equinox, this crossing point, was not fixed; both it and its counterpart on the other side of the celestial sphere moved slowly backwards. This discovery of the 'precession of the equinoxes' was, of course, of great importance for all later precision astronomy. He also measured the length of the year with great exactness, giving it as 365.2467 days (modern value 365.2422 days).

Hipparchos's measurements of the sizes and distances of the Sun and Moon were superior to those of Aristarchos. For determining the Moon's distance he showed his practicality, basing his observations not on a phenomenon difficult to measure accurately but on one he could determine with precision. He used a total eclipse of the Sun which occurred in 190 BC, and had it observed from two places on the same meridian – Alexandria and the Hellespont. At the latter the Moon completely obscured the Sun, but at Alexandria only 80% of the Sun was hidden. This was because the Moon is much nearer to the Earth than is the Sun, and observing it from two different places causes an apparent shift in its position in the sky. Knowing the different latitudes of the two places and taking account of the possible errors in measurement, Hipparchos was able to give the Moon's distance in terms of the size of the Earth. He found that the Moon's distance must be more than 59 times and less than $67^1/_3$ times the radius of the Earth. From a further study of eclipses of both Sun and Moon, he computed that the Sun's distance is 2,500 times the radius of the Earth, and was later able to give 60½ for the Moon's distance. These values were much superior to previous determinations, and although the distance of the Sun was some ten times too small, that of the Moon was virtually correct (modern value 60¼). His calculation of the size of the Moon was also reasonably accurate, though that for the Sun was, of course, far too small.

Ptolemy. There is no doubt that Hipparchos represented the best in Greek observational astronomy, and it is clear that his results stimulated the work of astronomers at Alexandria. Three centuries later his achievements were to be preserved for us by the Alexandrian astronomer Ptolemy, who also used them critically in his own vast astronomical compendium, which has come down to us with its Arabic-Latin title *Almagest*. In Greek the book was known first as the *Mathematike Syntaxis* (Mathematical Compilation), and later as *He Megiste Syntaxis* (The Greatest Compilation), and it was this which the Arabs called *Al* (The) *Majisti*, which later became corrupted to *Almagest*. But whatever we call it, the book marks the pinnacle of Greek astronomy, enshrining the results of the best of their work. Certainly it is not without its faults – no book ever is – but criticism that Ptolemy wilfully cheated in his astronomical measurements is without foundation, being based on a grievous misunderstanding of the evidence. In fact the *Almagest* became the basis of mathematical astronomy until the seventeenth century, and was successfully used by astronomers of the standing of Copernicus and Kepler.

Ptolemy was no relation of the Ptolemies who ruled Egypt – in fact the name Ptolemy merely signified that he came from Egypt – but it confused medieval and some later scholars who were moved always to portray Ptolemy wearing a crown! As far as can be determined his real name was Claudius and he was probably born in Ptolemais Hermiou in Upper Egypt about AD 100, but he seems to have spent most of his life in Alexandria; certainly it was there that he spent his working life and presumably where he died about AD 170. We have no details of his life, except that he may have studied (at Alexandria?) with the astronomer Theon of Smyrna.

Illustration page 87

Ptolemy's reputation rests above all on the *Almagest*, but before describing this monumental study it will be best if we look at his other achievements. He wrote a book applying mathematics to the art of constructing sundials and another dealing with the tricky business of representing the circles of the celestial sphere on a flat surface, the basic problem of all mapping. It was, of course, just such a map or projection that had been used by Hipparchos in the plane astrolabe. Ptolemy wrote also on astrology; his volume on this consisting of four 'books' or sections and, for that reason, becoming known in later antiquity as the *Tetrabiblos*, though its Greek title *Apotelesmatika* (Astrological Influences) was more explicit. Interestingly, he considered the influence of celestial bodies as something purely physical, deducing this from the obvious physical effects on Earth of the Sun and Moon. The book is also without fatalism – Ptolemy considered astrological influences only one of many affecting mankind – but the work does reflect the superstition and credulous belief prevalent in the Roman empire in the second century AD, which even minds like those of Ptolemy or Galen could not escape.

Ptolemy wrote also on optics; his original is lost but a later translation still exists. This contains the interesting idea that colour is an inherent property of bodies, and a study of stereoscopic vision, as well as the usual material on reflection and mirrors. Ptolemy accepted the erroneous Greek idea that vision was caused by something, a 'visual flux', emitted from the eye, but nevertheless he seems to have done some experiments to determine the size of our field of view and proved that vision 'travels' in straight lines. He also discussed the refraction of light, using experiments to demonstrate its existence; for example that a coin lying on the bottom of a vessel but hidden by the rim becomes visible when water is poured in. In this section he came close to the correct law of refraction which was not stated fully until 1621 by the Dutchman Willebrord Snel. Clearly Ptolemy was a physicist of considerable ability, and another book of his, on music, was a judicious mixture of mathematical analysis and the evidence of the ear; all along he never forgot the importance of empirical evidence.

Besides these books, and in addition to the *Almagest,* Ptolemy wrote a large and valuable treatise on geography. This, his *Geography* or *Geographike syntaxis*, was an attempt to map the known world, and the bulk of the text consists of a list of places with their latitudes and longitudes, a system of co-ordinates which dates back at least to the time of Eudoxos but which had never before been so extensively applied. The book was accompanied by maps, but since so many mistakes can be made when they are copied by hand, as they were bound to be in the days before printing, Ptolemy wisely gave instructions which would allow a copyist to reconstruct them anew. He also gave some sound practical advice on map projections, that is on how to represent the curved surface of the Earth on a flat chart, as we should perhaps expect from someone who had written a mathematical text on the subject. Of course the work contained errors and Ptolemy's maps were inadequate wherever they depicted areas outside the bounds of the Roman empire, but this in no way detracts from what was a monumental compilation, more comprehensive by far than anything before it.

Illustration page 88

For the achievements of the Chinese in map-making, see page 166.

In spite of the value and comprehensiveness of the *Geography*, the *Almagest* was Ptolemy's crowning glory. On the one hand it is a vast compendium of Greek astronomy up to his own day; on the other it contains the new results of his original work on the theory of planetary motions, as well as a catalogue of star positions and a new extensive table of chords. Using the observations of Hipparchos, and of Aristyllos and Timocharis, two Alexandrian observational astronomers, as well as evidence from other Greeks and Babylonian data, Ptolemy built up a detailed mathematical description of planetary motions and the motions of the Sun and Moon that was to act as the basis for all Western astronomy for the next fourteen centuries and more. Some of this new work was variable in quality; in working out a theory of the Sun's motion Ptolemy relied too much on the observations of Hipparchos (which in this particular case contained many errors) and discarded even his own when they disagreed too much. His work on the Moon's motion was much better, even though the Moon's behaviour is far more complex when one begins to consider the fine details. Based primarily on Babylonian data it shows a great improvement on the theory developed by Hipparchos, and by using it he was able, in conjunction with his theory of solar motion, to explain precisely how to compute the dates of future eclipses of both Sun and Moon.

As to the five planets, Ptolemy was working in an almost untouched field as far as the elegant and powerful geometrical tools of the movable eccentric and the epicycle and deferent were concerned. Apollonios had devised them, to be sure, but no one had applied them in such detail as Ptolemy did. In his lunar theory Ptolemy had modified the epicycle and deferent by using two epicycles; when it came to the planets he had to account for their apparent standstills and retrograde motions, and he invented an extra device, the 'equant'. This was an additional point about which the motion was uniform, but it was not the usual pivot point of the epicycle nor was it a point at the centre of the Earth. Indeed what Ptolemy really achieved, looked at from a modern point of view, was to give the planets slightly elliptical orbits centred on the Earth together with a close approximation to the variable motion which they display. Yet all this was achieved by using standard circular motions at a regular, unvarying rate, truly a masterpiece of mathematical artistry.

For Apollonios and his theory of epicycles, see page 111.

Illustration page 87

Ptolemy's success at 'saving the phenomena' was recognized by all later generations of astronomers and by his contemporaries. He was even prevailed on to write a popular abridgement of the theories given in the *Almagest*. But the full text is what his successors used, as well they might, for not only did it present a vast amount of original work, it also provided the best results the Greeks ever attained in a text which was a model of clarity and systematic presentation. Indeed it has few equals among the vast army of scientific books produced since that time.

The Fate of Greek Science

With Ptolemy we come to the last of the great figures of Hellenistic science, and the end of an astonishing intellectual development that was Greek in origin. It started, as we have seen, with philosophers who wished to make sense of the physical world in which they found themselves. They would not of course have called themselves scientists – the word is a nineteenth-century one – though they might have accepted the seventeenth-century name natural philosopher, yet it was science that they practised; not the mathematically orientated experimental science that we have today but science nevertheless, an attempt to rationalize the world of natural experience without recourse to divine intervention. This we have traced from Thales to its high development in Athens at the Lyceum and at the library and museum in Alexandria. It marks the first great concerted effort of western man to understanding the workings of Nature, and this immense achievement was recognized even in its own day. Roman science was Greek science consolidated and confirmed, but it contained no Roman contributions: Roman achievements were of another kind. Today we can look back and see in it the foundations of the scientific ideal – the pursuit of science untrammelled by political or religious restriction – and the basis of our present scientific culture. Yet it very nearly failed to reach us in anything but a very garbled version.

For the relationship between Greek and Roman culture, see pages 245–256.

Certainly, the library and museum at Alexandria continued after Ptolemy's time, though notable original research withered away. The library itself, originally developed under the Ptolemaic regulation which caused all books brought into Alexandria to be deposited for copying, contained some hundreds of thousands of papyrus rolls. Rumour had it that the collection was swollen in about 40 BC when Mark Antony gave the 200,000 rolls of the library at Pergamon to Cleopatra, but after the death of Ptolemy two centuries later little new material seems to have been added. Indeed the collection was soon to be damaged by invasion and insurrection. In AD 269 the library was partially burned when the beautiful but ruthless Septimia Zenobia, queen of Palmyra, captured Egypt, and it suffered even more severely in AD 415 in a mob attack. This seems to have been instigated, or at least condoned, by Cyril, the bishop of Alexandria, who championed orthodoxy against those whom he considered to be Christian heretics and against pagan learning. Hypatia, a woman mathematician, Neoplatonist philosopher, and head of the museum was brutally murdered by monks while an incensed mob burned the library itself. After this Alexandria never recovered its standing as a centre of learning, and it fell into complete oblivion after a final burning during the Islamic invasion of Egypt in AD 640. But long before that time scholars had fled, taking valuable manuscripts with them and, as we shall see (Chapter 5), it was in Islam that Greek learning was preserved for recovery in medieval Europe.

For the implications of this for Arabian science, see page 203; for the development of Neoplatonism, see page 206.

Chapter Three

Chinese Science

Until the 1960s little about the history of Chinese science was known in the West. A few ideas were general currency among historians of astronomy, and some information was available on Chinese botany and medicine, physics and engineering, but by and large appreciation of Chinese scientific culture was woefully inadequate. Certainly there were some widespread misconceptions which resulted either in attributing to the Chinese almost every early invention and discovery ever made or, on the other hand, throwing doubt on the claim that they made any notable contributions of any kind.

The problem has primarily been due to the language barrier. Until very recently those who knew Chinese were for the most part untutored in natural science – their specialist knowledge lay in other fields – but now, due to the efforts of Joseph Needham in Cambridge and his colleagues from all over the world, the situation is altering. The work of these scientifically trained scholars is still incomplete, but a brief resumé of their results so far will give us some insight into Chinese science up to the seventeenth century when it was irrevocably transformed, becoming merged with universal world science. In 1582 the Jesuit Matteo Ricci had arrived at Macao and almost twenty years later entered Peking. Here this learned scholar – linguist, mathematician, scientist – not only carried out religious missionary work but, in order to gain the confidence of the court, made known the first results of the new scientific revolution in the West. After Ricci's death in 1610 the Jesuit mission continued, and other scientifically trained missionaries took over his role. Indigenous Chinese science could no longer maintain a separate identity, and later Chinese achievements became part of the international scientific scene.

For the impact on European thought of the reports of missionaries such as Ricci, see page 400.

In looking at early Chinese science, historians find that they meet with one advantage that is not present in the study of the science of any other people, an advantage that is provided by the way the Chinese language is written. Still using ideographs to represent objects, written Chinese has not changed essentially from early times; it is as easy now to read an early text as a late one.

Why writing with ideographs has persisted in the Far East is not clear, but the very fact that an alphabetic script was never developed underlies something that we must keep in mind when we look at Chinese science; this is that we are seeing an explanation of the natural world build up in a quite different culture from our own. It is a culture that for a very long time remained somewhat isolated from the West, and developed almost entirely on its own. Certainly there were contacts, long before and long after Marco Polo in the thirteenth century, but they were not frequent and at times the Chinese did not encourage foreigners at all.

The isolation of China was, and is, in part geographical. Although it has a long coastline, stretching some 5000 km (3125 miles), most of this lies to the east and faces the Pacific, while what is in the south has its shores on the China sea, which is itself some 1600 km (1000 miles) wide. In very early times the sea formed an impassable barrier as far as the West was concerned. By land things were little better. To the north lies Mongolia and the Gobi Desert, to the west the vast mountain ranges of Tienshan and the Tibetan massif; indeed western exploration means climbing what is virtually a mountain staircase. From a geographical point of view, the wonder is not that China was almost isolated, but that there was any communication however, fitful, with the West.

As for the country itself, at a first glance it looks as if it is divided into three parts, the dividing lines being the two great rivers, the Yellow River (Huang-ho) and the Yangtze, which run from west to east. This, of course, is an oversimplification, because the land is so mountainous. However, in the area on either side of the Yellow River are the plains of Hopeh and Honan, as well as the sacred mountain T'ai Shan. More plains are to be found eastwards in Shensi, on the northern side of the river. Other rivers flow into the Yellow River; there is the Lo, on whose banks was the old capital of Loyang, and the river Wei in Shensi. Around the north of the Wei valley the soil is composed of dust blown down for ages from the deserts in the north; this area was the centre of the earliest Chinese civilization.

The Yangtze flows into the Pacific, just north of Shanghai. It is more readily navigable than the Yellow River, and one can travel up past the three great lakes of T'ai, Poyang and Tung-t'ing, and thence on to Szechuan province through gorges comparable with the Grand Canyon or the Great Rift Valley in East Africa. A general look at a map of China shows that, in spite of its two great rivers, it is still, like so much of Central Asia, an area where flat land is interspersed with mountainous regions, some of them particularly high and rugged. Great tracts of the country are not readily accessible, and the unification of Chinese culture was no simple task.

With a land of such contrasts there is naturally a wide range of climate, from the hot and humid south at a latitude of 20° to 25° north – on a level with central Burma and India, Egypt and the Sahara, Mexico and Cuba – to the more temperate north, latitude 40° – as far north as Turkey, Greece, Madrid, Philadelphia and Denver. In the west there are also the frozen wastes of the mountainous highlands, and in the east the arid mountains of Shantung where the winters are very cold. The valleys themselves vary, too; the North

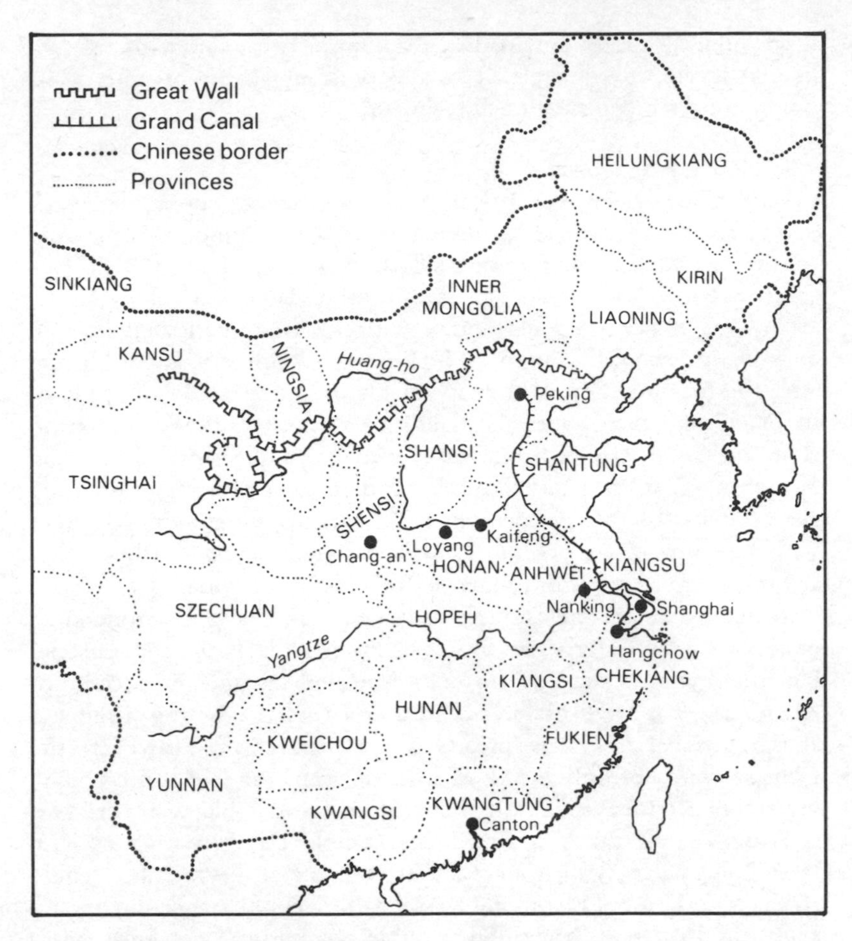

China plain adjoining Shantung is composed of silt from the Yellow River and is immensely fertile provided it is well manured. It is an area which is densely populated despite periodical flooding. North and north-east, in the provinces of Shansi, Shensi and Kansu, the dust-blown plains are exceptionally fertile, since the ground holds moisture well, and large harvests are the rule in spite of limited rainfall. The lower Yangtze valley can also be productive but it needs irrigation; here the Chinese genius for canals and waterways asserted itself very early on. Then in Szechuan there is a dense population in a basin beside the Yangtze, with a vast capacity for food production – sometimes achieving three harvests a year. In the far south the forests provide timber in abundance and there is much fishing; it is from here that the Chinese sailors traditionally came, but in early times the area was somewhat isolated from the rest of the country.

In brief, China is a land of climatic and scenic contrasts, a place where there are concentrations of population and wide tracts where only nomads eke out an existence. It may have been isolated, but it carried within itself a microcosm of the rest of the world. A vast area, larger than the whole of the United States, with a teeming population, it was indeed a self-contained civilization; a continent with its own highly developed culture that was astonishing in the breadth of its vision and its independent achievements.

The History of China

In spite of the evidence of 'Peking Man', whose fossil bones date back to 350,000 BC or earlier, and evidence of a Neolithic or Stone Age culture in China around 12,000 BC, the first civilization which concerns us is the Yangshao. This is of a much later date – about 2500 BC – which makes it contemporary with the Indus Valley civilization in India and the Old Kingdom in Egypt, though it is a millennium after the first building of the city of Ur in Sumeria. By the time of the Yangshao there were thousands of villages scattered in a band along the Yellow River from Kansu and Shensi to Shansi, Honan and Shantung. Textiles, painted pottery and an agricultural economy are characteristic of this period which, interestingly enough, coincides with a community of culture which stretched across northern Asia and northern America. Evidence for this is provided by various man-made objects, one example of which is a rectangular stone knife quite unlike anything to be found in the Middle East or in Europe, but found among the Eskimos and Amerindians and the peoples of Siberia and China. Probably migration across the Bering Strait lies at the basis of this widespread cultural migration. However, the Chinese were not only a part of this; they already had some unique inventions of their own. Of these the most notable were the two cooking vessels, the *li* and the *tsêng (zeng)*. The *li* was a pot with three hollow legs, which increased the surface area which heat could reach, and the *tsêng* a cooking vessel which could stand on top and allow food to be steamed. The double pot known as a *hsien (xien)* allowed more than one food to be cooked at a time, and when bronze-working began, it became widespread in China: in due course it formed the basis of the characteristic apparatus used in East Asia for distilling liquids.

A number of different systems exist for the transliteration or rendering of Chinese words into the Western alphabet. The most familiar until recently has been the 'Wade-Giles' system; but a new standardized and international Pin-yin system was introduced in 1962. In the present work, the more familiar Wade-Giles forms have been retained, but the Pin-yin spellings are included after the first appearance of each name in the text.

Illustration page 178

By 1600 BC there was a fully developed Bronze Age culture in China which lasted throughout the period of the Shang dynasty (1520 to 1030 BC). Knowledge of this period comes from archaeological digs at the capital, Anyang, in what is now Honan province. The evidence is in the form of 'oracle-bones', and their original discovery was a chance affair. Late in the nineteenth century odd pieces of bone were turned up by farmers in the Anyang area, and were sold to local apothecaries as 'dragon-bones' to be ground down for medicinal use. Fortunately, soon after this practice became common, Chinese scholars visiting the area were shown them and noticed to their amazement that there was writing on the bones. By 1902 it was realized that the bones, which were the shoulder-blades of cattle, had been used for foretelling the future. The technique, known as 'scapulimancy', which was done with turtle shells as well as shoulder-blades, was simple enough; a red-hot poker was applied to the blade and the ensuing

Illustration page 137

cracks were interpreted as the reply of the gods to specific questions. This was a particularly Shang method of divination, and its practitioners were extremely well organized, keeping records of their results on the bones. Written in a primitive form of ideographs, they show that much apparently legendary history was firmly based on real events, and so they are of great significance in illuminating earlier eras of Chinese history.

The Shang period of Chinese culture was notable for the beauty of its bronze work, for its extensive use of bamboo, which was sometimes prepared in strips and then made up into books, and for the use of cowrie shells as a form of currency. Where the cowrie shells came from is uncertain, but a possible source was the Pacific coast near the mouth of the Yangtze.

About 1027 BC the Shang were conquered by the Chou, a people who came from the west, from the provinces which are now Kansu and Shensi. The conquerors admired and respected the culture they found although this admiration did not prevent them deporting much of the Shang population to their own dukedoms – indeed it may have encouraged it – and they created a vigorous feudal society. However, in 771 BC the emperor was killed in battle, the capital was shifted eastwards to Loyang, and various vassal states gradually gained their independence during the three centuries following under princely rulers. This was a period of intellectual and technological development especially since it saw the start of a considerable amount of iron-working, while scholars and their 'disciples' took to a peripatetic existence, travelling from capital to capital to advise the feudal princes. Academies of scholars also arose, and at the capital of the Ch'i (Qi) state Prince Hsuan (Xuan) established one that provided quarters and maintenance for scholars from all over the country – this in 318 BC, some seventy years after the foundation of Plato's Academy in Athens. Various other technical and economic changes occurred at this time, including extensive use of irrigation, so that this period of the 'Warring States' is looked upon as China's 'classical' age.

The Unification of China

By 221 BC the feudal state of the Ch'in (Qin) had become more powerful and more efficiently organized than its neighbours and compelled their obedience. This 'Empire of All Under Heaven' was run by an elaborate feudal-bureaucratic system which set the pattern for all later Chinese governments; this divided the country into provinces, initiated large-scale standardization (of weights and measures, the width of roads, the size of carts, etc.) and connected up a number of small defensive structures to form the Great Wall, probably the largest building project ever carried out. Extending for more than 2,400 km (1,500 miles), it was designed not only to keep out the 'barbarians' from the pastoral economies of the north but also to keep the Chinese within the cultivated regions; it was as much a barrier to emigration as a bulwark against invasion. The emperor, Ch'in Shih Huang Ti (Di) carried his conquests into Japan but, to his credit, he

Illustration page 137

was not interested solely in military power. He travelled the whole of his empire, helped standardize the Chinese language and took a lively interest in alchemy. When he died and was succeeded by his son, who proved an ineffectual ruler, the kingdom crumbled. In 202 BC, power moved into the hands of the Han dynasties which were to last for the next four and a half centuries.

Under the Han the bureaucracy expanded, but was now recruited by way of competitive examinations, according to the principles of public service of K'ung Fu Tzu (Confucius). Of all the Han emperors, the greatest was Han Wu Ti (Di), who stabilized the country, ensured that the administration was efficient, and pursued an enlightened foreign policy. During his time Romans and Roman Syrians visited the country by sea, and a quite extraordinary diplomatic mission was conducted by an overland route to North Afghanistan, Tadzhikistan and Uzbekistan. This was to lead later to an expansion of the Chinese empire westwards and the establishment of a trade route to Iran, known as the Old Silk Road. The stability of Han rule encouraged science and technology and it was during this period that paper was invented, although the innovation was to take six and a half centuries to reach even as far as Central Asia, and twelve centuries to travel to Western Europe. It was during Han times, too, that Buddhism entered China from India; it consolidated its position when the country split into three kingdoms, since during this upheaval many people found a spiritual refuge in such an other-worldly religion.

In AD 265, following a period of instability, the Wei dynasty gained control of the country, though the ruling house was that of Chin (Jin), one of the Wei generals. But there was still no peace. There were invasions from the north, driving the Chin south of the Yangtze, and much attention was paid to the development of military technology, but those with little taste for this turned their energies more to speculation and to science, so scientifically this was a productive time. In the latter part of the fifth century, the country once again split in two and it was not reunited until a century later with the short-lived Sui dynasty followed by the T'ang (Dang), whose rule lasted until early in the tenth century. During the Sui the main links were built in the Grand Canal, which formed a vast communication network, cutting across the old battlefields where the forces of the Chin had contended. This canal was to prove of enormous benefit to China, in terms of supplying the armies, and of permitting the efficient movement of grain taxes – an important element in famine relief; yet it was built at the cost of great human suffering. Indeed, it is claimed that out of the five and a half million people involved in the work-force, some two million never returned.

Under the T'ang dynasty China's borders and influence began to expand. They moved into Tibet and became virtual rulers of Manchuria, Korea and part of Turkestan. However, their penetration westwards was halted in AD 751 in Uzbekistan at the Battle of the Talas River, where the Chinese confronted a Muslim people.

Although this marked the end of China's westward expansion, it also halted the Muslim march towards the east. It was one of the decisive battles of world history and, naturally enough, had repercussions in the Chinese empire itself, stimulating some satellites like Mongolia to struggle for and achieve independence. Moreover relations with Tibet deteriorated, but in the end the Tibetans became as troublesome to Islam as they were to China. The T'ang emperor eventually concluded a treaty with the Caliph Harun-ar-Rashid – the caliph immortalized in *The Arabian Nights*.

During the T'ang foreigners were once again welcome. Scholars came from Persia, Syria and other parts of the Muslim world, and it became the fashion for wealthy Chinese to employ a few foreign specialists (grooms and entertainers particularly) in their households. The Chinese themselves also travelled abroad. Foreign religions were imported – Christianity, Manichaeism and Zoroastrianism – while Buddhism flourished as never before. Indeed, Buddhism became so popular that it almost formed an alternative state and in 845 the T'ang authorities reacted: over a quarter of a million Buddhist monks and nuns were made to return to a secular life and more than 4,500 temples and 40,000 shrines were destroyed.

During the T'ang period art and literature flourished and Chinese law was codified, but there was little science. Alchemy was practised mainly by the Taoists, whom we shall meet shortly, and the Confucians were busy with map-making, but although a Buddhist monk, I-Hsing (I-Xing), made some advances in mathematics and astronomy, the general outlook was strongly biased towards the humanities. There was however technological progress – true porcelain was made for the first time and gunpowder discovered, though it was not used in war until 919, by which time the power of the T'ang had come to an end and China was once more divided.

Reunification came again in 960 with the Sung (Song) Dynasty, which survived until 1126 when a Tartar invasion drove them out, and the court and officials fled south, setting up their capital in Hangchow. Yet in spite of the upheaval both periods of Sung rule – a southern Sung regime lasted until the last quarter of the thirteenth century – were times of great cultural activity, with special emphasis on science and technology.

The Sung dynasty was finally overthrown in the thirteenth century in a clash between the nomadic culture of the far northern regions and the settled and intense agriculture of the main body of China. In 1204 the Mongol tribes were united and swept south in a surge of conquest, though it took them until 1279 to overcome the Sung. When they did they found a land of unbelievable agricultural wealth which their Chinese advisors persuaded them to enjoy rather than destroy. The territory that this Mongol conquest brought to the new Yuan dynasty was vast, covering great tracts of Central Asia, and posing problems in administration. Because they distrusted the Chinese, they gave the most senior posts to qualified Muslims or

Europeans if there were no Mongols suitable for the positions, and so this was a period when China became better known abroad then ever before. The Venetian Marco Polo (1254–1324) visited China and stayed for about seventeen years, acting as a court official.

The Yuan dynasty improved the roads and the waterways, and encouraged geographical exploration; it was also during this period that the great atlas, the *Yu Du* was produced by Chu Ssu Pen (Zhu Su Ben), and an important astronomical observatory established at Peking. But gradually the Yuan administration faltered; the scholar-gentry of the Chinese returned to administrative posts, Confucianism gained ground and the spirit of nationalism revived. By 1368 the Mongols were overthrown and the Ming dynasty established.

Illustrations page 139

By 1403 the capital had moved from Nanking to Peking and there followed a period of maritime exploration, Chinese ocean-going junks venturing to Sumatra, to Sri Lanka and to East Africa. This brought new geographical knowledge and a great variety of foreign produce, as well as a host of foreign animals such as ostriches and giraffes. Possibly the Ming purpose was geographical exploration of a specific area, for the voyages stopped as suddenly as they began, making no attempt to control the Indian Ocean which remained the province of the Arabs and, soon, the Portuguese.

There was some scientific work during the Ming, particularly in botany. Some members of the imperial household even took an interest in plants and a large botanic garden was set up near Kaifeng, while a book discussing the use of wild plants as food in times of famine also appeared. But perhaps the greatest achievement was the compilation of a vast pharmacopoeia, the *Pen Ts'ao Kang Mu (Ben Zao Gang Mu)* by Li Shih-Chen (Shi-Zhen). This appeared in 1596 and carried descriptions of about one thousand plants and another thousand animals, as well as containing more than eight thousand prescriptions based on them and, for good measure, a discussion on various medical matters.

The Ming were overthrown by a rebel leader Li Tzu-ch'eng (Zi-Cheng) in 1644, but the civil troubles that ensued led to him to request aid in restoring order from Manchuria, and once the Manchus were installed they refused to leave: they took over the government of the country and remained in power as the Ch'ing (Qing) dynasty for almost three centuries. With Manchu power our brief historical sketch comes to an end, for it was in the last decades of the Ming that the Jesuits arrived in China and brought European science to China. From the first decades of Manchu rule Chinese scientists were assimilating and then using the new knowledge, and China gradually became part of the worldwide scientific community. Thus to trace the development of post-Manchu Chinese science is to trace the development of ecumenical science, that attempt to probe the natural world which, from the seventeenth century onwards, has been a truly international undertaking.

The Chinese Outlook

With the Chinese civilization we come to an outlook on the world and on science which was different in many aspects from that characteristic of the West. We shall see this difference making itself manifest in various ways as we look at early Chinese science in the pages that follow. But it will be a help in understanding their achievements if we realize that even from very early times, the Chinese looked on the entire universe as a vast organism, of which man and the natural world were both a part. This had a profound impact on the way they explained phenomena that they observed; in some cases it helped them attain an understanding long before this was achieved in the West, but in a few instances it prevented them finding the true explanation for the way the world behaved. There was a second factor which also had an important part to play, and that was the Chinese rejection of – or lack of belief in – any kind of personal omnipotent deity as the ultimate power behind the whole universe. Some of the consequences of this will be considered later.

The Chinese always showed an outstanding practical sense, an immense ability to apply all knowledge to practical ends. Among all early peoples they were applied scientists *par excellence*, but this is not a history of technology, and so we shall not in general be considering their immense engineering developments in designs of efficient bellows and pumps, in iron and steel manfacture, in drilling deep boreholes, in shipbuilding and porcelain manufacture, or the many other aspects of mechanical ingenuity and inventiveness in which China pre-empted the West, in some cases by more than a millennium. Only two Chinese inventions – the mechanical clock and the magnetic compass – will fall within our brief. Nevertheless, as we shall see very clearly, it was not only in technology that the Chinese proved to be pioneers; they had some scientific views which were very advanced for the time although frequently these were formulated in practical terms.

Confucianism

The Chinese intellectual outlook – Chinese philosophy – was largely split into two camps, the Confucians and the Taoists, though there were those like the Mohists, the Logicians and the Legalists and Naturalists who also exerted some influence at times. The Confucians followed the teachings of Master K'ung or, in Chinese K'ung Fu Tzu (Kong Fu Zi), whom we know by the latinized version Confucius. Confucius himself was born in 552 BC in what is now the state of Shantung, and was related to the one-time ruling house of Shang. He spent his life developing and promoting his philosophy of harmony and justice in social relationships. For a time he was exiled from his home state and wandered from one feudal court to another with a small body of followers, but he returned home for the last three years of his life. Confucius died in 479 BC, mourned by his disciples and his family, but to the world an apparent failure, since his philosophical ideas remained unaccepted at large. Yet Confucius's teaching was to be taken up by subsequent generations until it became the dominant philosophy of government servants, of the entire vast bureaucracy that China was to develop. The later sobriquet 'uncrowned emperor of China' is not far short of the mark.

What was Confucianism? It has been aptly termed a doctrine of worldly social-mindedness; its purpose was to promote social justice, notwithstanding the fact that this was of course limited in the feudal-bureaucratic states of China in the sixth century BC. Confucius sought order in a country which was in political chaos and preached respect for the individual at a time when life was cheap. He wanted universal education, and he taught that administrative and diplomatic positions should go to those qualified in an academic, not a social, sense – an unusual and unacceptable suggestion at the time he made it. But academically inclined though Confucianism might be, it was not scientifically orientated. Confucius thought the universe had a moral order and it was man's duty to learn it and apply it. The proper study of mankind was man, not a scientific analysis of the natural world. Certainly, Confucian ideas were rational but they concentrated on social problems, with little importance given to the study of nature.

The most influential disciple of Confucius was Meng K'o (Ke), usually known by the latinized form Mencius. Born more than a century after Confucius, Mencius was an advisor to the states of Liang and Ch'i (Qi), and emphasized the democratic elements in Confucianism. Mencius had an exalted view of human nature; he thought every man had a natural tendency to do good, though not all later Confucians agreed with him, some believing the opposite, that men were born naturally evil in intent. Later, still, in the third century BC, it was claimed that man is born with elements of both good and evil, but that no man had an inherent tendency to behave in either way.

Another aspect of Confucianism was its 'ladder of souls'. As we saw in the last chapter, in the fourth century BC Aristotle had talked about various kinds of soul – vegetative, sensitive and rational – to account for various kinds of living things. The Confucians had a similar system: fire and water had a subtle spirit, but not life; plants and trees had life spirit but not perception; birds and animals could perceive, but it was only man who had a sense (or spirit) of justice. The Chinese system was subtly different from Aristotle's and, though a century later, it does not seem to have been derived from it. In China the doctrine was due to Hsuan Tzu (Xuan Zi) (third century BC) who displayed a very humanistic and sceptical outlook. He denied the existence of spiritual beings, demons and ogres, and although he did not approve of any formal scientific logic, he would today be considered an enlightened mind who rejected superstition. Yet, in the true Confucian tradition, he thought that to study Nature and neglect man was to misunderstand the universe. He therefore also led men away from science, emphasizing social studies too much and, in view

For Aristotle's classification of souls, see page 107.

of mankind's knowledge at the time, too soon.

During the Han, important sacrifices to the K'ung (Kong) family were offered in honour of Confucius by the emperor, and then such sacrifices were made more widely; Confucius became the patron-saint of officialdom and the religious cult of Confucianism grew up. A state religion, with the emperor as high priest of the whole nation, had been characteristic of an earlier age, but this was something quite separate.

Taoism

The greatest early scientific-philosophical movement in China was that which goes under the name of Taoism (Daoism). It was a mixture of religion and philosophy, magic and primitive science, and its name was derived from the aim of its followers to seek the Tao (Dao). What was the Tao? It is hard to define, since there is no equivalent word in English, but one can come near it by saying that it is a philosophical and spiritual term for 'The Way' or perhaps 'The Order of Nature' in the sense of the essential power of the universe; not, it must be emphasized, power of a divine all-powerful personal ruler, but the immanent power of that vast organism which is at once man and the universe. Better, perhaps, if we use the Chinese term Tao.

The origins of Taoism were twofold. It grew up partly among those philosophers who, during the wars of the eighth to fifth centuries BC did not attach themselves to the feudal princes to advise on government, but withdrew from most social life to contemplate and study the world of Nature. Yet its origins also owe something to the magicians, witchdoctors or shamans (magical priests) who believed in and prayed to a host of minor deities and spirits of the natural world to ensure good harvests and to cure sickness of every kind. From these influences it developed into a belief in the Order of Nature that brought all things into existence and governed every action by an inherent natural rightness, not an imposed force.

These beliefs stimulated a desire to ascertain the causes of things, to observe the natural world, and even to conduct experiments. The Taoist would contemplate Nature not out of idle curiosity, but in the belief that the knowledge so gained would bring inner peace. Instead of using this knowledge to gain mastery over Nature (as was the inspiration of so much Western scientific enquiry), the Taoist would never undertake what would be considered 'contrary' action to use force against Nature. Taoists were interested in cyclic changes and looked back to a simpler, 'purer' age when things were better – the age-old belief in a 'golden age' back in the distant past.

The Taoists were not concerned, as were the Confucians, with the social study of man alone nor were they preoccupied by ideas of social class. They admired the technology of the craftsman just as much as the art of the designer or the ideas of the theoretician, and it was doubtless this that enabled them to experiment and yet retain their standing as philosophers, something the Confucians could never do. They championed individualism and believed in the possibility

For more on Chinese alchemy, see pages 174–175; for a comparison with Islamic alchemy, see pages 237–239; and with alchemy in Europe during the Renaissance, pages 306–307.

of individual immortality, and immortality of the body at that. With this end in view they encouraged special gymnastic and similar practices, and sometimes they took drugs. The last were mineral in content (gold was considered especially efficacious) and this alchemy brought much, particularly biological, knowledge as it did in the West. (In China the main goal of alchemy was the discovery of the elixir of immortality; in the West it was the transmutation of base metals to gold.)

As with Confucianism, so with Taoism; both turned into religions. To begin with, the Taoist religion developed as reaction against the establishment of Confucianism as a religion. It was connected with a family, the Chang (Zhang) family, which was devoted to Taoism, and by the second century AD had developed into what can best be described as a Taoist church, which still exists. But after the eleventh and twelfth centuries the power of the Taoist religion waned, not least as a result of bitter disputes with the Buddhists. But the fate of the Taoist church is a side issue as far as we are concerned. The essence of Taoism – the seeking of the Tao – is the theme to be found running through many currents of Chinese thought.

T'ang Taoists and the Sung Neo-Confucians

The scientific side of early Taoism was not widely appreciated and this was the first part of Taoist teaching to wither from lack of support. Consequently, by the third and fourth centuries AD Taoist science was effectively dead and the Confucians, then in a position of supremacy, encouraged a religious and mystical interpretation of the rest of Taoist teaching. As a result mixed Taoist-Confucian schools of thought grew up, their names – the 'Philosophic Wit' or 'Pure Conversation' groups – giving the lie to their attitudes. But they nevertheless attracted the best minds of the day, minds that soon began to react strongly against the Establishment outlook which the schools enshrined, and took up the more liberal political ideas and outlooks of the old Taoists. Some of the writings of these intellectual rebels contained not a little alchemy and medicinal chemistry, and this was a side which had no small following. Indeed, between AD 389 and 404 the emperor of the Northern Wei established a professorship of Taoism and a Taoist laboratory for preparing medicines, in spite of objections from the orthodox Imperial Physician.

Much attention was also paid to the *I Ching* (*I Jing*, the Book of Changes) which will be described shortly, and during the T'ang there was a revival of old Taoist philosophy, with philosophers seeking after more fundamental explanations of the universe than the vague discourses of matter and form of the earlier Taoists. This new and deeper understanding took time to achieve, and only arrived in medieval China with the Sung Neo-Confucians.

It seems that this revival of Taoism left the Confucians feeling very acutely their lack of any ideas about the universe and the natural world, an intellectual lacuna emphasized by the Buddhists, who had their own views on matters of this kind. The consequence was that

Left A Chinese oracle bone of the Shang dynasty, from Anyang, inscribed with requests to the spirits for guidance; second millennium BC. Royal Ontario Museum, Toronto.

Below A section of the Great Wall of China, built in the 3rd and 4th centuries BC.

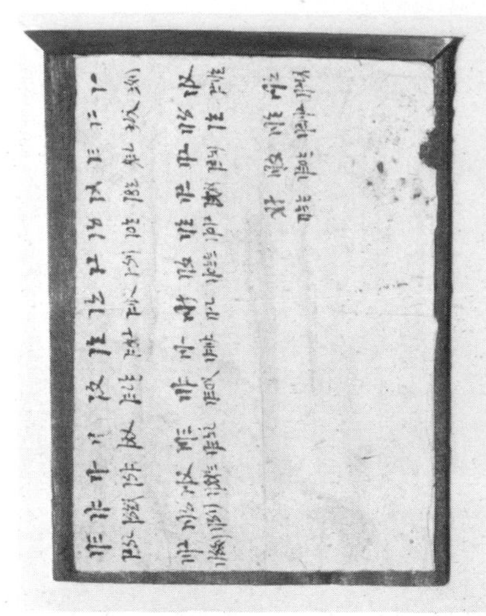

Right A Chinese abacus. Science Museum, London.

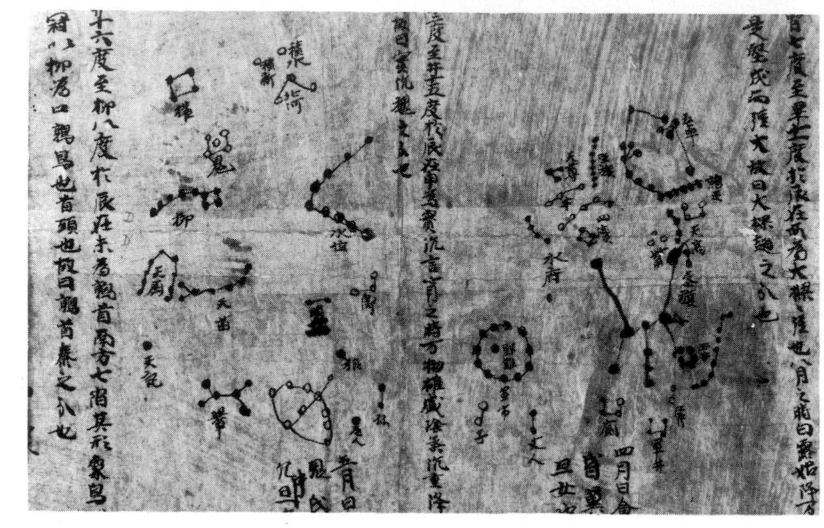

Right The Tunhuang manuscript star-map of AD 940 which uses a 'Mercator' projection. On the left are the constellations Canis Minor, Cancer and Hydra, on the right, Orion, Canis Major and Lepus. British Library, London.

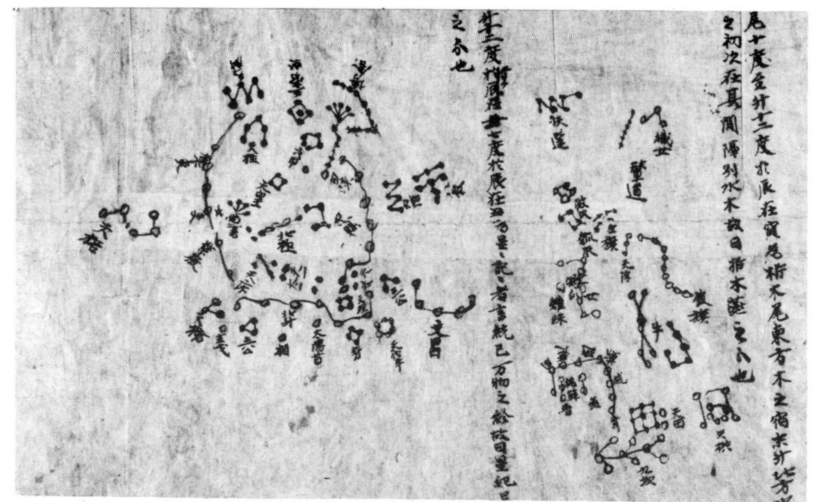

Right The Tunhuang manuscript star-map showing (left) the circumpolar stars and our constellations Ursa Major, on the right, Sagittarius and Capricornus. British Library, London.

Left The equatorial armillary sphere from the old Peking observatory of 1279. A sighting tube was originally fitted between the two rings of the declination circle. Now at Purple Mountain Observatory, Nanking.

Left and below The abridged armillary at Nanking, dating from the late thirteenth century. By separating the right ascension from the declination circles, it was more convenient for observing with than a conventional armillary sphere.

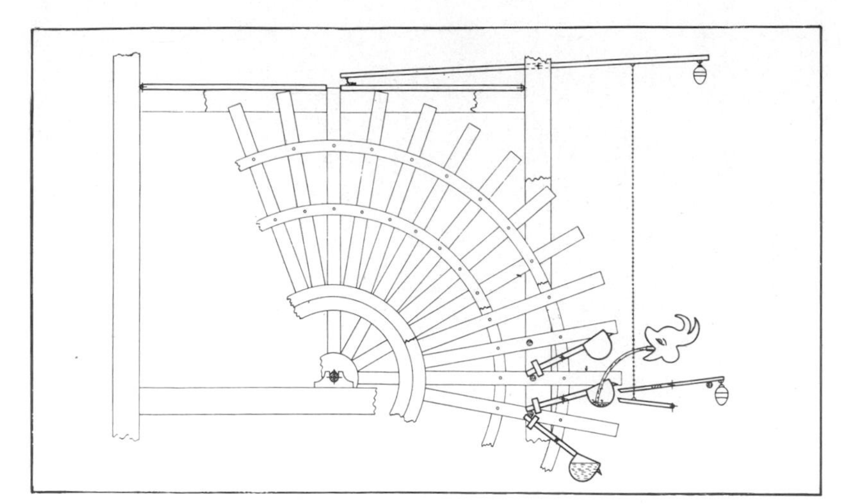

Right Astronomical clock-tower built at Kaifeng in about AD 1090. There is an armillary sphere at the top and the human figures below this indicate times and seasons. The boxes on the right of these are the siphon system designed to supply the water-escapement. *Above*, diagram of its escapement, the first escapement for a clock. The dragon's head (right) is fed by a constant supply of water.

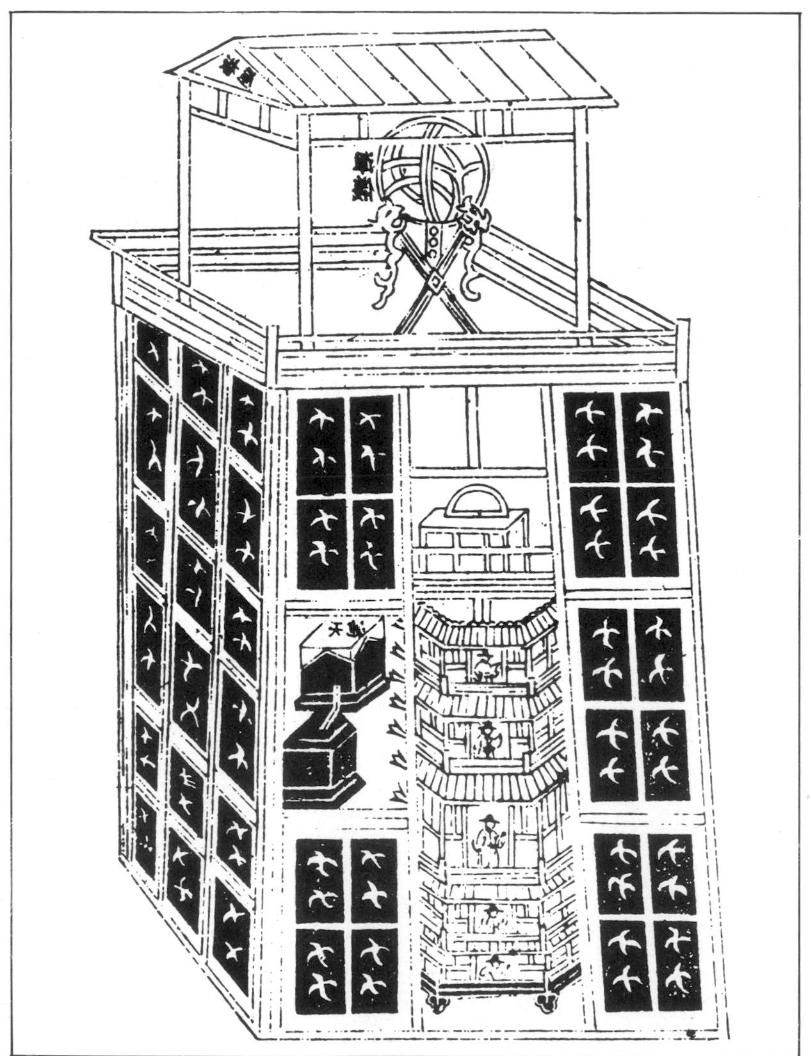

some Confucians began to try to expand their outlook, borrowing various concepts from both Buddhism and Taoism. This expansion of Confucianism was gradual, as might be expected, some thinkers being more steeped in Taoism than others, but it reached its peak in the Sung, with Chu Hsi (Zhu Xi). Born in AD 1131 and dying in 1200, Chu Hsi bent his considerable intellect not to making discoveries but to welding together Taoist and Confucian thought into one immense synthesis. His work was indeed a parallel to the kind of work which St Thomas Aquinas was to do a century later in the West when he fused Christianity with Greek and Arabic science. The outcome of Chu Hsi's work was to propagate a thoroughly naturalistic view of the universe against a mystical, magical view. It paved the way for other Sung Neo-Confucians to make an analysis of the natural world and, above all, helped to create the intellectual atmosphere for what may truly be called a golden age of Chinese science, in the Sung period.

For Aquinas, see pages 259–60.

The Mohists and Logicians

The Mohists, named after their founder Mo Ti (fifth century BC), and the Logicians were two early schools of Chinese thinkers who tried to work out a fundamental scientific logic. Mo Ti, who was probably an official of the Sung, preached universal love of man for man, and condemned offensive (but not defensive) war. His followers trained in the military arts to aid victims of oppression. Their concern with defensive strategy and fortifications led them to take an interest in the basic methods of science and they made some fundamental studies in mechanics and optics. Indeed it seems that the Mohists had a bias towards the physical sciences, just as the Taoists had towards the biological sciences.

In their analysis of experimental science, the Mohists discussed much basic logic. They tackled problems like the way the mind worked in ordering what it observed, and questions of cause and effect; in doing so they seem to have reached the two important thought processes of deduction (reasoning from the general to the particular) and induction (reasoning from particulars to generalizations). They also used the idea of conceptual models.

The writings of the Logicians, who were never very clearly differentiated from the Mohists, have been lost except for one partially preserved book, *The Book of Master Kungsun Lu*, and one containing paradoxes, *The Book of Master Chuang (Zhuang)*. The first is said to be the high point of Chinese philosophical writing, and discusses concepts (e.g. 'white', 'hard', 'horse', etc.) as distinct from particular things. It also contains a discussion of change – a very important aspect in biology. A series of paradoxes were prepared by Hui Shih, who lived in the fourth century BC, about a century later than the Greek philosopher Zeno. Probably reached independently, they were designed to shock, and so make the reader consider fundamental questions. Thus 'Fire is not hot. Eyes do not see' brings one face to face with questions of perception; fire is not 'hot' of itself, though

For Zeno, see page 79.

that is how we interpret it, and eyes do not see of themselves (despite the opinion of the Ancient Greeks). Again 'An egg has feathers' underlines the fact that an egg is a potential chicken. This technique contributed to the growing enquiries about the natural world, and so to nascent Chinese science.

The Legalists

As their name implies, the Legalists were an authoritarian school of philosophers. They lacked the humanity and love preached by Confucians and Mohists, nor did they possess the sensitivity and peacefulness of the Taoists. The movement began in the fourth century BC and reached its greatest influence a hundred years later; it was the school of advisers which was instrumental in helping the last prince of Ch'in (Qin) to become the first emperor of a unified China.

The attitude of the Legalists was simple enough. They thought that the collection of customs which the Confucians administered in a very paternalistic way were too weak; a more forceful and harsher system was needed based on laws already drawn up, not on common usage, which should be rigidly applied. Law, they taught, was what shapes people. If the law was strong, then the country would be strong. Punishments should be severe, to deter people from lawbreaking; if small crimes did not happen, great crimes would not follow. Indeed, it should be worse for people to fall into the hands of their own police than into the hands of an enemy in war, and they advocated a system of reporting and denunciation, even among members of the same family. The state required obedience, not virtue.

The harsh methods of the Legalists were not only unpopular; they brought in their train a natural revulsion, and some twenty years after Ch'in Shih Huang Ti had come to power, with Han rule firmly established, a milder rule set in with a return to Confucian ethics. But however severe and unbending the Legalists were, their efforts are important in the history of Chinese science, because it was they who began the custom of detailing everything precisely – of quantifying in numbers all conceivable matters, from the widths of chariot wheels to human conduct. There was nothing, they thought, which could not be specified and laid down in the form of a regulation, and nothing, from weights and measures to the emotions, lay outside their purview. In promoting this idea they were, indeed, doing something new, and it seems that they were stimulated to adopt it, in part at least, by new technological advances of the time. If it had been followed through, it might have introduced the quantifying of science: that attitude which expresses things in numbers not words. And this was an attitude which, as we shall see, was to be a great cornerstone in the European scientific revolution of the sixteenth and seventeenth centuries. Perhaps the scientific revolution might have begun in China seventeen hundred years before it arrived in the West. But it did not, and one of the reasons – though certainly not the only one – is that the Legalists failed politically and most of their teaching and attitudes died with them.

In stressing the rule of preordained law, the Legalists came close to the concept, so strong in Europe, of the Laws of Nature. It was a natural concept in the Western world, because there were always beliefs in personal guiding deities, mostly omnipotent. They ruled man and the world around him, so the idea that things, living or inanimate, behaved according to a divinely ordained law was to be expected. But this was not so in China, where there was no belief in a personal, guiding and lawgiving deity. The universe was an organism: it operated because everything fitted in to its natural place and acted according to its nature. When the Legalists fell, and their scheme went with them, the stimulus to express processes in numbers could not be replaced. Chinese law could not help because it was quite different from predetermined law in the Legalist sense where every crime and punishment was quantified; Chinese law knew only the 'natural law' of custom and usage tempered by what was fair and humanely desirable. That is not to say that China had no codes of laws, but they were codes with a humanistic bias. They reflected the Chinese attitude to the universe where everything happened because of an inbuilt universal rightness, harmony and well-established custom. Crime and legal disputes were looked on more as disturbances in man's relationship with Nature than anything else. Indeed, in the T'ang code of law it was stated specifically that it is dangerous to move from this kind of natural law to a law of legally fixed punishments.

With such an outlook there is no room for law in the Legalist sense, or in the sense epitomized in the West by Roman law. In China rights of individuals were not guaranteed by law; there were only duties and obligations. The supreme ideal was to demonstrate justness; not so much to decide responsibility on the basis of who has done something, but to assess the nature of what has happened. In the West there grew up the ideas of Natural Law, operating between all men, and a collection of Laws of Nature which the material world obeyed. In China no such concepts arose. The Taoists never developed a genuine idea of the Laws of Nature: natural forces, like the Ying and Yang (which we shall come to later) were sufficient for them.

Some Basic Ideas in Chinese Science

With the Chinese, as with the Greeks, there were some basic scientific concepts which were used in explaining the natural world. Of these the one to be found in both civilizations is that which concerned the basic properties of matter. The Greeks, as we have seen, had a few philosophers who favoured an atomic theory, but the general consensus of opinion was that this theory was unacceptable and that it was preferable to postulate four basic elements (earth, air, fire and water) and ally these to four qualities (hot and cold, wet and dry).

For Greek four-element theories and atomism, see pages 81–84.

The Chinese never developed an atomic theory, because it was not the kind of view which went with the concept of a natural universe that was an organism, operating according to the interplay of right and natural behaviour. The Mohists, it is true, seem to have flirted with an atomic outlook, but their views in this direction had no influence. Like the Greeks, the Chinese in general opted for a theory which used a small number of basic elements, in their case five not four. This Five Element theory goes back at least to between 350 and 270 BC and was stabilized and systematized by Tsou Yen (Zou Yan), sometimes called the founder of all Chinese scientific thought and the leading member of Prince Hsuan's (Xuan's) important Chi-Hsia (Zhi-Xia) Academy. Tsou Yen was a Naturalist and one of a sect of Chinese philosophers who did not shun the courts of princes as did the Taoists, but were nevertheless concerned with making sense of the natural world.

The five original Chinese elements were water, metal, wood, fire and earth, but they should not, of course, be thought of as mere substances (which they were not) but rather as active principles. The elements were concerned with processes found in nature, or in the laboratory. Thus water was characterized by soaking, dripping and descending, and was associated with the taste of saltiness; fire's characteristics were burning, heating and ascending and its associated taste was bitterness. Wood accepted form by cutting and carving, and sourness characterized it; metal accepted form by moulding when liquid or when remelted, and it was acrid. Finally, earth was characterized by the production of edible vegetation and was sweet.

Soon these elements were formed into a cyclic system, which had become very stylized by Han times. Various 'orders' of the elements were drawn up. One order set out the sequence by which the elements were supposed to come into being, with water being the primeval element. Another was the 'Mutual Production' order, which was believed to show how one element gave rise to another. Again, there was the 'Mutual Conquest' order in which each element was thought to conquer another. For example, wood conquers earth (since a wooden spade can dig up earth); metal conquers wood (it can cut it and carve it); fire conquers metal (it can melt it); water conquers fire (it can extinguish it). To complete the cycle, earth conquers water – it can dam it and contain it, as the Chinese with their efficient and often elaborate irrigation systems were very well aware. The Mutual Conquest order was used not only in science but in the political field, since it was widely believed that the behaviour of the prince or emperor and his court officials should, if good, be guided by the Mutual Conquest order of the elements, especially since these elements were associated with the seasons, and with manifestations of the natural world.

The Five Elements were associated with all experience. They were symbolic of change, of quantity (they were thought of as controlling a process depending on the amount of an element present) and, in due course, they were tied in with smells as well as tastes, with the cardinal points of the compass, with human mental and physical functions, and with animals. They were related, too, to the weather and the places of stars, the planets, and even aspects of government.

In short, the Five Elements were associated with all activity, both natural and man-made.

A second basic idea in the Chinese explanation of the natural world was that of the Two Fundamental Forces, the Yin and the Yang. These were used in a philosophical way at the beginning of the fourth century BC, Yin being associated with clouds and rain, with the female principle, with what is inside, cool and dark. Yang, on the other hand, is coupled with the ideas of heat and warmth, sunshine and maleness. They could not be found separately since each was the essential counterpart of the other; it was just that in every situation one or the other took precedence or (an idea parallelled much later in our own age in the terminology of genetics) one factor was dominant and the other recessive.

For the use of the ideas of dominance and recessiveness in genetics, see page 482.

The Five Elements and the Two Fundamental Forces could together present a multiplicity of associations within the natural world. They could cover everything that was susceptible to a fivefold arrangement, and those things that did not fit into the scheme were later fitted to other associations – into fours, nines, twenty-eights and so on. In other words the Chinese practised what is called 'associative' thinking; they looked for associations, for relationships between one thing and another.

Illustration page 160

The Five Elements and the Two Fundamental Forces helped Chinese science because they made it possible for relationships to be defined and, once defined, to be examined. They indicated how things could 'resonate' with one another or, as the scientist of today would put it, they allowed Chinese scientists to propose action at a distance between one body and another. What was not useful, however, was the mystical way of looking at these relationships which became widespread and was enshrined in the *I Ching*, the Book of Changes. This probably originated from a collection of peasant omens (sayings to do with unusual things noticed in man and in animals, unusual weather-lore and the like) and a mass of material used in divining the future, and it grew into an elaborate system of symbolic relationships and explanations in terms of them. The Five Elements and the Two Fundamental Forces were incorporated into its extensive appendices, although the idea of the two forces also frequently permeates the main text. To the main text itself is allied a collection of symbolic patterns, each composed of sets of three or sets of six full and broken lines. Each of these patterns is predominantly either Yin or Yang, and they are arranged to produce this result alternately. In brief this is a text full of magical overtones, and is the delight of those who dabble in mysticism connected with numbers.

Illustration page 180

The text probably dates from the third century BC, although its origins may be three hundred years earlier. If it had been treated as no more than a text to help with divining the future we should not be considering it, but the appendices, which were on a higher intellectual plane than the rest of the text, appeared to many to have real significance. In spite of their abstract nature, they seemed to contain so large a collection of relationships that it was assumed that they could throw light on every observed fact of Nature, and provide an explanation for every phenomenon. The *Book of Changes* seemed the natural reference book for those scholars of the Han who were trying to grapple with the problems of the tides or of the effects of magnetism; unfortunately it gave only pseudo-scientific explanations and misled rather than enlightened. Certainly discoveries of scientific value were made in spite of it, but often they were later declared to be present in the *Book of Changes* thanks to more than a little hindsight. This, of course, seemed to confirm its validity.

The real danger in the *Book of Changes* was that it tended to act like a sponge, absorbing every novel observation, as it was searching for the appropriate associations so that the new facts could be filed away. One effect of this was to discourage further observation, another was to stultify ideas. Yet in spite of it, Chinese science did progress. No other civilization seems to have suffered from an equivalent text to the *I Ching*, but perhaps this is not surprising; it was essentially an extensive and complex filing system of relationships and was bound to appeal to the Chinese mind which, after all, had developed the most extensive – and efficient – bureaucracy in the world.

The Sceptical Tradition

The Han period may have seen extensive use of the *Book of Changes* but it also saw the development of a sceptical tradition which was, of course, a great help to science both then and later. It was a movement of doubt and disparagement aimed against fortune telling and the host of superstitious practices that were then current. It did not begin in the Han – it went back at least five centuries before that – but it became a grand tradition among some Confucians in Han times. Its most important figure was Wang Ch'ung (Chong) AD 27–97) who produced an important book, *Discourses Weighed in the Balance*, which is essentially a classic of rationalism. In everything Wang was a sceptic: he attacked the prevalent belief in ghosts, in evil spirits, and a mass of other superstitions, but his book was not just a collection of attacks on superstitious ideas and practices; it also had a positive side. Wang talked, for instance, of the place of man in the universe and took an astonishingly 'modern' view: far from accepting the current belief that man was the centre of all things, he refers to men living on the Earth like lice in the folds of a garment. Certainly he had a great respect for man as the most intelligent of all creatures, but Wang did not consider his position unique. His arguments often involve statistical – or at least numerical – facts, and he managed to bring in an invigorating breath of logic at a time when magical and superstitious beliefs were widespread.

Wang also attacked those who believed that signs in the heavens, or storms and natural disasters, were reprimands sent from Heaven. It was claimed that if the emperor and his officials failed in their duties or followed wrong policies the universe would be out of joint and the heavens would display the fact, and even that disastrous events would occur. In practice, of course, this erroneous belief, known as Phenomenalism, had stimulated people to observe the heavens, but Wang's attacks on this attitude were necessary and timely. He was not alone in his criticisms, and the whole sceptical movement had some useful long-lasting effects. In later years it encouraged an interest in archaeology and the development of textual criticism, making China the first home of these humanistic sciences. So the sceptical tradition was no empty and destructive movement, and it brought Confucians to some fields of study they might otherwise never have known. Admittedly it did not foster science or a scientific and experimental approach to the world – that would not have fitted in with the Confucian ethic anyway – but it did give rise to a great and critical study of history, unequalled in the rest of the world until the last two centuries.

China and the West

To decide how much scientific knowledge was transmitted from the West to China and *vice versa* is not always easy, for independent but parallel lines of research and invention could and did occur in both parts of the world. For instance, it seems probable that the idea of a 'ladder of souls', which is to be found in Aristotle and Huan Tzu (Xuan Tzu), arose independently in Greece and China, for even though the ideas appeared within less than a hundred years of one another, they occurred at a time when conditions of travel between East and West were not easy. What is more, it is likely that the whole concept, which is really an expression of the differing complexity of living things, might be one which would be expected to occur quite naturally to someone concerned with the question of trying to explain and classify the world of Nature. All the same, communication between China and the West was not as rare as might be thought.

For Aristotle, see page 107; for the Chinese equivalent, see page 134.

As early as the Bronze Age (before 1500 BC, i.e. before the Shang) there was contact on a wide scale, as is proved from archaeological evidence in the form of ceremonial axes, swords, harness and other objects in strikingly similar designs. Again there are a number of plants thought to have mystical powers which appear among the beliefs of peoples as far apart as China, Gaul and Mexico. Clearly, information was diffused by migrations, conquests and, above all, by trade. Trade with China was established early, especially the export of silk to the West; indeed, China was known sometimes as Seres or Sina, names derived from the Chinese *ssu (si)* meaning 'silk'.

Travel to and from China could be made either overland or by sea routes via India; indeed, the first knowledge of China gained by the Greeks came by way of India in the fifth century BC. Maritime contact with China itself may go back to the third century AD, and certainly not long after this Chinese ships made their way to India and perhaps further. Greater contacts were made in the eighth century by the Arabs, the current masters of the seas, and in the ninth century the Arabs set up their own trading colonies at Canton and Hangchow.

For the role played by India in the transmission of Chinese ideas, see pages 188–196.

Later, the enormous Asian empire of the Mongols in the thirteenth century greatly facilitated contact between China, western Asia and even Christendom for a time.

The overland routes developed at the same time as the maritime. By 100 BC a regular overland silk trade was established over a number of routes, either out of China to the north-west and then west by way of Tashkent and Samarkand or from Lanchow south-west to Patna and then north-west to Merv in what is now the Turkmen republic of the USSR. This was the Old Silk Road, and it continued in use long after the sixth century when the secret of silk-making was smuggled out of China to Byzantium. But this we might expect, for silk was not China's only export, and the routes were used by immigrants from Roman Syria and from Persia, for embassies and even for transporting prisoners of war. On the whole, though, the Westerners seem to have been more adventurous than the Chinese, and there was more movement from the West into China than the other way round.

In spite, then, of China's relative isolation, it was not completely cut off from outside influences. Of course, as we have seen in the historical sketch, there were some times when foreigners were more welcome than they were at others; there were also periods when uprisings and wars in the countries of Central Asia or the Mediterranean and Arab countries were a deterrent to travel. But by and large information did filter out from China, as the Western adoption of some Chinese inventions shows. If this did not happen so readily in pure science, that is to be expected, but some ideas were exchanged, as a description of early Chinese science will show.

Mathematics

The Greeks, as we have seen (Chapter 2) had a genius for geometry. This the Chinese did not possess; instead they showed a flair for algebra and for ways of writing numbers. The writing of numbers is more important than might at first sight appear as, indeed, is the way one writes down mathematical operations like multiplication or division, not to mention the more complex kind of operations met with in higher mathematics. Thus, in eighteenth-century Europe, although both Isaac Newton and Wilhelm Leibniz invented the calculus independently, it was not in England that further progress was immediately made but rather in France and Germany, primarily because Leibniz wrote down the operations in a much more explicit way than Newton. The writing of numbers themselves is equally important for mathematical progress. Anyone who doubts this should try dividing CXLIV by XXIV using Roman numerals throughout; the cumbersome nature of the undertaking will be at once obvious!

The Chinese method of writing numbers developed gradually over the centuries, of course, but it had already reached a high level of simplification by the third century BC when place-value notation was used. Whether we realize it or not, place-notation automatically comes to us today when we write 1, 11, 111 and so on, using the places of the figures to denote tens, hundreds, and so on. It seems obvious. But not all civilizations found it so. And what about the numerals themselves?

Again, by the third century BC forms of numerals using straight lines were in regular use. These can best be thought of as collections of small rods, one rod for one, two for two, and so on, with the way they were laid changing direction with every power of ten. Thus the number which we should write as 11236 would appear in Chinese as |—||≡T. Notice that whereas Chinese is written from the top down the page, their numbers were written like ours, horizontally. The numbers apparently arose from the use of counting-boards. These are flat boards ruled with lines on which small rods or sticks are placed. The rods record a number and then, when a subsequent operation like adding, subtracting, multiplying or dividing is carried out, the appropriate rods are shifted in position, taken away or others added. It is a flexible system and one which in Chinese hands allowed considerable developments to occur not only in arithmetic but also in algebra. When it began is uncertain, but counting-boards may go back to 1000 BC, and the counting-rod numerals just described could have been in use a couple of centuries later.

A sign for zero, for an empty place on the counting-board, came later. It is first found in print only in the thirteenth century AD, but may well have been adopted a hundred years earlier. It has been suggested that it derived from India in the ninth century, although recent research indicates that the situation may not be as simple as this. Soon after AD 683 the first inscriptions using a zero appear simultaneously in Cambodia and Sumatra and for various reasons it is not unlikely that zero originated not in India itself but in an area bordering Indian and Far Eastern culture. Possibly the empty space on the Chinese counting-board may have played its own specific part in this, together with ideas like the 'void' of Indian philosophers and the 'emptiness' talked of in Taoist mysticism.

Types of Numbers

In most early civilizations there was a tendency to avoid fractions whenever possible but this never happened in China, and by the Han the Chinese had become adept at handling them. They were also familiar with a decimal system, which seems to have begun very early, for the rudiments are to be found in Mohist writings of the fourth century BC. Moreover, even earlier, by the time of the Shang oracle-bones, the Chinese were able to express very large numbers – something of which few early civilizations could boast – and by the second century AD they were using what we should today call 'powers of ten'. Thus 100 is 10^2, 1000 is 10^3, 10,000 is 10^4 and so on (the small number or 'index' indicating the number of noughts, that is the number of times ten is multiplied by itself). Of course, the Chinese did not write 10^4 just like that but they had characters which expressed this same idea.

In the West there was always difficulty when it came to 'surds' (a word derived from the Latin for 'stupid'). Such stupid, or irrational, numbers were those like the square root of 2 which cannot be expressed as a ratio of two whole numbers, as can a fraction like ½ or ¾. The square root of 2, expressed as a decimal, is 1.4142136 . . . but this incommensurability seems to have raised no question in the minds of the Chinese, whereas to the Greeks it seemed genuinely irrational because one could not make a straightforward ratio out of it. Again, the Chinese had no problem when it came to negative numbers; on their counting-boards red rods represented positive numbers and black ones negative numbers, and this at least as early as the second century AD. In India such numbers did not appear until the seventh century and in Europe not until the sixteenth.

For the difficulty felt by the Pythagoreans on this point, see page 74.

Arithmetic

As well as practising the basic processes of arithmetic and studying squares, cubes and roots as we have seen, the Chinese were intrigued by the relationships of the numbers themselves. In this they followed the pattern of many early civilizations. At a very early stage of Chinese history, odd numbers were thought to be unlucky and even ones lucky. Patterns of dots were studied, as they had been in Greece by Pythagoras and his followers, to derive sequences of numbers, and the Chinese discovered some numerical relationships unknown to the Greeks. For example, they knew that the sum of the sequence of odd numbers is always a square.

The Chinese also took an interest in 'combinational analysis', which shows up in their concern with 'magic squares' – squares with compartments filled up with numbers which all add up to the same total whether one takes all the numbers in a horizontal row, a vertical row or even a diagonal one. The magic square itself could become elaborate – three-dimensional squares were even devised – but in its simplest form it seems to go back at least to 100 BC or earlier, though it was not until nearly fourteen hundred years after this that the subject was really developed.

The Chinese were always adept in devising aids to calculation. Again like other early civilizations they counted on their fingers, and they employed the more complicated technique of allotting numbers to the joints of the fingers as well as to the fingers themselves. They also used a string with knots in it, similar to the *quipu* of Peru, though this was probably more for recording rather than for actually carrying out calculations. But their most efficient early calculating device was the counting-rod. These seem to have been used very early on and were ideal for semi-mechanical calculation; there are many stories of how quickly calculations could be performed with them by anyone who had practised using them. Indeed the eleventh-century astronomer Wei P'o (Po) is said to have moved them so quickly that they looked as though they were flying. But after the late Ming little more is heard of the rods because they had by then given way to the more efficient abacus.

Illustration page 58

Illustration page 138

The Chinese abacus is either called a *suan p'an* ('calculating plate') or a *chu suan p'an* ('ball-plate'); with its rectangular frame and its vertical wires each threaded with seven slightly flattened balls, it is probably familiar to most readers. The first description of it in any modern form did not appear until AD 1593 so it has sometimes been thought to have been unknown in China until the sixteenth century. Yet there are other texts which make it clear that the 1593 description is late: admittedly they do not actually speak of the abacus but what they describe is 'ball arithmetic', a type of calculation performed on a wooden trough with wires strung across it and balls travelling along the wires. This appears in a book of the sixth century AD, and the information it gives is claimed to have come from the end of the second century AD. Since there are other references to similar arrangements, it looks as though the abacus goes back at least to the sixth century in China and possibly even to the second century AD. In other early civilizations there are devices in which pebbles were moved in grooves, not balls on wires. The pebble method may well have come from India in the third or fourth century AD, being itself a development of the very ancient Mediterranean sand or dust trays used in calculating. These had been well known to the Greeks, the Egyptians and even to the Babylonians.

Geometry

Although the Chinese were not geometers – they never developed a deductive geometry as did the Greeks – they did concern themselves with some geometrical questions. Indeed in the fourth century BC the Mohists even went so far as to give some geometrical definitions of points, lines, and certain geometrical figures, at about the same time as Euclid was drawing up his *Elements* in Alexandria. But this was an isolated case. The Chinese never followed it up, though they did develop their own proof concerning the sides of a right-angled triangle, a proof that was not the same as the one invented by Pythagoras. By the second century AD they had also worked out the areas of all kinds of shapes and the volumes of many solid figures, probably using actual models to guide them, something which the philosophically minded Greeks could never have brought themselves to do.

A basic geometrical problem of all early civilizations was to find the area of a circle, because this is a quantity which enters into the solution of many other problems. Reduced to fundamentals, the problem is one of determining how many times the radius of a circle can be divided into the circumference. This ratio is a number which is now expressed as the Greek letter pi (π). The true value of π is 3.1415926536 . . . The Egyptians could not get an exact value nor could the Babylonians; the Greeks found it lay between 3.1408 and 3.1429, but the Chinese finally did better. In the first century AD they began to try to obtain an exact value, calculating the areas of the largest regular polygon they could fit inside a circle and the smallest one they could fit around the outside. The greater the number of sides of the polygons, the closer their areas approached one another

For the calculation of π by Eudoxos, see pages 90–91; by Archimedes, see pages 112–113; and by al-Kashi in Samarkand in the early 15th century, see page 223.

and the closer they came to giving a true value of π. A great leap forward was made in the fifth century AD with the calculations of Tsu Ch'ung-Chih (Zu Chong-Zhi) and his son Tsu Kêng-Chih (Zu Geng-Zhi), who finally obtained a value which lay between 3.1415926 and 3.1415927, a value which was checked nine centuries later by Chao Yu-Ch'in (Zhao Yu-Qin) who had to use polygons with up to 16,382 sides to do so. In the West this value was not arrived at until the seventeenth century.

Before leaving the subject of geometry, it is worth noting that the Chinese took the first steps in developing co-ordinate geometry, a form of geometry where lines and curves are represented by algebraic formulae. The Chinese invented the basic co-ordinates, probably derived from their map-making and their compilation of historical tables in which the entries were laid out in a system of co-ordinates as on a map. They realised too that a formula in algebra could express a geometrical relationship uniquely, appreciating this because it was just the way they frequently tackled geometrical problems. The Chinese had achieved this in the second century AD, but it was only in the seventeenth that such co-ordinate geometry was developed in the West, though it was then taken further than the Chinese had gone.

Algebra

Today, when we think of algebra, we think of the use of letter and number combinations, of equations like $x^2 - 5x - 6 = 0$. The Chinese practised algebra from early times, writing down their results fully in words, the words they used having a specific meaning in their mathematical context. Only rarely and much later did they use mathematical symbols. However, they also made use of the counting-board for algebra and, in Sung times, this was developed into a notation method that was so complete it took care of equations of quite high powers – equations containing x^9 could be handled. But excellent though all this was, – and there is no doubt but that it was the final peak of Chinese algebra – afterwards progress ceased since the Chinese had no general theory of equations, such as was developed in the Western world, to carry them further.

The Chinese lack of algebraic theory did not, however, inhibit their ability to tackle a great many problems using algebra itself. By the Han, they were able to solve simultaneous linear equations (equations with two, three or more unknown quantities) and later, in the fourth century AD, even indeterminate equations (where there are more unknowns than there are equations to solve them directly). Quadratic equations (equations with x^2) could also be solved early on and they were aware too of something we today call the Method of Finite Differences, a method which can be used, as it was by the seventh-century Chinese, to solve problems in connection with the apparent motion of the Sun in the sky. By the fourteenth century they had developed the method to a degree not reached in Europe until three to four centuries later.

The Chinese algebraists studied mathematical series (series of numbers linked with one another in a specific way) and also looked into mathematical ways of expressing permutations and combinations of things. Moreover, in solving their higher-power equations they made use of what we now call the Binomial Theorem. This is concerned with two-term (binomial) expressions like $(x + 1)$, which if multiplied together give a series of terms. If we take a couple of examples we can see how this happens: for instance $(x + 1)^2 = (x + 1)(x + 1) = x^2 + 2x + 1$ and $(x + 1)^3 = (x + 1)(x + 1)(x + 1) = x^3 + 3x^2 + 3x + 1$. Thus it becomes clear that the more multiplications (or powers) we have, the greater the number of terms in the series. But if one looks at the numbers before the x's (that is, the coefficients), one can see that there is a pattern among them. Those for power one [i.e. $(x + 1)$] are 1, 1; those for power 2 [i.e. $(x + 1)^2$] are 1, 2, 1; for power 3 [i.e. $(x + 1)^3$] are 1, 3, 3, 1; and so on. In fact they can be laid out in the form of a table:

power 1					1		1			
power 2				1		2		1		
power 3			1		3		3		1	
power 4		1		4		6		4		1
power 5	1		5		10		10		5		1
etc.										

Since the seventeenth century AD this array of numbers has been known as 'Pascal's Triangle' because Blaise Pascal drew it up, though it had appeared a century before in a book by Peter Apian. The triangle is of use in the mathematical analysis of probabilities: thus the second row (power 2) expresses the total number of ways or permutations in which 2 coins may fall (there is one way of getting 2 heads, two ways of getting a head and a tail, and one way of getting 2 tails); row 3 (power 3) the number of ways 3 coins can fall, and so on. However, the interesting fact for us at the moment is that the Chinese knew of it at least five centuries before Pascal, for it was expounded by the Sung mathematician Chia Hsien (Jia Xien) in 1100 and probably appeared first a little before this. In China, though, the triangle of numbers appears on its side, a clue for historians that the advanced Sung algebraic methods were indeed derived from the use of counting-rods on a counting-board.

For Pascal, see page 372.

Illustration page 157

What then can we conclude about Chinese mathematics? Firstly, it seems clear enough that the Chinese never evolved any rigid proofs such as did Euclid in his *Elements,* possibly because they did not develop formal logic according to strict rules such as had been laid down in Greece by Aristotle; their minds do not seem to have worked in that way. In China mathematics was tied to the solutions of particular problems – it was utilitarian. The Chinese never pursued mathematics for its own sake. But this does not mean that the Chinese made no mathematical achievements; they certainly did. They could extract square roots and cube roots, they dealt with fractions (writing the numbers in vertical columns as we do) and also with negative numbers. The Chinese proved the right-angled triangle relationship and determined areas and volumes of many kinds of geometrical figures; they devised a way of finding proportions (the 'Rule of Three') and the 'Rule of False Position' to solve equations; they practised indeterminate analysis, calculated Finite Differences and knew of Pascal's Triangle.

Astronomy

The idea that the heavens were affected by the behaviour of man, or rather of his rulers and their administration, was all part and parcel of the Chinese view of the cosmos as an organism; the illness or well-being of one part must affect the rest. This, therefore, acted as a stimulant to the administration to establish astronomical observatories and to appoint official astronomers to observe the skies and record what they saw. But there was another reason why the administration needed astronomical aid, and this was so that the calendar could be correctly determined. The acceptance of an official calendar was deemed part of the obligations of those who paid allegiance to the emperor. But, of course, this calendar needed to be reasonably correct; dates had to keep in step with the seasons. So, for these two reasons, astronomy was always an official science – a 'Confucian science' it has sometimes been called – unlike alchemy, for instance, which was an 'unorthodox' or unofficial science with very strong Taoist connections.

Chinese astronomy, as reported by Matteo Ricci and his colleagues about 1600, seemed by and large a poor thing, far inferior to Western astronomy, even Western astronomy of the late sixteenth century. But this was due to a number of misunderstandings, not least the fact that in describing the positions of bodies in the sky, the Chinese used a different method from that used by Western astronomers. That this was equally valid did not seem to temper the Jesuits' opinions. Ironically enough, the Western system which Ricci saw as the only correct one has long been discarded – indeed, it was even on the way out when he was making his criticisms. The 'Chinese' system has been adopted universally, though, it must be said, not because the Chinese approach was imported into the West – it never was – but because it was discovered independently and proved to be superior for precision observing.

The difference between these two methods can be understood by a simple diagram. Imagining the stars as fixed to the inside of a sphere – and this is still the most convenient way to deal with things when we are making position measurements, because those measurements are always made as angles – there are two ways of specifying position. One is to measure position with reference to the Sun's apparent path in the sky, that is to the ecliptic. The other is to specify positions with reference to the celestial equator (which is really the Earth's equator extended to the celestial sphere). The two circles, the celestial equator and the ecliptic, cross at two points (A and D in the figure).

These points, when the Sun is on the celestial equator, are those times when day and night are of equal length – the equinoxes – the point where the Sun moves up to the northern regions of the celestial sphere being the 'Spring equinox'. Suppose we now wish to specify the position of a star at X. We can do so by using co-ordinates based on the ecliptic, as Ptolemy and the Greeks did. We shall then say that X has a celestial longitude of so many degrees, measured from A along the ecliptic to C. The celestial latitude is then the distance from C up to X. The alternative is to measure along the celestial equator, from A to B, then up from B to X. This is what astronomers do now. They call the distance AB 'right ascension' not celestial longitude (even though CA is the celestial equivalent of terrestrial longitude), and the angular distance BX (the equivalent of terrestrial latitude) is known as 'declination'. The Chinese used this modern method, though instead of specifying a star's declination, they used the 'north polar distance', that is the distance NX. This was because they set great store by the celestial pole.

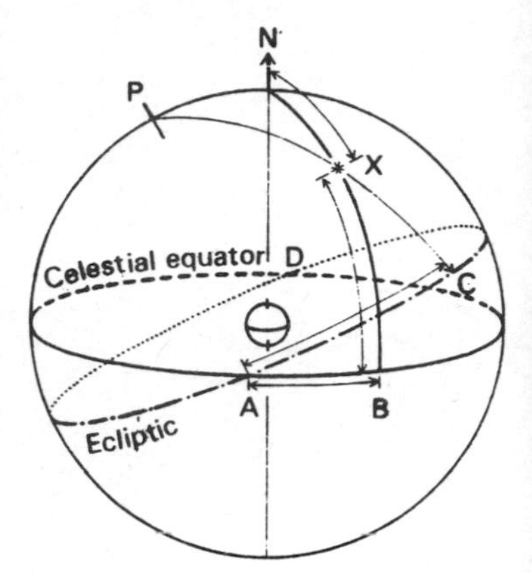

Methods of describing the position of a celestial body (X).

To the Chinese the north celestial pole typified the emperor, set at the centre of his government. It was the pivot of the heavens just as the emperor was the pivot of the empire. The north celestial pole is, of course, always in the sky to a country like China which is in the northern hemisphere. It cannot be seen by day, certainly, but it is there nevertheless. So, too, are the circumpolar stars. These never set, always remaining above the horizon. From such considerations the north pole of the heavens and the circumpolar stars assumed supreme importance, and this coloured the way the Chinese looked at all the constellations and the way in which the Sun's position in the sky was measured.

For determining a seasonal calendar knowledge of the Sun's position is crucial, as we saw in the first chapter. As we also saw, the early Mediterranean civilizations did this by observing heliacal risings and settings, while the peoples in more northerly European latitudes used alignments of stones. But there is a third quite different way of determining the Sun's position among the constellations, and that is to observe what stars are visible due south at midnight, because they will be lying directly opposite to the Sun in the sky. This is the method adopted by the Chinese. What they did was to use a set of 28 constellations or 'celestial mansions' through which the Moon appears to pass (both Moon and Sun follow nearly the same apparent paths in the sky). Once having determined these, they then tied in the right ascension positions of the beginning of each mansion with a particular circumpolar star which has the same right ascension. Often such a circumpolar star would be dim, but this did not matter; it could always be observed on a clear night.

For these two techniques, see pages 22 and 49.

The Chinese had a lunar calendar, as had all early civilizations, but they used a solar calendar for the seasons. By 1400 BC they knew that the length of the solar year was 365¼ days and that a lunation (New Moon to New Moon) was 29½ days. They used a 12-lunation cycle

(354 days) and added an extra month of 29 or 30 days from time to time to keep in step with the seasons. They also later developed a 19-year cycle (a period sometimes known in the West as the Metonic cycle, since it was developed in 430 BC by the Greek astronomer Meton of Athens in conjunction with Euctemon, another Athenian). This cycle consists of 235 lunations and gives a period of 12 years of 12 lunar months and seven years of 13 months; after this period solar and lunar calendars are almost in step (they are, in fact, only five days out, or, put another way, the average error is only a little over a quarter of a day per year). The Chinese use of this cycle seems to predate Meton's work by a century. This 19-year method was superior to the first and by and large had replaced it by the third century BC. These calculations were supplemented by a 'meteorological' cycle of 24 points – 'Spring Begins', 'Rain Water', 'Excited Insects', 'Vernal Equinox' and so on; each point meant a movement of the Sun of nearly 14° in right ascension, and just over 15° along the ecliptic. If a lunar month failed to contain one of the meteorological points – and this could happen from time to time – then this was taken to indicate that an extra month should be inserted. Thus the Chinese had an effective luni-solar calendar.

Besides a luni-solar calendar the Chinese also used a simple day-count. Their method for this did not depend on either Sun or Moon, but was based on a combination of 12 'earthly branches' with 10 'heavenly stems' (a fortune-telling system), giving two cycles of 60 days, used as far back as the Shang. The cycle could be broken down into 6 periods of 10 days, and a 10-day week was common in early China. The 7-day week came rather late, being introduced into China in Sung times some time about AD 1000 by Persians or by merchants from Central Asia.

The movements of the planets were observed and noted in China from very early on. The planets – Mercury, Venus, Mars, Jupiter and Saturn (the only ones then known) – were associated with the Five Elements but the Chinese never formulated any theory of planetary motion as did the Greeks. However, they took a particular interest in Jupiter because the time it takes to complete one orbit of the stars is 11.86 years, i.e. almost 12 years, and this seemed to fit closely with the 12 cyclical 'earthly branches' and also with the number (12) of lunations in a year. Other cycles were recognized, as was the case in other civilizations, and the Jupiter 12-year period was frequently involved. The largest Chinese cycle, the 'Supreme Ultimate Grand Origin', combined all the others, and amounted to no less than 23,639,040 years; it was the period after which all variations of the relative positions of all celestial objects would be repeated.

Recording Observations of the Heavens

Besides their observations of the planets, the Chinese made and recorded a wide variety of astronomical phenomena. Today their observations are being found of considerable use by astronomers when they need to trace back cyclic events like eclipses, or the appearance

Left 'Pascal's Triangle' as depicted by Chu Shih-Chieh in *Precious Mirror of the Four Elements*. AD 1303.

Left Rear of a bronze mirror of the 8th century AD, showing the Earth, the four cardinal points and the five elements, as well as the primordial ocean. Seattle Art Museum, Washington, Eugene Fuller Memorial Collection.

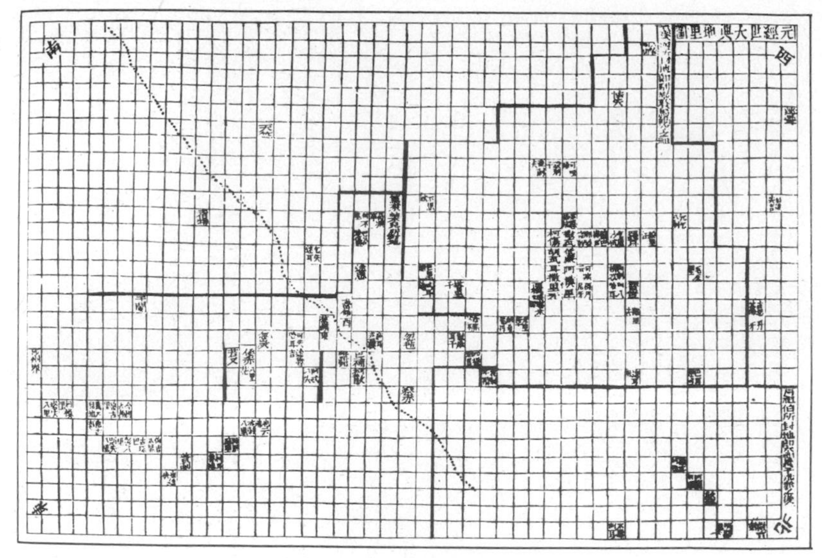

Right Schematic grid-map of AD 1329, from the *History of Institutions of the Yüan Dynasty*. It shows north-west China, with north in the bottom right-hand corner.

Right A Han-dynasty (c.200 BC–AD 200) mortuary jar with the representation of mountains on the lid. As such it was probably a precursor of the relief map. Museum of Fine Arts, Boston, Massachusetts.

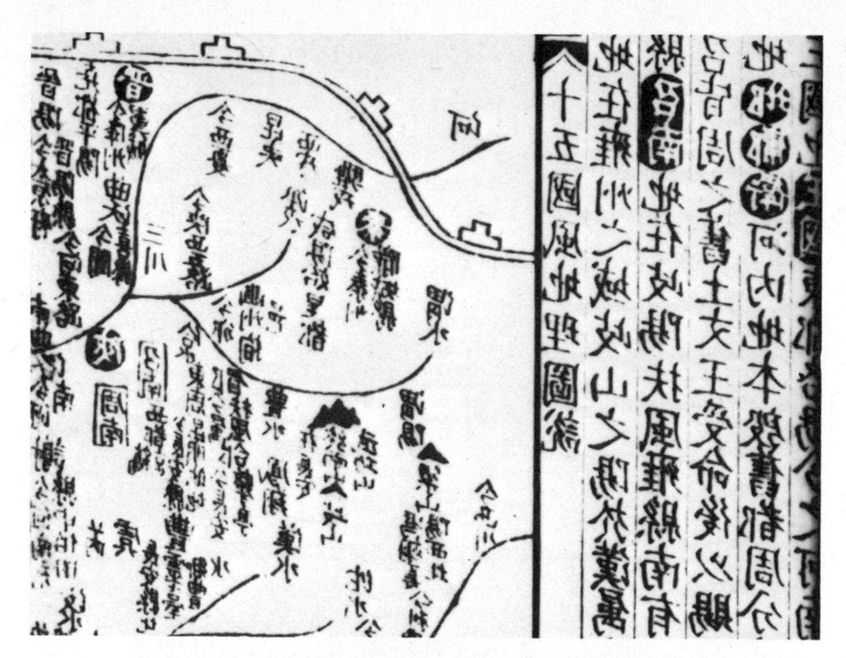

Left The oldest printed map in the world produced in AD 1155 and showing west China. From *Illustrations of Objects mentioned in the Six Classics*.

Left Reconstruction of the seismograph of Chang Heng (7th century AD).

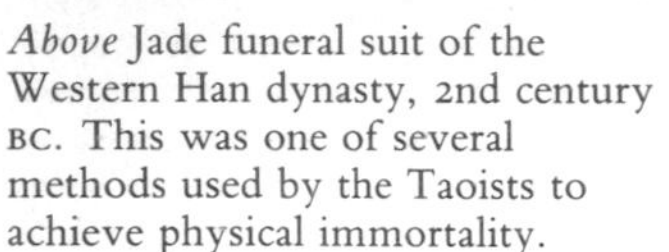

Above Jade funeral suit of the Western Han dynasty, 2nd century BC. This was one of several methods used by the Taoists to achieve physical immortality.

Above right Bronze hand-bell, possibly of the 6th century BC. Danske Kunstindustrimuseum, Copenhagen.

Right A Ming-period engraving of one of the Five Holy Mountains, showing an interest in dramatic geological formations.

of comets, or even rarer events like exploding stars. From very early times, eclipses of the Sun were considered of great significance, and Chinese eclipse records go back to 720 BC, some 300 years earlier than those listed by Ptolemy in the *Almagest*; they are very reliable and remarkably detailed. However, since an eclipse was thought to be a warning from heaven about bad government policies, eclipses that were observed during reigns popular with court officials tended not to be recorded, an attitude of mind that explains why astronomical records were considered state secrets.

If the Chinese had inhibitions about recording eclipses for posterity in popular reigns, they were not inhibited in what they might observe and record by any preconceived ideas about the stability and changelessness of the heavens, as was the case in the West. Thus the Chinese observed and recorded sunspots, which were never recorded in Europe until the seventeenth century because of the prevalent belief there of the perfection of celestial bodies, a perfection which would not allow the Sun to appear spotted. The Chinese sunspot records, which start in 28 BC, are the most complete we have.

Another example of the Chinese lack of inhibition in observation compared with the West is the case of novae and supernovae. These we now know to be stars that explode, creating vast envelopes of hot, brightly glowing gas. Thus they may appear where no star has been observed before because, of course, they were previously too dim to be observed – hence their name of *nova* (new). After the middle of the Han, the Chinese very appropriately used the term *k'o hsing (Ge xing)* or 'guest star'. In the West no such stars could be admitted to exist, at least officially, since the heavens were perfect and complete; new stars were a theoretical impossibility. The very large explosions – the supernovae – are rare, but Chinese records cover appearances in AD 1006, 1054, 1572 and 1604. The last two appeared after the European belief in a perfect unchangeable cosmos had gone, and were recorded in the West as well as the East. Some Western accounts of the 1006 supernova also exist for it was exceptionally bright and seems, by and large, to have been thought to be a comet; that would have calmed consciences, for comets could suddenly appear, as they were thought to be flaming vapours in the upper air, something 'below the sphere of the Moon' and so transitory. However, only in China and Japan does the supernova explosion of 1054 seem to have been recorded. Its remains, visible even today nine centuries and more after the explosion, are known to modern astronomers as the Crab nebula and attract intensive study; for this the Chinese records are invaluable. They are valuable, too, because they do not confine their attention only to supernovae; the less exciting and noticable novae, which are so much dimmer than the supernovae, are also recorded. Indeed, the Chinese observed and noted down details of no less than 75 'guest-stars' between 352 BC and AD 1604.

The Chinese observed comets with great care whenever they

appeared, and this is very helpful because records from other civilizations are far from complete. Their total list of comets covers appearances between 613 BC and AD 1621, a total of well over 22 centuries. Of all these, appearances of Halley's comet are the most important (being fairly frequent and very striking), and it is to the Chinese records that today's astronomers look when they wish to trace that comet's earlier movements. Moreover, modern astronomy has shown that there is a link between comets and 'shooting stars' (meteors), which appear for only a fraction of a second or so as they flash into view on a dark night. These are lumps of rock and metal from outer space which heat by friction as they rush through the Earth's atmosphere, usually becoming completely vaporised in the process, though sometimes they are too large to be completely destroyed and land on the ground; they are then known as meteorites. Old records of such falls are important and fortunately Chinese records were kept of them and, when possible, meteorites were examined, scientific details of them being given long before any such objective accounts are available in the West. Periodical showers of meteors were also recorded in China.

For Halley's comet, see page 349.

Mapping the Skies

From what has been said already it will be no surprise to learn that the Chinese seem to have been the first to compile systematic catalogues of the stars. As early as the fourth century BC three astronomers – Shih Shen, Kan Te (Gan De) and Wu Hsien (Wu Xien) – compiled catalogues in such detail that these were in use a thousand years later. In the West the first catalogue was compiled in the second century BC. The three Chinese catalogues were used independently for a long time but were combined in the fourth century AD when Ch'en Cho (Chen Zhuo) constructed a star map based on them. Star charts were in fact begun a century earlier although even this does not mark the beginning of drawings of constellations: Han reliefs are known. It was the Chinese, too, who invented the 'ball-and-link' method of showing the constellations, a method which is still in use today in popular star atlases.

One of the most astonishing of all early star maps first appeared in the year AD 940. Based on the star map by Ch'ien Lo-Chih (Qian Lo-Zhih), the fifth-century AD Chinese 'Astronomer Royal', it is exceptional not because it was a detailed chart – those had been available for centuries – nor because it was in three colours to differentiate between the stars in the catalogues by the astronomers Shih Shen, Kan Te and Wu Hsien, but because of the way in which the stars were actually mapped. To draw out correctly on a flat surface the stars as they are observed on the inside of a sphere – the equivalent of trying to represent the curved surface of the Earth on a flat sheet – has always been a problem. There are various ways of getting round the difficulty, various forms of 'projection' that can be used, as Ptolemy knew and explained in the second century AD. What is surprising about the 'Tunhuang' star map of Ch'ien Lo-Chih is the form of projection he adopted: a 'Mercator' projection. The reader will doubtless be familiar with this method from current maps and atlases; it was devised in AD 1596 by the great Flemish cartographer Gerhard Mercator, yet this is six centuries after the Tunhuang star map. Whether Mercator knew of such Chinese use – after 940 the projection was certainly used widely in China – we do not know, but the West definitely has no priority in this matter as was once thought.

Illustrations page 138

Mapping the heavens was also carried out with globes and planispheres. A planisphere is a circular, or near-circular, chart with the celestial pole at the centre; it is a map of the celestial globe looked at as if one were hovering over the celestial pole, a different projection from Mercator's. The Chinese knew of this projection too and their most important planisphere appeared in the twelfth century, carved on stone; it depicted the pole, the ecliptic, the equator and the celestial mansions, and it also showed the 'white road' of the Moon and gave a correct account of eclipses of both Sun and Moon. The true reasons for eclipses – the Moon blotting out the Sun's light in a solar eclipse and the Earth intercepting the Sun's light for a lunar eclipse – were known to some in China a millennium before the stone planisphere, although for a time some Chinese philosophers continued to prefer an explanation depending on the waxing and waning of the Two Forces, Yin and Yang.

A celestial globe on which the stars could be represented is an ancient device: it can be traced to Greek times and may even have originated in Babylonia. A solid globe with the stars drawn on it was to be found in China in the fifth century AD, about the time of Ch'ien Lo-Chih. Perhaps it goes back before this, though previously the Chinese may have found that an armillary sphere would do all they needed. An armillary sphere is a sphere of rings (Latin *armillae)*, each ring depicting a circle on the celestial sphere such as the ecliptic, the equator and so on; the Earth is thought of as situated at the centre. In brief, it is a skeleton celestial globe. If the rings are graduated and a bar with mounted sights is fixed to a movable ring, the instrument can be used for measuring positions.

Illustration page 139

Observing Instruments

The Chinese used a wide variety of observing instruments, some of similar design to those in use in other civilizations, such as the gnomon (the vertical stake) and the armillary sphere. They were also guided later by the Arabs in building huge astronomical instruments in masonry. Thus at Kao-ch'eng (Gao-cheng) in Honan province there is 'Chou (Zhou) Kung's Tower for the Measurement of the Sun's Shadow'; this is a tower some 12 metres (40 feet) high with a level masonry scale more than 36 metres (120 feet) long. A shadow was cast along this scale by a 12-metre (40 feet) gnomon. Restored in Ming times, the tower was first set up in AD 1276. But not all Chinese instruments were derivative; they had a number which were entirely of their own design.

One of their earliest was the gnomon shadow template, a graduated standard piece of jade or terracotta material to set against the shadow cast by a gnomon. These date from at least the second century AD. Another was the 'circumpolar constellation template', derived from the jade disc pierced by a central hole (the *pi)* which was a device that could be used for identifying some of the stars close to the celestial pole and picking out the celestial pole itself. It could also indicate the angles at which these stars appeared, thereby acting as an instrument to key in to their 'celestial mansions' by observations of circumpolar stars. Such templates go back to at least 600 and possibly to 1000 BC.

The Chinese also turned their attention to timekeeping – an important adjunct to astronomy – and raised the clepsydra or water-clock to a new stage of development. The clepsydra itself was not a Chinese invention – its origins go back to Mesopotomia and Egypt – but the Chinese turned it into an instrument of some precision. They fitted its water tanks with siphons or used a number of tanks feeding one another, or even adopted both methods in the same clock to ensure a regular flow of water at all times. They also developed a 'steelyard' clepsydra using containers for the water hung on the arms of a balance arm pivoted near one end (a steelyard) and, for measuring short periods of time, they even built small clepsydrae of jade and used mercury instead of water. But their most significant development was the invention of a mechanical clepsydra which had an 'escapement'. The escapement – the device that allows the rotation of a shaft to occur only in regular steps, by permitting a train of gears to 'escape' by only one tooth at a time – is the essence of all mechanical clocks. It was first devised by the Chinese Buddhist monk and astronomer I-Hsing (I-Xing) and the engineer Liang Ling-Tsan (Liang Ling-Zan) about AD 723. In China its most notable astronomical application was in the eleventh century by the astronomer Su Sung who built a large 'clock tower' at Kaifeng. This showed the time and, astronomically rather more important, it carried at the top an armillary sphere which was driven by the water-clock escapement so that it followed the apparent motion of the Sun and stars in the sky. Yet Su's clock tower was not the first structure to be fitted with a clock-driven armillary sphere. In the eighth century Chang Heng (Zhang Heng) applied the escapement of I-Hsing and Liang Ling-Tsan in this way. The idea, it seems, was not so much to make observing easier by providing an instrument which would automatically depict the daily rotation of the heavens, but rather to provide a standard against which the actual rotation (observed with another armillary sphere) could be checked. In this way discrepancies in the heavens could easily be noticed, discrepancies which it was the duty of astronomers to record. In all this, the Chinese pre-empted the West: mechanical clocks did not arrive in Europe until early in the fourteenth century, seven hundred years after the invention of the escapement in China, while mechanically driven observing instruments were not actually made in the West until the eighteenth century.

Illustrations page 140

Chinese concern with co-ordinates based on the celestial equator, rather than the ecliptic as used in the West, had two effects. On the one hand it meant that the Chinese armillaries were often simpler than those in the Western world because they omitted a ring for the ecliptic; more importantly it led them to devise the 'equatorial mounting'. When one looks up at the heavens they appear to rotate around the pole, but since the pole is never overhead (unless the observations are made from one of the Earth's poles), all celestial bodies appear to move in a curved path across the sky. To follow this movement with a sighting instrument – an astrolabe, for instance – the observer must turn it in two directions, one sideways parallel to the horizon and the other straight upwards or downwards. What the Chinese realized some time in the thirteenth century AD was that if one tilted the axis of the instrument over on its side until it was lined up on the celestial pole, only one movement is necessary for following the daily motion of the stars. Just such an instrument, mounted 'equatorially', is Kuo Shou-Ching's (Guo Shou-Zhing's) 'equatorial torquetum' of AD 1270. Although the equatorial mounting is now used in the West for most large telescopes it did not arrive until towards the latter part of the sixteenth century for an armillary instrument, three hundred years after its appearance in China.

Illustration page 139

Theories of the Universe

Before leaving early Chinese astronomy, it is important to consider their views about the universe as a whole. There were three main theories, the first and oldest of which seems to have been inherited from Babylonia. This was called the Kai T'ien (Gai Tian) or 'hemispherical dome' theory, which was just what it describes; a dome for heaven with a domed Earth beneath it, surrounded by a circular ocean. The second idea, the Hun T'ien (Hun Tian) or 'celestial sphere' theory, though later, was nevertheless known as early as the fourth century BC. A step forward from the hemispherical dome concept, it was ideal as a description of how the universe appears; with it was linked the idea of the Earth itself as a sphere.

The third Chinese idea, the Hsuan Yeh (Xuan Ye) or 'infinite empty space' theory, was the most advanced and imaginative of the three. It is associated with Ch'i Meng, who lived sometime during the later Han (AD 25 to 220). What this view propounded was that the heavens were empty and void of substance . . . having no bounds'. The Sun, Moon and stars floated freely in space. What drove them in their paths? Here the Chinese called on a concept of the 'hard wind', something derived from the Taoists and possibly originating in the powerful air blasts of bellows used in smelting. At all events, this whole theory was a totally novel idea. Indeed, an infinite empty universe with bodies floating in it, driven by a hard wind or not, was a very advanced concept, and advanced not because it fits in more with our present views of the universe (which it does), but because it was less restrictive than the only other concept of the time, the rigid Greek belief in solid spheres. It was a grand view of the cosmos and it encouraged the Chinese to develop a broader outlook on the whole of Nature.

Earth Sciences

The Earth Sciences (geology, geophysics, meteorology, oceanography) are all recent studies, going back no earlier than the eighteenth century. Geography is older, certainly; it was practised by the Greeks, but really as more of an art than a science. Nevertheless it is interesting to see that the Chinese were able to bring some semblance of science to these subjects long before modern times.

Map-making

In geography this rational approach of the Chinese showed itself in the way they tackled the problem of mapping the Earth. In the West there was a hiatus between the time of Ptolemy and about AD 1400, an interval in which European map-making suffered an almost complete eclipse. Maps of the world became the vehicle for religious ideas, and bore little resemblance to the real world, yet in China a tradition of scientific map-making was carried on. In the Han, around AD 100, Chang Heng (Zhang Heng) introduced a grid system – a system of lines at right-angles to each other – for specifying the positions of geographical features, and in the T'ang the growth of the empire saw an extension of accurate mapping within wider boundaries. During the Sung also much mapping was done and in the twelfth century two magnificent maps were produced. Both were carved in stone, one with a precision grid, and were so accurate that they do not compare unfavourably with modern charts. Interestingly, north is at the top, a convention universally accepted today but unique at that time. The greatest of all Chinese map-makers was Chu Ssu-Pen (Zhu Si-Ben) who worked during the Mongol period. He not only made magnificent and accurate maps but also warned about the dangers of inserting countries for which precise detailed evidence was not available, advice which was not universally accepted even in China. It is also to the Chinese that credit must be given for the first relief maps. In Sung times these were made of wood, but they appear to have originated long before the tenth century. They were used for military purposes in the Han during the first century AD, but their original derivation is in some doubt. Relief models of population statistics were used very early on in China and there were also 'hill censers' – instruments for burning incense fashioned as sacred mountains – in the first century BC, these having Buddhist and probably Taoist connections; perhaps, then, these were the forerunners of the later maps.

Illustrations pages 243, 289

Illustration page 158

Illustration page 158

Meteorology and the Tides

Chinese interest in the physical nature of the world and its environment led them to make studies both of the weather and of the behaviour of the tides. Weather, of course, had connections with the behaviour of the heavens and the conduct of government affairs; after all, meteorological phenomena do occur in the sky above the Earth.

This connection, as well as the day-to-day administration of the state, which needed to know of special and disastrous weather conditions such as floods, snow blockages, droughts and so on, led to a widespread system of meteorological recording. Therefore although weather prediction in China, as elsewhere in the ancient and medieval world, never progressed beyond the stage of the countryman's weather lore, their weather records were quite another matter. Temperature was regularly recorded at least from Han times – not on a temperature scale, of course, that was a seventeenth-century idea – but excessively hot summers and cold winters were noted and provide records that are useful to today's meteorologists. Rainfall and wind were always recorded, at least from Shang times, and records exist back to 1216 BC; these give information on rain, sleet, snow and wind, with details of rainfall and of direction of the wind itself. Snow gauges were in use; these were large bamboo cages placed beside mountain passes and on highlands, and officials were supposed to report to the central government to assist in calculating what repairs and maintenance of dykes and other public works were needed.

Humidity was assessed: the first hygrometer to give a measure of the water content of the air was devised in the second century BC; it weighed dry and damp charcoal and then compared the results. Predictably, thunder and lightning were explained in terms of Yin and Yang at a time when no scientific explanation was possible, but it is an indication of Chinese rational thinking that they were well aware of the water cycle in the late fourth century BC. This idea that rain fell, was evaporated to form clouds and was then recycled back, was also known in Greece in the sixth century, so perhaps this was a case of the transmission of ideas, though difficulties in the exchange of information at this time make it unlikely.

Many other meteorological, or seemingly meteorological, phenomena were observed and recorded by the Chinese. Rainbows and haloes round the Moon were noted, as well as those much more picturesque haloes and mock-suns which are sometimes seen near to and around the Sun. These last were observed a millennium before records of them were kept in Europe. Again the aurora borealis or 'northern lights' were duly noted and records of these go back to the third century BC.

But the phenomenon into which the Chinese seem to have shown the greatest insight was the movement of tides. What doubtless drew particular attention to this subject was the very considerable range of tides which China experiences; there is a range of over 3 metres (10 feet) at the mouth of the Yangtze in the Spring while the Ch'ien-T'ang (Qian-Tang) river near Hangchow has a tidal bore unequalled anywhere in the world except on the Amazon. By the second century BC enough was known for high tides to be expected at Full Moon, but it was not until two centuries later that it was realized that the Moon actually caused the tides. By the eleventh century it was appreciated that the Sun also played a part, while soon after this the engineer and astronomer Shen Kua recognized what we call the 'establishment of the port', that is the delay in high tide due to local irregularities in the shore line.

For an early study of the same phenomenon in Europe, see page 252.

The tidal ideas of the Chinese must be seen in perspective. About 200 BC the Greek philosopher Antigonus of Carystos had suggested that the Moon was the main tidal influence and Greek understanding of tidal phenomena was ahead of China. But Europe did not develop this knowledge, and even the recognition of the establishment of the port came a century later than in China. In the West the Moon's influence as a tidal force was not fully accepted until five hundred years after that. Yet if the Chinese were not the first to connect the Moon with the tides they were pioneers in the preparation of tide-tables; their systematic compilation of these goes back at least to the ninth century AD.

Geology

Chinese art shows that they had a very clear appreciation of geological features. Indeed, Chinese artists were extremely faithful to Nature and for many hundreds of years guides to the delineation of hills and mountains, outcrops of rock and expanses of water were among their standard textbooks. And no later than the twelfth century AD the Neo-Confucian scholar Chu Hsi (Zhu Xi) understood that mountains had been elevated from land that was once below the sea. This was something that was not to be appreciated in Europe until the beginning of the nineteenth century. But Chu Hsi (Zhu Xi) was not the first to realize the fact; the idea can be found in China much earlier than the eleventh century, back in fact to some time between the second and sixth.

Illustration page 160

The Chinese seem to have been the first to recognize fossils for what they are – the remains of once-living material. Certainly they were in the van of every nation when it came to appreciating the true nature of plant fossils. Knowledge of fossilized pine trees appears to go back to the third century AD, and later, in the eleventh century, fossil bamboo was discovered and described, while between the two, other identifications were made, although not always correctly because there was sometimes confusion between fossilised plants and rocks with veins of material that looked very like fossil plant remains. The Chinese also recognised fossil animals. They were not always correct over their identification of species; there was, for instance, confusion between a bird – the swallow – and the extinct mollusc *Spirifer,* because the mollusc's shells look very like birds' wings. 'Stone fishes' were recognized as early as the sixth century AD and described in minute detail, but in the first century BC the Chinese had recognized fossils as the remains of living creatures, a view which, though it had been suggested by some Greeks, was totally forgotten or ignored in the West until after the Renaissance.

For the gradual awareness in Western science of the implications of fossil remains, see pages 389–390, 423 and 425.

Mineralogy

From very early times all civilizations took an interest in stones, ores and minerals. The Chinese were no different but they also made an extra and worthwhile contribution. It was widely recognized that minerals must be due to slow changes occurring in the Earth's crust, and mechanisms to account for them involved 'exhalations' from the Earth; Aristotle suggested two, one from moisture within the Earth and one from the dry Earth itself. The Chinese had a similar idea around the same time, and it is tempting to suppose that both derived from an earlier common source, possibly Babylonia.

Minerals were classified in China by hardness, colour, appearance and taste. Metals were distinguished from stones. It was also noticed, as Theophrastos had observed, that some could be fused by heat, others not. Attention was also paid in China to the association of ore beds with different kinds of rock. Of course, Chinese mineralogy was helped by the special interest in minerals of the Taoists who had always specified mineral as well as herbal drugs either to help bring about physical immortality or for more orthodox medical use. In the West, in the second century AD, Galen had firmly set his face against any but herbal drugs, but in China there were no such inhibitions.

The Chinese recognized and used a number of mineral substances. Alum was used in dyeing and in a host of other industrial processes, as well as in medicine. So was sal ammoniac (ammonium chloride) and, much later, borax. The Chinese knew and used that strange fibrous mineral asbestos; they were aware of the touchstone (jasper) as an aid in assaying gold and silver and, of course, they knew of precious stones. However, it is strange to find that although the Chinese were familiar with diamonds and recognized their hardness they were not acquainted with them as cut and polished stones until the Portuguese brought them in the sixteenth century. The Chinese decorative mineral *par excellence* was jade, which is a silicate of sodium and aluminium. In spite of its hardness, it was already being carved in Shang times and later cut using abrasives. Rotating circular cutters were in use by the twelfth century AD and it seems likely they appeared long before this.

An interesting facet of Chinese mineralogy is what we should today call geo-botanical and bio-geochemical prospecting. In ancient times miners had to locate the presence of ores by their own intuition and experience. They noticed the lie of the land, the appearance of certain outcrops of rock and so on. In China these factors were recognized too but, in addition, they observed that certain plants were associated with certain minerals; a kind of shallot indicated that gold might be found, ginger gave a clue to tin and copper. They appreciated, too, that the condition of the plants was important, and as we now know that some kinds of plant actually take up certain metals, we can appreciate that they were pioneering a technique which lay far in the future. Indeed the Chinese knew that some metals could actually be extracted from certain plants. One might expect that miners in other civilizations would also have been aware of these facts, but if they were there is no evidence of it. In the West geo-botanical prospecting was unknown until 1650 or later.

Seismology

China forms part of one of the world's great earthquake regions and, as we should expect, the Chinese kept extensive records of all disturbances. One of the earliest recorded was in 780 BC when three river courses were interrupted, while the records as a whole show that there were twelve peaks in frequency of occurrence between the end of the Sung and the beginning of Manchu rule. One of the worst earthquakes in history occurred in China in AD 1303 when more than 800,000 people perished.

Although no notable progress was made in China in devising a theory of earthquakes, nor was there in any other ancient or medieval civilization. But the Chinese do have one special claim to fame in seismology: they constructed the ancestor of all seismographs. Designed in the early second century AD by Chang Heng (Zhang Heng), this 'earthquake weathercock' contains the basic element of such a device, a very heavy pendulum which, while not affected by light local disturbances, will respond to the deep, very low-pitched rumbles of an earthquake.

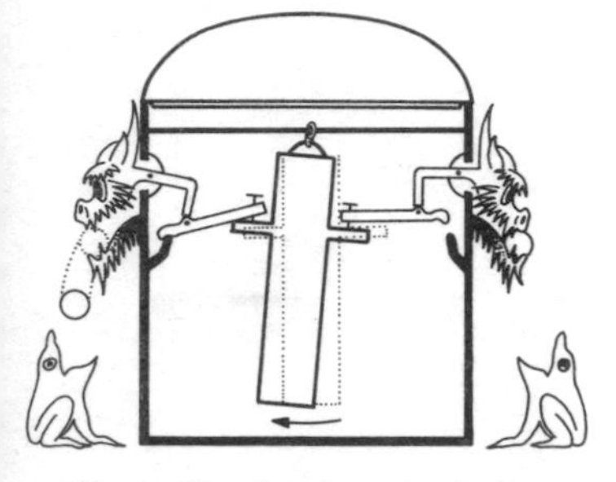

Chang Heng's seismograph. In the event of an earthquake, the hanging weight would swing in the direction of the tremors, and this swing would be recorded by a falling ball. See also the illustration page 159.

Chang Heng's remarkable instrument consisted of a bronze wine jar about 2 metres (6 feet) across, and fitted with a cover. Around the jar's surface were eight dragon heads, each holding a ball. On the floor around the jar were placed eight bronze toads with open mouths. In the event of an earth tremor, one of the dragon mouths would open and drop its ball into the mouth of the appropriate toad. The dragon which disgorged the ball indicated from which direction the tremor had come, and since the apparatus then automatically locked itself so that no other balls could be released, a permanent indication was given. The apparatus proved useful to the administration, and there is evidence that similar recorders were made in later times. Such instruments seem to have been known at the Maraghah Observatory in Persia in the thirteenth century, but as this observatory was in close touch with the Chinese, there seems no doubt that the Persian instruments were imported from China.

Physics

Chinese physics was never concerned to any serious degree with atoms or theories of atoms. In the fourth century BC the Mohists seem to have had some leanings that way, and in the first century AD Buddhism brought with it atomic ideas, but the general Chinese outlook was one which concentrated on the mutual growth and decay of the Two Powers, Yin and Yang. It was a cyclic view, and one which led to a universe permeated by continuous wave-like changes. Today, in modern Western science, cyclic wave-like motion is part and parcel of our explanation of change within the universe. It is only one aspect of an elaborate and powerful atomic theory, but that does not alter its significance in scientific thinking. It is interesting, then, to realize that from very early times the Chinese, too, used a form of wave theory to account for change throughout the natural world;

as Yin waxed so Yang waned, and then as Yin later waned, so Yang grew again. So powerful is this Chinese concept that some modern atomic physicists are turning to Chinese imagery in an attempt to explain their theories of sub-atomic particle behaviour.

The Chinese were also adept at practical measurement, as perhaps we should expect of a nation which was so successful in developing every field of engineering. They had a metric system very early on, and were also among the earliest when it came to problems of statics – forces, weights, levers, balances, etc. – for the Mohists were already making worthwhile contributions to this in the fourth century BC. They attempted also to deal with moving bodies, an endeavour that was not followed up in China as it was in the West with the early Greek studies of motion. It was the Mohists, moreover, who began the study of optics in China, discussing shadows and early appreciating the fact that light travels in straight lines. They also experimented with the 'camera obscura', and knew how the image of a distant scene is turned upside-down when light passes through a pinhole. However, in China, as in Islam, the camera obscura only came into its own in the eighth century AD. Flat mirrors and concave mirrors were objects of study, and the Mohists knew what are today called 'real' and 'virtual' images given by a concave mirror; in all this they seem to have been ahead of the Greeks, who, as we saw in the last chapter, had some fundamentally wrong ideas about light and vision. It is true that at about the same time the Mohists were at work in China, Euclid in Alexandria was writing out a series of theorems on optics; but what he had to say about mirrors is lost, and probably it did not reach the stage of Mohist understanding owing to the prevailing Greek misconceptions.

For Ptolemy's views on this, see page 122.

In China, concave burning mirrors were put to practical use, and in the Han really large metal mirrors were common but, as in the West, glass mirrors were unknown; they were to be a nineteenth century invention. The Chinese were also familiar with burning lenses. By the tenth century they had fashioned lenses of various shapes, and knew that while some magnified images others could give a reduced image (diminishing lenses), though this did not lead them to wear spectacles nor to develop the telescope. Chinese lenses were made from naturally occurring rock crystal, though it seems likely that glass was also used, at least from Han times onwards, for China had a glass industry as early as the sixth century BC.

In all early civilizations music acted as a stimulus to the study of sound, though how scientific such a study became depended on attitudes of mind to enquiries of this kind; they could remain on a purely artistic level. But in China, as in Greece, it was a study particularly concerned with careful measurement. However, there was a difference; whereas the Greeks were concerned mainly with analysing sound, the early Chinese tended to look into relationships, since this was sympathetic to their way of thinking, more in line with their organic view of the universe, where everything bore some

For the Pythagoreans' views, see page 72.

relation to everything else. Thus, whereas the Greeks would examine why the shape of a pipe would alter the pitch of its sound, the Chinese were more concerned with sympathetic sounds or, in modern terms, sympathetic resonance. Such sounds are heard, for instance, when plucking a string of a musical instrument sets up vibrations in other strings. To the Chinese this was merely an example of a natural relationship in the natural world; there was nothing marvellous in it. Since the notes to which the strings were tuned were in specific relationship with one another they would be bound to have sympathetic resonances, for that is the way the world was constructed.

In the earliest times in China sounds had non-scientific relationships too; the Chinese linked sound with taste and colour in some religious ceremonies. But this did not prevent a parallel scientific attitude being maintained also. Sounds were classified by timbre and by pitch, and various musical scales were specified. These, in their turn, demanded precise tuning. This was achieved partly by resonance; a properly tuned bell could be used as a standard and would set another bell ringing in resonance when the second bell was correctly tuned. Again, vessels filled or partly filled with specific quantities of water were used as resonators to give standard tones, and bells were sometimes used to help tune strings. None of this was unique to the Chinese, but what was peculiar to them was the precision they achieved: this was partly due to the use of millet seeds as a measure of the capacity of pipes of similar bores, which may have been derived from a basic desire to measure capacities of containers of various kinds, but certainly from the Han onwards had become a standard method for tuning pipes. The Chinese also recognized that sound is vibration; they were not alone in this but a special contribution they did make, and which arose out of their scientific work on sound, was the evolution of an 'equal temperament scale'. This is the kind of scale now adopted in the West; it allows one to move readily from one key to another. It was used – one might almost say publicized – by J. S. Bach in his *Well-Tempered Klavier*, for such a scale was comparatively new even in the early eighteenth century, having been first introduced in the West about 1620. Yet it was known and worked out in China by Chu Tsai-Yu (Zhu Zai-Yu) in 1585, and there is reason to think that it was from China that the Western scale was derived.

Illustration page 160

For its publication by Simon Stevin in the early 17th century, see pages 314–315.

Of all the Chinese work in physics, and it is evident that it was a field in which they made many important contributions, their most significant was the invention of the magnetic compass. This is particularly interesting, since it is a prime example of processes of magic leading to a scientific discovery. It apparently began in the third century BC with the use of diviner's boards. These were operated by those fortune-tellers concerned with state affairs, and they consisted of two parts, an upper and a lower. The upper, which represented the heavens, rotated on a pivot over the lower disc, which represented the Earth. The upper board had the north circumpolar constellation of the Plough or Big Dipper, and both boards were marked with 'compass points' or directions, each 15° apart. For divination, 'pieces' symbolizing various objects were thrown on to the board, and the diviner would then read the future from their positions. One of the symbolic pieces was in the shape of a spoon, which itself symbolized the Dipper. It was able to rotate quite readily and, at some time not later than the first century AD, such rotating spoons completely replaced the upper part of the divining board.

In company with many other early civilizations, the Chinese knew of the natural magnetic properties of lodestone (a form of iron oxide sometimes known as magnetite), and in early days its power to attract iron was considered magical. In either the second or first centuries BC, this led fortune-tellers to make some of their divining-board pieces out of lodestone and the spoon itself began to be constructed of this material. As soon as this had happened, and the spoon had replaced the 'heaven' plate, its position, which was now in the centre of the board, made it clear to diviners that it was a truly magical object, for its handle always pointed in the same direction. In due course it became known as the 'south-pointing spoon'. Of course, such a spoon moved jerkily – friction would see to that – and so the practice grew up of making a wooden spoon with a piece of lodestone inside it; later still the piece, no longer by now resembling a spoon, was mounted on a pin or even floated on water. This, then, was the essence of the magnetic compass, pointing to the south rather than the north, and a familiar sight by the first century AD. By the sixth the Chinese had discovered that small iron needles could themselves be magnetized by being stroked with a piece of lodestone, and later still, in the eleventh century, they found how to magnetize iron by raising it to red heat and then cooling it while it was held in a south-north direction.

The 'magic' magnetic compass was first used away from the divining board in laying out public works and private buildings. In due course it was adopted by sailors and was common on Chinese shipping perhaps as early as the tenth century, and certainly by the eleventh; Chinese navigational usage thus preceded its adoption in the West by at least a hundred years. But this is not all. The use of a magnetic needle brought about greater precision in observing direction magnetically, and as early as the T'ang dynasty it led the Chinese to discover that magnetic north and south do not coincide with geographical north and south, a discovery that was not made in the West for a further seven hundred years.

Illustrations page 177

For the first European investigations into magnetism, see pages 315–316.

Chemistry

As a scientific discipline chemistry is a fairly recent study; only in the West in the seventeenth century was scientific chemistry developed and it was another hundred years before it reached China. Over the ages, of course, the Chinese acquired a vast amount of practical chemical knowledge, as did the peoples of other civilizations, and this knowledge should not be despised. With its techniques and its applications to medicine it formed an essential background without which no science of chemistry would ever have developed.

Early Chinese chemistry – or perhaps we should call it 'protochemistry' or even just 'alchemy', though it was more than that – made a number of valuable contributions to the basic knowledge of what was to be chemical science. It began, as probably did chemistry everywhere else, as a development of the art of cooking, and was a study which was very congenial to the Taoists; it had a mystical side, at least in the way they practised it, and it allowed them not only to philosophize but also to use their hands. Chemistry is nothing if not a practical, laboratory-based science, and the practical work it demanded meant that the Taoists could clearly demonstrate the difference between their outlook and that of the Confucians, who took a superior attitude to any practical, artisan practices. But there was more to it than this. The Taoists' main goal was the ambitious one of physical immortality; they sought some means by which ageing could be prevented. To achieve this they advocated a number of methods, which included gymnastics, breathing exercises and the taking of special medicines, often of mineral content. They also paid particular attention to the way bodies were buried.

Illustration page 160

Immortality ever eluded them, but in seeking it they amassed much chemical knowledge. One aspect of this showed up in some recent archaeological work in China. Excavations of a tomb in Honan province brought to light a coffin which, when opened, was found to contain the body of a woman, the 'Lady of Tai'. Although she had died about 186 BC – over two thousand years ago – the body was like that of someone whose death had occurred only a week or so previously; the flesh, for instance, was still elastic enough to return to normal after being pressed. The body was not, however, embalmed, nor was it mummified, tanned, or even frozen; it was preserved by being immersed in some brownish coloured liquid, containing mercuric sulphide, inside a coffin which was itself inside a second coffin, sealed tightly with layers of charcoal and sticky white clay. The atmosphere in the coffins was mainly methane and it was under some pressure. Thus the burial had preserved the body in what we should today call anaerobic conditions; it was air-tight and water-tight and the burial chamber had ensured that the temperature remained reasonably constant at some 13°C (55°F). There are many stories of Taoists actually achieving bodily incorruptibility, and the evidence of the Honan excavation makes it clear they are not all myth; knowledge of chemical preservation was clearly at an advanced state, even by the second century BC.

In practising their mystical alchemy the Taoists were at one with the proto-chemists of Alexandria, of India and, indeed, of every civilization where attempts were made not only to investigate the chemistry of natural substances, but also to turn metals which were cheap and abundant into gold, which was not only rarer but far more beautiful. The word 'alchemy' is, of course, derived from the Arabic but, interestingly enough, recent research indicates that the Arabic was itself derived from Chinese, not from Greek, Egyptian or even Hebrew sources as was once thought. The Taoists, then, may have had an influence far outside their immediate circles; the general alchemical attitude which we find everywhere, an attitude which took an 'organic' view of many substances, which conceived of experiments as copying their gestation within the womb of the Earth, may perhaps have owed something to them. Certainly it was an outlook which fitted in well with the Chinese view of the universe as an organism. But the Taoists were helped, too, by other aspects of Chinese philosophy; the Five Element theory helped them to classify various substances and experiment purposefully with them, while the doctrine of the Two Forces led them to an idea of flux and reflux, a sense of cyclic change where, as soon as one process reaches a peak, its opposite must begin to assert itself.

For the possible influence of Taoist ideas on alchemy in India, see pages 193–194; for the alchemy of Islam see pages 237–239; and of Renaissance Europe, pages 306–307.

Their experimentation led them to design a variety of special chemical apparatus, which included such items as special stoves and furnaces as well as vessels in which chemical reactions could take place under isolated conditions. Often such reactions meant that high pressures were set up and strong metal containers were frequently used, sometimes bound with wire to prevent the whole retort from exploding. And although the Chinese never invented thermometers as such, their alchemists and proto-chemists certainly appreciated how important it is that some reactions occur at a certain heat; they therefore devised water-baths and other temperature stabilizers. Steelyard balances were used for weighing and, most ingeniously, bamboo tubing was used for connecting one piece of apparatus to another.

Perhaps, though, the most significant piece of apparatus which they devised was the still. It was derived, basically, from the neolithic cooking pot, the *li*. This had three hollow legs and was later developed into a special type of double steaming-vessel, the *tseng* (*zeng*), which had, in effect, a second vessel standing on top of the first, separated by a perforated grating. For chemical purposes the second vessel was capped with a basin of cooling water so that evaporated substances would be cooled and thus condensed; they then dripped downwards where they were collected in a small cup. This design, which came to be used throughout East Asia, was different from the type of still developed in Alexandria, where the distilled material was brought out at the side of the apparatus and passed along a tube to a collecting vessel; such cooling as there might be was achieved only by the air surrounding the external tube. The basic design of the Chinese still is with us today in the modern molecular still, used for the extraction of small quantities of complex compounds, but it may possibly have been a later development than the Alexandrian or Hellenistic type. The latter dates from some time before AD 300 whereas the Chinese goes back probably to the fourth century AD, though as it is impossible to locate the original invention, it could be earlier.

Illustration page 178

What is not in doubt, however, is that distillation was widely practised in seventh-century China during the T'ang. Moreover, the immediate cooling of the distilled material which the Chinese still achieved was important chemically; such a cooling process was not available in the West until four or five hundred years later.

One of the techniques which the Chinese still enabled the proto-chemists to exploit was the distillation of alcohol; a cooling system is imperative for this, otherwise the alcohol is lost. They also practised a special freezing-out process; a method in which water is frozen out first, to leave alcohol behind. This technique, which requires no still, gives a very concentrated form of alcohol which the Chinese seem to have known as early as the second century BC.

For the use of the same technique by Paracelsus in the 16th century, see page 310.

As time passed, so chemical knowledge was amassed. Some minerals were prepared in forms suitable for medicinal use – the sulphides of arsenic were one such example – and this long predated their use in the West, where minerals were not used in medical treatment until the sixteenth century. Industrially the Chinese became adept at the extraction of copper by precipitating it out of solutions, and they also used weak nitric acid to obtain substances which were not soluble under normal conditions. This work brought them into contact with potassium nitrate or saltpetre, which they used in experiments in combination with charcoal and sulphur, a substance which had been mined for a very long time. The experiments may have been, and probably were, done with a view to obtaining an elixir to help in achieving immortality, but whatever the original purpose, they led the Chinese to the discovery of gunpowder. This was used for fireworks and for military purposes, being first adopted in battle in the tenth century during a period when the country was once again split up into warring factions. During the next two hundred years it played a regular part in military actions in China, but it did not become known outside until the thirteenth century, when it was used in Islam; it reached Europe in the fourteenth.

What then can we say, in summary, of Chinese chemistry? In its more mystical or magical aspects it led to unique methods for preserving the dead, and in its more practical aspects it brought about industrial, military and medical advances. Scientifically the Chinese also took some valuable steps forward, because they realized quite early on that chemical reactions could provide not only mixtures but also totally new substances, while their proto-chemists also drew up tables of substances and the way they reacted, thus anticipating the Western idea of chemical affinity which evolved in the seventeenth century. In addition, Chinese chemistry seems to have made much of weighing and measuring the proportions of substances which took part in reactions, and thus they gained some understanding of what today's chemist would call combining weights and proportions, an important aspect of modern research. Moreover, their concern with precision was an attitude which presaged what, in eighteenth century Europe, was to help lead to the birth of modern chemistry.

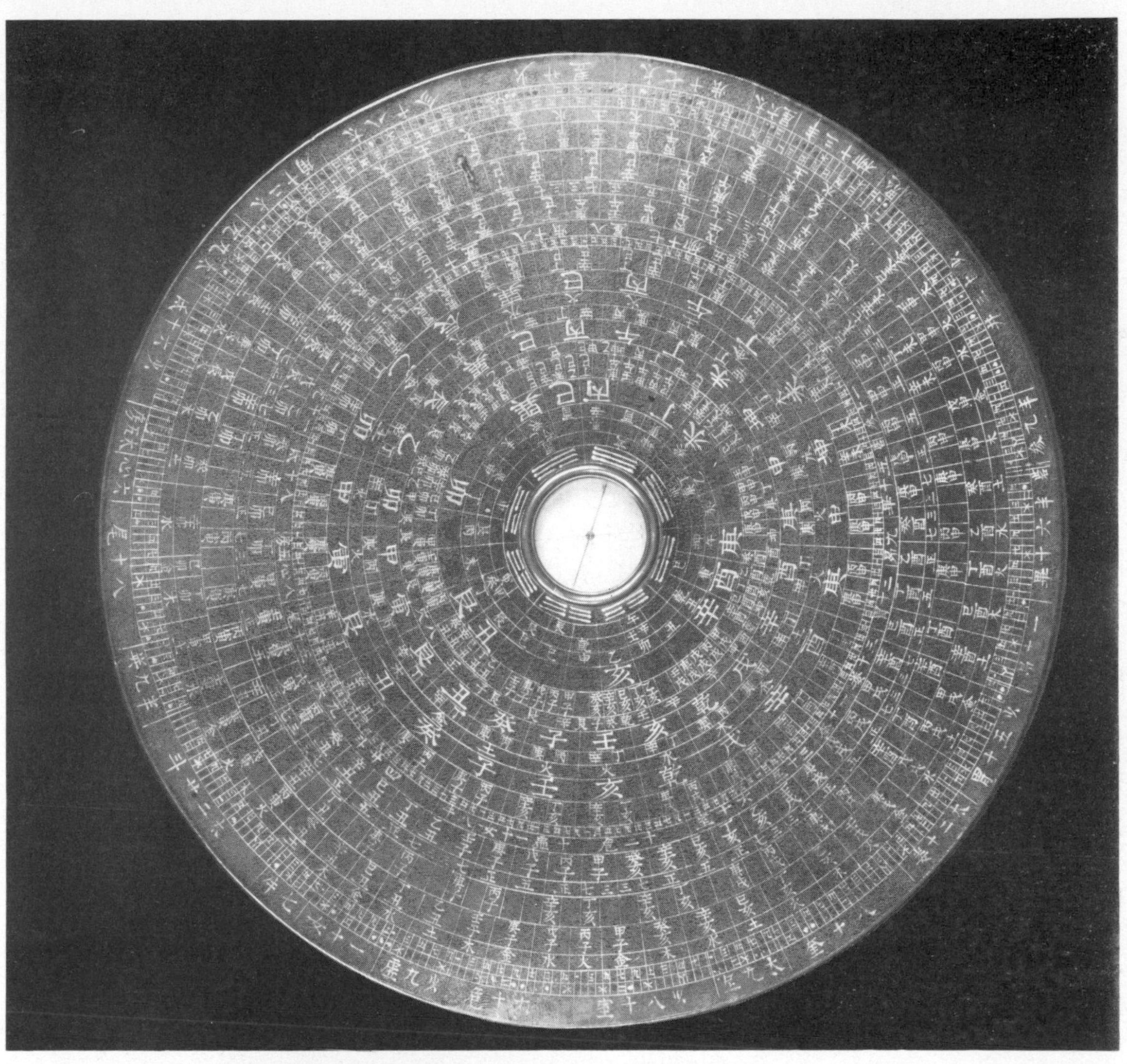

Above A Chinese magnetic compass used for geomancy and the laying out of public buildings. Geomancy was a means of divination by figures or lines. Science Museum, London.

Left Chinese mariner's compass of the 19th century. Science Museum, London.

Right A three-legged bronze ritual wine vessel from the Shang dynasty, made of lead alloyed with antimony. Rijksmuseum, Amsterdam.

Below and below right The refining or cupellation of silver from lead and other impurities. This technique dates from at least the 3rd century BC, and is one of several chemical processes of which the Chinese had a good practical experience very early. From *The Exploitation of the Works of Nature* AD 1637.

Left Early 15th-century painting on silk of two mynah birds, emphasising how, in addition to the accurately observed anatomy and botany, the Chinese showed a great interest in animal behaviour and in relationships between individual phenomena. Fogg Art Museum, Harvard University, Cambridge, Massachusetts. Gift – Dr Denman W. Ross.

Left A painting of two cranes, dating from the early 15th century. Accurate detailed observation is matched by great artistic sensibility. Museum für Ostasiatische Kunst, Cologne.

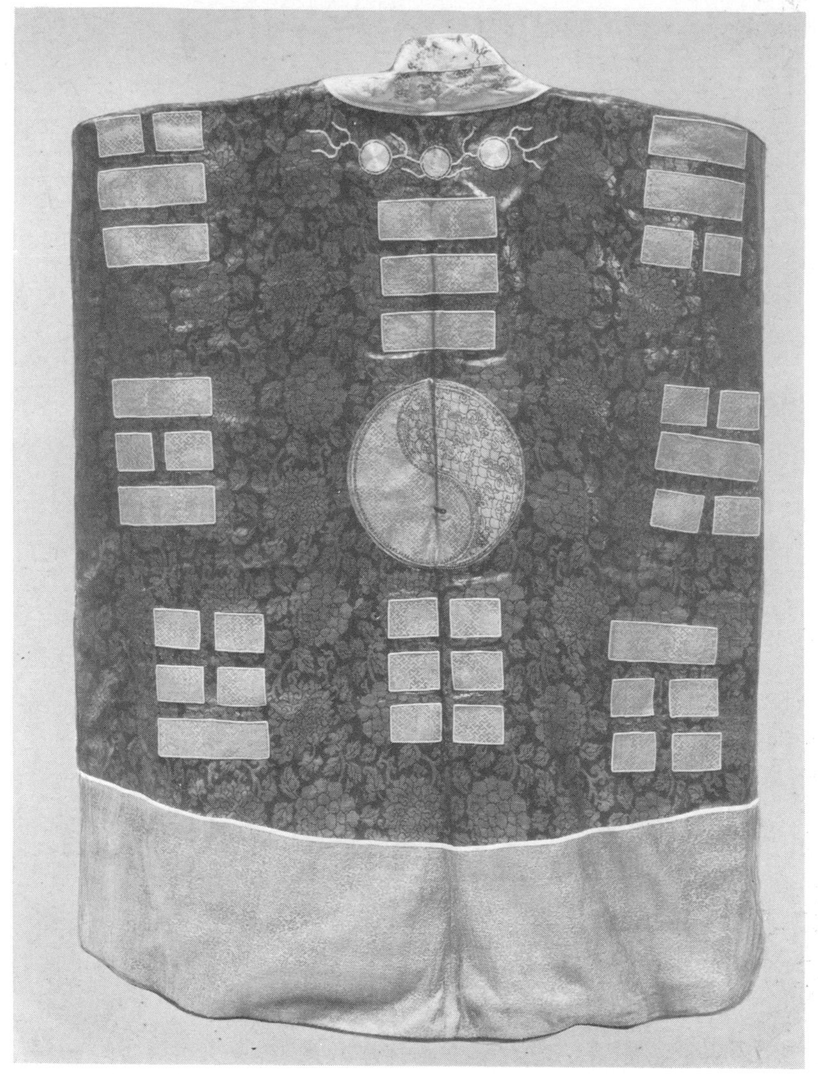

Right Taoist priest robe of the 9th century AD. Its motifs include in the centre the Yin-Yang symbol, the Two Fundamental Forces, and the Eight Mystic Tangrams deriving from the *I Ching* and linked with the Five Elements. Metropolitan Museum of Art, New York. Gift of Joseph J. Asch, 1936.

Right Painting of the Five Holy Mountains. These were associated with the Five Elements, and the Four Mountains of the Cardinal Directions were supposed to have been laid out in an exact square. National Palace Museum, T'aipei.

Biology and Agricultural Science

Much research still needs to be done on the biological sciences in China, but fortunately enough information is already available for something of what they achieved to be appreciated. The Chinese had, of course, a very wide variety of animals in their vast country, among them rare species such as the giant panda, the giant salamander, and the gibbon. From early on they knew enough to undertake selective breeding; thus it was they bred the water-buffalo, which was used for ploughing in the paddy field and, for a higher social milieu, the Pekinese dog and a wide variety of goldfish. The Chinese were also interested in horses; they bred the Mongolian pony, and the Han emperor Han Wu Ti persisted in his conquest of Fergana (now the Uzbek region of the USSR) in order to capture some of the 'blood-sweating' horses of the region. These were fine horses, possibly the forerunners of the famous Arab breeds, but were infected by a parasite which caused a skin haemorrhage; the emperor wanted them for mystical reasons rather than as steeds for his cavalry.

The Chinese wrote widely on animals and descriptions abound, especially in the many encyclopaedias which were compiled from as early as the fourth century BC onwards, and also in a special class of literature which can best be called pharmaceutical natural histories. These began a couple of centuries later and collated and described the growing Chinese knowledge of the natural world. There were also special short tracts on specific animals and, of course, on those creatures particularly associated with the Chinese, like the silkworm. Silkworms were bred from Shang times (around 1500 BC) onwards though it was not until a little later in the Chou dynasty that a really well-organized breeding programme began. It remained a Chinese monopoly until the mid-sixth century AD when its secrets were smuggled out of China and introduced into Europe. However, the silkworm was not the only insect domesticated by the Chinese. In Szechuan they also bred scale insects which flourished on ash trees, because these insects are covered with white waxy scales, while the female also exudes wax. The wax was used medicinally as well as for candles. Other scale insects, like the cochineal, were bred for the colouring material they provided which was useful in dyeing.

Indeed, the Chinese seem to have had a special predilection for insects. They bred crickets for sport, keeping them in cages and releasing them for fighting, rather similar to cock-fighting. And, of course, they also kept bees, though the honey obtained was used mainly for medical purposes. But undoubtedly the most unusual Chinese use of insects was to protect crops from ravages by other pests. A book written in the third century AD describes how in the south the citrus farmers would hang bags of ants on their trees as a protection against spiders, mites and other pests. This is the first known example of biological plant protection and there is no doubt about its effectiveness; the actual species of ant has now been identified and the practice continues even today.

The Chinese study of plants was at least as extensive as their work on animals, and it was continuous. In the Western world there was little development after the Greeks until Renaissance times, but in China there was no such hiatus. Moreover, the Chinese had the advantage of an immensely richer variety of plants than was ever available to Western botanists, from the coniferous forests in the north to the deciduous forests and woodlands further south, as well as areas of desert and scrub. And if this were not enough, Chinese botanists were stimulated in their studies by the administration, for the bureaucrats took more than a passing interest in the use of land, and needed to know the different species of plants and trees and the environments in which they flourished best. In its turn, this policy encouraged the Chinese to make a careful study of soils, a subject which seems to have had no counterpart among the Mediterranean civilizations. Such a study came early in China, for soil conditions were being written about in the fourth century BC, and there are indications that primitive soil science goes back three hundred years before that.

Throughout their history the Chinese have had to face the problems brought about by a large and continually increasing population. This had its affect on agriculture, but it also affected Chinese botany, because it encouraged attempts to try to grow crops in regions outside their normal habitats, and even to breed strains which would be successful under different conditions. Much useful work was done on this from the tenth to fourteenth centuries AD. Another study, brought about by the ever-present danger in some regions of famine, was what is sometimes known as the 'esculentist movement'. Covering the period from the fourteenth to seventeenth centuries, this was a movement which studied action to be taken under famine conditions. It tackled the problem by listing and describing in detail all the wild plants which could be used for food and was, in a very real sense, a forerunner of all the attempts being made today to enlarge the natural sources of mankind's food supplies.

The writings of this esculentist movement were part of a much wider range of botanical literature which the Chinese produced. Botanical material appeared in their encyclopaedias and in the pharmaceutical natural histories already mentioned, but there were also, as in the case of zoology, a number of specific works on particular plants, these far outnumbering those on various animals. This was, of course, partly due to the great need for details of plants which could be used medicinally. When printed words became available in China, as early as the ninth century, the texts multiplied, and three hundred years later, when the art of preparing wood blocks for illustration had matured, they carried very careful and accurate illustrations of plant species.

For the spread of printing in China and the West, see page 271.

Scientific botany needs a system of plant classification, as we shall see when we come to consider the work of Linnaeus in eighteenth-century Europe, but some important steps along this path were taken long before that in China. The Chinese language itself was largely responsible: in primitive times the pictorial symbols lent themselves to representing things like trunks, branches and stems as well as various kinds of leaves and even fruit, and when these symbols developed into the ideographs of written Chinese, this facility was not lost. New signs were invented where necessary, and the Chinese never had to resort to another language, as Western botanists later had to do; indeed, they had no 'dead' language to which to turn. All this meant that botanical names grew up naturally; there were, of course, common-or-garden names as well, but as early as the third century BC, the Chinese were using two-word technical names for plants. This helped them in their ability to classify and to recognize natural families of plants not very different from those that were drawn up in the West.

The greatest of all Chinese botanical writers, Li Shih-Chen (Li Shi-Zhen), was born in 1518 and worked under the Ming dynasty. A man of wide reading and scholarly outlook, in 1583 when he was nearly seventy, he put the final touches to a vast pharmaceutical natural history (the *Great Pharmacopoeia*) in which he presented all his facts accompanied by a careful critical assessment; he gave his own preferred technical name for every plant mentioned, followed by a list of other names, and in this really anticipated modern nomenclature.

The Chinese were also great cultivators of garden flowers, and many of those which now grace Western gardens we owe to China; the rose, the chrysanthemum and the peony are examples. To the Chinese the garden was a place of quiet and contemplation, and it was designed and cultivated with this in mind. Yet it was for their gardening techniques that the Chinese have sometimes been criticized, at least as far as their agriculture is concerned; but if the methods they used for large-scale food production did seem more suitable for market gardening rather than farming, this gave them certain advantages when it came to intensive cultivation. For the Chinese need to produce enough food led them to methods which, from Han times onwards, produced a far greater yield per acre than anything known in the West. This was partly because crops were sown so that each plant could be reached and given individual attention, while in South China, during the later Han, the seed was sown in seed beds, followed in due course by transplanting. Much use was, of course, made of irrigation, and round the irrigation pools for the rice paddies plants such as water chestnuts, beans and cucumbers were grown and duck were raised. In addition mulberry trees were planted for silkworm breeding and in the shade of these the water buffalo tramped around, levelling the ground and keeping it firm as they did so. Nothing was wasted.

The Chinese also practised crop rotation. Unlike the Romans, who left the ground fallow each alternate year, after Han times the Chinese did this only as a last resort. Human as well as animal manure served as a fertilizer, while mud dredged up from the Yangtze was also used. As the population grew, so land reclamation became more intensive and from the ninth to the thirteenth centuries AD terracing was practised and lakes were drained and surrounded by earth walls – a technique later applied in Europe with great success by the Dutch – while the Chinese also used bamboo rafts covered with water-weeds and earth as an additional way of increasing the area that could be cultivated. As well as all this, the Chinese were not averse to importing new crops from surrounding countries; wheat and barley were early imports from western Asia; in the sixth to eighth centuries AD there was cotton from India, and special strains of rice, imported from the ancient kingdom of Champa in the eleventh century, led to double cropping and, with selective breeding, to a strain that would ripen in 60 days or less. As further assistance, the Chinese developed some well-designed agricultural machinery which, because of their market-gardening methods, could be built with the simple technology available. Because European farming techniques demanded much heftier machinery an industrial revolution was necessary before agricultural machinery became really useful in the West.

Medicine

The Chinese developed intensive agriculture to feed a constantly increasing population. But how did they keep them in health? There are aspects of Chinese medical treatment and Chinese ideas about the way the body operates which deserve at least a brief comment, particularly now that acupuncture is accepted by the non-Chinese medical profession, and is occasionally used as a means of treatment in other countries.

The origins of acupuncture go back far into Chinese history. The first documentation dates from about 600 BC and from that time the technique has been continued, though now used in China in conjunction with modern Western medicine, for instance as an anaesthetic during surgery. The technique is now being investigated in the light of modern knowledge, but it is a subtle system and appears to be a practical method of stimulating the natural responses of the body to the onslaught of a disease. In China it was used not only in human medicine but in veterinary treatment as well. The idea of acupuncture is to be found in the Chinese belief in the close link between men and earthly things (the microcosm) and the universe at large (the macrocosm). In part it echoed their view of the entire universe as an organism, resulting in the original acupuncture points bearing a specific relationship to the compass points and arrangement of the heavens; it also reflected the view of a kind of 'vital spirit' or 'air' which moved within living things and whose motion was facilitated by the implantation of needles. Indeed 'vital spirit' played an important part in Chinese medicine, for it was later developed into the theory of a 'pneumatic' circulation within the body, powered by the lungs. This circulation was, however, only one of two types which the Chinese recognized; the second was powered by the heart and concerned the blood, which was thought to carry a vital 'juice'. Both circulatory ideas seem to have been based on a careful examination of the body, its nerve branchings, its veins and its arteries. Such ideas probably originated in the early Han, and they therefore preceded ideas of the circulation of the blood in the West by more than 1600 years.

For an Islamic account of the capillaries, see pages 236–237; for Galen's explanation of the blood system, see pages 248–249; and for William Harvey's demonstration of the circulation of the blood, see pages 395–396.

The Chinese interest in the 'two circulations' led their physicians to set great store by the pulse rate of a patient. But the pulse rate was not all that concerned the Chinese physician; he was also concerned with the patient's general medical state, including the smell of the breath, the cleanliness and colour of the tongue and the beating of the heart. To guide him the physician had the great medical textbook, the *Manual of Corporeal Medicine*, the equivalent both in authority and comprehensiveness of the Greek Hippocratic corpus. In this he would find his treatments well explained, and among them was another unique Chinese method, moxibustion. This word is derived from the burning of wormwood (*Artemisia moxa*) which could be burned in various ways, close to the skin or on it in certain cases, the application taking place at specific points, as in the case of acupuncture. Moxibustion may first have been used in China to relieve rheumatism in the joints and for other pain-relieving purposes, but this was not its only function. Chinese physicians used it also as a form of treatment for certain disorders, and it is now known that it does have some specialized healing powers especially in connection with skin diseases. Moxibustion and acupuncture ran, of course, in harmony with prescriptions of drugs.

As might have been expected in a strongly bureaucratic state such as China, medical practice was a strictly regulated profession. Examinations had to be passed, in general education as well as in medicine, and by the fifth century AD there were already academic positions in medicine, while by the T'ang there was an Imperial Medical College. It seems to have been in China, too, that the idea of hospitals first arose. They came before the Han, but with the arrival of Buddhism their number increased. Originally Taoist and Buddhist foundations, they were taken over by the state from late T'ang times onwards. There were also quarantine regulations, certainly as early as the fourth century AD if not before, and two hundred years later leper colonies had been established.

Conclusions

Like so many other aspects of Chinese science, medicine advanced to a high level in the prevention and treatment of disease. In this the State played its part by imposing effective bureaucratic regulations. Generally, it was this involvment of the State in science in China – in assisting the collection of facts, in organizing and helping to implement the knowledge so gained throughout the enormous country, as well as in using the perceptions of the scientists to improve its revenues and its ability to wage war – that made the history of

Chinese science very different from that of early science in the West. Chinese studies were far more comprehensive, far less sporadic and ideas were more likely to be seen through to completion than the work of the relatively isolated scientists of medieval Christendom.

Despite this, and despite the fact that in so many fields Chinese science reached at a very early date a level of knowledge equal or superior to that of Europe in 1500, there was no 'scientific revolution' in China: the breakthrough into the era of powerful modern science occurred in Europe not in the East. Why should this have been? It is plainly impossible to give a categorical answer to such a question, but it may in part be connected with this same close association of science and the State bureaucracy. In China, the urgent impulse to exploration, to new discovery for its own sake never developed as it did in Renaissance Europe, and there was no aspiration to break the mould of existing orthodoxies as inspired men such at Galileo. And one of the reasons for that must lie with the prevalence of the efficient but traditional bureaucracy of China, its rules and outlook defined by Confucius many centuries before. Even so, it is impossible to look at Chinese science without a sense of debt for its insights, and many of its ideas and methods are being vigorously developed still. Equally, the West is in some cases beginning to turn to these insights, particularly the Chinese holistic attitude to the universe, which is now seen to be less destructive than aspects of the Western approach to Nature.

Chapter Four

Hindu and Indian Science

The history of Hindu science – the science of the subcontinent of India before the advent of Islam with the Mughal dynasty of the sixteenth century – still requires much research. The detailed modern work that has been done for China really needs to be done for the civilization that lay to the south-west for China, but the subject is fraught with difficulties. There are serious problems of dating and problems over the availability of literature; and in a field where so much is obscure, the questions of the culture and scientific contact with other civilizations, or of the independent development of ideas, became very hard to answer. In view of this uncertainty, this chapter must be kept brief, although is should be possible to outline several aspects of science in India which are interesting in their own right, and others that were important in the interchange of ideas between China, Islam and the West before the scientific revolution.

Early History

The history of civilization in India dates back as far as that of the Near East of China, as has been revealed in the excavations of the chief cities of the so-called Indus Valley civilization. These cities primarily Mohenjo-Daro and Harappa, situated in what is now Pakistan – date back to the period 2300–1750 BC, and exhibit a complex town-planning and engineering knowledge, and a well-developed economy. There were public baths and a drainage system, the streets were paved and in many respects their cultures were more advanced than those of the same date in Mesopotamia. As well as an as-yet undeciphered written language, the people of these cities have left a vivid record in a great quantity of seals, which display a rich awareness and knowledge of animal and plant life.

Illustrations pages 197, 198

The origins of the more or less continuous culture of Hindu India can be found in a series of invasions of Iron Age Indo-European cultures from the north in the eighteenth century BC. These cultures eventually invaded the entire sub-continent, and introduced the precursor of the caste system, by dividing society into four rigid groups, the warriors, the priests, the merchants and the labourers. This was also the period in which the Vedic beliefs, which formed the basis of later Hinduism, were developed and written down; since the tenor of these beliefs has some bearing on the future development of Indian science, it may be useful to explain them at this point.

The basis of Hinduism is an animistic religion with a large number of different gods not unlike those of the Egyptians, and the religious activities of the Hindu are devoted to ritual observances which permit every aspect of life to come into tune with these various gods and spirits. For it is accepted that the universe is in danger of being destroyed by chaos, and the sacrifices offered by men could help to strengthen the gods. This idea is linked with a belief in the correspondances between microcosm and macrocosm, that man and earthly things are a reflection of the wider universe and vice versa. This belief has been important in the mingling of physical and spiritual studies as can be seen in the widespread practices of yoga and other forms of asceticism. The Hindu religion has encouraged spirtual enquiry, and has developed lofty concepts of the unity of Nature and of the Divine Principle; there is also the doctrine of *karma* or force generated by a person's actions and influencing both future human actions and the form in which that person's spirit would be reincarnated after death.

The ideas of Hinduism combine a reverence for all forms of life with a certain disdain for the material aspects of existence. These tendencies were developed in an extreme form by the teachings of Siddhartha Gautama (564–483 BC), who is known as the Buddha, or Enlightened One. He taught that the sole aim of life was the achievement of *nirvana*, or the release from the cycle of rebirth and of the suffering that is the inevitable core of life. Nothing in the material world has any reality equal to this fact, and the renunciation of the world, and of individuality are both essential for his followers.

Illustration page 199

Buddhism was no more conducive to enquiry than was Hinduism, though some of its beliefs, such as the idea of cosmic cycles, had an effect on mathematics. But it does have a place in the history of science, for it was to achieve a far wider appeal than Hinduism, and attained great importance throughout south-east Asia, expecially in China Japan and Korea. As a result, it served as a channel through which ideas and knowledge might pass from one culture to another.

The native rulers of India had never been able to achieve a permanent or total unification of their enormous country – the most notable attempt being that of Asoka in the third century BC and based on the Ganges valley – and the subcontinent for a long time was prey to invaders from the north. Although India was not absorbed in the first great expansion of Islam in the eighth century AD, an Islamic dynasty conquered virtually the whole country in the sixteenth and seventeenth centuries. These Islamic rulers brought a much closer contact with Persia; and as their rule declined, the Europeans, and particularly the British, moved in and overlaid many aspects of Western culture and science on to the indigenous traditions.

Astronomy and Mathematics

During the Vedic period, which ran from about the fifteenth century BC to the eleventh century AD, there was some observation of the sky, and the universe was divided into three distinct regions (the Earth, the starry firmament, and heaven) and each of these was then broken down into three further subdivisions. The Sun's path was observed, probably, like the Chinese, by noting which stars were south at midnight and therefore opposite in the sky to the Sun, while the Moon was also observed, and calendars were worked out on the basis of the motions of both bodies.

There seem to have been two ways of reckoning the month, one counting it from New Moon to New Moon, the other from Full Moon to Full Moon again. Then, about 1000 BC, a year of 360 days was used, being divided into 12 months of 27 or 28 days: on this reckoning what must have been observed was the Moon's path against the background of the stars (27.32 days). Of course, 12 × 27 gives a total which is 36 days short of the 360-day year, but if a reckoning is taken from Full to Full (or New to New) Moon, a 30-day month would have been more likely and would have fitted in with the 360-period. Vedic hymns give the first two values (27 and 28), but it would seem that the period was altered as the years passed, for in 100 BC a Vedic text 'concerning the luminaries' refers also to the 'theoretical' 30-day month. Even this would give a calendar which was short of the solar year by 5¼ days, and the Vedic Hindus had two methods open to them, either to add an extra month at regular intervals or to add 5 or 6 days to one or more months. They tried both and in the end settled on the first alternative.

The planets do not seem to have aroused very much interest, but there is one intriguing point about them. Five bright planets are visible to the unaided eye, but the Hindus imagined that there were also two other 'bodies', Rahu and Ketu, which they introduced to account for solar eclipses. Since such eclipses can take place only when the Sun is at a point where its apparent orbit (the ecliptic) crosses the Moon's orbit, Rahu and Ketu were thought, presumably, to lie at these points, although the precise meaning of the terms is difficult to determine since the word *ketu* is also used to refer to unusual phenomena like comets and meteors.

The astronomers of ancient India do not appear to have been very interested in the stars themselves; they did not prepare star catalogues, as did the Greeks and the Chinese, and seem to have looked on the stars only as a guide to solar and lunar motion which, of course, they needed for purposes of calendar determination. Thus the stars that concerned them were those which lay along the ecliptic, and these they divided into 28 *naksatras*, each having a length of about 13°. Nevertheless, in spite of this utilitarian approach, they did recognize some star groupings and named some of the brighter stars: for instance, the Pleiades, Castor and Pollux, Antares, Vega and Spica.

The views so far mentioned were modified by the Jains. These were the followers of the Jaina religion which was founded in the sixth century BC by Vardhamana Mahavira, as a protest against orthodox early Vedic ritual. It aimed at the perfection of man's nature mainly through a monastic and ascetic life, rejected the idea of a creator god and preached non-injury to all living creatures. A dualistic religion, it saw reality as constituted of two entities, and in astronomy its followers thought of two Suns, two Moons and two sets of *naksatras*; a consequence of their belief that the Earth was really a series of concentric rings of land separated by concentric rings of ocean. The innermost circle or *Jambudvipa* was divided into four quarters, with the sacred Mount Meru in the centre; India was the southernmost quarter and the Sun, Moon and stars were believed to take circular paths around Mount Meru as pivot point and move parallel to the Earth. Theoretically the Sun should give daylight to each quarter in succession but since a day lasted for 12 hours, it could only cover two of the quarters each 24 hours. For this reason two Suns, and two Moons and two sets of stars were required.

Lest it be imagined that all ancient Indian astronomy was somewhat vague and imprecise, and that calendar computation was all that concerned their astronomers, it must be emphasized that they did take an interest in applying numerical methods and measurements to the heavens. Late in the fifth century BC when the Persian Achaemenid dynasty controlled the north-west of India, Mesopotamian astronomy and literature flooded in; in the second century AD there was an influx of Greek astrology, and later other Greek (Alexandrian) astronomical material arrived. This allowed tables of planetary positions to be drawn up and Greek planetary theory to be worked on, while attempts were also made to measure the sizes and distances of both Sun and Moon. This more mathematical approach developed strongly from the sixth century AD onwards, and its leading figure seems to have been Aryabhata I, who was born in 476 and worked in the region of Patna. (He is known as Aryabhata I to distinguish him from another astronomer, Aryabhata II, who lived and worked at the close of the tenth and the beginning of the eleventh century). Aryabhata I's attempts to make his measurements seem to have been based on the methods of Hipparchos and were, presumably, derived from the *Almagest*. The values he obtained were not dissimilar, being just a little too large for the Moon, but far too small for the Sun, in fact too small by a factor of almost 28 times, and even some later measurements by Bhaskara II, who was born some six hundred years after Aryabhata, still showed errors; indeed they were not as good as Aryabhata's for the Moon, though their error for the Sun was only 19 times too small. Again, Ptolemy's scheme of planetary motion was adopted during the first few centuries after the *Almagest* was written, though Aryabhata I did put forward the idea of a rotating Earth.

For Hipparchos, see pages 119–121.

The observing instruments used by Hindu astronomers were those used throughout antiquity – the gnomon, the circles and half circles for finding distances of the celestial bodies above the horizon and along the ecliptic, the armillary sphere, and water-clocks – while they adopted the astrolabe and giant instruments built in masonry which they later inherited from Muslim astronomers. In observing techniques, therefore, they were not great innovators; indeed the famous and beautiful observatories with masonry instruments built at Delhi and at Jaipur under the guidance of Jai Singh in the eighteenth century were to a great extent anachronisms. They followed a tradition more than three centuries old and took no account of European celestial measurements using a telescope, measurements which gave greater precision than could be obtained with masonry instruments, however large.

Illustration page 199

One other aspect of Hindu astronomy which deserves at least a brief mention, was their concern with long time cycles. One of these was the *mahayuga*, a period of 4,320,000 years; it is four times 1,080,000 which is the least number of years that contain a whole number of civil days, taking the year to have a length of 365.25874 days. (This is close to the modern figure of 365.25964 days for the year measured from the nearest point of the Earth's orbit to the Sun back to the same point.) Later, Aryabhata I used the value 1,728,000 to give what was known as the 'Golden Age', 1,296,000 years for the 'Silver Age', while half and quarter the Golden Age gave other cycles. The last of these periods, 432,000 years – the 'Iron Age' – was thought to have begun on 17 or 18 February 3102 BC when the planets were last all in conjunction (together in the sky) and the period was believed to be a cycle, at the end of which the planets would all be in conjunction once more.

The Buddhists also used long time cycles, theirs giving periods for the cyclic destruction and rebirth of the universe. They also conceived of a plurality of universes, each constructed on the pattern of the Babylonian universe: an Earth surrounded by an ocean beyond which was a mountain chain supporting the sky. But whether the cycles were Buddhist or Hindu, they involved very large numbers and the writing and handling of these was one of the requirements which the Indian astronomer demanded of mathematics.

Indian mathematics was to a great extent numerical and algebraic, rather like Chinese mathematics, although some geometrical work was also done, mainly on the volumes of various solids. At the start, Indian mathematics was purely practical; weights and measures were regularized in Mohenjo-Daro and presumably all the cities of the Harappan culture had similar, if not the same, standardization. The first written numerals they used were vertical strokes, which were gathered into groups, but these 'counting-rod' numerals, if this is what they were, do not seem to have undergone a systematic change at ten, though counting in tens was certainly adopted by the Vedic Hindus. They had specific words for very large numbers – up to 10^{12} or a million million – multiples larger than this being given more than one word, as we have just done in describing 10^{12}. However Jainism and Buddhism called for larger numbers still, and special terms were devised for 10^{29} and 10^{53}, because these recurred in the cyclic rebirth of the universe.

Again, like the Chinese, the Hindus seemed to have had no difficulty over irrational numbers, and they calculated the square roots of two and three to a number of decimal places; they were, of course, well aware that their values could not be exact. Hindu mathematicians also knew the relationship between the diagonal of a square and its sides; in other words they were familiar with the Pythagorean relationship between the sides of a right-angled triangle. It is claimed, too, that they knew about binomial expressions and the coefficients that arose, and were able to write these down, using long and short syllables, as early as the third century BC. It is, therefore, sometimes said that they knew of Pascal's Triangle as early as this, but no text showing this diagrammatically seems to exist, so the priority for the triangle, if not for recognizing the pattern of coefficients, still rests with the Chinese.

For Chinese irrational numbers, see page 150; for the discovery of Pascal's Triangle, see page 153.

As in Hindu astronomy so in Hindu mathematics there was much progress in the sixth and subsequent centuries. As discussed in the previous chapter the Hindus had a sign for zero around this date, though they probably did not actually invent this. Decimal value notation was introduced and Sanskrit digits assumed a very convenient form which came close to our present way of writing numbers. These Hindu numerals were adopted in Muslim mathematics by al Khwarizmi in the ninth century AD, and three hundred years later entered Europe when Adelard of Bath began translating Arabic works into Latin; it was for this reason that they became known as Arabic numerals, though their origin was really Hindu. A number of notable mathematicians worked during this time, especially Aryabhata I and Brahmagupta, who lived a century after him. Aryabhata calculated the value of π to four decimal places, and prepared tables of chords and arcs of a circle for lines inclined to each other by different degrees. These were useful, particularly in astronomical calculations, though with the later Arabic development of trigonometry they were to be replaced by the even more convenient trigonometrical quantities sines, cosines and tangents, which still do duty today. Aryabhata and his successors also considered the relationships between triangles drawn on a sphere rather than a flat surface and came close to the subject of spherical trigonometry.

For the origin of the concept of zero, see page 149; for the transmission of Hindu numerals to Islam, see page 225; and to Europe, page 321.

Brahmagupta produced a considerable amount of mathematical work and is perhaps the best known of all Hindu mathematicians. His main claims to fame are the rules he devised for finding the volume of a prism and of four-sided figures inscribed within and around circles and, above all, his summing of series. For the latter, he worked out rules for finding the totals of the sums of squared numbers and cubed numbers, and the sum of any number of terms of a simple arithmetical progression, where the first term is 1 (for instance a progression like 1, 2, 3, 4, 5 . . .). Thus it did not matter how many terms one took, Brahmagupta gave a formula to calculate the answer if the first and the last terms were known, and the difference between one term and the next.

Hindu mathematics had a bias towards numbers rather than shapes, towards arithmetic and algebra rather than geometry, and Hindu astronomy was concerned primarily with the practical results of the theoretical work of its astronomers. Tables were required for determining the calendar and for astrology. With such an aim it should not, perhaps, be a matter for surprise that no astounding original discoveries were made. What was done was to digest and modify the astronomical knowledge received from other civilizations outside India, and, in course of time, to hand it on to Islam where it was put to great use.

Chemistry and Physics

The earliest appearance of chemical knowledge in India was in purely practical matters. Pottery was made and fired and pigments were prepared, but the most significant of these early uses of chemical lore was in the smelting of iron, which probably began in India between 1050 and 950 BC. A millennium and a half later, Hindu foundrymen were able to cast some famous iron pillars. One of these, still at Delhi, has a height of well over 7 metres (24 feet) with another half a metre below ground, and a diameter ranging from 40 cm (16 inches) to just over 30 cm (12 inches); it weighs over 6 tonnes, is made of wrought iron and would have been considered impossibly large to cast in Europe until comparatively recent times. But perhaps the most remarkable thing about this and the other pillars of its kind is their lack of decay or of any sign of rust. Why this is so is not certain even now, though it seems likely to be due to the formation of a film of magnetic oxide of iron over the surface, resulting from the original surface treatment it had received.

Illustration page 197

Nothing so far indicates any attempts at chemical research; with iron smelting, pottery, dyeing, glass-making, the manufacture of pigments and all the other practical uses of chemical knowledge there was no underlying theory, no attempt to enquire into the nature of the processes. Interest centred on the product and the product alone. Things seemed to change, however, in the seventh century AD, when the Tantric Buddhists were finding support from all strata of society, for it was then that alchemy entered the scene; this was very late compared with other civilizations and it was clearly an import from outside. Nevertheless Hindu and Buddhist minds gave their own particular slant to alchemy, and the subject had a rapid growth, concentrating on the one hand on male-female symbolism and on the other on the importance of mercury. Hunting for an elixir of immortality does not seem to have appealed to Indian alchemists as it did to the Taoist alchemists in China, although such ideas did appear in Indian medicine, but considerable effort was spent in preparing substances for alleviating the diseases which afflicted humanity. It is interesting to note that although minerals were used extensively in alchemy their use in medicinal preparations had, the Hindus believed, always to be tempered by herbal ingredients which 'digested' the

For Taoist ideas, see pages 135–136.

minerals. Alchemy was accompanied by the alchemical laboratory with its furnaces, retorts and, above all, its stills for the extraction of essences, and it is, perhaps, significant that it was the East Asian still that the Indian alchemists seem to have used rather than the Alexandrian type. This may well be crucial evidence for the origins of Indian alchemy; there had been contact between India and China through the institutions of Buddhism since the first century AD.

Illustration page 262

From the fourth century AD until about the eleventh, Indian science made its greatest headway and it was towards the latter part of this period that Jainist and Buddhist ideas stimulated what was a new concept in Indian science, an atomic theory. A four-element theory coupled with a fifth heavenly essence had long been held – it was an import from the Greeks – but now the formation of the bodies to be found in the natural world was described in an atomic context. The Indian atomic theory postulated that each of the four elements had its own class of atoms, all of which were indivisible and indestructible. Dissimilar atoms could not enter into combination, but similar atoms could do so provided they were in the presence of a third. Two atoms could cause an 'effect' (a *dyad*), while three of these 'effects' could produce an 'effect' of another kind (a *triad*). Thus cause brought about effect, but was absorbed immediately into the effect to which it gave rise, which in turn assumed the role of a cause, so that the sequence continued. The way the first 'effects' (*dyads*) were arranged in a *triad* gave rise, it was thought, to the different qualities of a substance.

In the West, as we know, an atomic theory was proposed by Democritos and Leucippos and was promoted with great flair by Lucretius, but the Indian theory, with its dyads and triads, was both more complex and more subtle. In its description of causal effects it was unique among early atomic ideas, and it attracted Indian thinkers and scientific men until well into the eighteenth century.

For the Greek atomists, see pages 82–84.

Another aspect of Indian physics that must be mentioned is the theory of impetus, which was proposed to account for the continued motion of a body. This was one of the problems which the Greeks tackled and with which they had less than their usual success. Because of his concepts of natural and violent motion, Aristotle had been forced to consider the pressure of the air as the means by which the motion of a body continued once it had been given its initial push. What the Indian view suggested was that when a body first experiences the force that sets it moving, the very application of this force imparts a quality, *vega* or impetus, which causes the body to continue to move in the same way. When a body meets an obstacle it either comes to rest or continues moving, but more slowly; how slow depends on how much the obstacle has neutralized the impetus: complete neutralization results, of course, in a stoppage.

This doctrine of impetus was a notable contribution to thoughts and explanations about the motion of bodies. In the West the Aristotelian doctrine, for all its faults, was held until the fourteenth century AD although, it is true, there were a few brave spirits who dared to question it. In the fourteenth century a theory of impetus developed, but its debt to the Indian theory is not clear. It *is* clear, however, that what the Indians proposed was a forerunner of what was later developed mathematically in the West during the scientific revolution.

For a late-medieval critique of Aristotle's theory of motion, see pages 265–266.

Biological and Medical Sciences

As far as medicine was concerned, the Indus Valley people set great store by hygiene, as excavations at Mohenjo-Daro bear witness, and after this the Vedic physicians developed techniques of dealing with a wide variety of ailments, presumably taking old teaching a few stages further. They believed that diseases often had a hereditary cause, though they also taught that seasonal changes brought about some forms of illness and, more interestingly, that other diseases were due to minute organisms within the body; unfortunately there was no attempt to classify diseases. Treatment involved herbal remedies, tempered sometimes with minerals and with animal parts, while of course ritual, charms and incantations also had a place. In addition, from the second century BC onwards the practice of yoga was also an accepted form of physical healing. Yet in all this it is clear that although there may have been little formal medical systematization, those who practised medicine made many observations, inventing their own technical terms, so that in due course there was enough material to set out the great basic Hindu medical treatise, the *Ayurveda*, which seems to have been compiled approximately 2000 years ago.

In the *Ayurveda* the idea that disease is an imbalance in the body is a central theme, but the book has many modifications made in the light of later experience and is really a compendium of medical practice, a Hindu Hippocratic Corpus. Treatment of disease is a two-way process, the elimination of those ingredients within the body causing lack of balance and their replacement with harmonious ones. The text also shows some understanding of the digestive system – food is taken in, burned (by 'fire in the stomach'), and then transformed into blood, muscle, fat, bone marrow or semen – and perhaps it was this view which gave rise to the idea of treating disease by the removal and replacement of deleterious substances. The text moreover includes details of surgical treatment and many operations were carried out, particularly of the abdomen and of the bladder (for the removal of stones), while removal of cataract of the eye was also performed. Moreover, the Hindu physician knew how blood vessels should be sealed after cutting and even performed cauterization. Indeed, it seems that it was in surgical treatment that Hindu medicine excelled.

There is ample evidence that in the wider field of biology the Vedic Hindus amassed a considerable collection of facts about the form, structure and internal structure of plants. They had names for the root, for shoots, for the stem, leaves, flowers, fruits and branches, and divided plants into three broad groups, trees, herbs and creepers. Herbs, which were of such use medicinally, were further subdivided into seven classes based on their detailed form, structure and other characteristics. And what was done for plants was also applied to the animal kingdom, for Vedic literature contains over 260 named mammals, birds, reptiles, fish and insects, and lists also those species which are poisonous and cause diseases in cattle and man. Domestic animals included the cow, buffalo, horse, sheep, goat, cat and dog and, uniquely to the Indian subcontinent, the elephant.

Illustrations page 200

Later biological knowledge included further studies of plant and animal life, much of the evidence for this being found in poems and dramatic literature as well as the more usual sources such as encyclopaedias and works with a more philosophical bias. The germination of plants was studied, and in the fifth century AD Prasastapada suggested a plant classification based on whether reproduction was sexual or not. Later still, in the thirteenth century, more descriptions of animals were published, stimulated by the royal passion for hunting, while from the early sixteenth century the Mughal emperors, who were fascinated by all kinds of animals both domesticated and wild, caused a vast amount of zoological knowledge to be accumulated and encouraged experiments in breeding. Plants, too, came in for their share of this increased biological awareness.

Conclusion

In the period before the scientific revolution, Hindu science made a number of original contributions that were to be importantly developed in China, in Islam or in Europe. Nevertheless, perhaps because of the prevailing religious tone of the Indian civilization, it never developed into a fully-fledged science, and over the past 200 years science in the Indian subcontinent has had a primarily Western flavour. India has made some important contributions to science in the twentieth century, notably the mathematical work on the theory of numbers by Srinivas Ramanujan (1887–1920), and the work of the physicist Chandrasekhara Raman (1888–1970). Raman's studies on the scattering of light in crystals opened a new insight into the behaviour of their molecules, while he himself did much to foster the growth of modern science within India.

Left The ruins of Mohenjo-Daro, now in Pakistan. This 4000-year old Indus Valley city was carefully laid out, its streets paved and its houses drained.

Left The iron pillar at Delhi still unrusted since the 4th century AD. It is not clear whether this was a freak or represents an unusual degree of metallurgical knowledge. It is, however, far larger than anything similar that could have been produced in the West at such a date.

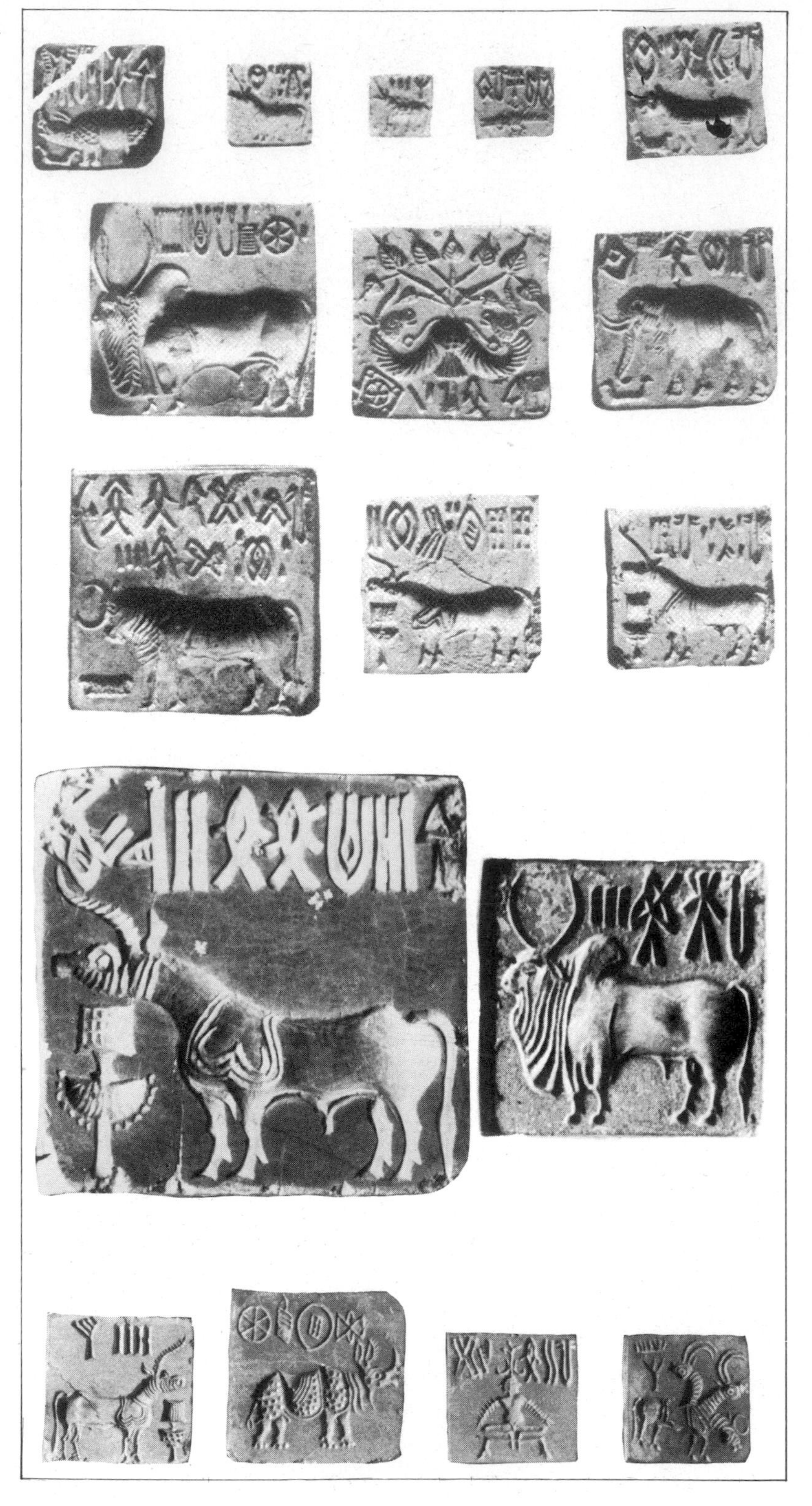

Right A selection of seals from Mohenjo-Daro, showing a range of detailed, naturalistic and realistic representations of animals. (Above, National Museum of Archaeology, Karachi. Below, Museum of Central Asian Antiquities, New Delhi).

Left Statue of the Buddha preaching, from the 2nd century AD. Within the scientific field, Buddhism contributed only some medical knowledge and an atomic theory. Indian Museum, Calcutta.

Below One of the stone quadrants at the observatory of Jai Singh at Jaipur, built between 1728 and 1734. Other instruments there included a large sundial accurate to within 15 seconds, and two 5.4m diameter bowls to help determine the positions of stars and planets.

Chapter Five

Arabian Science

The history of science in Arabia is largely the history of science in Islam. Yet it is not wholly so, for science began some centuries before the time of Muhammad in the land which was to be the cradle of Muslim civilization, and some of its scientific practitioners were men of other creeds and races.

The religion of Islam originated in the Arabian peninsula, and has been permeated by Arabian culture and the Arabic language throughout its history. Nevertheless, it today covers a wide area, including the Middle East, central Asia, north Africa and parts of south-east Asia, while at other times it has flourished in Spain and in the Balkans. Islam has consequently played a crucial role in world history, both as an important civilization in its own right, and as an intermediary between the civilizations of antiquity and those of the early modern world. This twofold aspect of the contribution of Islam is reflected in the history of science.

For the present purposes, we mainly concentrate on the so-called golden age of Islam, from the eighth century AD to the eleventh, when Islamic cultures flourished in Spain, north Africa, Syria and Iran. The end of this golden age was marked by the revival in Christendom that drove the Arabs from Spain in the *reconquistà*, beginning about 1000 and completed by 1500. This was the period at which Islam's most creative scientific work was done.

The early history of Islam

The origins of Islam were interwoven in the history of the Arabian peninsula, which with the decline of the western Roman Empire formed an important crossroads of the trade-routes between the Mediterranean and the Indian Ocean, and between north Africa and western Asia. Its society comprised a number of disunited nomadic tribes, and a series of trading bases, often with a cosmopolitan flavour. Among these were Mecca, already a centre of pilgrimage, and Medina.

The Islamic religion emerged when Muhammad, the son of a merchant family, saw a vision of the angel Gabriel, assembled his revelations in the Qu'ran and accepted the role of Prophet of the one true god, Allah. For a variety of reasons, his teachings were unacceptable to the ruling elite of Mecca, and in AD 622 he left that city for Medina. This journey, known as the *hijrah* or *hegira*, is traditionally considered to mark the beginning of the Islamic era. In Medina

Opposite above Four deer carved on the wall of a Buddhist cave-temple at Ajanta, in Maharushta state, India, excavated between the 1st and 7th centuries AD.

Opposite below A wall painting of two bulls fighting, from Ajanta.

he made many converts, and when he marched back to Mecca in 630, he took it virtually without bloodshed. Muhammad died two years after this, but his followers built on the momentum of his teachings, and quickly achieved the formidable task of uniting the disparate tribes of the Arabian peninsula. They then undertook an extraordinary expansion into Syria, then through western Asia and north Africa, contriving to appeal to those subject peoples exhausted from the long struggles between the Byzantine Empire and the Sassanid dynasty in Persia. This expansion was fired by Muhammad's idea of a *jihad* or holy war; and by AD 750 the Muslims controlled a continuous empire from Spain to the Indus.

Despite their missionary zeal and their often fiercely puritanical teachings, the Islamic conquerors were prepared to accept a degree of tolerance for the native cultures of their new conquests, and the courts that they established consequently saw a remarkable fusion of the indigenous arts and knowledge and the Arabian styles of Islam. In this way they inherited, among other things, the science of the Greeks from many Hellenistic cities, as well as the culture of Sassanid Persia. A golden age of Islamic culture emerged after the age of expansion was over, though another flowering was later to occur in Istanbul after the Ottoman Turks had taken over the old Byzantine capital in 1453.

Islam was never an entirely unified entity, either religiously or politically. At an early date the creed had divided into two main sects, known as the Sunnis and Shi'ites, initially in a dispute over who should succeed Muhammed as *caliph*, or leader of the faith. This division later acquired a theological and a geographical dimension, as the Shi'ites became primarily confined to Iran, where they were the inheritors of the old Persian culture. There were further divisions, after the Umayyad dynasty which had led the great expansion of Islam beyond Arabia was usurped by the Abbasids, who built a new capital at Baghdad. Meanwhile, an Umayyad dynasty still flourished in Spain, and a Fatimid dynasty ruled in Egypt. Later, the Mongol invaders from central Asia in the thirteenth century also espoused Islam, as did the Ottoman Turks, who rose to prominence shortly after.

Despite these differences, Islam maintained an essential unity in its cultures, preserved partly by the religion itself, and partly by the consciousness of the shared Arabian origins of the movement, which were preserved in the use of the Arabic language and script throughout the Islamic world. This language was to prove a highly flexible and appropriate medium for the expression of scientific concepts.

The Earliest Science in Arabia

There are two aspects of Islamic science, on the one hand the scientific ideas which were imported from outside and, on the other, the contribution of the Arabs themselves to the sum of scientific knowledge. This latter, the Arabs' own contribution, has often been neglected or skimmed over in favour of the more exciting advances which were to come in Western Europe from the sixteenth century onwards. Too often science in Arabia has been seen as nothing more than a holding operation. The area has been viewed as a giant storehouse for previously discovered scientific results, keeping them until they could be passed on for use in the West. But this is, of course, a travesty of the truth. Certainly the Arabs did inherit Greek science – and some Indian and Chinese science too, for that matter – and later passed it on to the West. But this is far from being all they did. They interpreted what they inherited, commented on it and added valuable analyses of what it contained, and, above all, they made many original contributions of their own. Indeed, Arabia produced some original scientific minds; it nurtured them and encouraged them to make their own individual contributions. So when we think of the West's indebtedness to Arabic culture, it is important to appreciate both aspects, the original work as well as the transmitted ideas of an earlier age.

Since Ptolemy's work at Alexandria in the mid-second century AD Greek science had largely stagnated. Only two mathematicians of note seem to have flourished at Alexandria: Diophantus, about a century after Ptolemy, and Pappus, a hundred years later still. Diophantus had written his *Arithmetica*, a computational arithmetic designed to solve practical problems in which great ingenuity was displayed in solving equations of all kinds, including those where the unknowns exceeded the equations devised for their solution. Pappus for his part wrote a handbook on Greek geometry and arithmetic, adding some new material such as a method of dividing an angle into three equal parts and discussing geometrical solutions of problems connected with curves – work that was to be taken up in the West some thirteen hundred years later. But these were before the very severe destruction which the library and museum suffered in the second decade of the fifth century. Although it was a Christian mob which destroyed the Alexandrian institute, it was partly due to Christian scholars that the losses were not more severe and that something of Greek learning was salvaged. This happened because of two major factors. One was that scholars at the library had long been aware of the possibility of destruction – the attack by Queen Zenobia in the middle of the third century had made evident the library's vulnerable position, and the mob incited by Bishop Cyril to violence against pagan learning confirmed it – so that scholars of all creeds and outlooks began to move away, taking material or copies of material with them.

For the destruction of the library of Alexandria, see page 124.

The other factor was the establishment of a college at Edessa, a Graeco-Roman city in the south of Turkey, not so very far north of the Syrian border. Set up as a theological college primarily for Syrians, it had become a haven for Nestorian Christians. Loyal to the teachings of Nestorius, the fifth-century Patriarch of Constantinople (Istanbul), who affirmed the separateness of Christ's human and divine natures, they were condemned as heretics by the orthodox at the Council of Ephesus. However, the Christian Church in Iran accepted the doctrine and rejected the Ephesian condemnation, so the Nestorians found safety at Edessa. The importance of the Nestorians to us is not only that they were among the scholars who helped Greek scientific teaching to survive, but also that they gave it greater currency by translating a number of works into Syriac. Later they helped translate these versions into Arabic for, when Islam was established, their reputations as scholars still held.

When Edessa was shut down in 489 some Nestorians moved to the great Iranian intellectual centre at Gondeshapur. This had been established by the Sassanian monarch Shapur II for the Greek scholar Theodorus who had written a medical treatise in Pahlavi (an Iranian writing system devised in the second century BC), but it was the influx of Nestorians that gave it what was to be an enviable reputation. Other Christians from the Eastern churches also made contributions: in the sixth century Sergius, a Monophysite priest and technically another 'heretic' (the Monophysites held that Christ had only one nature), translated into Syriac the philosophical works of Aristotle as well as those of the philosopher Porphyry (third century AD), the works of Galen and various treatises on agriculture. Then, in the seventh century Severus, a Syrian bishop, wrote praising Indian astronomy and remarked on how excellent their calculations were because they used nine different signs for the first nine digits; in other words, because they used Hindu numerals. This was indeed important; it marked the introduction into Arabia of what later came to be called Arabic numerals – the type of numerals we use today – although their acceptance into Arabian mathematics and astronomy by Al-Khwarizmi had to wait for another two centuries.

For the development of Hindu numerals, see page 192.

Severus also explained how eclipses of the Moon were caused by its moving into the Earth's shadow, and in another Syriac treatise, possibly by a different author, is a description of a Greek astrolabe. A century later another Syrian bishop, George, wrote about the calendar. But this was the century that saw the advent of Muhammad and the start of the Islamic era, a period which began not with cultural advance but with religious revelation, conversion and, under the first Muslim dynasty, the Umayyads, the beginning of a holy war that swallowed up Syria and the whole of the Middle East, as well as extending still further east and west. But, of course, such expansion could not continue for ever, and when the Abbasids came to power, there was a return to the cultivation of the arts of peace.

The chief architect of this revival was the second Abbasid caliph al Mansur, who destroyed the last vestiges of Umayyad resistance and is usually looked upon as the real founder of the Abbasid dynasty. Described as a tall, lean man with a sparse beard, it was he who, in 762, laid the foundations for the new capital at Baghdad. His successors included the famous Harun al-Rashid. During this formative period of the caliphate fresh mathematical astronomical texts arrived from India; they too contained the Hindu numerals and were to exert a strong influence when later translated into Arabic. It was Harun al-Rashid's second son, al-Ma'mun, who saw to this.

Al-Ma'mun came to power in 813 and was a judicious ruler. After some troubles at the beginning of his reign, he settled at Baghdad and showed favour to the Mu'tazilite movement, a group of Muslim supporters who believed the Faith could be supported by rational arguments and based their modes of reasoning on the methods used first by the philosophers of Greece and Alexandria. To further the Mu'tazilite cause there was a need for translations of more Greek and Alexandrian works, so al-Ma'mun set up in Baghdad the Bayt al-Hikmah or 'House of Wisdom' where a whole body of translators, many of them Christians, were recruited. Where manuscripts of more important works did not exist, al-Ma'mun had them imported from Byzantium. He also set up observatories so that Muslim scholars could verify the astronomical knowledge contained in the ancient texts. With al-Ma'mun we come, then, to the beginning of that cultural renaissance in Arabia which was later to prove so important to the West and thus to the development of the modern scientific outlook.

A wide range of work besides translations was done at the House of Wisdom, and one of the earliest and most influential of the broad-minded scholars to labour there was Abu Yusuf al-Kindi, sometimes referred to as the 'first Arabic philosopher'. Born about 801 and descended from a noble branch of the Kinda tribe of the Yemen, he came to the attention of al-Ma'mun probably because of the latter's abiding concern to have good translations of Greek and Hellenistic works, since al-Kindi wanted the study of philosophy to be improved, and advocated complete access to the accumulated scientific wisdom of the ancient world. Al-Kindi did not confine himself to pure philosophy, and took a great interest in various branches of science. He studied optics, though rather from the geometrical aspect – tracing the paths of light beams – than the experimental, and seems to have accepted the old Greek views although he did emphasize the fact of linear propagation – that light always travels in straight lines. He also made studies in geography, geology, meteorology, astronomy and astrology, as well as investigating clocks and astronomical instruments, and even took a great interest in sword-making. He wrote about medicines and their composition, remarking that their efficacy was proportional to their component parts and, an interesting point, their effect could not be reduced to the behaviour of only one of the constituents to the exclusion of the others. Was this an appreciation of 'side effects'? At all events al-Kindi appreciated that qualitative effects were due to the quantities of the constituents used, and his teaching on this was to have a great effect in Western Europe during the Middle Ages. Of course, al-Kindi could not pursue in depth all the subjects that interested him, but he encouraged others and, though no expert in Greek, he knew enough to help develop an Arabic terminology of Greek technical words. He was a believer in the infinite extent of the universe and in the study of mathematics – which to him seems primarily to have meant geometry – as a necessary preliminary to acquiring any other knowledge.

Al-Kindi's greatest forte lay in the field of pure philosophy. Having read Aristotle and Plato, he also studied Plotinus, a philosopher of the third century AD and the founder of what has become known as 'Neoplatonism'. This was the final form of Greek pagan philosophy and was a somewhat one-sided development of ideas to be found in Plato, spiced with elements from Aristotle and from the Stoics (who had held the universe to be both governed by fate and yet completely rational), and tempered with a peppering of gnosticism (an early form of theosophy which considered matter as evil). Plotinus himself seems to have been opposed to the full gnostic creed, but certainly elements of this outlook were included in his philosophy. Neoplatonism exerted an influence not only in the Arabic world, but it may be as well to state its main tenets here. The philosophy – or that form of it which Plotinus devised – taught that there is a hierarchy of spheres of being, the lowest of which exists in time and space and can be perceived by the senses. The other spheres, which are derived one from another, all lie outside space and time. Each establishes its own reality by turning back to its superior in contemplative desire, this desire being implanted in it by its superior. Thus the Neoplatonic universe is characterized by a double movement – an outgoing and a return. There is also a decreasing degree of unity as one moves from the highest sphere to the lower ones, for each sphere is an image, on a lower level, of the one above, and as one goes down the scale a greater degree of multiplicity, of separateness and increasing limitation, becomes evident; at the lowest level the separation into atoms of our spatio-temporal world appears. The highest sphere of being, from which everything else is derived, is itself derived from the ultimate principle. This transcends everything else; it is 'beyond being', and can be called 'the Ultimate Good'. It is absolutely simple and is devoid of any specific traits. It can be known only when it raises the mind to union with itself, and cannot be imagined or described.

For the role of Neoplatonism in Hermetic thought during the Renaissance, see pages 273–277.

The Neoplatonic outlook had an immediate appeal, and al-Kindi welded it firmly and surely to Islamic ideas. In the House of Wisdom and in Baghdad, where the learning of the Greeks was now being revived, a marriage of philosophies was badly needed if there was to be no opposition from orthodox Muslims to the assimilation of pagan knowledge. Some welding together of the diverse outlooks had to be made, and this al-Kindi achieved. As tutor to the son of the succeeding caliph al-Mu'tasim his influence was immense in court circles, and though at the end of his life petty jealousies ousted him from his pre-eminent position, by then his work was done. He may only have been what one historian has described as 'an innovator with an archaic streak' but he managed to help set going the great intellectual movement that was to blossom into Islamic science.

Astronomy

With the establishment of observatories as well as the House of Wisdom by al-Ma'mun, and the influx of Greek astronomical texts, the stage was set for the serious development of an indigenous Islamic astronomy which was closely coupled with Islamic advances in mathematics. Indeed, as soon as the House of Wisdom was founded astronomers were at work. Some like Habash al-Hasib and al-'Abbas al-Jawhari, were more concerned with the mathematical side, although al-Jawhari did do some observing. Among the most important of the early ninth-century astronomers was Abu Ja'far Muhammad ibn Musa al-Khwarizmi. His main contributions were to mathematics, but he did write on astronomy and was well acquainted with Ptolemy's *Almagest*. He prepared a set of *zij* (that is astronomical tables) of future planetary and stellar positions called the *Zij al-sindhind* since they were based on some Hindu tables or *siddhanta* that arrived at Baghdad. Influenced by Ptolemy's original tables, they are the first Islamic astronomical work to survive in anything like its entirety. Al-Khwarizmi wrote also on the Greek astrolabe, an instrument that Islam was to make its own. Built of brass, it was flat and circular. In the centre was a disc engraved with indicator lines whose positions were worked out mathematically. This disc rotated in a holder, one side of which had a fret of thin pieces of brass ending in points that represented the stars. By rotating the inner disc it was possible to find rising and setting times for the celestial bodies, and determine the occurrence of other astronomical events. In this respect the astrolabe was a graphical computer. The other face of the holder carried scales and a sighting arm. Thus by its aid the user could determine the altitude of a celestial body and, by holding the instrument horizontally, the body's azimuth (i.e. its position along the horizon measured from true north). Such a way of measuring celestial positions, using altitude and azimuth, was a specifically Arabian system and, indeed, our word 'azimuth' is itself a word of Arabic origin.

Illustration page 209

Another of the early Baghdad astronomers was Abu-al-'Abbas al-Farghani, who also wrote on the astrolabe, this time a substantial book which is an improvement on al-Khwarizmi's, giving not only the mathematical theory behind the instrument but also correcting faulty geometrical constructions for the central disc which were current at the time. Al-Farghani also wrote a more general book on astronomy, a critical commentary on al-Khwarizmi's *zij*, and a commentary on the *Almagest*. This last was most important since it gave, in Arabic, a thorough account of Ptolemaic astronomy in a clear and well-organized text which enjoyed considerable popularity.

If al-Farghani was more a theoretician than a practical observer, so too was another of the House of Wisdom's astronomer-mathematicians, the Mesopotamian Arab Thabit ibn Qurra. He was more mathematician than astronomer, and we shall come across him in that context, but he also wrote on astronomy as well as displaying a sound knowledge of Greek, Syriac and Arabic. In his young days he was a money-changer in Harran, the Mesopotamian town in which he was born, but ibn Shakir, an Islamic mathematician passing through the city, was so impressed with his abilities that he persuaded ibn Qurra to go to Baghdad. There ibn Qurra wrote about the sundial, and made a careful study of the Sun's apparent motion across the sky, noting particularly its acceleration and deceleration at different times of the year. He also studied the Moon's motion across the background of the stars and came to the conclusion that there was a hitherto undiscovered movement in the Sun's path. This affected both the precession of the equinoxes and the angle between the Sun's path (the ecliptic) and the celestial equator. The changes ibn Qurra thought he had discovered could be described as saying that the equinoxes made a small circle in the sky once every 4,000 years; thus the ecliptic appeared to tremble, and it was for this that the effect became known as 'trepidation'. With the trepidation circle having a diameter of 8°, this new factor substantially affected all subsequent astronomical tables, not only in Arabia but in Western Christendom during the Middle Ages. Only in the late sixteenth century when fresh and far more precise astronomical observations were being made in Denmark by Tycho Brahe was ibn Qurra's trembling of the ecliptic shown to be a chimera.

For Tycho Brahe, see pages 334–336.

Of all the early Arabian astronomers the greatest and justly the most famous was without doubt Abu 'Abdallah al-Battani. He too seems to have been born in Harran, and came from a family of Sabians, that is believers in an ancient religion which contained much of the old Mesopotamian astral theology and star lore. Thabit ibn Qurra was also of this religious persuasion, and it says something for the tolerance of the Muslim rulers that this was one of the many religions which they permitted to co-exist within Islam; Sabianism died out, however, during the eleventh century. The claim that al-Battani came from a noble or even royal household has not been confirmed by recent research.

For Babylonian astronomy and astrology, see pages 42–44.

Al-Battani made astronomical observations from al-Raqqa on the north bank of the Euphrates some 160 km (100 miles) east of present-day Aleppo. He made observations of eclipses and other celestial phenomena, but his reputation really rests on his *Kitab al-Zij* or Book of Astronomical Tables. It was prepared, as he said in his preface, because the errors and discrepancies he found in other *zij* led him to try to improve theories about celestial motions and the inferences drawn from them by basing them on new observations, just as Ptolemy had done using the observations of Hipparchos. To this end he constructed a sundial, a novel type of armillary sphere known sometimes as 'the egg', a large quadrant fixed to a wall (a 'mural' quadrant) and a device which came later to be called a 'triquetum' – an arrangement of straight arms acting like a quadrant without its

Left A fine example of the astrolabe, designed in Iran in the 18th century. Developed by Arabian astronomers, the astrolabe could be used to calculate the altitude and azimuth of Sun, Moon, stars and planets, to calculate the time, and to measure distances and heights. Museum of the History of Science, Oxford.

Below left Astronomers, using various instruments for measuring the position of celestial bodies, at the observatory founded in 1575 at Istanbul. The observatory was closed down soon afterwards because of the unpopularity of its astrological predictions. Istanbul Üniversitesi Kütüphanesi.

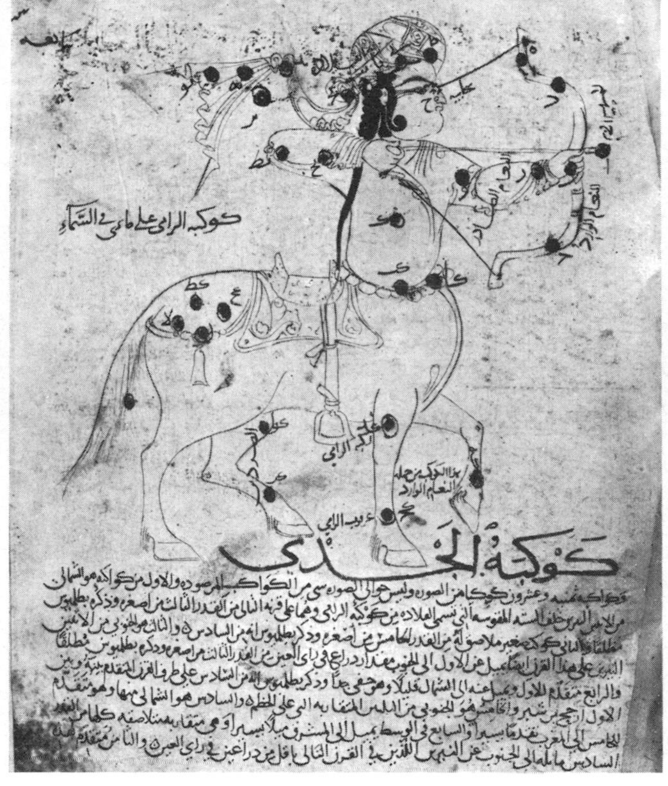

Below The constellation Sagittarius from al-Sufi's *Book on the Constellations of the Fixed Stars*; 10th century. British Library, London.

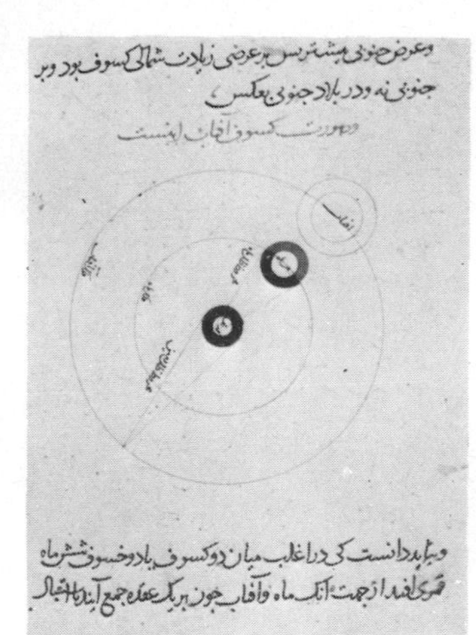

Above Diagram in explanation of a solar eclipse by al-Tusi, from a 14th-century manuscript in the Central Library, University of Teheran.

Right A giant sextant with a radius of 40m, hewn out of rock at the observatory of Ulugh Beg at Samarkand, built in the 1420s.

heavy scale and derived from an instrument used by Ptolemy. Indeed, he was so concerned with practical observational matters that sometimes his explanations of planetary theory are less than perfect. But his corrections of errors of observation in the *Almagest* were very valuable, particularly those on planetary motion, while al-Battani also pointed out that Ptolemy had been wrong to suppose that the angle between the ecliptic and celestial equator, the obliquity of the ecliptic, always remained the same, and that the point in space where the Sun appeared at its most distant – the solar apogee – was fixed. And of course these were vital points for the future of precision astronomy. But al-Battani did more than point out errors; he made observations of his own and obtained better values for these important constants and, in writing them up, not only gave his results but explained very clearly how he achieved them.

Although so good on planetary motions and movements of the Sun and Moon, and more successful than anyone before him in measuring their apparently changing diameters throughout the year, al-Battani was content to accept Ptolemy's positions for the fixed stars, merely adding a correction to bring them up to date. Perhaps this is not surprising because it was on the moving celestial bodies that early astronomers primarily concentrated their attention, and here al-Battani's contributions were so valuable, not only in themselves but also in the mathematical techniques he used for deriving them, that they were taken up by later generations. In Western Europe they were to be quoted from the fifteenth to the seventeenth centuries by such astronomical giants as Copernicus, Kepler, Tycho Brahe, and Galileo. And nearer his own times the famous Hispano-Jewish scholar Moses Maimonides (1135–1204), the intellectual leader of medieval Judaism and physician to the Egyptian sultan Saladin, closely followed al-Battani when it came to astronomical matters.

Besides straightforward astronomy, ninth-century Arabia also saw some serious work on astrology; indeed it was at this time that its greatest Arabic proponent al-Balkhi Abu Ma'shar laboured in Baghdad. Known later to the West as Albumasar, Abu Ma'shar was born at Balkh, an ancient city in the far western regions of Iran, in 787, dying in Iraq almost ninety years later. Balkh was an outpost of Hellenistic civilization which had become a multiracial city where Chinese, Indians, Graeco-Scythians and Syrians intermingled with the Iranian population, and later communities of Buddhists, Hindus, Jews, Manichaeans, Nestorians and Zoroastrians co-existed. When the Abbasids had come to power, Balkh and the surrounding Khurasan area provided the new caliphs with their army, their generals, and a host of intellectuals who played no small part in translating and commenting on Greek works at the House of Wisdom in Baghdad. Abu Ma'shar was a third-generation member of this intellectual elite, and throughout his life had a sense of Iranian intellectual superiority over colleagues from other parts of Arabia.

Abu Ma'shar's intellectual traditionalism led him to quarrel with

For Plato's views on astronomy, see pages 97–98; for Aristotle's, see pages 104–105.

al-Kindi and to formulate his own ideas, which were coloured by the ideas then current in Baghdad at the time as well as by his traditional beliefs. They owed something to Neoplatonism and to Aristotle. Abu Ma'shar believed in an outermost sphere of divine light and eight ethereal celestial spheres, with our own (ninth) sublunary sphere in the centre. But Abu Ma'shar's primary interest was astrology, a subject in which his cosmological picture nevertheless played a part. He taught that all knowledge came from a divine source and that every science contains a modicum of revelation; the three spheres of influence – the divine, the ethereal and the sublunar – all interacted and made astrology a real science. He himself prepared horoscopes and gained an immense reputation, both in his own time as well as later, and was described as 'the teacher of the people of Islam concerning the influences of the stars'.

Astrology seemed logical enough in the ninth century. The Earth was the centre of all things, and with Abu Ma'shar's hypothesis of divine revelation, the influence of the stars did not seem far-fetched. The surprising thing is rather that so few of the Arabians concerned with the heavens spent much time on astrology. They seem to have been concerned more with perfecting what the Greeks had begun rather than applying Greek knowledge to the art of fortune-telling; their mathematical abilities were bent to matters of scientific fact rather than prognostication.

The last great representative of the astronomical and mathematical school that had grown up after the founding of Baghdad around the beginning of the ninth century was the Iranian Abu'l Wafa' al-Buzjani, who was born in the city of Buzjan in 940. Primarily a mathematician, he followed tradition by writing a complete textbook on astronomy from the mathematical point of view, his mathematics making his solution of the astronomical problems particularly elegant and explicit. Another Iranian based in Baghdad between AD 970 and 1000 was Abu Sahl al-Quhi; he carried out observations of summer and winter solstices which were made in Iran at Shiraz, and also observed the movements of the Moon and planets. Al-Quhi seems to have been noted for his observing skill, since his work displayed the greatest precision attainable at the close of the tenth century.

It was around this time that some objections to Ptolemy's ingenious use of the equant to explain irregular planetary motion in regular terms were raised by the great Islamic physicist al-Haytham. Al-Haytham believed Ptolemy had missed some basic aspect of the movement of the planets and went on to show that, in his opinion, Ptolemy's theory of the Moon's motion was impossible from a practical point of view. In one sense, of course, al-Haytham was right; Ptolemy's 'errors' were due to assuming regular motion in a circle as the basis for all celestial movements, but unfortunately no Muslim astronomer ever abandoned it. A totally new outlook had to wait until the seventeenth century and then it was in Europe, not in Islam, that this was done.

So far the Arabian astronomers had concentrated on planetary motions, but in the late tenth century there was one notable exception, the Iranian Abu'l-Husayn al-Sufi, of whose life and career little is known. He was renowned for his observations and descriptions of the stars themselves, and his *Book of the Constellations of the Fixed Stars* became a classic of Islamic astronomy, and later found its way to the West, where its author's name was translated as Azophi. What al-Sufi did was to make the first really critical revision of Ptolemy's star catalogue, adding evidence from his own careful observations. Moreover, he set out his results very clearly, constellation by constellation, discussing the stars in each – their positions, their magnitudes (brightness) and their colour. Two drawings of each constellation were provided, one as seen from the outside of a celestial globe and one from inside, i.e. as seen in the sky. There was also a table of all stars with their positions and magnitudes, and identified, too, by their Arabic names, some of which we still use today (for example, Aldebaran, Altair, Betelgeuse and Rigel). There is no doubt that al-Sufi's book met an important need, and it was put to good use by generations of Arabian and Western astronomers.

Illustration page 209

Al-Sufi also constructed instruments, as might be expected from so practical an observer, and he wrote a book on the astrolabe as well as one on astrology. Astrology was also accepted by his contemporary, Abu Rayhan al-Biruni who was a native of Armenia. Born in 973 he took to science very early and at seventeen had graduated a ring dial with half-degree divisions with which he observed meridian altitudes of the Sun. When civil war broke out in 995 he fled abroad, to return home two years later when peace reigned again. While away he stayed at Rayy and visited the mountain observatory where there was a large sextant, but on his return he became 'compelled to participate in worldly affairs', and was at the court of successive monarchs. Nevertheless, he managed to do some observing, co-operating with al-Buzanji, though the latter was some 30 years his senior. Al-Biruni was primarily concerned with astronomical geography, using eclipses to determine longitudes of places on the Earth, and making astronomical observations to determine the distance of a degree of the meridian. Yet astronomy was not al-Biruni's sole interest; he wrote on mathematics and geography proper, on optics, on drugs, on gems and on astrology. Indeed his prodigious literary output amounted to something like 13,000 pages of highly technical material. A staunch Muslim, he is said to have invented an instrument for determining times of prayer, but since it used Byzantine months, a religious legalist abused him as an infidel. 'The Byzantines also eat food', al-Biruni is supposed to have replied; 'Then do not imitate them in this!'

At the close of the tenth century we also find at work Abu'l-Hasan ibn Yunus, one of the greatest of all Islamic medieval astronomers. Born in Egypt, ibn Yunus came from a respected family and as a young man saw the Islamic Fatimid conquest of Egypt and, in 969, the foundation of Cairo. From 977 to 1003 ibn Yunus made astronomical observations under the caliphs using, among other instruments, a large copper astrolabe nearly 1.4 metres in diameter. There was also a giant armillary sphere with nine rings and weighing nearly a tonne, large enough 'for a man to ride through on horseback', though this may have been built soon after ibn Yunus had died.

Ibn Yunus is remembered primarily for his *zij*, *The Large Astronomical Tables of al-Hakim*, named after the caliph. This contained no less than 81 chapters and was much larger than al-Battani's *zij*, with twice as many tables. It differed from all previous *zij* by beginning with a list of observations, either of ibn Yunus himself or his predecessors, and gave details of all kinds of astronomical phenomena from eclipses to planetary conjunctions. Its mathematics and mathematical tables were good. The *zij* also contained some astrology, for ibn Yunus was famous as an astrologer as well as an astronomer, as might be expected considering the predilection of the caliph al-Hakim for the subject. It is said that ibn Yunus predicted his own death seven days before it happened, proceeded to clear up his personal business and then locked himself in his house, washed the ink off his manuscripts, and recited the Qu'ran until he died – on the day he had predicted! True or false, the story clearly underlines his reputation as a man whose predictions were always reliable. Astrology notwithstanding, the *zij* of ibn Yunus was a very valuable contribution to Arabian astronomy, its reputation no doubt enhanced by the general dearth, during the eleventh century, of any successors capable of preparing so vast a work.

Ibn Yunus also prepared a second major work. This was concerned with the astronomical determination of times of prayer. An astronomical guide was needed because the Prophet had laid down that evening prayers – the first prayers of the day – must be made between sunset and nightfall, morning prayers between daybreak and sunrise, noonday prayers when the Sun was on the meridian, and afternoon prayers when the shadow of any object was equal to its midday shadow length plus the length of the object. This, of course, all meant that the apparent daily motion of the Sun must be known accurately and ibn Yunus's tables, based as they were on a sound mathematical footing, were excellent and extensive, containing more than 10,000 entries of the Sun's position throughout the year. Indeed they were so good that they remained part of the corpus of tables used in Cairo until the nineteenth century.

Arabian astronomy entered the doldrums for a century after al-Biruni and ibn Yunus. The only astronomer of note was Abu Ishaq al-Zarqali, who came from a family of artisans at Toledo. He made astronomical instruments and clocks, though he is probably best known for his *zij*, *The Toledan Tables*, which were similar to the tables of al-Khwarizmi but gained a great reputation in the Western world. They contained results which included the effects of 'trepidation', a subject on which al-Zarqali wrote a book. He wrote too on scientific instruments and, in particular, on those concerned with drawing or 'projecting' the sphere on to a flat surface. In his own day, however, he seems to have been most famous for his complex water-clocks, some of which even showed the movements and phases of the Moon. These were made at Toledo, which al-Zarqali left in 1078 because of the disturbances caused by the repeated attacks of Christian armies under Alfonso VI of Castile. In 1133, after the city was conquered by the Christians, Alfonso VII gave permission for the clocks to be dismantled by a craftsman, Hamis ibn Zabara, to discover how they worked. But ibn Zabara was not equal to the task; he took them to pieces, failed to discover how they worked, and was quite unable to reassemble them. By then al-Zarqali was dead and details of his methods of construction were therefore lost.

Illustrations page 219

The eleventh century was a time of social and political upheavals, but things settled down again in the twelfth century under the Almohads, a military Islamic power centred on Spain, and scholars began again to be encouraged. Yet there was one notable Muslim astronomer and mathematician who had lived and worked for the most part in the eleventh century, and that was Ghiyath al-Khayyami, better known to the western world as the poet Omar Khayyam. Born in 1048 in Iran at a time when his country was overrun by the Seljuk Turks, he spent his boyhood and formative years at Balkh in Afghanistan and then moved on to Samarkand, where he not only took a great interest in astronomy but also began to write on mathematics and music. To be a scholar in those days meant either being rich or having a patron and in Samarkand al-Khayyami received support from the chief justice. Clearly news of his abilities spread, for in 1070, when he was still only 22 years of age, he was invited by the Seljuk sultan and his grand vizier to go back to Iran, to the capital at Isfahan, and take charge of the observatory there. Here he remained for the next eighteen years during which time he produced his own set of *zij*, the *Malikshah Tables*. Unfortunately most of this work is lost, and all that remain are some stellar positions and a list of magnitudes of the hundred brightest stars. Al-Khayyami also had a plan for calendar reform which advocated a system that would have been in error by no more than one day in 5,000 years.

While al-Khayyami was at Isfahan he had also to act as court astrologer, but this was not work in which he believed at all; he merely looked on it as one of the less desirable aspects of his duties. But al-Khayyami was not only a disbeliever in astrology; he was also something of a free-thinker in other matters, as some of the quatrains of his famous *Ruba'iyat* bear witness. Indeed, his apparent lack of religion angered orthodox Muslims and when, after the sultan's death in 1092, he fell into disfavour at court, funds for running the observatory dried up, and al-Khayyami had to take definite steps to free himself from charges of atheism. He went on a pilgrimage to Mecca, and it seems that the late philosophical writings he prepared were done with the same end in view. Finally al-Khayyami moved to

Merv, the new Seljuk capital, where he died in 1131.

It was in Merv that another astronomer, Ab'ul-Fath al-Khazini spent his early years and his working life, which coincided with al-Khayyami's later years. Originally a Byzantine slave-boy owned by the court treasurer, and probably a eunuch, al-Khazini was given a first-class education and made a niche for himself in Islamic physics; in astronomy he prepared a *zij*, based on his own observations and containing a particularly notable set of studies of eclipses, and a book on Islamic astronomical instruments. Also in twelfth-century Islam, at Damascus, the mathematician and astronomer Sharaf al-Din al-Tusi flourished. Al-Tusi is now remembered mainly for his invention of the 'linear astrolabe', a simple device consisting of a graduated wooden rod with a plumb line and a double cord. Used for making angular measurements, it was essentially a reproduction of the meridian line on an astrolabe. It could, al-Tusi claimed, be made in half an hour by an amateur and was extraordinarily simple to use. He himself measured the altitudes of stars with it, as well as the direction of Mecca and the Qu'bah, though its accuracy did not equal that of the ordinary astrolabe.

Illustration page 210

In spite of the work of al-Khayyami, al-Khazini and al-Tusi, the chief work in astronomy during the twelfth century took place in western Islam, in Spain and Morocco. Later, it was to be from Spain that the fruits of Arabian scholarship were to be absorbed by Europe and so we find that the leading astronomers there had their names latinized for easier reference in western Christendom. Thus Jabir ibn Aflah was known as Geber (a Latin name sometimes wrongly applied to the alchemist Jabir ibn Hayyam), al-Bitruji al-Ishbilt as Alpetragius and Abu'l-Walid ibn Rushd as Averroës. Jabir ibn Aflah of Seville worked there during the first half of the twelfth century and his claim to fame is that he published a *Correction of the Almagest* in which he criticized Ptolemy, especially over the positions adopted for Mercury and Venus; ibn Aflah placed them beyond the sphere of the Sun, a scheme which was widely adopted in the West as well as in Islam. He also simplified the *Almagest's* planetary mathematics by using some new developments in Islamic trigonometry.

A more surprising figure was ibn Rushd, known in medieval Europe as 'the Commentator'. An able physician and logician, he was also a sound observational and theoretical astronomer who exerted a considerable influence through his writings, especially on astronomy in the West. Born in Cordoba in Spain in 1126, he spent the greater part of his working life in Marrakech in Morocco, where he died in 1192. Ibn Rushd was given a sound Muslim education, with a bias towards law, and seems to have been endowed with a good logical mind. He studied theology and seems to have taken a mean road between extreme doctrines, but he later turned against the whole subject since he found arguments could not be followed to their logical conclusions. Nevertheless, this did not prevent him occupying the post of cadi (religious leader) at Seville for a short time. But ibn

Rushd's love of logic and his mastery of it led him to take up the natural sciences and, as a starting point, he studied medicine, receiving his training from two eminent practitioners. He himself became physician to Prince Abu Ya'qub Yusuf (who reigned as Almohad caliph in Marrakech in the 1170s and 80s) and wrote a comprehensive medical textbook to replace the one prepared by ibn Sina (Avicenna), which was not popular in Andalusia.

For ibn Rushd's medical work, see page 236; for his importance in medieval Europe, see page 255.

It was Prince Yusuf who commissioned ibn Rushd to prepare explanatory commentaries on Aristotle and this gave full reign to his logical mind. Indeed ibn Rushd seems to have succeeded in thoroughly understanding Aristotle, many of whose philosophical writings are obscure, and he made many corrections to the commentaries of ibn Sina. But he was not content to act only as elucidator, and developed his own theory of the intellect which was tied up with the doctrine of forms; he believed that man thinks by abstracting the forms behind things and that the human intellect is the receptacle of these 'intelligible' forms. Ibn Rushd's views exerted an important influence on Western medieval thinking and brought him a considerable reputation in his own day. He also wrote extensively on religion, discussing questions of revelation and of free will and predestination; while he was particularly interested in demonstrations of the existence of God, of His attributes and His unity, in the origin of the universe and in the principle of causation.

It is clear that ibn Rushd was not only a great thinker, but also a man who could present his views clearly, and it is little wonder that what he said was listened to with respect. His views on astronomy, which were theoretical, also commanded attention. He first reviewed previous knowledge and ideas, and did not confine himself to those of Aristotle and Ptolemy, but also discussed the opinions of his predecessors. After due consideration he then allied himself with those who called for a return to Aristotle, though he did not follow Aristotle's views slavishly. Recognizing three kinds of planetary motion – that observed by the eye, that visible to observers only over very long periods of time, and that which was recognizable only in theoretical terms – he concentrated his own efforts on the last, the theoretical side. He believed fervently in regular, even motion for all moving celestial bodies – physics demanded it, he thought – and he set his face firmly against Ptolemy's eccentric motions; indeed he claimed that the ingenious device of the equant did 'not accord with the nature of things'. He accepted the idea of concentric spheres which Eudoxos had originally proposed and which in Aristotle's time had numbered 55 to account for all the various movements observed. By ibn Rushd's time, Arabian astronomers had managed, by a careful selection of motions, to reduce these to 50, but he was able to better this, requiring only 47 to achieve the necessary variations.

For the system of spheres proposed by Eudoxos, see pages 100 – 101.

Ibn Rushd's contemporary al-Bitruji, who worked in Cordoba some time around 1190, was also a great Aristotelian. As far as Ptolemy's interpretations of planetary motions went, he believed these to be no more than mathematical constructions; he was sure they could not be physical, because they conflicted with Aristotelian physics. What he did therefore was to adopt Ptolemy's parameters for the moving celestial bodies and then make a clever adaptation using the homocentric sphere theory which accounted for trepidation as well as everything else. Its somewhat unusual results included a spiral motion of the stars, but it recommended itself to many who favoured Aristotle at the expense of Ptolemy, and 'Alpetragius' was much quoted by late medieval Western scholars such as Albertus Magnus, Robert Grosseteste and Roger Bacon.

With the thirteenth and fourteenth centuries we have moved past the peak of Arabian astronomy and are coming towards its close, which occurred in the 1440s. Indeed, the thirteenth and fourteenth centuries were somewhat in the nature of troughs; in neither were there any great astronomical figures. Some astronomical constants were reobserved, some books written and *zij* prepared, but the only real progress was a new planetary system using extra epicycles in place of Ptolemy's equant and his other constructions. The author was the fourteenth-century observer 'Ala' al-Din ibn al-Shatir, but although ingenious his new ideas were more aesthetic than scientific. It was only at the very end of the fourteenth century and the first half of the fifteenth that there was a brief resurgence of astronomical activity near Samarkand; it centred on the figure of the great ruler, Ulugh Beg (1394–1449).

The name Ulugh Beg means 'great prince' and was a title used in place of this ruler's original name, Muhammad Taragay. Born in Sultaniyya in Central Asia, Ulugh Beg was brought up in the court of his grandfather, the Mongol conqueror Timur, more familiar as Tamerlane. A Muslim, Tamerlane made great conquests in the 1360s to 90s to rule over an area from the Mediterranean to Mongolia. He died in 1405 while on an expedition to conquer China. His tomb – the Gur-e-Amir – in Samarkand is one of the great monuments of Islamic art. Tamerlane's grandson, who in 1409 became ruler of Maveramnakhr (chief city of which was Samarkand), was not interested in conquest but had a penchant for the sciences. In 1420 Ulugh Beg founded a *madrasa* or institute for higher learning in Samarkand, with astronomy as its chief subject. Four years later he built an observatory; this was a three-storey building with a giant sextant, the largest astronomical instrument of its type in the world, having a radius of no less than 40 metres (132 feet). The sextant was used for observing the transits across the meridian of the Sun, Moon and planets whose changing positions, as well as the precise length of the year and important astronomical quantities such as the angle between the Sun's path and the celestial equator, the observatory was built to record. The instrument was made large in the interests of accuracy, for on its masonry scale one degree occupied over 70 cm (28 inches) and one arc minute 12 mm (½ inch). This meant that a precision of something between 2 to 4 arc seconds could be obtained, a remarkable

Illustration page 210

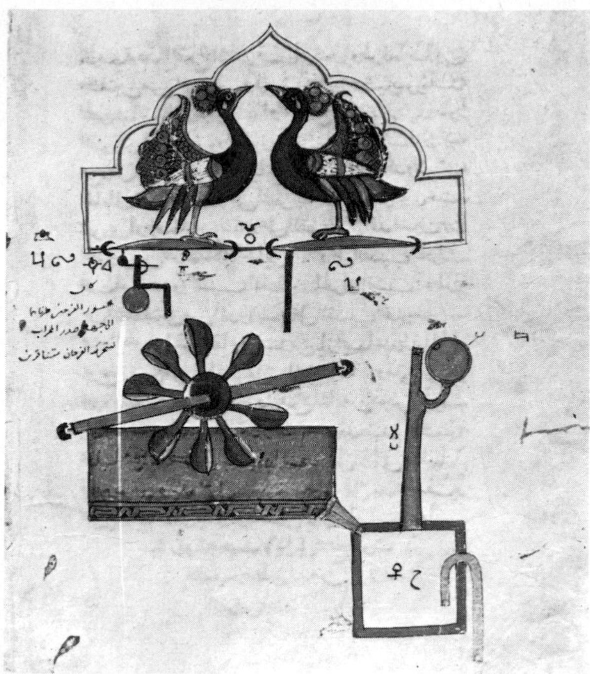

Above and above left Illustrations from a treatise of 1206 by Abu al-Jazari on mechanical contrivances, showing a water-driven clock with jacks, illuminated glass balls and revolving zodiac. The Arabs inherited their interest in such devices from Alexandria. *Above left*, Museum of Fine Arts, Boston, Massachusetts. Goleubew Collection. *Above*, Metropolitan Museum of Art, New York. Rogers Fund 1955.

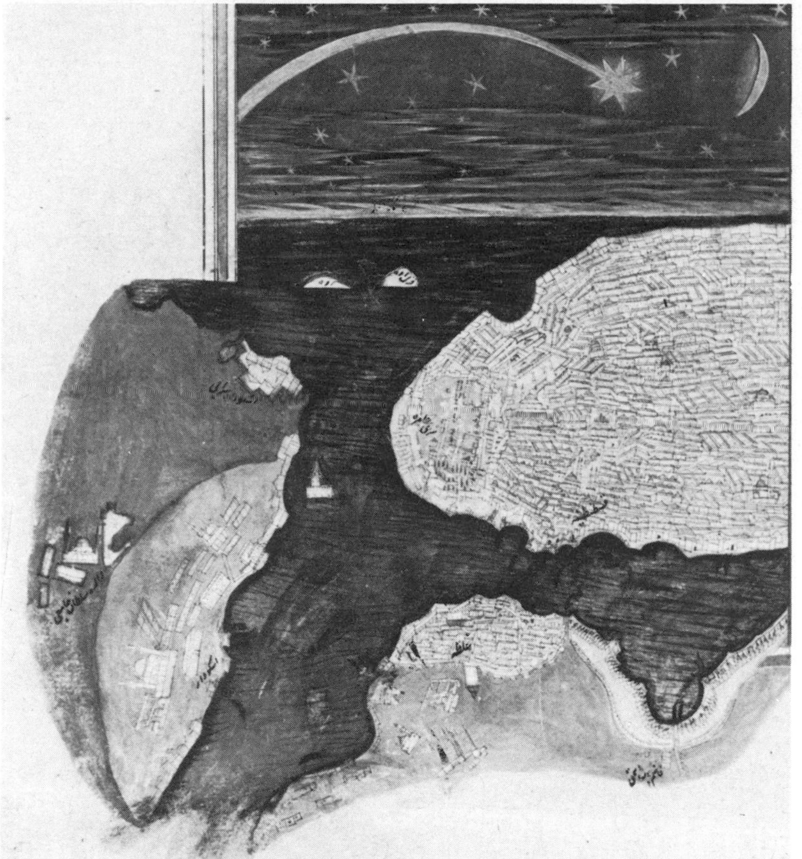

Left The comet of 1547 shown over the Bosforus. Istanbul Universitesi Kütüphanesi.

Right Inlaid doors from Egypt c. 1300, demonstrating the Islamic fascination with complex and subtle geometrical patterns. These were intended to explore the principles underlying reality, and as such were similar to the mathematical interests of the Pythagoreans. Metropolitan Museum of Art, New York. Bequest of Edward C. Moore, 1891.

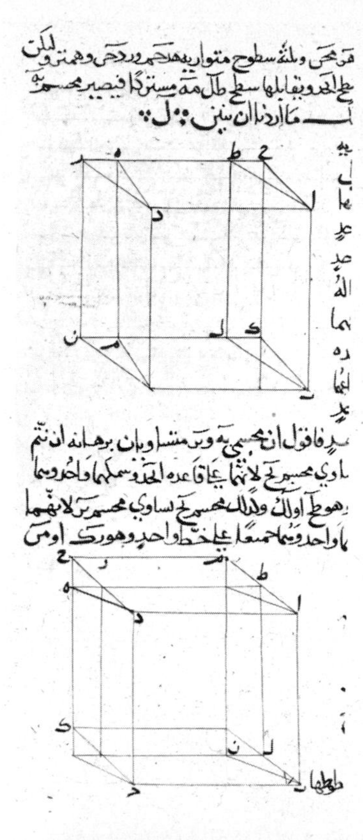

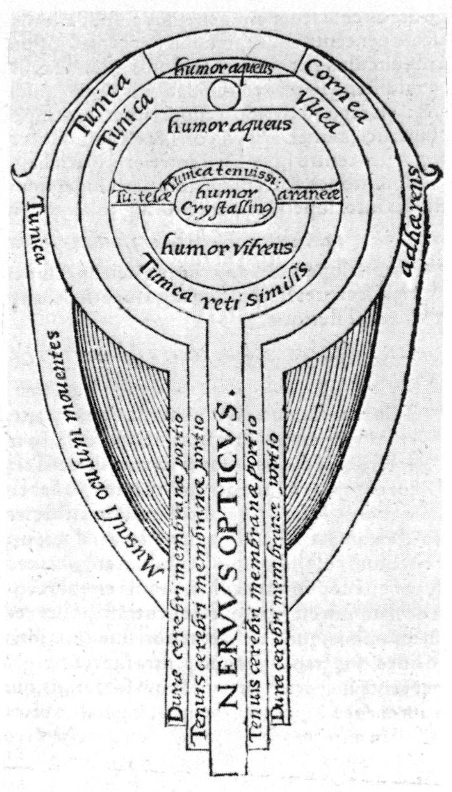

Far Left A 9th-century manuscript of Euclid's *Elements*, in the version revised by Thabit ibn Qurra. Bodleian Library, Oxford.

Left A 16-century Western edition of al-Haytham's study of the structure of the eye. This work and his other studies in optics stimulated much research during the Scientific Revolution. *Opticae Thesaurus* ed Risner, 1572.

Left The Brethren of Purity, an esoteric Islamic sect whose *Epistles* reveal a deep study of many scientific subjects, including animal biology, mineralogy, music and mathematics. Süleymaniye Kütüphanesi, Istanbul.

Right The eastern Mediterranean and western Asia, in a map by Ibn Hawqal, reflecting the sacred geography of much early Islamic mapping. Süleymaniye Kütüphanesi, Istanbul.

Right The world-map compiled by al-Idrisi in the early 12th century for the Norman king Roger II of Sicily. It has been described as the 'height of cartography in Islam': al-Idrisi was unusually well-informed about regions as far apart as Scandinavia and the Niger (north is at the bottom of the map).

achievement for the fifteenth century when one realizes that 4 arc seconds are equivalent to the width of an ordinary wooden pencil 1.4 km (7/8 mile) away. The observatory had, of course, other instruments, including an armillary sphere and an astrolabe.

As we might expect, the observatory produced its own *zij*, sometimes referred to as the *zij* of Ulugh Beg, sometimes as the *Zij-i Gurgani* (Guragon was a title used by Ulugh Beg). They were very accurate as far as the planets were concerned but not so good on the stars, some values for which were taken from al-Sufi's *zij*. The mathematical tables were, however, extraordinarily precise; so good, in fact, that they can bear comparison with similar tables today.

The director of Ulugh Beg's observatory was Ghiyath al-Din al-Kashi. Born in Kashan, Iran, at some unknown date, al-Kashi seems to have lived for a time in penury, trying to muster up patronage for his astronomical and mathematical work, rather than take up his second study of medicine and practise as a physician. When Ulugh Beg established his 'university', al-Kashi moved to Samarkand, and obtained a secure and honourable position at court, later becoming the first director of the observatory. Ulugh Beg thought very highly of him; he excused al-Kashi's bad manners and lack of court etiquette because, as Ulugh Beg himself put it, he was 'a remarkable scientist, one of the most famous in the world, who had a perfect command of the science of the ancients, who contributed to its development, and who could solve the most difficult problems'. Al-Kashi was indeed remarkable; he calculated the value of pi (π) to 16 decimal places, and other mathematical ratios with equivalent accuracy, while it was probably his work that caused the *Zij-i Gurgani* to have such extraordinarily good tables. He also wrote an elementary encyclopaedia on practical mathematics for astronomers, surveyors, architects, clerks and merchants. When he died in 1429, five years after the observatory's foundation, his loss was mourned, not least by Ulugh Beg himself.

For the Greek calculation of π, see pages 112–113; for the Chinese, see pages 151–152.

Al-Kashi was succeeded at the observatory by Qadi Zada al-Rumi, who had been born in Bursa in the west of Turkey in 1364. Trained in mathematics and astronomy, he moved to Samarkand in 1383 to continue his studies, and when Ulugh Beg decided to build his university Qadi Zada was appointed rector. When al-Kashi died, he became director of the observatory in his place, and seems also to have spent time calculating mathematical ratios for mathematical tables. He was obviously a competent scientist – Ulugh Beg would hardly have appointed him otherwise – but does not seem to have been of al-Kashi's stature.

Ulugh Beg himself used to make astronomical observations, and was a dynamic force behind the cultural life of Samarkand. Unfortunately, with his death by assassination, the intellectual glory of Samarkand faded and its great traditions of learning went into eclipse. In the sixteenth century the observatory was razed to the ground by religious fanatics and not rediscovered until our own times.

Mathematics

If Arabian astronomy mainly consolidated and perfected a science which was essentially an inheritance from the Greeks, Arabian mathematics was quite different. Certainly it was a means whereby Hindu numerals were transmitted to the West, but above all it brought into the mathematical art two powerful techniques – algebra and trigonometry – which are as valid today as they were when the Arabs introduced them.

The first great Arabian mathematician was Thabit ibn Qurra, whose astronomical work at Baghdad has already been described. He made many contributions in all areas of mathematics; as a translator he wrote Arabic editions of all the works of Archimedes and the work of Apollonios on conic sections (the ellipse, parabola and hyperbola) as well as Euclid's geometry. He wrote on the theory of numbers and extended their use to describe the ratios between geometrical quantities – a step the Greeks never took – and he discussed the question of where, if anywhere, parallel lines can meet. Ibn Qurra also prepared a *Book of Data*, a geometrical book halfway between Euclid and the *Almagest* in difficulty; this was to have a vogue in the West in the Middle Ages.

Illustration page 221

Another of the Baghdad astronomer-mathematicians was al-Battani, and his notable achievements were, first, that he gave up the old Greek system of chords of angles and adopted the far more convenient trigonometrical proportion known as the sine; he also used its converse ratio, the cosine. However, he did not use the other ratio, the tangent (and its inverse, the cotangent), and so some of his formulae were still a little cumbersome. With Hipparchos and Ptolemy the Greeks had come close to trigonometry but they had never finally settled on the ratios that al-Battani adopted, and which, because of their simplicity and convenience, were to revolutionize the mathematics of triangles used so much in astronomy and surveying, divesting it of some of its previous difficulty. Al-Battani's second achievement was his use of trigonometry, and the projection of figures from a sphere on to a plane, to allow him to get some new and elegant solutions to astronomical problems. His methods were copied in Western Europe in the fifteenth century by the astronomer Regiomontanus.

For Regiomontanus, see pages 320–321.

The ninth century saw other mathematical advances in Islam: al-Jawhari developed some methods of calculating life expectancy using astrological data, and Kamal al-Din worked in algebra using high-degree equations with ease and, in all equations, handling irrational quantities, such as the square root of 2, without difficulty and thus extending the field over which such mathematics could be applied. But of all the mathematical practitioners of this century, perhaps the most important was al-Khwarizmi, for it was he who wrote a treatise on practical mathematics, something to show 'what is easiest and most useful in arithmetic', in which he used algebra in our modern sense. He did this when he explained how it is possible to reduce any

For the development of algebra in the West, see pages 322–323.

problem to one of six standard forms, using two processes, the first known as *al-jabr*, the second as *al-muqabala*. *Al-jabr* was concerned with 'transferring terms' to eliminate negative quantities (so that, for example, $x = 40 - 4x$ becomes $5x = 40$); *al-muqabala* was the next process, that of 'balancing' the positive quantities that remain (thus if we have $50 + x^2 = 29 + 10x$ *al-muqabala* reduces it to $x^2 + 21 = 10x$). In his book al-Khwarizmi used no symbols as we do now – these came later – and he expressed his mathematics in words; moreover he did not invent algebra as a technique, for he derived it either from Greek or, more likely, from Hindu sources. But his achievement was to make the technique clear and, by explaining it so well, promote its use. It was al-Khwarizmi, too, who wrote glowingly about the Hindu numerals and so encouraged the use of these as well.

For the origins of Hindu numerals, see page 192.

The tenth century saw more mathematical research and development, some of it geometrical, some algebraic and trigonometrical. Thabit ibn Qurra's grandson, Sinan ibn Thabit ibn Qurra, concerned himself with geometry, and al-Quhi devised a compass, with one leg which changed length, for drawing ellipses and other conic sections. Trigonometrical work concentrated on preparing tables of sines – ibn Yunus compiled these to four decimal places – and the 'sine theorem' was discovered. This theorem is used for triangles drawn on a spherical surface such as are met with in astronomy when triangles are measured on the celestial sphere, and is particularly important. No one is certain of the inventor; it may have been Sinan, or al-Khujandi, or perhaps Abu Nasr al-'Iraq, but certainly it is to one of these Muslim mathematicians that we owe it. In algebra al-Karaji developed the use of binomials and defined algebra's proper task as the 'determination of unknowns starting from known premises'.

Illustration page 220

But the great allrounder of tenth-century Arabian mathematics was Abu'l-Wafa'. He wrote a good manual on practical arithmetic, *A Book on What is Necessary From the Science of Arithmetic for Scribes and Businessmen*, and a similar one on geometry, *A Book on What is Necessary From Geometric Construction for the Artisan*. This last gave solutions of two and three-dimensional problems using only a compass and straight edge – a form of practical geometry that would have upset the Greeks, to whom geometry was a solely theoretical art – and Abu'l-Wafa's constructions were so eminently serviceable that they were widely circulated in Europe during the Renaissance. In trigonometry he prepared new tables and developed ways of solving some problems of spherical triangles.

As in astronomy, so in mathematics, the eleventh century was a time of virtually no development, but things improved in the twelfth. The poet and astronomer al-Kayyami wrote a commentary on Euclid and on algebra, using some of the ideas of Abu'l-Hasan al-Nasawi, who had worked at least a century earlier and whose methods seem to have owed much to the Chinese. Al-Khayyami also discussed finding the roots of the fourth, fifth, sixth and higher powers by a method he had discovered that did not involve using geometry and

For Pascal's Triangle in China, see page 153.

may have involved Pascal's Triangle. If so, then his discovery was contemporaneous with that in China, but since al-Khayyami's method is lost, it is not possible to be sure whether he made a parallel discovery or not. The astronomer al-Tusi wrote a large algebraic treatise and did so not long after al-Khayyami's death; this too dealt with finding the roots of equations and discussed changes of variables in them, something that proved to be unique in Arabian mathematics. Again it was in the twelfth century that the physician ibn Yahya al-Samaw'al showed his mathematical precocity by writing a book *The Dazzling* when only nineteen years old. In it he discussed the multiplication and division of powers (i.e he dealt with the 'power series', x^4, x^3, x^2 . . . $\frac{1}{x}$, $\frac{1}{x^2}$, $\frac{1}{x^3}$ etc.), he adopted the convention that 1 can be expressed as power zero (i.e. as x^0) and in fact enunciated our present methods (although he did not use the same notation). Moreover he not only went one step further than al-Karaji by writing algebraic results in a rather more symbolic form, but he was also the first Arabian mathematician to display an understanding of negative numbers, choosing to treat them as separate identities. Thus al-Samaw'al was able to subtract numbers from zero, devising rules that only appeared in Europe 300 years later.

After al-Samaw'al little more development of Arabian mathematics occurred. In the mid-thirteenth century the Spanish Muslim Muhyi 'l-Din al-Maghribi recalculated pi (π) and values for sines, and gave new proofs of the sine theorem, but a hundred years later these calculations were surpassed by al-Kashi at Samarkand with his values correct to sixteen decimal places. It was al-Kashi, too, who introduced methodical ways of dealing with decimal fractions. And it is in the fifteenth century that we come not only to the last Spanish Muslim mathematician but also to the last development of Islamic mathematics; this occurs in the work of Abu'l Hasan al-Qalasadi who is known to us now for a book on algebra written in verse. However it is not the fact that algebraic rules became poetry in his hands that concerns us, but the fact that his text made many algebraic symbols more widely known. They were not his invention but rather the work of ibn Qunfudh and Ya'qub ibn Ayyub a century before: al-Qalasadi's importance was that he made them more familiar, so that they were later to stimulate Western mathematicians when they inherited the superb mathematical legacy of Arabia.

Physics

For the ideas of the Mohists, see pages 141–142.

Physics underwent little development after Aristotle and Archimedes until late medieval times in Europe. Certainly the Mohists in China took some useful steps and other specialized work was done there, but little appears to have been achieved in Arabia except for the study of optics, where there were advances made, most notably by ibn al-Haytham in the late tenth and early eleventh centuries; there was also a little done on balances and the subject of equilibria by Thabit ibn Qurra and al-Khazini.

As a translator of the works of Archimedes Thabat ibn Qurra must have known the basic physics of the lever, of pulleys and, of course, the problems connected with weighing and the use of the balance. Indeed it is for his writings on the principle of the balance and on the equilibrium of bodies, including the equilibrium of a heavy beam, that ibn Qurra as a physicist is remembered today. His work does not, however, seem to have been followed up in any particular way for well over 200 years, until the time of al-Khazini, whom we have also met briefly. An ascetic, mystic and teacher in his adult life, al-Khazini wrote extensively on weights and weighing and, having a good practical turn of mind, wrote the substantial *Book of the Balance of Wisdom*. Well aware of the works of his predecessors, Archimedes, ibn Qurra and al-Asfizari (who had lived a generation before and whose balances had been destroyed by the sultan's treasurer out of fear), al-Khazini's book is essentially concerned with the hydrostatic balance. His design used a steelyard with reference markers fitted at various points so that weighings could be made using different liquids, and it was widely adopted for determining the adulteration of precious metals and the assessment of precious stones. To help in using the instrument, the book also included tables of what we should call specific gravities. But worthy though all this work was, it is completely overshadowed by the achievements of the greatest Islamic physicist, ibn al-Haytham.

For his influence in medieval Christendom, see page 254.

Known to Western mediaeval scholars as Alhazen, al-Haytham was born at Basra in Iran in 965 and moved to Cairo during the reign of the Fatimid caliph al-Hakim, the ruler who patronized the astronomer ibn Yunus and founded a library in Cairo that was almost as famous as the one at the House of Wisdom in Baghdad. At Cairo al-Haytham had what was outwardly a rather disastrous career. It seems that when he arrived in Egypt and witnessed the regular annual inundation of the Nile, he assumed that this occurred because there was no proper hydraulic control of the river. He therefore obtained the caliph's patronage for an engineering expedition to the south of Egypt, with himself in charge. The aim was to find the high ground he assumed to be at the source of the Nile and take the appropriate steps to control the river. Alas, as he travelled upstream and saw the vast buildings, especially south of Aswan, he began to realize that if anything could have been done the ancient Egyptians would have done it, and the success of an irrigation scheme such as he had imagined was doomed to failure. He was forced to return home and admit his error. As a first step he was given a lowly government post, but he feared for his life, as the caliph had a reputation for being both eccentric and murderous. Al-Haytham therefore decided to feign madness, and he remained under house arrest until the caliph was dead, after which he returned to sanity!

When once again free to pursue learning, al-Haytham was in his late fifties and he seems by then – or soon after – to have come to the conclusion, after extensive enquiries into religion, that truth was to be had only in 'doctrines whose matter was sensible and whose form was rational'. These doctrines he found in Aristotle and in the fields of mathematics and physics, and it was in the latter that he made his mark with his original optical studies. Though al-Haytham's optical work does contain some Greek elements, particularly from Ptolemy, he rearranged and re-examined everything in such a way as to produce results that were entirely new. Especially was this so in his theories of light and vision which are totally his own, owing nothing either to antiquity or to previous Islamic ideas. Light, claimed al-Haytham, is something emitted from every self-luminous source; it is a 'primary emission'. He also considered a 'secondary emission', from what he termed an 'accidental source'. Light from such a source is emitted 'in the form of a sphere' (i.e. in all directions). This concept of a secondary 'accidental source' meant that from every point of the specks of dust in a beam of sunlight or an illuminated opaque object, light is also emitted. Such light would, like light from a primary source, travel in straight lines (as al-Kindi had pointed out) but it would be weaker. This was original thinking indeed, really enshrining the principle of secondary wavelets proposed six centuries later by the Dutchman Christiaan Huygens.

Illustration page 221

For Huygens's idea of secondary wavelets, see page 376.

Al-Haytham described colours as being real and distinct from light, coloured bodies radiating their light in all directions in straight lines; colours are always present with light, mingled in with it, and never visible without it. Admittedly this is not a view we could accept today, but in the eleventh century it was a valiant attempt to explain an intriguing phenomenon which was to wait for its final solution for centuries. Certainly it was like a breath of fresh air after the less than satisfactory views of the Greeks. Al-Haytham was consistent, too, when he discussed reflection at polished surfaces; in keeping with his theory of light he claimed that such a surface does not 'receive' light but sends it straight back, and quoted experiments to prove his case. It is to al-Haytham, too, that we owe the introduction of that extremely useful concept, a 'ray of light'; indeed he had a truly physical picture of such a ray.

The whole Greek idea of vision as something emanating from the eye was rejected by al-Haytham as absurd, and he drew upon his physical theory of rays and mathematical constructions of their paths to explain what is seen. He believed, like the Graeco-Roman surgeon Galen, that sight is stimulated first at the crystalline lens of the eye, and there is no talk in his work of the formation of an image inside the eye as in the camera obscura. He did however discuss the optic nerve and its connection with the brain. Here al-Haytham's views were not correct, but once again they were a great step forward and showed for the most part a completely new approach. All this and more was discussed in his major book, *Optics,* which like a modern physics treatise, takes a mathematical and experimental approach, citing no authorities but the authority of empirical evidence.

Al-Haytham's work was important; how important may be judged not only from its continual quotation by later medieval scholars in the West, but also by the fact that his conclusion that the refraction of light is caused by light rays travelling at different speeds in different materials, as well as his laws of refraction, were used in the seventeenth century by Kepler and Descartes. Al-Haytham, then, represented, in embryo, the modern physical scientist. His work marked the high point of Arabian physics.

For optics in Renaissance science, see pages 317–319.

Geography

It was in the mid-thirteenth century that Baylak al-Qibaji of Cairo was the first to write in Arabic about the magnetic needle as a ship's compass, but Arabian navigation and geographical knowledge goes back far beyond this. Itineraries and routes had, naturally, been drawn up so that diplomatic missions could be sent to distant lands – to China for instance – and for military campaigns, while merchants would have had some knowledge of caravan routes. But no organized scientific geography seems to have begun until the early ninth century, the time of al-Ma'mun and the establishment of the House of Wisdom at Baghdad. Here al-Farghani made Ptolemy's *Geography* known to the Arab world, while al-Khwarizmi wrote his *Book of the Form of the Earth*. The latter was primarily a list of the latitudes and longitudes of places which included the old Greek 'climata', seven strips of latitude in each of which places were supposed to have the same length of daylight on their longest day. The map of al-Khwarizmi differed substantially from Ptolemy's world map in some places, due perhaps to using the different longitudes and latitudes collected at the House of Wisdom.

For the Chinese development of the magnetic compass, see pages 172–173.

Geographical research continued in the tenth century with geographers like ibn Khurdadhbih and ibn Ya'qub Ibrahim, who did not, however, work in Baghdad. Abu'l-Qasim ibn Khurdadhbih (sometimes Khurradadhbih), who was of Persian descent, was chief of posts and information at al-Jibal, a city on the Tigris. A close companion of the cultured caliph al-Mu'tadid (the third and last of the Abbasid dynasty), ibn Khurdadhbih wrote on wines and cookery and also prepared an economic and political geography, the *Book of Roads and Provinces*. The book's mathematics were poor, but it organized very well the vast amount of material it contained. As to ibn Ya'qub Ibrahim, he was an Hispano-Jewish merchant well known for his travels throughout the breadth of Europe either on business or in diplomatic missions. He visited Jewish communities, and noted descriptions given to him of the areas in which they lived, and though little now remains of what he wrote, what there is gives a good description of the Slav countries and southern Russo-Arabic territories, and is an excellent source of details about contemporary life. But of all the Arabian travellers of the tenth century, the most notable was Abu'l-Hasan al-Mas'udi who left Baghdad about 915 and spent his life travelling all over the Islamic world as well as in India and East Africa. He only settled down in his later years, dying in Cairo at about 956. Al-Mas'udi believed true knowledge could only be obtained by personal experience and observation; he was a prolific writer with 37 works to his credit though unfortunately only two have survived.

To al-Mas'udi knowledge accumulated with time, and he disagreed with those who accepted the ancients as final authorities and minimized the value of contemporary scholars. 'The sciences', he said, 'steadily progress to unknown limits and ends.' He openly challenged the 'traditionalist' outlook which, two centuries later, was to exert a dead hand on new learning and in due course lead to a decline in Islamic science and Islamic society in the Middle Ages. A good historian, who advocated always going back to original sources and who tried to take a scientific and objective view of the past, Al-Mas'udi conceived of geography as an essential prerequisite of history, and a geographical survey preceded his own world history. He stressed the point that the geographical environment strongly affected a region's animal and plant life, and was able to sort out many contemporary confusions in geography. He did not subscribe to the Islamic school of thought which took Mecca as the centre of the world and made geography conform to the concepts of the Qu'ran; he was equally critical of geographers of the past and did not accept Ptolemy's belief in a *terra incognita* in the south; he accepted the views of sailors who told him that there were no limits to the southern ocean.

Al-Mas'udi has sometimes been called the Islamic Pliny because, like the Roman Pliny, he took a wide interest in the world around him, and tried to discover things for himself; certainly he was one of the most original thinkers of medieval Islam. But not all Islamic geographers followed his example; map-makers were still drawing rather formal maps of only Islamic areas, but more telling, perhaps, is the fact that the chief eleventh-century geographer, the Hispano-Arab Abu 'Ubayd al-Bakri who compiled details of land and sea routes and lists of place names, never himself travelled outside the Iberian peninsula and relied almost entirely on reports from others. Indeed, Muslim geography seems to have reached its peak with al-Mas'udi and, with two exceptions, to have declined after his death.

Illustration page 222

One of these exceptions was Abu al-Idrisi, a Muslim from a noble house which laid claim to the caliphate. Born at Ceuta in Morocco in 1100 he was educated at Cordoba in Spain, and although he died in Ceuta in 1166, he spent his working life outside Islam. At the age of 16 he started to travel through Asia Minor, Morocco, Spain and the south coast of France, and even visited England. Then al-Idrisi was invited by Roger II, the Norman king of Sicily, to come to live at Palermo for his own safety, since he would be in continual danger of assassination attempts if he remained in Muslim circles. Thus it was that he went to Palermo, a meeting place of Arabic and European cultures, and achieved what was to be one of the great examples of Arab-Norman co-operation. Roger II, being dissatisfied with Greek

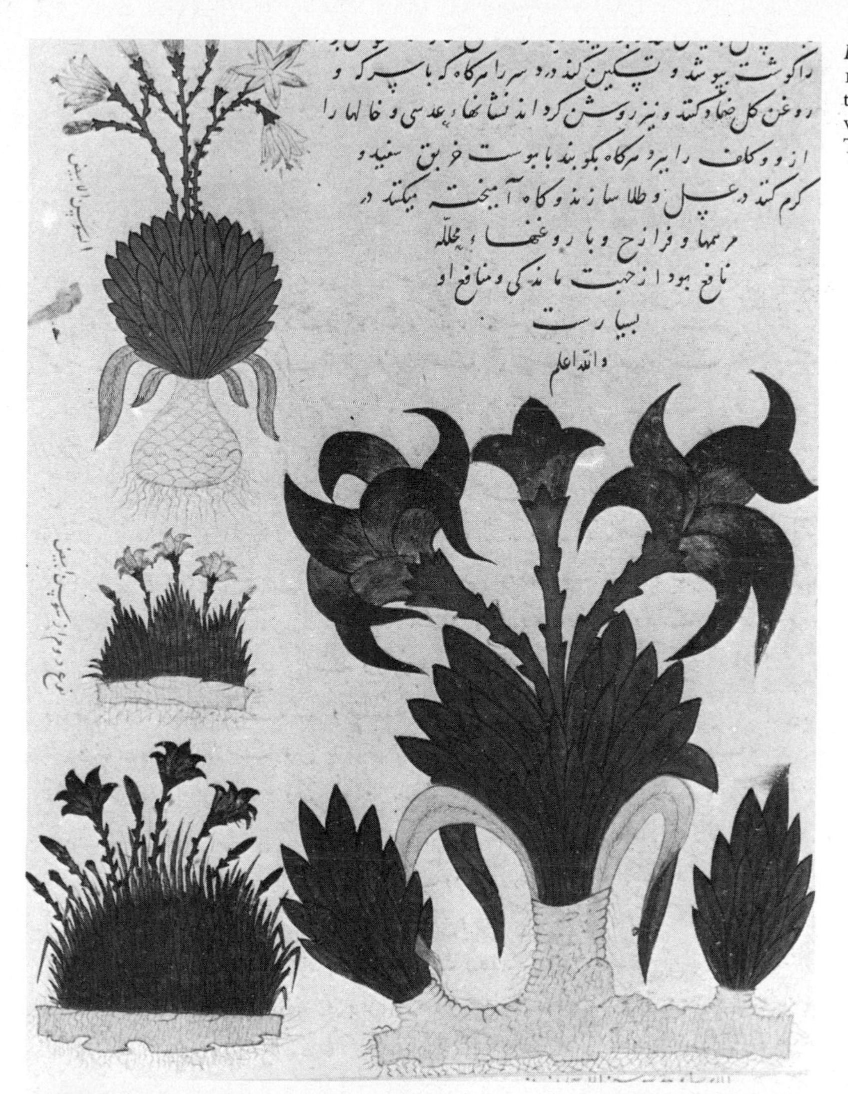

Left The iris and the white lily in a 15th-century Persian botanical treatise. The best Islamic botanical work was done in Persia and India. Topkapi Sarayi Müzesi, Istanbul.

Left The lion and the jackal from the 15th-century *Kalilah wa Dimnah*, a Persian collection of moral tales about animals. Topkapi Sarayi Müzesi, Istanbul.

Above A goose on a clutch of eggs from a 14th- or 15th-century edition of Abu al-Jahiz's *Book of Animals*. Biblioteca Ambrosiana, Milan.

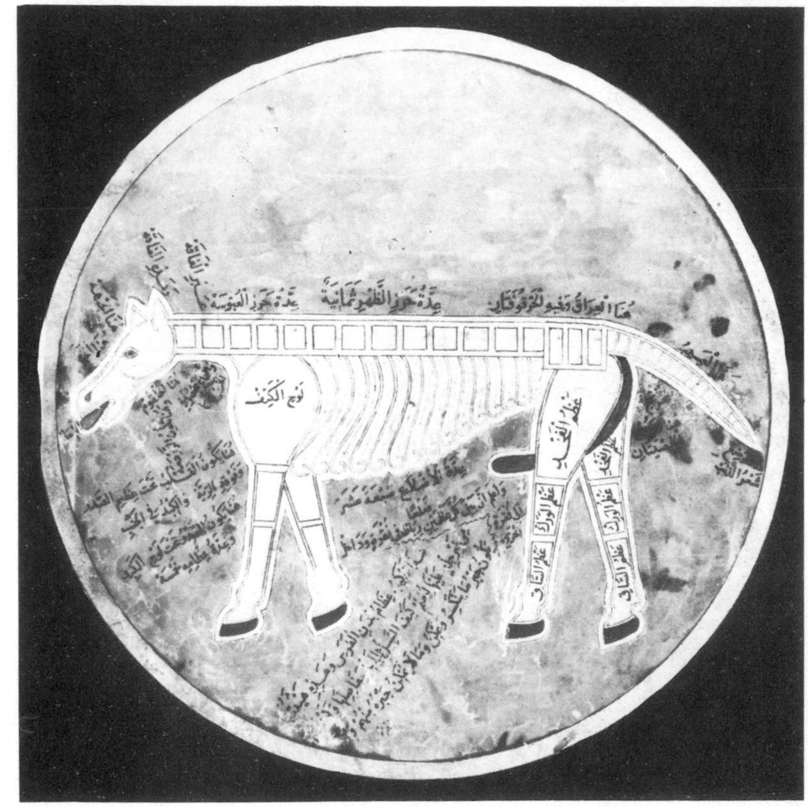

Right An anatomical drawing of a horse from a 15th-century manuscript. The study of horses formed a central part of Islamic biology. Istanbul Üniversitesi Kütüphanesi.

and Arabian maps, decided to commission a new one, with features to be shown in relief and the whole to be engraved on silver. Al-Idrisi was put in charge of the project, and envoys were sent overseas to collect information. In due course the project was completed, though nothing of it now remains except a geographical compendium compiled by al-Idrisi and containing sectional maps. The maps show the 'inhabited world', are mainly of the northern hemisphere and are divided into climata. They contain no evidence of originality of thought – they were based on Greek and Arabian conceptions of the world – but within their limitations give evidence of a thoroughly workmanlike job.

Illustration page 222

The other exception in the general decline was the thirteenth-century geographer Zakariya al-Qazwini, whose wide interest in science was strongly coloured by the Islamic faith, and who was somewhat given to metaphysical speculation. He wrote well on geography, basing his work on the results of his own travels throughout Asia Minor. He also has the distinction of being the first to explain the rainbow correctly, and he played a major role in observing at the observatory established at Maraghah in Iran, thus being instrumental in the preparation of the famous *Ilkhani zij*. Indeed al-Qazwini's work and writings were to do much to help the renaissance of science and philosophy in Iran.

Biology and Medicine

In biology and the medical sciences the Arabians inherited a vast amount of material from the Greeks, Romans, Persians and Indians. To Arabian philosophers, plants were primarily studied for their use either in agriculture or in medicine, an attitude taken at the beginning by Jabir ibn Hayyan (whom we shall meet again under alchemy) which set the seal on much that followed. At Baghdad some of the early translations were of botanical books mistakenly believed to be by ancient Greek authors such as Aristotle and Theophrastos, and those that were genuine, like the works of those army medical men Dioscorides and Galen of the first and second centuries AD, also confirmed this medical bias. However, although Dioscorides was author of by far the largest pharmacopoeia in Western antiquity, his book contained other useful botanical information. Some of it was purely practical – the effects of storage for long periods and suggestions for suitable containers – but not all, for he did discuss each plant in relation to its habitat. Yet by and large the medical bias of botanical study remained; we find it in the writings of the eleventh-century doctor ibn Sina, in those of the Spanish Muslim school and in the works of Arabian scholars in Andalusia in the twelfth and thirteenth centuries. It continued too in the encyclopaedic works of the fourteenth century.

There were two exceptions to this approach to botany. In the ninth century Abu al-Dinawari, an Iranian historian who also made astronomical observations in Isfahan, wrote a large compendium, the *Book of Plants*. Admittedly this contained much philosophy and history but it was full, too, of botanical details and was often quoted in later times. The second exception originated in the second half of the tenth century in Basra with the Ikhwan as Safa or Brothers of Purity, a secret radical Islamic confraternity affected by Neoplationism and Manichaeism (a dualistic Persian religion which preached the release of the spirit from matter by asceticism). Taking an esoteric interpretation of the Qu'ran open only to initiates, they tried to neutralize Greek philosophy and followed alchemy and the occult sciences, though in their 'Epistles' they taught enlightenment first by a study of natural science. In botany this led them to examine the form and structure of plants and their growth. They then went on to discuss the number symbolism of the various parts and their place in the cosmic order, but it was their studies of growth and morphology which proved a valuable contribution to knowledge.

Illustration page 221

As far as zoology was concerned, the Arabian peoples were familiar with the life and habits of all the domesticated animals which even now provide the basis of living for nomadic tribes. In pre-Islamic times much specialized knowledge was amassed about the camel and the horse, while in Islam animals assumed religious significance since they shared man's destiny and were thought to provide lessons about God's wisdom and man's duties on Earth; indeed, religious laws laid down certain responsibilities over the way animals were treated. Thus in Islamic literary works there are frequent references to animals, which are used as symbols of cosmic qualities, but such references often display an intimate knowledge of animal behaviour.

If eighth-century zoological writing was mainly concerned with camels and horses, by the ninth the Mu'tazilite theologians (whose name implies they 'stood apart' in the quarrels over leadership in the Muslim community and who had a great sense of the rationality of the Islamic religion), and the scientist and mystic ibn al-A'rabi, were concerned with underlining the way the animal world gives evidence of God's wisdom. Abu al-Jahiz, known for his masterful Arabic prose, was a Mu'tazilite and he, al-A'rabi and others classified some 350 animals into four categories, these categories being based on how the animals moved.

During the ninth and first half of the tenth centuries we also find some descriptions of animals written by al-Kindi and by al-Farabi, but the most important zoological contributions came from those who were compiling general encyclopaedias and from others whose interest centred on natural history rather than on philosophy; these were objective descriptions and included some of the exotic animals of India. But after the middle of the tenth century the more philosophical works took over. These discussed the 'chain of being' rather along the lines of Aristotle's ladder of nature, and described the habitat, methods of reproduction and the number of senses animals possessed. They provided anatomical descriptions of the internal organs, but all done from the point of view of divine design. Ibn Sina in the eleventh century and ibn Rushd in the twelfth also wrote in this vein, though ibn Sina went as far as to give some consideration to animal psychology. Encyclopaedic works with large sections on zoology continued to appear in the thirteenth and fourteenth centuries, and during the thirteenth al-Qazwini brought in a new animal classification based on an animal's means of defence. Later on in the fourteenth century Kamal al-Din al-Farisi wrote *The Great Book on the Life of Animals*. This proved to be the foremost late Muslim work on zoology; in it al-Farisi systematized all previous studies and so produced a large compendium of zoological knowledge, which proved of great popularity and was translated into Turkish and Persian, because of the religious as well as the factual material it contained. This interest in animals continued throughout Mughal times, and the Mughal emperor of India, Jahangir, himself devoted sections of his own *Book of Jahangir* to careful descriptions of plants and animals, with pictures painted by miniaturists at his court, especially the seventeenth-century master Mansur.

Illustrations pages 231, 232

In the medical field Arab culture owed much to the work of Galen (second century AD) whose practical knowledge as an anatomist was gained as surgeon to gladiators and soldiers but whose general medical outlook was based on earlier research carried out by Herophilos, Erasistratos and the teachings of Aristotle. Galen's works were among the first Greek texts to be translated and commented upon, especially by Thabit ibn Qurra in Baghdad, while other medical men, notably Qusta ibn Luqa and Ishaq ibn Hunyan also made translations from Syriac and Greek, again mainly of Galen. But ninth- and tenth-century Islamic medicine was not just a revival of Galenic medicine; straddling the two centuries like a medical colossus was the figure of Abu Bakr al-Razi.

Known to the medieval West as Rhazes, al-Razi was born at Rayy in Iran about 854, and died there between 925 and 935. Little personal information about him has survived, but he was for some time director of the hospital in Rayy and later of the one in Baghdad. His philosophical outlook, which has to be gleaned from his critics, was unusual. He held egalitarian views, rejecting the idea that men can be sorted into strata according to their innate capabilities. But even more to the point in such a society as Islam, he attacked religion; men, he said, were not only equal, they also had no need of a discipline imposed by religious leaders in order to manage their affairs. As far as miracles were concerned, he thought they were impossible, and even wrote a book (now lost) called *The Tricks of the Prophets*; men of science like Hippocrates and Euclid were, he thought, much more important than religious leaders. Indeed, al-Razi claimed that religion was positively harmful as it led to fanaticism which engendered religious wars.

He took an equally critical and non-authoritarian attitude to science. Believing it to be a subject characterized by continual progress, a view he strongly defended against the Aristotelians of his day, to whom the summit of science had already been reached or would soon be, al-Razi was quite prepared to criticize ancient authorities, whoever they were, and even wrote a book with the title *Doubts concerning Galen*. He distrusted scientific dogmas, whether they came from Galen or anyone else. As for the natural world, al-Razi held atomic views not very different from those of Democritos, and it was the atoms, in his opinion, that gave rise to the four basic elements.

Al-Razi was a most successful physician and had considerable reputation as a medical author, for he wrote many comprehensive manuals, as well as specialist works on smallpox and measles. However his non-medical views, his radical rationalism and his anti-religious attacks made him unpopular, a situation summed up later by ibn Sina when he said that al-Razi should have confined himself to boils, urine and excrement and have refrained from dabbling in matters beyond the range of his capacity!

Ibn Sina has sometimes been called the 'Galen of Islam' because of his encyclopaedic *Canon* of medicine, which met with great praise and was thought to be impossible of improvement, except in Muslim Spain where other medical works were preferred. Its circulation did much to keep Islamic and European medieval medicine in a static condition, but it would be unfair to blame ibn Sina entirely for this; such a state of affairs was partly due to the general over-developed reverence for authority in the eleventh and later centuries, coupled with the fact that al-Razi's progressive ideas were generally unacceptable. Ibn Sina seems to have been as much a philosopher as a medical practitioner; in fact it would probably not be an exaggeration to say that, technically, he was a better philosopher than al-Razi, but that al-Razi was the better physician. Most of ibn Sina's philosophy was influenced by the Qu'ran and by Aristotle, though it would be wrong to suppose he did not allow science to help form his opinions, only it tended to be traditional science.

For the astronomical work of ibn Rushd, see pages 216–217.

In the century after ibn Sina we come to ibn Rushd, whom we have already met in connection with astronomy. Ibn Rushd's comprehensive medical textbook was much preferred in Andalusia to ibn Sina's, but he too seems to have been a practitioner rather than an innovator. However, Arabian medicine reached new heights in the thirteenth century because of two notable medical men. One was ibn al-Quff who was born in Karak, Jordan, in 1233, son of a Christian Arab who held an important post under the Ayyubid caliphs. A fine teacher, al-Quff moved to Damascus and became surgeon at the headquarters of the Mameluk army in Syria. While here he wrote profusely, producing a commentary on the scientific views of ibn Sina and a number of medical books, and was editor of the largest Arab medical text on surgery, a subject in which he had much practice since the Christian Crusades kept him busy. Today al-Quff is remembered because he described the capillaries and explained the use of the cardiac valves. This was long before the days of the microscope and four centuries before the detailed description of the capillaries by

Illustrations page 242

Malpighi in Europe, but al-Quff's description was much more than a lucky guess and gives evidence of very careful observation.

The other notable thirteenth-century medical man was the surgeon ibn al-Nafis. Born near Damascus, he studied medicine at the great Nuri hospital founded in the previous century by Nur al-Din (Nureddin). Later he moved to Egypt, and although the hospital at which he taught and worked is not known, he was also personal physician to the Mameluk ruler al-Zahir, a post which gave him disciplinary powers over other medical practitioners. Since he was an expert in religious law and lectured on jurisprudence, he would seem to have been an ideal candidate for the post.

Al-Nafis wrote his *Comprehensive Book on the Art of Medicine* when he was in his thirties, a book compiled from eighty volumes of notes which, nevertheless, appears not to have drawn on the whole of the astonishing total of 300 volumes which he possessed. Among the unpublished material was much on surgery, particularly on surgical techniques and post-operative care, as well as detailing the duties of surgeons and their relationships to both patients and nurses. Al-Nafis also wrote a commentary on Hippocrates's *Nature of Man*. Yet the main reason why this prolific writer is remembered today is his discovery of the lesser circulation of the blood, i.e. the circulation between heart and lungs. In doing so he boldly pointed out that Galen was utterly mistaken in assuming that blood travelled from one side of the heart to the other. Al-Nafis's discovery was an important one, and it was made three centuries before that circulation was described in Europe by Servetus (1553) and Colombo (1559), some thirty years after al-Nafis's ideas had reached the West.

Illustration page 241

For Servetus and the lesser circulation of the blood, see page 305.

Alchemy and Chemistry

Before finishing this description of Arabian science we must glance at the subject of alchemy, that mixture of science, art and magic that gradually blossomed into an early form of chemistry. Alchemy was concerned with the transformation of the substance of things in the presence of a spiritual agent, often called the Philosopher's Stone. The alchemist himself was not untouched by such transformations. Metals and minerals were used, but were thought to participate not only as material bodies but also as symbols in man's cosmic world, hence their correlation in alchemical manuscripts and drawings with astrological signs, where, for example, the sign of the Sun indicated gold, that for the Moon silver, while Mercury signified mercury (quicksilver) and Venus copper. It was a 'science' that embraced the cosmos and the soul, where nature was a sacred domain that gave birth to minerals and metals. Thus although it inspired much mysticism it also stimulated a careful study of minerals and metals that was later to be of use to legitimate science. In Arabia this last involved both taking over Greek, Indian and Persian descriptions and also looking at some substances anew. Many Arabian scientists were involved in this, above all, al-Biruni, who wrote a vast compendium

For alchemy in China, see page 136; in India, see pages 193–194.

on mineralogy *The Book of the Multitude of Knowledge of Precious Stones,* and ibn Sina, in whose *Book of Healing* and *Canon* there are classifications of minerals and metals and descriptions of the way he believed them to be formed.

The greatest alchemist of all was Jabir ibn Hayyan whose work spanned the late eighth and early ninth centuries. Ibn Hayyan (or Jabir as he is usually known) had a total philosophy of nature based on the microcosm-macrocosm concept and on a deep belief in the interplay of cosmic and terrestrial forces. The mineral kingdom had special significance in his scheme of things, which included such phenomena as the transmutation of base metals into gold. He accepted hylomorphism – the Aristotelian doctrine of the four elements and the four qualities – and from the four qualities (hot, cold, dry and wet) he obtained two basic principles, mercury and sulphur, which were to run through all subsequent alchemy, both Arabian and European. These two principles were not the actual substances we know as mercury and sulphur but principles of action, like the male and female principles or the Chinese Yin and Yang. It was the 'wedding' of these two principles which gave rise to all the different metals found in nature and which differed only in the proportions of mercury and sulphur they contained and in the celestial influences under which their principles had been brought together. Jabir believed the celestial influences came into it because of the 'unnatural' and 'extraterrestrial' nature of all metals; they were the insignia of the planets on the terrestrial plane (the Sun and Moon were among the planets in the sense Jabir used the word). He also believed in numerological relationships between the metals. Thus, when applied to metals, each of the four qualities (hot, cold, dry, wet) had to be divided into four degrees and each degree into seven parts, making a total of 28, a number equal to the number of letters in the Arabic alphabet. There were also four natures which could be expressed by the series 1, 3, 5, 8, which totals 17, the key to understanding the structure of the world. The number 17 was also related to a magic square whose component numbers were themselves related to the Pythagorean scale of musical notes, Babylonian architectural proportions and the symbolic Chinese shrine to heaven, the Ming Tang (Hall of Light), recently erected by the Empress Wu in 688.

For Yin and Yang, see page 145.

Jabir's scheme sought to make order out of the host of separate substances in nature, but it did so by seeking correspondences between the natural and supernatural worlds. It was a descendant of the alchemy which had grown up in the second and third centuries in Alexandria, most notably in the hands of Zosimus, and of other elements derived from Pythagorean mysticism and Persian allegory. Yet it was not just a scheme devised to bring order out of chaos, but a system for developing techniques whereby spiritual forces could be used to transcend the cosmos. And to Jabir and other Islamic thinkers of his day this cosmos was no mere physical realm as science visualizes it now; it was rather a domain with a variety of levels of existence,

Illustration page 262

illuminated by Islamic revelation. It was a conglomeration of spheres, of the four elements and of the signs of the zodiac, in which the twenty-eight divine names played their part, and the peak was the supreme heaven of the divine throne: it was a place where the Prophet was the symbol of all that was positive.

The transmutation of base metals to gold was looked on not merely as a physical process but the intrusion of a higher principle operating in the natural world, and was bound up with the idea of an elixir, which was itself concerned with the alchemical concepts of death and resurrection, dissolution and coagulation. But was transmutation really possible? Did it ever occur? These were questions which were debated throughout Islamic history. Theologians in general did not accept it, nor did they like alchemy and other occult studies, though there were exceptions. The Mu'tazilite Qadi Abd al-Jabar supported the idea and most philosopher-scientists and medical men accepted transmutation, though ibn Sina, while he accepted the alchemical concept of mercury and sulphur, inveighed against it.

Predictably that hard-headed rationalist al-Razi rejected much of the mysticism of alchemy and concentrated more on the experimental results which alchemists had obtained. Nevertheless he still tended to use much of the language of alchemy – he wrote *The Book of Secrets* and *The Book of the Secret of Secrets,* titles which underline the esoteric and arcane side of alchemy – though he described clearly and without mystery many chemical processes such as distillation and calcination (when materials are heated to a high temperature, without fusion, in order to obtain changes such as oxidation or pulverization). He also classified substances according to 'kingdom' as animal, vegetable or mineral – then a useful pharmacological scheme – and was interested in the medical uses of chemical compounds. The tradition that al-Razi was the first to make medicinal use of alcohol is incorrect, but it was he who began the transformation of Islamic alchemy into a science of chemistry.

In the tenth century both ibn Sina and al-Farabi wrote on elixirs and some other subjects related to alchemy, but not on alchemy itself, and a century later we find Abu'l-Hakim al-Kathi writing a useful guide to alchemical apparatus. Indeed, alchemy was carried on side by side with the more pragmatic approach to chemical reactions by those who rejected its mysticism, and though alchemy was one of the legacies of Arabian culture to the West, it was also accompanied by the proto-chemistry that had been begun by al-Razi and others.

Illustration page 242

For alchemy in the West during the Renaissance, see pages 306–307.

The Final Stages of Arabian Science

From what has been said here, there can be no doubt that the philosophers and scientists, geographers, natural historians and medical men of Arabian culture contributed materially to the sum of man's knowledge about the natural world. This was part of their bequest to the late medieval West: the other, as we saw, was the whole corpus of Greek science, sometimes filtered through the sieve of Islamic culture, sometimes not. Yet although the early Arabs and the whole Islamic world studied science and made notable contributions, their achievements came to an end; they never extended to modern science.

Islam extols the value of revelation above all else: it is the supreme authority. That is not to say that reason is discredited, far from it; the use of the human intellect is prized as one of God's gifts, but it must be for ever under the control of revelation. The Mu'tazilites, who emerged about 700, were aware of this; indeed, they set such store by reason that they said it could fathom even the deepest profundities of religious belief. On the other hand, the Asharites, whose views first appeared a couple of centuries later, condemned the over-zealous use of reason and its 'adulteration' of religious dogma, and for nearly two centuries the rival schools wrangled with each other until during the twelfth century the Asharite arguments carried the day. There then developed the attitude of passive acceptance. This attitude was inevitably inimical to independent scientific thinking, as intellectual traditionalism won the day. Islam never separated religion and science into watertight compartments as we do now, and the torch of science had to be carried on by others.

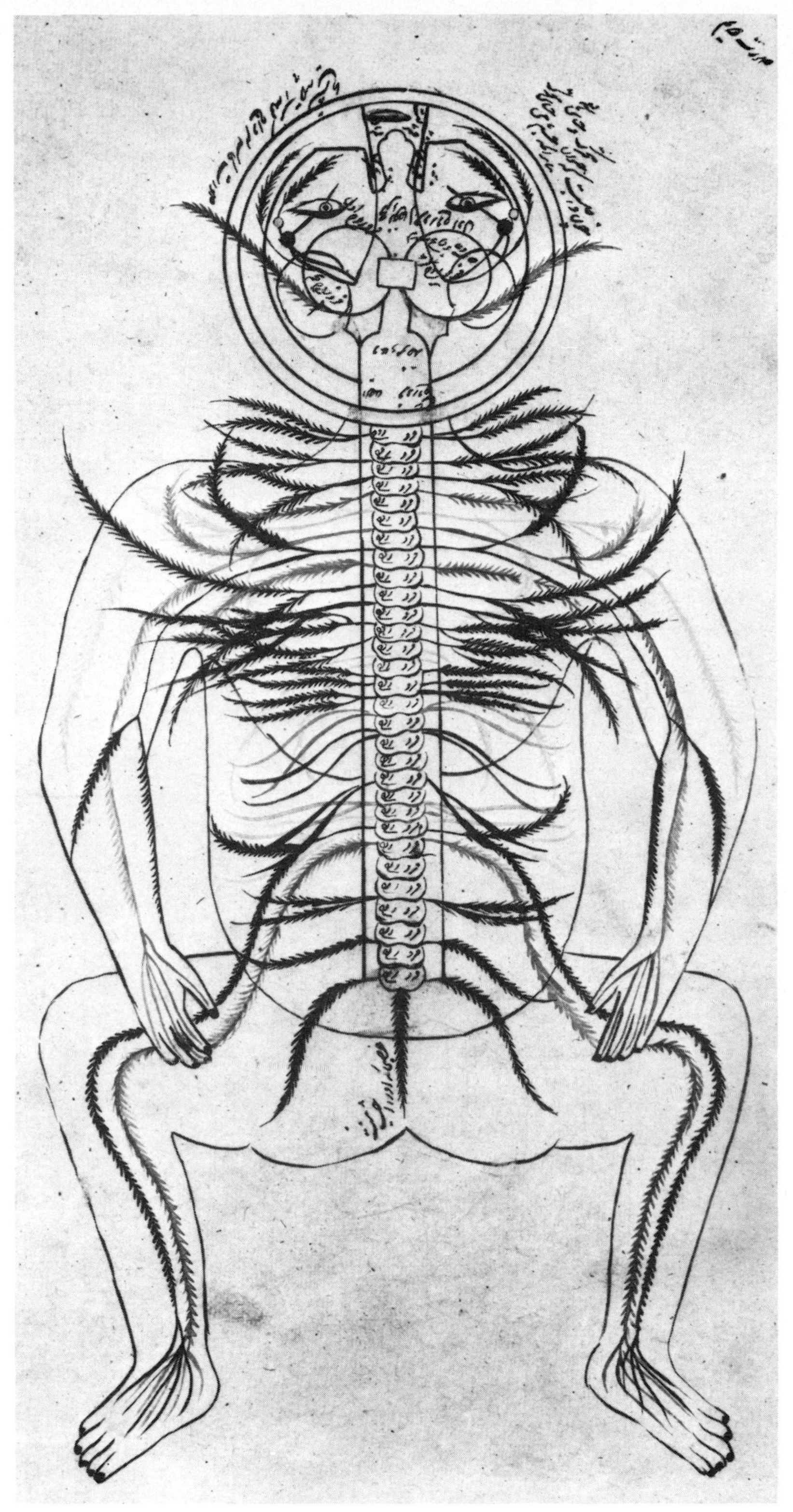

Left Anatomical drawing of veins and arteries from a 17th-century manuscript by Mansur ibn Ahmad. Wellcome Institute of the History of Medicine, London.

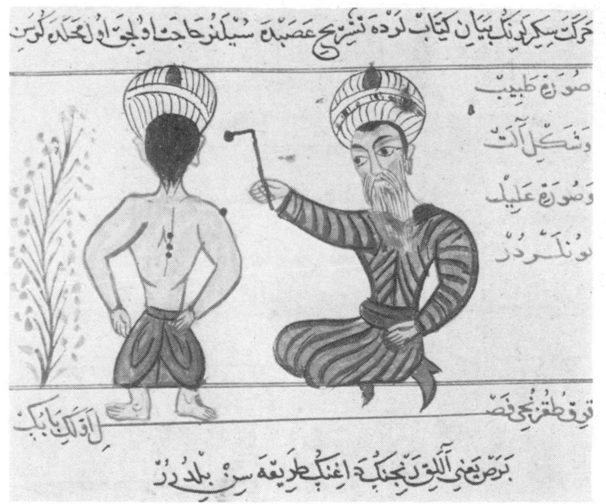

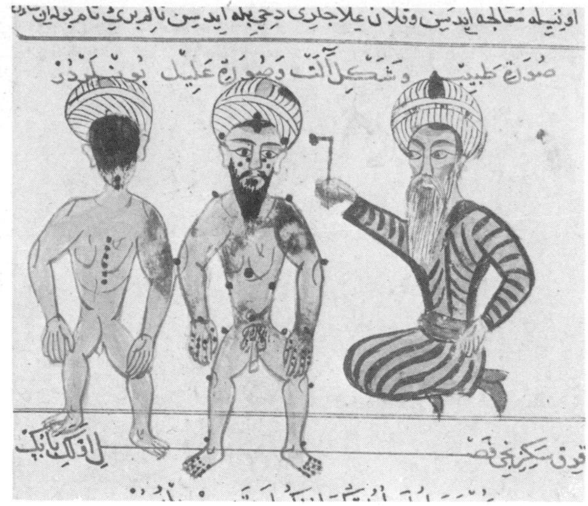

Above The cauterization of leprous lesions, from a Turkish version of a Persian manuscript of c. 1300. This was a widely used practice in Islam as in the medieval West. Bibliothèque Nationale, Paris.

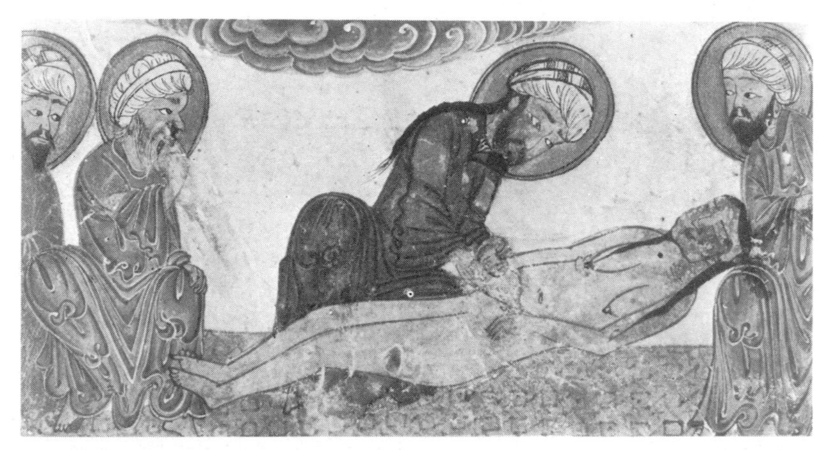

Right A caesarian operation from a 15th-century Islamic manuscript by al-Biruni. Edinburgh University Library.

Right A 10th-century glass alembic used for distillation. Islamic alchemy developed skills in chemistry combined with a spiritual emphasis on the relation between matter and the cosmos. Science Museum, London.

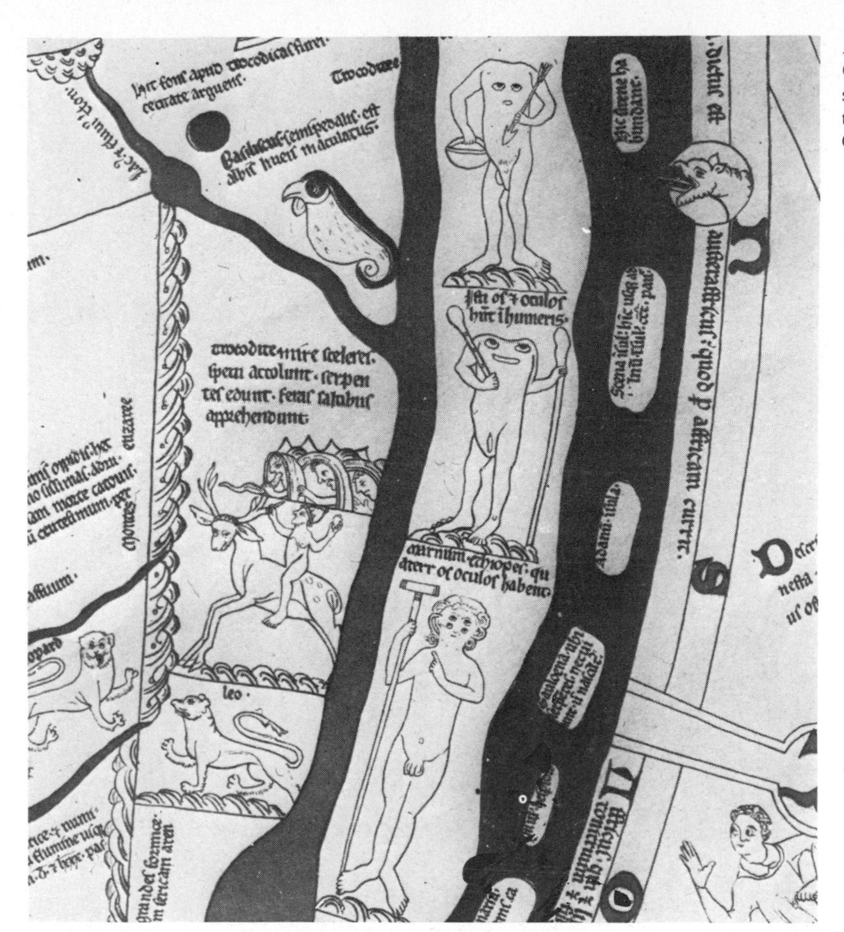

Left A detail from the Hereford Cathedral *Mappa Mundi* c. 1300, showing three of the grotesque tribes thought by medieval Christendom to inhabit Africa.

Left How to pluck a mandrake with appropriate ritual. The forked root of the mandrake (a plant of the order Scrophulanales) was thought to depict the human form and to be in the power of dark earth spirits: no human hand could do the plucking. From a 12th-century manuscript of a Latin translation of the Greek medical botanist Dioscorides. British Library, London.

Right A late 14th-century English drawing of animals, some with formal, heraldic representations and others more naturalistic. Magdalene College, Cambridge.

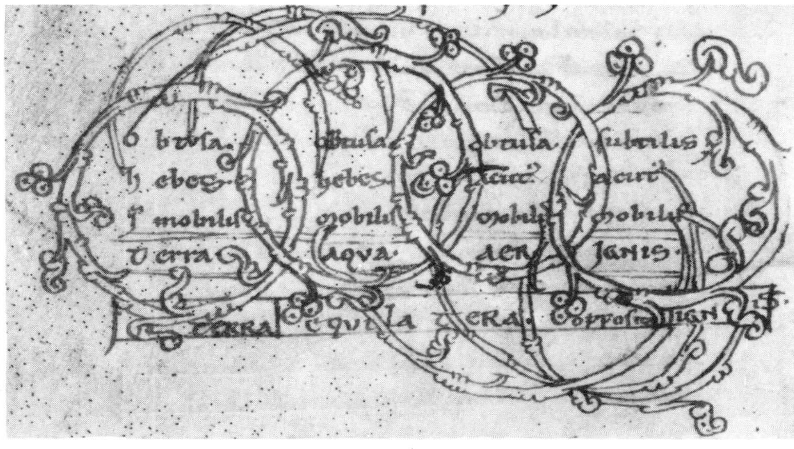

Right Diagram of the tour interlinked elements, fire, air, earth and water, c. 1100. Gonville and Caius College, Cambridge.

Chapter Six

Roman and Medieval Science

We must now return to the early centuries of the Christian era to consider the development of science in the West. Whereas in the Middle East it was the influence of Islam that had the greatest bearing on the preservation of Greek thought, in the West it was first the attitude of the Romans to the work of the Greek scientists and thinkers, and, later, the dominance of Christianity over all aspects of European intellectual activity.

Roman Thought

As early as the third century BC, at the time when Eratosthenes at Alexandria was measuring the size of the Earth, Rome was already ruler of Italy. Two hundred years later the Romans controlled almost the entire Mediterranean, including the Greek world, and they set up an empire that was to provide an unprecedented degree of peace, cohesion and law to a region from Egypt to Britain. The cultural unity imposed by this empire, and enshrined in the Roman towns and roads built everywhere, was responsible for the transmission of Mediterranean arts and learning to the previously backward areas of northern Europe. The unity of this empire became threatened in the third century AD, and in 285 it divided into an Eastern empire, based on Byzantium (which was to survive until the fifteenth century), and a Western, based still in Rome. The West was increasingly subject to barbarian invasions, but the cultural dominance of Rome was maintained in the institutions of the Christian church.

It is often said that the Romans were a practical and technological people not much given to intellectual speculation; for abstract thought they looked to the Greeks for inspiration. Even so, considering the complex formulations of Roman law and the architectural daring of their great aqueducts and basilicas, it is still surprising that they undertook so little theoretical scientific work. It is perhaps less surprising that what work was carried out tended to be a gloss on Greek ideas. Thus even the poet, orator and philosopher Cicero, who lived in the first century BC, drew heavily on the ideas of Epicurus and Democritos, while his cosmological views show the strong influence of Ptolemy. Others of the notable thinkers of the Roman world were really Greek by origin and culture; the physician Galen, sometime surgeon to the gladiators at Pergamum, was one such.

Yet it would be quite wrong to write off the Roman civilization as an intellectual or a scientific disaster. Even though Roman thinkers were not much given to scientific speculation, or Roman patrons to the support of scientific experimentation, the respect in which the Romans held the Greek world was striking. Even in the very last days of the Western empire, the works of Plato, of Aristotle, and of Homer, were being taught as ideals of clear thinking and good writing; and in general the logical and non-metaphysical nature of Greek thinking appealed to the Roman mentality. Thus Rome played a vital role in keeping Greek ideas alive, a role that was to be much appreciated by the scholars of the Renaissance, a thousand years after the eventual fall of the western Roman empire.

Pliny

Pliny, or Gaius Plinius Secundus (sometimes known as Pliny the Elder), was born in Como, Italy, in AD 23, of a well-to-do family and was educated in Rome. As far as can be determined he never married and certainly he adopted his nephew Pliny (Pliny the Younger) as his heir. This nephew became a senator, thus entering the highest order of Roman citizenship, but Pliny himself was a knight (*eques*), the next order below, who therefore began his duties with military service. It was while away on military duties that he began to find his flair for writing, which started with a book on the use of javelins by the cavalry. By his mid-thirties, with his military obligations behind him, Pliny settled in Rome, where he probably practised law as well as writing, and in due course became a friend of the emperors Vespasian and Titus. Pliny is now remembered mainly for two works: by art historians for his history of the fine arts, which is the earliest extant history of its kind in the West, and by the scientific world for his *Natural History*.

Pliny's *Natural History* is a monument to the enthusiasm and perseverance of a man whose efforts were indefatigable, who was able to do with little sleep, and whose watchword was 'To live is to be awake'. Dedicated to Titus, the book is said to be a compilation based on 100 principal authors who, Pliny claimed, supplied him with 20,000 useful facts; yet it seems that his sources actually covered no less than 473 authors in all, and the facts he amassed totalled nearly 35,000. Designed to be of practical use, it contains some excellent and painstaking descriptions and there is much imaginative as well as eccentric writing; unfortunately, though, Pliny was not always reliable in his quotation of sources and he was also somewhat uncritical, so that the fabulous appears side by side with the prosaic. Nevertheless, the work is a magnificent compilation in spite of all its faults; it may not be science but for centuries it did stimulate wide interest in the world of Nature.

Pliny's last assignment was as commander of the fleet in the Bay of Naples, whose duty it was to suppress piracy. It was while on this tour of duty that in AD 79 he was told of an unusual cloud formation (due, it was later learned, to the eruption of Vesuvius) and went ashore to investigate, as well as to reassure the terrified population of the area. Unhappily he was overcome by the fumes, and died as a result.

Galen

No such accident occurred to Galen, whose importance in the history of medicine and in medieval medical thought both in Arabia and the West can hardly be overemphasized. He was born some time in AD 129 or 130 in Pergamum. When aged sixteen he decided to study medicine, and while still a student he began to compile medical books; it is as a medical writer that he was later remembered. From Pergamum he moved to Smyrna and then to Corinth and Alexandria to complete his medical training, returning to Pergamum for a time when he was 28 as official physician to the gladiators, an appointment which brought him useful knowledge of nerves, muscles and tendons. In 161 he moved to Rome where he set up in a medical practice which soon flourished because of some startlingly successful cures of influential patients.

At Rome, Galen was introduced to Flavius Boethus, an influential member of the government who stimulated the young man both to write anatomical and physiological books and also to give public anatomy lectures and demonstrations, which were then very popular. This brought him to the notice of other senior officials and in due course he became a physician to the imperial family. Clearly he knew how to make the best of his opportunities, but it seems that he was also a more than competent physician. Although poor at prognosis (forecasting the course of an illness) he appears to have been a brilliant diagnostician. He took a very real interest in medicines and strongly supported experimental tests of their efficacy, though he himself was not always very scientific in his approach to them. Perhaps because of his experience in Pergamum, Galen had a distaste for surgery – his own surgical experience was primarily confined to sewing up wounds – and disagreed with the followers of Erasistratos who were trying to put surgery on a sounder basis. And in spite of his public lecturing and his writing, Galen seems to have felt no desire to impart his knowledge personally; he took on no students.

For Erasistratos, see page 116.

Galen's work in anatomy and physiology was hampered by the lack of human dissection; when he wanted to carry out dissections he was forced, like others of his profession, to use the bodies of animals. Being well aware of the need for physiological experiment, he must have found this very irksome, and it forced him into a certain amount of speculation. As a practising physician he followed Hippocratic tradition – and herein perhaps lay much of his success – but his father had instilled into him a sound philosophical training, and when it came to questions of anatomy and physiology, it was to Aristotle that Galen turned. This was the teaching he was to promulgate and, in consequence, it was the Aristotelian viewpoint that was to permeate Western medicine right down to the seventeenth century.

Aristotle had drawn attention to the importance of the blood in forming the tissues of the developing embryo, and concluded that it was blood which nourished and maintained the flesh of the adult. This Galen accepted. Again, Erasistratos had linked the digestion with the blood by way of the liver; food digested in the stomach and intestines passing as chyle by way of the mesenteric veins to the liver, where it was converted into blood. From the liver it went by a large vein (the *vena cava*) to the right side of the heart, and thence to the rest of the body. This again Galen accepted, for it allowed him to adopt a scheme whereby Aristotle's 'vegetative soul' could operate by assuming that the liver was the vegetative centre of the body, attracting the blood-forming elements of the chyle. Furthermore, Galen also agreed with Erasistratos in thinking that blood from the *vena cava*, which entered the right ventricle of the heart, nourished the lungs and was prevented by valves from returning to the body.

Yet there were difficulties facing him when he came to some of the other ideas of Erasistratos and Aristotle. For instance, in studying the distribution of blood in an animal, Aristotle had said that it should be strangled before being dissected so that the blood was retained inside the body. This was unfortunate, because it meant that the left side of the heart and arteries were largely drained of blood, the arteries looking just like empty tubes, and this misled many medical men, including Erasistratos. Indeed, Erasistratos went so far as to base part of his physiology on the fact, explaining that 'pneuma' or 'spirit' passed from the lungs through the 'empty' arteries and thus to the whole body. Galen's difficulties arose because his own observations made it clear that blood was normally present in all arteries; he could not, therefore, use Erasistratos's concept of the circulation of the pneuma, and had to devise new theories about the functions of heart and arteries. These led him to make a fundamental mistake.

Galen began by assuming that the main function of breathing was to draw in air in order to cool the blood, and even the heart itself, by way of the pulmonary vein (which runs from the lungs to the left ventricle of the heart). The heated air then returned to the lungs. He knew of the pulsation of blood in the arteries, but put this down to the fact that when the heart was dilated a wave of dilatation passed along the walls of the arteries – an idea originally due to Herophilos – and then claimed that the arteries drew into themselves blood from the veins and pneuma from the surrounding air by way of the pores of the skin. Thus the whole body breathed and innate heat penetrated everywhere. If, of course, this arterial pulse was cut off from a limb – by a tourniquet for instance – then the limb became pale and cold because its innate heat was no longer receiving nourishment from the pneuma. All this seemed logical enough, but it meant that Galen had to modify Erasistratos's teaching about the valves of the heart – heated air could, he said, return past the valve between the left ventricle and auricle, but on the right-hand side the valve, he thought, could not be by-passed. This, then, posed a problem. Blood entering the right ventricle from the *vena cava* was unable to return; true, some of it

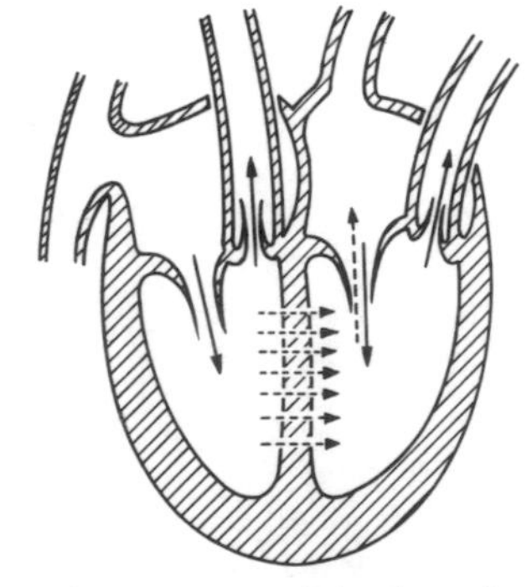

Galen's concept of the flow of blood through the heart.

could go through the pulmonary artery but not enough (because the pulmonary artery was narrower than the *vena cava*). Galen therefore concluded wrongly that some blood seeped through from the right side of the heart to the left.

The whole Galenic system was further complicated by his belief in Plato's tripartite soul – nutritive, animal and rational – but it was an ingenious system. In essence it proposed that the liver and veins supplied the body with nutrition; the lungs, left ventricle and arteries maintained pneuma and innate heat throughout the whole body, while brain and nerves controlled sensation and muscular movement by way of a special psychic pneuma. Later systematizers recognized three forms of spirit: natural spirits which were formed in the liver, vital spirits formed in the heart and arteries and animal spirits in the brain. This, then, was the basic physiological scheme of medieval medicine. It was adopted as the fundamental scheme in Muslim medicine, as we have seen, and it informed medicine throughout the West. Indeed, so basic was it in the West that when medical students went to dissections at the university, the works of Galen were read out aloud by the 'Reader', and the parts of the anatomy pointed out by the 'Demonstrator' – we still retain these academic titles – and there was little or no idea of independent investigation or dissection by the students themselves. But such reliance on authority was not confined to medical studies; it permeated every branch of learning.

For the ancient Mesopotamian idea of the importance of the liver, see page 38; for Islamic anatomical knowledge, see pages 236–237; for the overthrow of the Galenic orthodoxy by Vesalius, see pages 285–286.

Illustration page 298

Christianity and Medieval Science

From its obscure beginnings as a minor Jewish sect Christianity had become by the early fourth century the official religion of Rome after the emperor Constantine the Great's conversion in 312. As a result its priests and bishops were vested with an authority and power previously denied them.

What was the attitude of this new religion to science? Was it going to encourage study of the natural world created by God, or was it not? As so often in matters of this kind, some Christians took one view, some another. On the one hand a good case could be made out for ignoring all secular studies – scientific or otherwise – and concentrating all attention on the important matter of the salvation of souls. And since science at least meant going back to Greek sources, to pagan learning, it might indeed be only prudent to leave it alone lest the mind become contaminated with dangerous ideas, damaging to Christian souls. There was, on the other hand, a diametrically different approach. If God had made the world, and seen that it was good, then studying His handiwork through science could only bring an increasing sense of wonder at such divine wisdom, and a sense of awe at such marvels as the Creator allowed man to see. Ranged on the side of science, strong in the belief that the contemplation of the works of God could only bring increased awareness of the deity's omnipotence and wisdom, was Aurelius Augustinus, later to be canonized as Saint Augustine.

Saint Augustine

Augustine was born at Tagaste in the Roman province of Numidia (now Algeria) in 354; his father was a minor government official and his mother a Christian. Educated in the classics and in rhetoric, he became a teacher of rhetoric at Carthage when aged 21, later teaching at Rome and Milan where he was a professor of rhetoric.

Augustine was successively attracted by Manichaean religious doctrines, the Greek philosophers and Neoplatonism. Clearly, at this stage, Augustine was only seeking some kind of intellectual satisfaction, but suddenly he experienced a dramatic conversion, centred on the epistles of Saint Paul. He took almost literally the text (Romans 13:14) to 'put on the Lord Jesus Christ, and make no provision for the flesh, to gratify its desires', and gave up the world, retiring to a monastic life. In 387 he was baptized.

Augustine's change of heart was permanent, and after due meditation he decided he must return to North Africa, where he began a prodigious career as an author, writing on philosophy and metaphysics and doing so with a convert's conviction, illuminated by the certainty and reality of his inner experience. The result was so impressive that his works exerted an immense influence. Soon, however, he had to forsake his monastic life, for he was consecrated Bishop of Hippo in Numidia, then one of the major centres of Christian influence in North Africa, and was obliged to spend much time in pastoral duties and in travelling. He died in 430.

Augustine wrote nothing on science, nor did he make any scientific observations of any kind, yet he marked a crucial stage in the development of that pattern of thinking and system of values in which science in the Western world was to emerge. Greek philosophers had made the bold claim that true knowledge of the universe could be grasped by sciences which man himself constructed – they believed the science of geometry was proof of their case – but now, when Greek science had largely been forgotten, the Church came with a message which, while religious, clearly had philosophical overtones. In the hands of Augustine, science had a part to play in the Christian religion; he believed that anything, including the natural world, which depended solely on God for its being must be essentially good. The universe, whose creation was clearly the act of an intelligent Creator, must be good. Its study could only be good and would lead to a greater appreciation of God's wisdom.

Augustine stood at a parting of the ways between Eastern and Western views of Christianity; he turned away from Greek and oriental views of history as a cyclic process, and saw time as a one-way progression, believing history to be moving from a beginning towards an end. He also rejected astrology because from his Manichaean experience he thought it to entail a denial of human freedom, and man, he knew, had the freedom to choose between good and evil. To make the right choice man needed divine help and a strong faith, but then Augustine put faith above all things. Indeed, he believed

Illustration page 264

faith had priority even over knowledge, yet he nevertheless still thought it desirable for man to acquire knowledge, epitomized in his designation of theology as 'the queen of the sciences'. Thus at a crucial time, Western Christendom found in Augustine a champion of science and learning. His theory of the divine nature of knowledge dominated Western thinking for generations; in fact it was not until the thirteenth century that it was challenged at the universities of Oxford and Paris in the light of the newly rediscovered teachings of Aristotle. But that was another age.

Yet if Augustine was on the side of those who would seek out knowledge, it becomes clear that for centuries the Christian Church gave little thought to science. It was concerned with other matters. Questions involving science arose, as they were bound to do in a community which sought rules for the determination of Easter, and so introduced a date based on a lunar calendar into civil dates determined on a solar calendar, and in a Church with a care for those facing agricultural problems such as the failure of crops or epidemics amongst sheep and cattle. But such science as was involved concentrated on practical problems, not theoretical questions; theory had been left behind with the Greeks.

The Venerable Bede

Typical of this attitude was the work of the Venerable Bede, a monk of Jarrow in the north of England. Born either in 672 or 673, he was a pupil of Benedict Biscop, a learned cleric who founded monasteries at both Jarrow and Wearmouth, and encouraged visits from British and European scholars so that his religious foundations were something of a cultural centre. Bede obviously enjoyed the intellectual atmosphere surrounding him, though, unlike Biscop, he himself never travelled abroad to visit other centres of learning; indeed it is said he never travelled more than 50 miles (80 km) from the monastery. However, he had little need to do so, for there was a quite exceptional selection of reference material at his disposal.

We know of Bede because he became a prolific writer, the chief voice of what, for a short time, was a flowering of English culture. Half of what he wrote concerned Bibilical exegesis, an art of explanation and critical analysis at which he excelled, while the remainder was mainly concerned with material useful for teaching in the newly emerging monastic schools. Some of the latter contained science, particularly his *computus*, a study of the subject of keeping time and calculating dates, which were both vital matters for monastic communities. Bede spent considerable time on calendar computation; he was the first in the West to have created, or at least to have recorded, a calendar on the basis of the 'Metonic' 19-year lunar cycle, preparing and tabulating a 532-year cycle of dates for Easter. In this it seems he built upon previous work (an earlier 84-year cycle and another of 532 years) managing to reconcile them. And it was Bede who first used as his anchor date the birth of Christ, thus initiating the use by historians and later others of the Christian era, still denoted today by

For the Metonic cycle, see page 156.

the letters AD, *Anno Domini*, meaning 'in the year of the Lord'. Bede also interested himself in the tides, for they were a notable feature at Jarrow and Wearmouth, both places lying at the mouths of rivers which disgorged into the North Sea. Bede noticed that the times of high tides at these places were affected by local conditions, thus drawing attention to that phenomenon which was later to be recognized as general in seaports over the whole world and known as 'the establishment of the port'.

The Revival of Greek Learning

Bede represents the learned man of the early – though not the earliest – centuries of the Christian era in Western Europe. Science was but a small part of his scholarship, which was centred on the faith as seen by the Church and revealed in the scriptures. All scholars, like Bede, were clerics, whose minds were inclined more to matters of salvation and the glorification of God than to enquiry into the details of the natural universe. They lived in an atmosphere inimical to independent thought and enquiry, and it is little wonder that in subsequent centuries there was no scientific speculation; such knowledge of the natural world as was current was based on commentaries which imparted something of original Greek ideas, though only in a garbled and truncated form. It was an unsatisfactory situation, but one which was a natural consequence of the ecclesiastical sense of priorities, and it remained unchanged until the intellectual bombshell of original Greek learning burst upon the West from the twelfth century.

This sudden influx of Greek learning appeared first as Latin translations of Islamic texts, either of original Arabian works such as al-Khwarizmi's *Algebra* and ibn al-Haytham's optics, which reached the West as *Opticae Thesaurus* (Treasury of Optics), or as Arabic translations of, and commentaries on, Greek texts by Aristotle. Later on, Greek texts were to become available in Latin and in Greek. Much of the first work was done in Toledo, a city of Islamic Spain, by such men as Adelard of Bath, Gerard of Cremona and Michael Scot. Adelard, the future tutor of the English Henry II, was responsible for many scientific translations into Latin, including one of an Arabic version of Euclid's *Elements* which was to do duty for centuries in the West as the chief text on geometry, while Gerard was another prolific translator with some 80 works to his credit, including Ptolemy's *Almagest*. Michael Scot (1175-1232), famous as an astrologer and wizard and mentioned in Dante's *Inferno* and by Boccaccio, finally settled at the court of the Holy Roman Emperor Frederick II, whose kingdom in Sicily was another important route through which Arab scholarship reached Christendom. Scot too was a linguist, translating works from Hebrew as well as Arabic, and was responsible for Latin versions of commentaries by ibn Rushd on a number of Aristotle's scientific works and an astronomical text by al-Bitruji which contained a description of the celestial spheres. Yet these scholars were not alone; others were also at work on this new treasure house of knowledge.

Illustration page 221

Among the first to be affected by this recent knowledge were the new universities of Paris and Oxford. These universities were the archetypes of the Western European university, Paris being founded about 1170 as a development of the cathedral schools of Notre Dame, closely followed by Oxford which also grew out of the schools founded in the ninth century by King Alfred. Paris quickly became a great centre of Western Christian theology, and by the 1220s the mendicant orders – the Dominicans and Franciscans – began teaching there. At Oxford the Franciscans were to be found in considerable force and produced two great scientific men, Robert Grosseteste and his pupil Roger Bacon, both well versed in the host of translations then available from Arabic sources.

Robert Grosseteste and Roger Bacon.

Grosseteste was born in the east of England, probably in Suffolk, some time about 1168, and seems most likely to have been at the University of Paris between 1209 and 1214. Until his death in 1253 he was the central figure in England in the important intellectual movement of the first half of the thirteenth century. He held various ecclesiastical appointments and carried out some lecturing in Oxford. (He was certainly there for some five or six years before his consecration as bishop of Lincoln in 1235.) At Oxford he was chief lecturer in theology to the Franciscans, who first came to the university in 1224, but his influence was broader than it might appear from this appointment alone. For it was Grosseteste who not only directed the English Franciscans to a study of the scriptures and of languages, but also set them about learning mathematics and natural science; in brief, his influence was immense at a time when the new knowledge of Greek science and philosophy was having a profound effect on the whole of Christian philosophy.

Grosseteste himself had a great curiosity about natural things and wrote important texts on astronomy, on the cosmos, on sound and, particularly, on optics. His thorough knowledge of Aristotle's works also stimulated him to write about the nature of scientific enquiry. Science, he said, began with man's experience of phenomena, which were usually complex. Its aim was to discover the reasons for the experience – to find what its causes might be. Then, having uncovered the causes – the 'causal agents' – the next step was to analyse them, breaking them down into their component parts or principles. After this, the observed phenomenon must be reconstructed from these principles on the basis of a hypothesis and, finally, the hypothesis itself must be tested and verified – or disproved – by observation. These were important views and his recommended procedure was valuable, for it contained the essential basis of all experimental science.

He also made an analysis of the 'causal agents' as a starting point for an Aristotelian procedure, a classification of the sciences to show how some were dependent on others. Thus he claimed that optics and astronomy were subordinate to geometry, because both used geometrical techniques to explain the behaviour of rays of light reflected from mirrors or refracted through glass or water, and the motion of celestial bodies. On the other hand he held that mathematics itself could only provide a formal causal agent for a phenomenon, because material and, in Aristotle's terms, 'efficient' causes could only be provided by the physical world itself.

Grosseteste's ideas about the sciences were significant, especially in view of his influence as a teacher. Yet of equal significance was his own scientific work which, although it included sound – he was a great music-lover and enjoyed working out mathematical relationships between notes – and astronomy, was mainly on optics. But it is worth noting that in astronomy he made the interesting and, at the time, very novel suggestion that the stars themselves were composed of the four terrestrial elements; this was an idea which had all kinds of implications, but does not seem to have been followed up. He also wrote about the calendar, showing that a year of 365¼ days and a 19-year lunar cycle were not accurate enough, and advocated a more precise measurement of the length of the year and its relationship with the month.

Illustration page 244

Optics, Grosseteste considered, was the basic physical science. Light, he thought, was the first 'form' of prime matter to be created, and he considered it to be a physical substance propagating itself outwards from its source point in a sphere, thus giving rise to the three dimensions of space, an original and interestingly scientific interpretation of the creation text in the first chapter of Genesis, 'Let there be light'. But Grosseteste did not confine himself to philosophical speculation about light; stimulated by al-Haytham's work on optics, he discussed in detail the behaviour of rays of light. He discussed direct visual rays, reflected rays and refracted rays, and the formation of the rainbow. On refraction he is very interesting, for when he comes to mention the images formed by lenses and mirrors he says that all this was 'untouched and unknown among us until the present time' and that:

For al-Haytham, see pages 227–229

This part of optics, when well understood, shows us how we may make things a very long distance off appear to be placed very close, and large near things appear very small, and how we may make small things placed at a distance appear as large as we want, so that it is possible for us to read the smallest letters at an incredible distance, or to count sand, or grain, or seeds, or any sort of minute objects.

This appears in his *On the rainbow or on refraction and reflection*, and reads like a description of magnifying and diminishing lenses and of the telescope. Yet, as we shall see later, it is generally thought that the telescope was not invented for another 350 years. Perhaps, though, Grosseteste had experimented enough with concave mirrors and magnifying and diminishing lenses to satisfy himself that such a device was possible, even though the optical qualities of his lenses would have given only poor results. But this we shall come back to in a moment.

Illustrations page 263

Grosseteste also discussed the law of refraction of light through a lens, though he did not reach a complete understanding; that was only to come in or after 1621. Nevertheless his optical work was a great achievement and a first glimpse of the stimulation to scientific and experimental work which the influx of books on Greek science was to have.

As a university teacher Grosseteste would have had countless students, but his most important disciple, at least from the scientific point of view, was Roger Bacon, some fifty years his junior. From 1241 Bacon lectured at Paris but he was back at Oxford by 1247, where he was introduced to Grosseteste. Grosseteste's influence on Roger Bacon was immense, indeed he must have fired the younger man's imagination in the most extraordinary way for thereafter Bacon spent the rest of his life promoting the study of languages, of mathematics and, especially, of optics and immersing himself in experimental science.

When nearly forty years of age Bacon became a Franciscan, but this seems to have brought him nothing but trouble. He was soon in conflict with Saint Bonaventure, the minister-general of the order, mainly over alchemy (the question of the transmutation of base metals) and astrology, both of which Bacon supported and Bonaventure detested, a disagreement exacerbated by the minister-general's insistence that there should be no publication by members of the order either on subjects of this kind or on theology without his express authority. Although Roger Bacon circumvented the regulation by getting Guy de Foulques (soon to become Pope Clement IV) to request copies of his works, his intellectual views continued to get him into trouble. After he had written condemning the teaching practices of both Dominicans and Franciscans, his unpopularity with the authorities grew and it is perhaps not surprising to find that in the end he was imprisoned for some years, mainly on account of his 'Averroist teaching', a condemnation which at this time meant that he regarded reason and philosophy as superior to faith and revealed knowledge (Averroës was the name by which ibn Rushd was known in Christendom).

For ibn Rushd's work in astronomy, see pages 216–217.

Roger Bacon was an important figure, not because he rebelled against authority but because of the positive virtues of his scientific outlook, flavoured though it was with alchemical and astrological overtones. As far as Bacon was concerned there were four obstacles to grasping the truth of things: (i) 'frail and unsuitable authority'; (ii) long custom; (iii) uninstructed popular opinion; (iv) concealment of one's ignorance in a display of apparent wisdom. That some, if not all, of these criticisms are still valid – especially the last – is a measure of the clarity of Bacon's thinking. He did not class the authority of the scriptures in his category (i), though he did claim that scriptural authority must be informed by reason, and that reason, for its part, must be confirmed by experience. What kind of experience? Here again Bacon was specific, dividing experience into two kinds, experience obtained through inner mystical experience, and knowledge obtained by way of exterior causes, aided by instruments and given precision by the use of mathematics. Thus he confirmed Grosseteste's belief in the significance of observed evidence, an attitude which was to assume cardinal importance some three centuries later.

Bacon went on to claim that natural science led not only to a knowledge of things, but also to knowledge of their Creator, both types of knowledge forming a unity under the guidance of theology. Thus he was of the opinion that men should study languages, mathematics, optics, experimental science, alchemy, metaphysics and philosophy. Clearly, there was no conflict in his mind between philosophy, metaphysics, religion and experimental science (*scientia experimentalis*). All were part of man's knowledge, revealed or observed. In this Bacon's outlook was more medieval than modern and, of course, his *scientia experimentalis* was not our modern experimental science; rather it was what later became called *magia naturalis* or natural magic. This was knowledge gained from experience of the natural world, using one's senses alone or aided by instruments and leading to 'wonderful' discoveries and descriptions, but it was not knowledge obtained by specially designed experiments to answer a specific question; that was something which belonged to a much later age. Nevertheless, Bacon's work was a step along this road, as his important work in optics makes clear.

For al-Haytham's description of the workings of the eye, see pages 227–229.

Illustration page 263

Following Grosseteste, Bacon drew on the remarks of Euclid, Ptolemy and al-Haytham and also emphasized that lenses were of use not only for burning but also for magnification, to aid failing sight. He described al-Haytham's account of the eye from the point of view of an image-forming device, and went on to give eight original rules of his own for classifying convex and concave lens surfaces with respect to the eye. Again, like Grosseteste, his remarks about their optical effects are of interest, and at one point he wrote in his *Opus Majus* (Major Work), which was completed in about 1267:

> For we can so shape transparent bodies and arrange them in such a way with respect to our sight and objects of vision, that the rays will be bent in any way we desire, and under any angle we wish; we may see the object near or at a distance. Thus from an incredible distance we might read the smallest letters . . . So also we might cause the Sun, Moon and stars in appearance to descend here below.

For della Porta, see pages 318–319.

Had Bacon an experimental telescope? This is not outside the bounds of possibility, though if he had it would seem likely, considering the lenses available at the time, that its magnification would be small and the optical image it produced very poor. Moreover, given the general outlook of the medieval mind, it would have been unlikely to be pursued as a scientific observing instrument; it would have been looked on as a curiosity and an example of natural magic, as indeed it was to be as late as 1589 when Giambattista della Porta published a Bacon-like description in the enlarged version of his *Magia Naturalis*.

Bacon believed in astrology, but it would be a mistake to make too much of this, for it was a widely held view current at a time when the Earth was still the centre of the universe and when the medical profession prescribed its treatments for astrologically auspicious times. In general Roger Bacon had a sound scientific sense. It was he who in his writings introduced the concept of the 'laws of nature'. Admittedly he did not invent the term; it is to be found in the works of Saint Basil (fourth century AD) and further back still in Lucretius, but Bacon's revival of it was important at a time in the thirteenth century when changes in philosophical thought were constantly occurring. However, perhaps Bacon's main importance was that he always tried to balance one authority against another, always insisted on good translations of Greek and Arabic works and emphasized the importance of observation of the natural world. His writings, in fact, show the virtues rather than the vices of scholasticism, the mixture of religious dogma and philosophy which was the hallmark of Western scholarly thinking between the ninth and the fifteenth centuries.

Albertus Magnus

With Robert Grosseteste and Roger Bacon we have come closer to what we should recognize today as Western science, but they were not alone either in taking this approach or in promoting new ideas stimulated by the discovery of the works of Greek authors. A notable contemporary of similar outlook and influence was Albertus Magnus or 'Albert the Great', sometimes known as the 'universal doctor'. Born at Laningen in Bavaria in about 1200 of a good family, he went to the University of Padua where he studied the liberal arts and then, much against his parents' wishes, became a Dominican friar. Albertus Magnus remained teaching in the German states until some time around 1241, and then moved to the University of Paris where he held the 'chair for foreigners'. Seven years later we find him in Cologne establishing a *studium generale*, for in spite of his formal degrees in theology, Albertus was a scholar with broad interests and a penchant for teaching. Indeed, though he became principal of the German Dominicans for three years and bishop of Regensberg for three more, most of his life was spent preaching and teaching.

Albertus Magnus played an important role in introducing Greek and Arabic science to Western European universities, and doing so against no little opposition. By and large such science as there was in the university curriculum had been culled either from encyclopaedias, which were more often than not a strange melange of fact and fable, or found in theological works that discussed the six days of creation; it was not the newly rediscovered science of Aristotle and other Greek philosophers. Moreover, the Church in general was opposed to Aristotle's ideas, especially those in natural science, and in 1210 the ecclesiastical authorities in Paris had condemned Aristotle's scientific works so that anyone teaching them, publicly or privately, was in danger of excommunication. In the days when most teaching was in

For Albertus Magnus's involvement in Arabic science, see pages 217–218.

the gift of the Church and when there was a real and terrifying belief in the torments of hell, this was a serious and frightening threat, and though the ban was revoked in 1234 it still inhibited any real spread of Greek science within the universities or schools.

It was in 1240, while he was at Paris, that Albertus Magnus became acquainted with Aristotle's works, and he was so impressed with them that he let himself be persuaded by fellow Dominicans to paraphrase the science which they contained. This turned out to be a monumental task for Albertus did not stop at Aristotelian science; he also paraphrased Aristotle's logic, mathematics, ethics, politics and metaphysics. It was an astonishing achievement and in the eyes of some scholars put Albertus Magnus on an equal footing with the famous Arab commentators and even Aristotle himself; hence the title 'universal doctor'. Into these paraphrases Albertus imported a certain amount of Platonic thinking – a trait that became a habit with many medieval Aristotelian thinkers – and he was never afraid of making it clear whenever he disagreed with Aristotle; unlike some later scholars, he never fell into the trap of thinking Aristotle infallible. Indeed, he emphasized the importance of knowledge based on observational evidence; science, Albertus taught, does not consist merely of believing what one is told but involves enquiring into the nature of things.

This attitude towards science can be seen in his own scientific work. He was unhappy about Aristotle's treatment of the motion of arrows and other projectiles, he speculated that the Milky Way was composed of stars and that the Moon was not a perfect unblemished body, suggesting that its dark markings, far from being shadows created by the Earth, were actually due to surface features. All this was new. So, too, were his suggestions about the way the elements made up chemical compounds and his benign glance towards the atomic theory of Democritos. He did not dismiss the idea of transmutation, so beloved of the alchemists, though he did not believe they had ever achieved it. But, like Aristotle before him, his greatest work lay in his own observations of nature. He classified more than a hundred minerals, many of which he found in the mines and excavations he visited during his pastoral trips as bishop, and even made time not only to study animals and plants but also to write books about them. His treatise *On Animals* does, admittedly, contain some descriptions of fabulous creatures, but all the same Albertus was wise enough to reject many unfounded popular myths, such as the widespread belief that the pelican opens its breast to feed its young. More positively he observed insect mating and dissected crickets and examined some of their reproductive parts; he also opened eggs at various intervals after they were laid to observe development of the chick; he observed the development of fishes and mammals and had specific ideas about nutrition of the foetus. He also described many northern animals unknown to Aristotle and speculated that if there were any which inhabited the polar regions they would have thick skins and their fur would be white.

On plants Albertus Magnus was no less original. He gave excellent descriptions, made a systematic classification by shape – bell-form, bird-form, star-form, etc. – and drew a distinction between thorns and prickles on the basis of their formation and structure. He also made a comparative study of fruits and was the first to observe the importance of temperature and light on the growth of trees, to establish that sap in the roots is tasteless but becomes flavoured as it ascends and, incidentally, he was also the first writer in the West to mention spinach. But perhaps the most interesting thing about Albertus Magnus's botany was his belief that some existing types of plant could be transformed into others and that, by grafting, new species could be developed.

With Albertus, as with Grosseteste and Roger Bacon, we see the reactions of great minds to the stimulation which Greek science brought to a culture previously lacking in any systematic study of the natural world. But of course to the Christian theologian this influx of Greek philosophy had its dangers: its outlook was pagan and inimical to revealed religion, for it had been nurtured in a culture where lip-service was paid to a pantheon of gods and goddesses, about whose existence the intellectual had mental reservations. Moreover there were many aspects of Aristotelian philosophy which were in conflict with the scriptures accepted as authentic by the Christian Church, as Grosseteste himself had realized. What then was to be done? It might be safe to let men of the calibre and orthodoxy of Grosseteste and Albertus Magnus have access to the originals but what of others, whose Christian doctrine might become heretical under this pagan influence? Clearly there was a need for some acceptable synthesis of Christianity and Aristotelianism. Certainly, it would be a formidable task but the right man appeared to do just this – to weld Christian doctrine and pagan thought into an acceptable amalgam. His name was Thomas Aquinas.

Thomas Aquinas

Born near Monte Cassino about 1225, Thomas was the youngest son of a rich Italian family, then supporting the Holy Roman Emperor Frederick II against the papacy and consequently in a precarious political position. He was sent away to be educated, first to a monastery and thence, at the customary age of fourteen or fifteen, to the new University of Naples where he studied grammar, logic and some natural science. At the age of twenty and against family opposition, Thomas, like Albertus Magnus, became a Dominican friar; almost at once he was sent to Paris to receive further education, for in the thirteenth century the Dominicans were in the forefront of intellectual life. At Paris he may possibly have become a pupil of Albertus Magnus; what is certain is that Thomas came under his influence at the *studium generale* in Cologne.

Thomas Aquinas soon showed himself an able theologian, and between 1256 and 1259 held a Paris professorship in the subject.

For ibn Rushd, see pages 216–217.

Then, after a decade back in Italy, he returned to Paris to hold the chair in theology for a second time; this was unusual but due most probably to the disputes over Aristotelianism which had arisen there between those of traditional Augustinian orthodoxy and the Averroists, whose leanings were more towards the independent thinking of ibn Rushd. One of the key issues was the question of the creation of the world, and here Aquinas was careful. He taught that reason alone could not decide whether the world was created at a moment in time or not, since there was nothing from the purely philosophical point of view that prevented it from being eternal. Obviously this was a case where revealed knowledge from the scriptures was vital in reaching a decision. Yet in spite of Aquinas's attitude he became tarred with the Averröists' brush and was included in a condemnation of them; seven years later in 1277 a further condemnation was brought against him for certain writings. But Aquinas had died in 1274, having returned to Naples the year before to erect a Dominican *studium* near the university where he lectured and wrote, attacking his detractors with polemical treatises which allowed him to weld a still closer synthesis of pagan philosophy with Christian theology.

Although no scientist himself, in Paris, the centre of these intellectual battles over the newly discovered learning, Aquinas stood up for the cause of reason. Almost alone he was able to make the theological faculty change course and come to terms with Aristotelian teaching. He set his face against the notion that all knowledge comes by divine illumination; the invisible things of God could, he believed, be seen through His visible creation. Reason could bring truth and certitude. The world of nature was a book written by God. And all this, coupled with his careful theological interpretations of Greek science, meant that orthodox Christians need have nothing to fear from pagan philosophy. When it came to science, the Greeks revealed God's world; when it came to matters of salvation then the Church and the scriptures were the revealing authority. After his death and the 1277 condemnation, it began to be appreciated just how far Aquinas had gone in reconciling Christian orthodoxy and Greek science, and within fifty years he was being looked on by the Western Catholic Church as its most representative teacher; in 1323 he was canonized.

John Duns Scotus and William of Ockham

The attitude which Saint Thomas Aquinas promoted was developed by two other important philosophers, the Franciscans John Duns Scotus (born at Duns in Scotland about 1266) and William of Ockham born near London, twenty years younger. Duns Scotus made things 'safer' for the scientist by separating experiment and scientific reasoning from theology, claiming that higher knowledge was only attainable through inner awareness – an awareness that did not involve the senses – and stressing that God in his infinite being was unknowable by physical observation. Moreover, he claimed that God could override physical and natural laws should He wish to do so; the one limitation was that He could never do anything that was self-

Left An early 9th-century drawing of cauterization, a form of treatment that was thought to work by drawing noxious liquids from the body through ulcers. Biblioteca Medicea-Laurenziana, Florence.

Left Large alchemical alembics for distillation. Western alchemy was far more concerned with transmutation of base metals into gold than was that of the East.

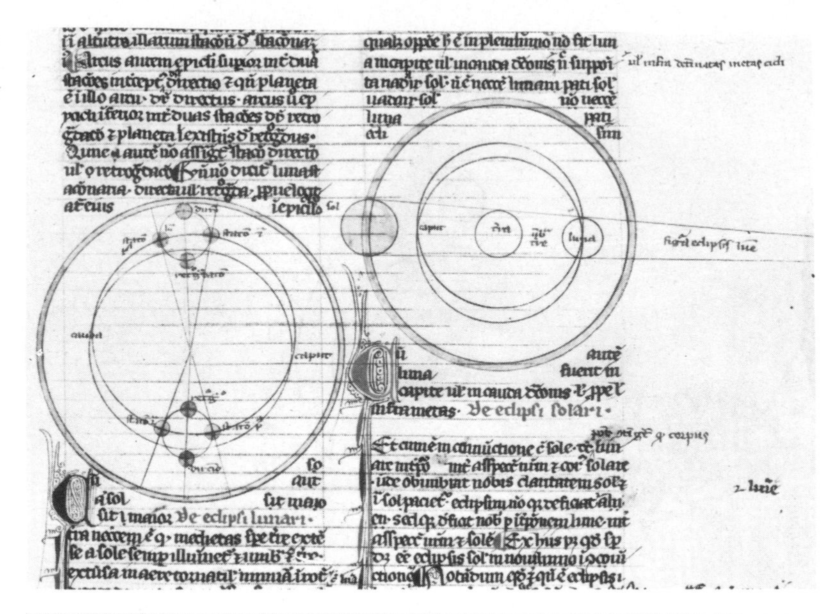

Right Eclipse of the Sun and Moon as shown by John Holywood (Sacrobosco), *De Sphaera* originally written about 1200. Universitäts Bibliothek, Basel.

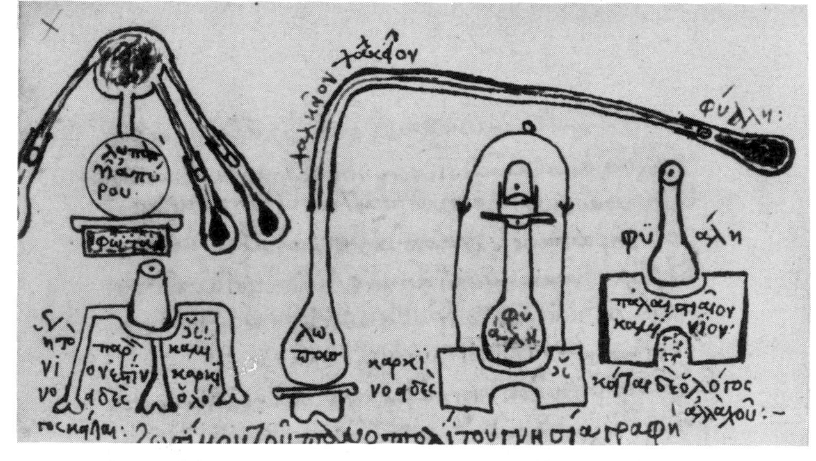

Right Byzantine sketch of an alchemical study, with equipment known by AD 300. Alchemy and astrology were the only fields of science to flourish in Byzantium, particularly before the 5th century and after AD 1000. Bibliothèque Nationale, Paris.

Right The 'formula of the Crab', an alchemical formula for the transmutation of base metal into gold, by the 3rd or 4th century Byzantine scholar Zosimus. It indicates a sequence of materials and processes comprehensible only to initiates. Biblioteca Nazionale Marciana, Venice.

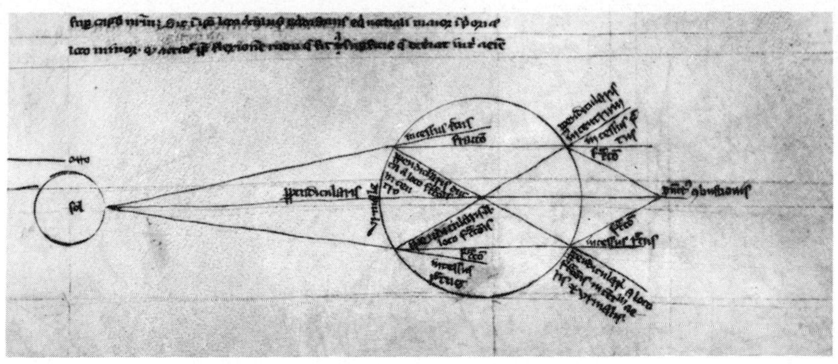

Above Portrait of a Dominican friar wearing spectacles by Thomasso a Modena, 1352; the earliest known illustration of lenses used for this purpose.

Above left A study of the diffraction of light by a spherical lens, made by Robert Grosseteste in the mid-13th century. British Library, London.

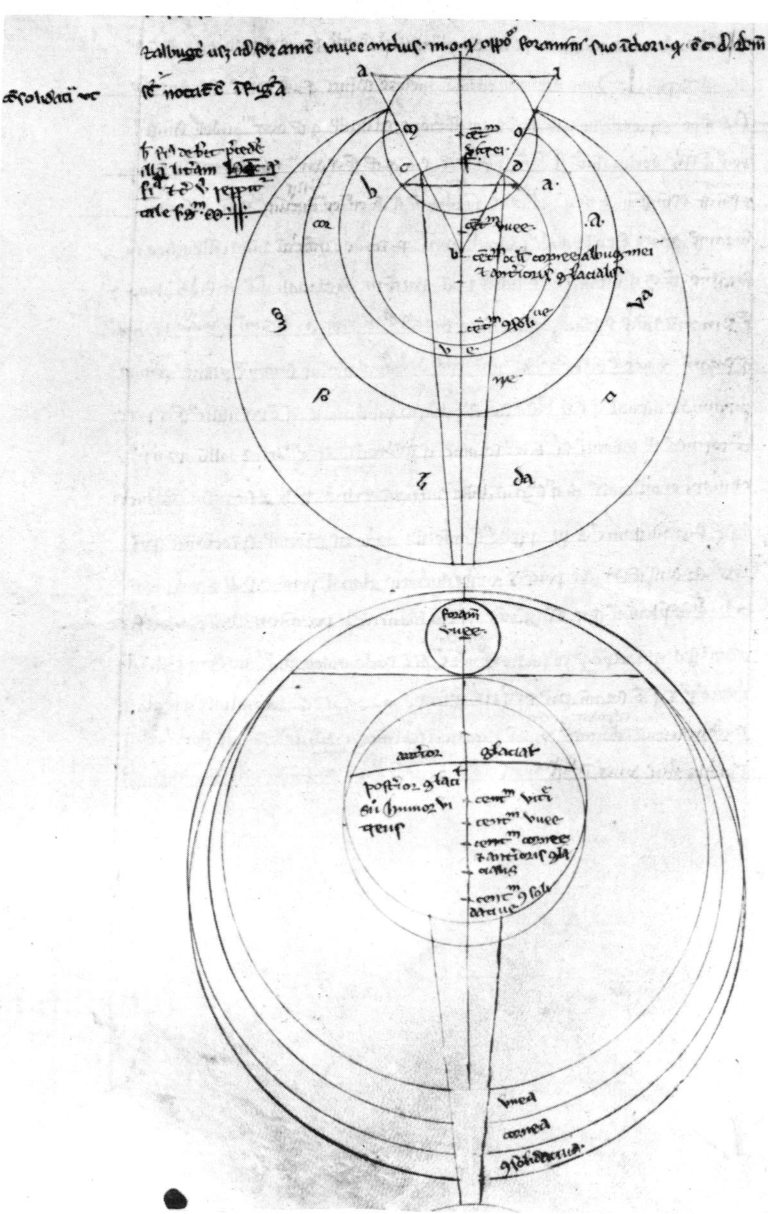

Left Roger Bacon's study of the internal structure and optics of the eye, from *Opus Majus*, 1268. Compare this with the drawing by al-Haytham, page 221. British Library, London.

Right The tower of philosophy, with theology, 'the queen of sciences', presiding over the hierarchy of learning which includes geometry, astronomy, logic, music and poetry. From *Margarita Philosophica*, by Georg Reisch 1508.

Right A waggon drawn by horses equipped with the inefficient medieval lateral trace harness. From the Luttrell Psalter c. 1338. British Library, London.

contradictory. Duns Scotus died in 1308.

William of Ockham initiated the philosophical doctrine of nominalism that was to dominate thinking in the universities of northern Europe during the fourteenth and fifteenth centuries. It was a doctrine about 'universals', and really concerned the question of essence and existence. Its niceties do not concern us, though in passing it may be mentioned that Ockham proved that there could be no real distinction between the two, an attitude that was still to exert an effect in the seventeenth and eighteenth centuries on philosophers like John Locke and David Hume. Ockham, now probably best remembered for 'Ockham's razor', his philosophical principle of economy *entia non sunt multiplicanda praeter necessitatem*, 'entities are not to be multiplied beyond necessity', carried still further the separation between scientific knowledge and revealed knowledge which Duns Scotus had promoted a generation before. Though Ockham disagreed profoundly with Duns Scotus's general philosophical outlook, he was at one with his attitude to science.

Motion

This separation of science from revealed religion was stimulated by, if not wholly due to, the 1277 condemnation of Aristotelian philosophy; coupled with the later reverence for Thomas Aquinas's interpretation of those aspects of Aristotle's teaching which impinged on Christian dogma, it encouraged those interested in science to speculate and even to look at Aristotelian science with a critical eye. Nowhere was this more in evidence than in Aristotle's teaching on moving bodies. Indeed, one of the few scientific questions which Saint Thomas Aquinas had himself discussed was the difficult problem of the motion of a projectile. This was one of Aristotle's less successful essays on motion, and it arose because of his belief that for a body to take on any 'unnatural' motion (i.e. motion different from the 'natural' kind of motion such as falling to the ground), a force must constantly be applied. Should the application of this force cease, then the 'unnatural' motion would cease. Thus in the case of an arrow shot from a bow, if the force causing its 'unnatural' motion towards its target should be withdrawn, the arrow would fall directly to the ground. This whole concept raised many problems.

For Aristotle's theory of motion, see page 105; for an alternative, Hindu, view, see page 194.

First of all, its basic principle was that a constantly applied force caused a regular or uniform motion; it did not result in any acceleration (as we should now expect it to do), because it involved a body moving against resistance (the resistance of the air). Should the resistance be removed, the body would move faster and faster until it reached an infinite velocity. This was one of the reasons why Aristotle thought that a complete void, which would, of course, offer no resistance, was impossible. If one accepted a void, a vacuum, then the whole question reached absurd proportions.

Secondly, there was the question – and this indeed was the main one – of what kept the body in motion. When the bowstring was released it pushed against the arrow and started it off on its journey.

But once the arrow left the bowstring why did it not fall straight to the ground? What kept it moving? Aristotle suggested that the air, parted by the moving arrow, closed in behind it and so propelled it along. But the air was the substance resisting motion, so how could it be the propelling force as well?

One of the first to tackle anew the question of motion through a resisting medium was Thomas Bradwardine. An Oxford scholar who was to become archbishop of Canterbury for only a month in 1349 before he fell victim to the Black Death (a form of plague which killed one-third of Europe's population between 1347 and 1351), Bradwardine wrote an important *Tract on Proportions*. Essentially this was a discussion of the problem of the precise mathematical relationship between the magnitude of Aristotle's moving force, the strength of the resisting medium and the velocity reached by the moving body. What Bradwardine did first was to show that all current formulae were inadequate because none completely satisfied the basic Aristotelian teaching that motion could only occur when the motive force exceeded the resistance; they all broke down under extreme conditions. Having shown up the inadequacies of previous formulations, Bradwardine then offered his own solution: this involved the 'powers' of the variable quantities involved, a step which opened up the problem to deeper examination and was used by those who later tackled the question, especially Nicole Oresme, a pupil of the famous and influential teacher Jean Buridan.

Buridan was born at Béthune in the north of France in about 1295, was a secular cleric (i.e. not a member of a religious order), twice rector of the University of Paris, and a man who exerted immense influence on the study of natural science in the first half of the fourteenth century. Stimulated by Ockham's work, Buridan wrote on logic as well as on physical science; indeed he is probably best remembered today for his logic and the allegory known as 'Buridan's ass'. This concerns the problem of choice between equally desirable alternatives, and the allegory tells how an ass, placed between a pile of straw and a container of water, dies of thirst and starvation because it cannot make up its mind which to consume first. The source of the allegory is rather obscure, because Buridan's example involved the behaviour of a dog, not an ass, the choice was between two equal amounts of food, not food and water, and the dog did not starve but realized that its choice in such a case must be made at random.

Buridan's importance to us is that he spoke openly against the use of supernatural explanations for natural phenomena and was a strong believer in the principle of cause and effect in the natural world; above all, it was he who brought the solution of the motion of a projectile nearer by using the concept of 'impetus'. Albertus Magnus had mentioned the idea of impetus when he had discussed what we may call the projectile problem, but Buridan looked on it almost in the way in which we should consider inertia, as something inherent in a body. At all events, Buridan's attitude was of great significance.

Interestingly, Buridan applied the doctrine of impetus to planetary motion. He suggested that God had perhaps originally imbued the planets with impetus, and if this seems a supernatural explanation, it was nevertheless an improvement on the Aristotelian idea that each planet was moved by a guiding intelligence. It was a significant step forward towards a physical explanation of the behaviour of the entire universe, and it clearly influenced later thinkers, not least Buridan's pupil Nicole Oresme.

Nicole Oresme

Oresme was born about 1320, studied at the University of Paris and in later life became a friend of the dauphin of France (the future King Charles V) and bishop of Lisieux, where he died in 1382. A dynamic preacher, friend of princes, subtle in disputations and a skilful translator from Latin into French, Oresme was an influence to be reckoned with, and his scientific ideas had wide currency.

A strong opponent of astrology, Oresme was the first to use the metaphor of the heavens as a clock – the mechanical clock had recently reached Western Europe – though he did not accept Buridan's argument about a divine gift of impetus to the planets, and still hankered after the guiding intelligences. To Oresme there was still a basic difference between terrestrial and celestial motions. Again he did not accept Buridan's idea that impetus was something permanent in a body, rather he thought of it as self-expending. This was an interesting and important development. It made impetus synonymous with a transient form of the impressed force which carried the projectile along after it had received its initial push. It would last, though, only for a time, and when it was expended the projectile would cease to move and would drop to the ground.

Illustration page 288

Oresme applied mathematics to planetary and other motion, and in doing so extended Bradwardine's work because he used not only 'rational' numbers for the powers of the variable quantities in the projectile problem, but also brought in 'irrational' powers. Since irrational numbers were those which could not be expressed by simple ratios [e.g. whereas a rational number like one half (0.5) can be expressed as ½, an irrational number such as the square root of 2 (1.4142136. . .) cannot be written as a simple ratio], Oresme believed there was a certain amount of numerical indeterminacy in expressing the behaviour of the universe. This provided him with ammunition against astrology which, he claimed, was based on a rigid determinacy.

Oresme also applied the concept of the 'centre of gravity' to bodies in the universe. In this he followed Buridan, who thought that the Earth slightly altered its position in space as erosion of its surface altered the position of its centre of gravity, which must lie at the centre of the universe. He also talked about the possibility of other inhabited worlds in space, an idea that had the most immense religious implications. The fact that he could even suggest this is a measure of how science and religion had become separated, even though it was obligatory for Oresme and others, when expressing radical or unorthodox ideas in the name of science, to make it clear that these were speculations, not descriptions of reality.

The motion of falling bodies was another aspect of the basic laws of the physical world which was tackled by the fourteenth-century natural philosophers. At Merton College, Oxford, in the 1330s, a method of measuring uniform acceleration by the average speed had been devised; it was a method to be adopted by Galileo, but Oresme did not use it. He suggested that the velocity of descent for bodies on Earth depended on the time for which they had been falling rather than the distance they had travelled; he was clearly thinking in terms of equal increases in velocity in equal times, thus presaging our present expression of the rate of acceleration of 981 centimetres per second every second (or 32 feet per second per second). It was Oresme, too, who wrote on geometrical and numerical ways of analysing motion, work that was to bear fruit centuries later, as we shall see.

Conclusion

The late medieval scientific movement concentrated on physical science because it was a subject in which it was possible to exercise precision of thought and freedom of speculation which would have been more difficult or even impossible in other fields. It was work which was to be continued in the following centuries, in the time that has come to be called the Renaissance, and on into the period of what is often called the Scientific Revolution. And it is in the physical sciences that we see most clearly the emergence of modern science, which was to a great extent based on the enquiring attitudes of late medieval scholars.

Chapter Seven

From Renaissance to Scientific Revolution

Throughout the history of science there have been many revolutionary theories about the natural world, many changes in the paradigms which man has adopted to make sense of the way the universe works. Such 'scientific revolutions' have differed in intensity. Some, like those major revolutions which saw the introduction of mathematical paradigms to describe planetary motions, first among the Babylonians and later among the Greeks, have turned out to be significant strides forward in our whole way of looking at the universe with a new understanding; others, such as Aristotle's paradigm of a scale of nature as a method of classifying the myriad creatures which populate the world, were mini-revolutions, important but not radical enough to cause a complete reorientation in man's approach to the subject. But the revolution in outlook and basic ideas about Nature which stood head and shoulders above all others, the one that engendered the modern scientific approach, was that which began in the fifteenth century and carried on to the end of the sixteenth. Its consequences were in fact so great that with good reason it is often called 'The Scientific Revolution'.

It was not the first or the last revolution in the history of science. In recent decades, much discussion has taken place on the manner in which absolutely fundamental scientific ideas and even scientific 'facts' can change as a new theory emerges to account for facts that the older ideas cannot explain. Such a new theory can revolutionize its field. The theory will be tested for both in generating new, more detailed insights. Such changes in theories, or paradigms, may vary in intensity or breadth: some may only affect a single field of science for a short time. The scientific revolution of 1500–1600, though, not only affected every field of science but changed the techniques of scientific investigation, the goals that the scientist set himself and the role that science might play in philosophy and even in society itself. So profound a change could not simply happen of its own accord, but was an offshoot of the more general change in the way in which man viewed himself and the world in which he lived. This was the change known as the Renaissance.

The Renaissance began in Italy in the fourteenth century, and can be seen reflected in the works of the poets Petrarch (1304–74) and Boccaccio (1313-75), whose humanism and belief that their culture was the heir of classical antiquity laid the foundations for the Renaissance outlook. Underlying the Renaissance was the influence of classical literature, original manuscripts and translations. These opened up a new perspective on classical antiquity which inspired authors and poets, painters and sculptors, and imbued them and their patrons with a humanistic outlook. The effect began slowly but gathered momentum and then began to challenge the general medieval approach which surrounded it. Thus in the 1450s we still find Lorenzo the Magnificent not only pressing his scholars for translations of the most recently discovered classical texts, but also asking a friend to comb the bookshops for so typical a product of medieval scholarship as a commentary on Aristotle's *Ethics*. Yet gradually the change came. The humanism stimulated by classical learning penetrated every aspect of cultural life, broadening it and extending its boundaries far beyond the confines of the religious symbolism which was so dear to the medieval mind. It began increasingly to secularize men's attitudes, encouraging them to recognize beauty in the natural world, not only in a world limited to sacred imagery.

Illustration page 292

Why did the Renaissance begin in Italy which, after all, was no more than a loose conglomeration of city-states with no common cultural unity. Why was it here that a new appreciation of humanistic values should blossom first? There seems to be no simple answer: there was the natural genius of the people, the advent of economic development and entrepreneurship – the growth in fact of capitalist expansion and an element of political independence, as well as other influences less easy to define. Moreover it did not begin at the same time all over Italy; it started in Tuscany, the centre of Italian banking and a province with a tradition of learning that permeated court as well as cloister which was therefore particularly sensitive to any new intellectual stimulus.

The Renaissance of the fifteenth century saw not only a rediscovery of classical antiquity but also a new discovery of the world itself. It was an age of geographical exploration. Since Ptolemy had drawn his map of the world in the second century AD geography in Europe had seriously deteriorated. Map-making had ceased to be scientific, specifying places by a system of co-ordinates, and moved to religious cosmography in which the world was represented by a disc divided by a few partitions into continents that bore little resemblance to their true shape or even their correct positions. By the fourteenth century, however, a number of sea charts of areas of the Mediterranean began to appear and of necessity these showed the shores of the land-masses properly, with harbours and ports. The Portuguese then began a series of voyages of exploration, which led to the establishment of an overseas empire for Portugal and to imports of goods, especially spices, that were of economic importance. The first expedition took place in 1418 and led to the occupation of the Azores, Madeira and the Canaries; later expeditions resulted in various settlements down the west African coast as far south as Sierra Leone and the annexation

Illustrations pages 289, 290, 291

of the Cape Verde islands. By 1500 they had reached India.

Of all the explorers around this time, the most famous was surely Christopher Columbus. Columbus's claim to fame is his discovery of the Americas, though this was neither his aim nor that of his sponsors, the Spanish court. What he really set out to do was to find a western route to the Far East. His calculations were based on a study of the geographical writings of classical antiquity, as well as Hebrew, Arabic and European sources, and contained two basic errors, though without them he would never have dared attempt his voyage. One was a belief that the land mass of the Far East extended much further eastwards than it does, the other was Columbus's calculation of the distance he would have to travel. In 1492 he set sail and reached the New World.

Illustration page 293

The discovery of new sea routes and new areas of the world, especially the totally unexpected 'New World' in the west, had the most profound repercussions on contemporary outlook. They underlined the fact that the ancients, in spite of the greatness of their civilization, had not known everything there was to be known about the world; this implied that man had only to look and he might be expected to make totally new discoveries in other fields.

The voyages of exploration were not sponsored by Portugal and Spain as an adventure, but for reasons of national prestige and, above all, for commercial gain. These, however, were not the only commercial enterprises to have an effect on the Renaissance; there were also the important inventions of paper and printing. Both had originated in China, and came to the West, by way of Islam. Paper, developed at the beginning of the second century AD by Tshai Lun (Cai Lun), had moved to the oases of central Asia within the next hundred years or so, but after this its progress was slow. By the eighth century it was being made in Samarkand, and by the tenth was being used for wrapping goods in Cairo bazaars, but not until 1150 was it available in Europe. The transmission of printing followed a somewhat similar course. Printed books appeared in China as early as the ninth century and a book trade flourished there almost at once; a century later parts of the Qu'ran were being printed in Cairo. The first printed items on sale in Europe were playing cards, which did not appear until 1377 in Germany. Germany was also to be the centre from which printing began to spread after the invention of movable type during the first half of the next century by the goldsmith Johannes Gutenberg. This was a crucial development, for it meant that books of every kind could be set up from a stock of separate alphabetical letters.

For the invention of printing in China, see page 182.

Once the invention had been made, the process spread quickly within Europe To those who had been forced to depend on handwritten manuscripts it was a revolution of the greatest significance. No longer were commentaries and originals liable to be marred by copyists' errors; and whereas the making of copies had been slow, tedious and costly and, for the most part, confined to monasteries now once the printed page had been set in type, the more copies that could be made the less expensive the final product became. And since only a comparatively small body of scholars could read Latin, printers began to issue books in the vernacular. This was to spread the new thinking of the Renaissance far beyond the confines of cloister and university.

Illustration page 291

In Europe the invention of printing from movable type coincided with yet another invention, which was the technique of printing illustrations from engraved metal plates. Originating in the Rhine valley and in northern Italy in the 1450s, and again the product of men who were trained as goldsmiths, this too helped the spread of knowledge in Renaissance times and was to have special importance in some scientific fields. The development of printing from delicate woodcuts was also to prove significant in advancing science. Naturally enough the importance of these printing inventions was not lost on many scholars and there grew up a new phenomenon, the advent of the scholar-printer and scholar-engraver. Artists, too, took up the art of engraving. Albrecht Dürer of Nuremberg was a notable example of a creative painter who appreciated the material success to be gained from making engraved copies of his works and who, incidentally, combined this with the printing of his own books.

Illustrations page 294

The Reformation

Besides science, another field in which printing was to play a very significant part was religion, an area where the humanism and independence of thought that the Renaissance encouraged was to have the effect of fragmenting Christianity. Certainly religious revolt was nothing new, but now there was dissatisfaction with Rome's autocratic rule. In Bohemia in the 1420s John Huss wanted a Czech liturgy in place of the Latin as well as other changes, while he also objected strongly to the poverty of many clergy and wanted to appropriate Church property. His followers were grievously persecuted. In England, John Wycliffe, one-time master of Balliol College, Oxford, had preached somewhat similar reforms a generation earlier. He also had set about translating the Bible into English. Once printing was developed and thousands of copies of vernacular translations of the scriptures could appear, the movement of Church reform had an impetus that was to carry it forward and even further from what many saw as the yoke of Rome.

Looking back now with hindsight it is not difficult to see how the established Church had difficulties in keeping to the 'golden age' of 'pure faith' of the early Christians, the recovery of which was the aim of the reformers. For to survive and make converts the early Church had been forced to make some compromises, accommodating itself in some degree to the thoughts and behaviour of the converts themselves. Thus it was that certain pagan customs became incorporated into Christianity; what had been fertility rites became Church feasts and festivals. And when whole tribes were converted, as in the Germanic lands, the previous local custom of commuting the penalty for a crime to a payment of money was taken over complete. Then, as the Church's organization became more centralized and regularized, this gave rise to the sale of indulgences, a practice that became fraught with abuse.

Thus it was that although the Church had set out with the best intentions, over the centuries it had become corrupt. Reforms were needed and, although a simple cleansing of abuses should have done all that was required, other factors, spiritual, political and social, which had been at work in the 300 years before the dawn of the Renaissance decreed otherwise.

In brief, the stage was set for an act or a series of acts of regeneration, but when these arrived in the early sixteenth century the Church could not contain them; western Christendom was torn asunder by a movement that has since been named the Reformation, and Protestantism was born. Needless to say the Papacy was forced to react, and there was a Counter-Reformation. Certain internal reforms took place around the mid-sixteenth century, and the Church also revived the Inquisition, which had been established in medieval times to enquire into heresy, witchcraft, sorcery and alchemy.

All this – the Reformation and the Counter-Reformation – was to have a profound effect on the growth and practice of science during the Renaissance and for long after, as will become clear as we trace the progress of science from the fifteenth century onwards. That happened because of the ethics of the emergent Protestantism. On the one hand, the Protestant attitude to work encouraged in northern Europe (and especially in Germany) the growing capitalism of the time, and on the other hand it stimulated scientific research. The scientific stimulus was caused by the wish to use discovery to create an orderly and coherent picture of the universe, with a view to uncovering more of God's handiwork. This helped satisfy a need felt by those to whom God's ways with man were to be discerned more in the Bible and in Nature than in the mysteries of the sacraments and of the Church.

Hermetism

Research in the last two decades has shown that during the Renaissance and for a short while after it, there was another movement in thought and belief which, like orthodox religion, was to exert a deep influence on studies of the natural world. This was the semi-religious and quasi-magical set of ideas supposedly due to Hermes Trismegistus, and known as Hermetism. These were based on writings which were thought to have originated in Egypt at the time of Moses, and to have been derived from or inspired by the Egyptian deity Thoth, the god of reckoning and learning and advisor to the other gods of the Egyptian pantheon. His Greek equivalent was Hermes and, with the inflated reverence of the Greeks for the Egyptian mysteries, by Hellenistic times he had received the additional appellation of

Trismegistus, or Thrice Great.

The ancient provenance of the Hermetic writings was to be disproved in the seventeenth century, but in the fifteenth century when they and other works from antiquity found their way to Italy, no one questioned their authenticity. After all, they came with what seemed to be unimpeachable credentials. The early Church Father Lactantius, the 'Christian Cicero' as he was known in the Renaissance, whose *Divine Precepts* was the first systematic Latin account of the Christian attitude towards life, had no doubt about their genuineness. Indeed, it is clear that he thought Hermes Trismegistus was a pagan ally of Christianity, a veritable example of the importance of using pagan learning in support of the Christian faith, and even quoted him in the *Precepts*. But Lactantius went even further than this, for the use by Hermes Trismegistus of the phrase 'the Son of God' and his reference to the 'Son of God and of the word', led Lactantius to regard Hermes as the most important of all the seers and prophets of antiquity who foresaw the coming of Christianity. But one had not to rely only on Lactantius. The credentials of the Hermetic works were vouched for by no less a figure than Augustine of Hippo. Admittedly Augustine had condemned what 'Hermes the Egyptian, called Trismegistus' wrote about how the Egyptians animated their 'idols' by drawing spirits into them, but all the same he too believed that Hermes had predicted the advent of Christ.

In about 1460 one of the agents employed by the Tuscan ruler Cosimo de Medici bought for him a Greek manuscript from Macedonia. This contained fourteen of the fifteen treaties that made up the Hermetic writings, and its arrival caused no little excitement. Cosimo was now in his seventies and was determined to read what he could of these magical and mystical philosophical works before he died, so he instructed the scholar Marsilio Ficino, whom he employed to translate and interpret the works that had come from the Platonic school, to turn his attention to the Hermetic corpus. Ficino was a powerful intellectual figure of the time; in 1462 he was appointed head of the Platonic Academy in Florence which, situated in the Medici villa of Careggi and endowed with an enviable collection of Greek manuscripts, was a forerunner of later Italian learned and scientific academies. For this scholar to forgo his work on Plato and concentrate on the Hermetic writings was not, however, as great a switch as it might seem, for the Hermetic works contained much Neoplatonic thinking, a circumstance thought at the time to confirm the belief that Plato himself had once sat at the feet of Egyptian sages. Ficino, of course, was well aware of the standing of the Hermetic writings and far from objecting to Cosimo's instructions seems to have welcomed the opportunity to exercise his skill on so worthy a task, which he approached with a mixture of awe and wonder.

For Neoplatonism, see page 206.

There is nothing surprising about Ficino's respect for the Hermetic works. They had not only fired the imagination of his master, Cosimo, but, when he died in 1464, still stimulated the curiosity of his grandson Lorenzo the Magnificent, who not only urged Ficino to carry on but even commissioned Sandro Botticelli to paint a picture, the *Primavera*, full of Hermetic symbolism. Yet veneration for Hermetism was not confined to the Medici and their circle. Bernardino di Betto di Biago (Pinturicchio) was commissioned in the 1490s by Pope Alexander VI to decorate a suite of six rooms in the Vatican – the Borgia Apartments – with Hermetic frescoes, and in the floor at the west end of Siena cathedral is a mosaic pavement depicting Hermes Trismegistus, placed there sometime in the 1480s. And elsewhere, outside Italy and even, later, in the Protestant world, the Hermetic writings were revered.

Illustrations page 293

What, then, was the attraction of Hermetism? Doubtless this was partly due to its magical content but above all it seems to have been because of its immense antiquity. This was a passport to respectability in the Renaissance of the fifteenth century, when the more ancient a culture the more standing it possessed. But what about the mystical aura that enveloped it? This veneration was certainly due in part to its apparently prophetic insight into the coming of Christianity, but there was more to it than this. It was a corpus of teaching that mixed magic and metaphor, that mingled Neoplatonism with mysticism; it contained mysteries which only the initiate, the Magus, could understand, and that only after a period of study and meditation. Its universe was the Aristotelian-Ptolemic universe of spheres, but guided by divine beings and operated upon by magic, astrology, alchemy and the other occult 'sciences'. Its underlying philosophy was a form of gnosticism, teaching that man is able to uncover the divine elements within himself by attaining a mystical sympathy between the world and mankind, and it emphasized the doctrine of microcosm and the macrocosm. Of course not all the works in the collection teach precisely the same thing; some parts are concerned with the Cabala – that esoteric Jewish mysticism based partly on an oral tradition believed to be given by God to Moses and partly on elaborate numerology – other parts with various occult practices, but uniting them all is what has been termed a 'pervasive and intense piety'. The whole Hermetic corpus was concerned with individual attempts to reach an intuitive understanding of God and of salvation, helped throughout by the figure of Hermes. It was this that impressed Ficino and the other Renaissance scholars.

Illustration page 298

The effects of Hermetism on science will become clear when we look at the revolution wrought by Copernicus and his heliocentric view of the universe, but it can be pointed out here that Hermetism was soon acting as a stimulus to scientific observation. Immediately these powerful and respected writings became widely available in Ficino's translation they exerted a wide effect which was particularly noticeable in the seventeenth century, influencing experimenters and mathematicians alike: experimenters because practical trials and tests were all part of the stock-in-trade of the magician and so even of the sophisticated Renaissance Magus, and mathematicians because

For the Hermetic overtones in Copernicus's writings, see page 328.

mathematics was a necessary adjunct to alchemical operations. But mathematics attracted the Magus for another reason and one of even greater significance for the new scientific spirit. This was the presence in the Hermetic writings of a Pythagorean version of Platonism which extolled the quantitative approach to the universe, encouraging the use of mathematics to show relationships and demonstrate essential truths about the whole of creation. In the late seventeenth century when Hermetism had virtually faded from view, this way of tackling problems was to remain and proved the scientists' most advanced tool for understanding the cosmos.

For the Pythagorean interest in a mystical approach to numbers, see pages 72–77.

For more than a century Hermetism continued to seduce churchmen and lay scholars, for in spite of the Church's opposition to magic, alchemy and the occult, many prelates thought the Hermetic writings were hallowed because of their Christian prophecy. Even so staunch a Catholic as Philip II of Spain had more than 200 Hermetic texts in his library. How long all this magical philosophy would have held sway is a moot point, for in 1614 Isaac Casaubon proved that the Hermetic writings were not what they were supposed to be. A Protestant from Geneva, Casaubon was one of the most brilliant Greek scholars of his day – 'the most learned man in Europe' a contemporary called him. In 1610 he was invited to England and, with the encouragement of James I, set about making a critical study of a recent Roman Catholic history of the Church, *Annales Ecclesiastici*, written by Caesar Baronius, the Vatican Librarian. Baronius's twelve-volume history claimed to prove scriptural authority for all the Church's practices, and it was therefore important for the Protestants to disprove its claims. Casaubon used his immense scholarship to achieve this and, embedded in his detailed critical comments to the history, he also showed quite conclusively that the texts of the Hermetic works could not have been written in very early times. They were certainly not contemporaneous with Moses; their style and the quotations they gave showed that the texts must have been prepared well after the beginning of the Christian era, not before it. The advent of Christianity to which the texts referred was no wonderful example of divinely inspired prophecy, but a mere reference to events that had already occurred.

What had come as a revelation to Renaissance man – a set of texts with an impact at least equivalent to the discovery in our own time of the Dead Sea Scrolls – was now shown to be nothing more exciting than an everyday collection of Neoplatonic writings. There were, of course, those who did not know of Casaubon's demolition of the accepted date of Hermetism, or if they were aware of it preferred to ignore it and continue breathing their accustomed magical atmosphere. But gradually his criticism fell on fertile ground and began to turn men away from this particular form of Renaissance mysticism. Casaubon's textual criticism alone did not give rise to the new breed of natural philosophers like Galileo and Newton, who were to transform man's view of the cosmos; other factors were at work as well.

Illustration page 297

But it is probably true to say that it played a part, helping to complete the process of weaning Renaissance scholars from magic so that those in the seventeenth century were able to examine the natural world without recourse to magical ideas or the Cabala.

Leonardo da Vinci

Mention of Isaac Casaubon has carried us outside the Renaissance and forward to the seventeenth century, but now we must return to the fifteenth to consider a man who typifies the period with his broad interests and humanitarian outlook: the artist, engineer and proto-scientist Leonardo da Vinci (1452–1519). Born in Tuscany in the foothills above Florence, he was the illegitimate son of Pietro da Vinci, a Florentine notary. Leonardo's artistic talents soon became evident and his father, who still took an interest in him, was persuaded to apprentice the boy to the studio of the artist Andrea del Verrocchio. Verrocchio was one of Florence's leading artists and so his studio undertook a wide variety of work, and it was here that Leonardo received most of his informal education. Here he would have learnt the art of perspective, and here his mechanical abilities would have been stimulated, for Verrocchio not only painted but was also a goldsmith and sculptor; since the latter involved moving heavy weights, the block and tackle and the winch would have been every-day equipment.

Leonardo's notebooks are full of mechanical devices, most of which he would not have seen in Verrocchio's studio, but this does not mean that he invented them all or even that he invented any. The Renaissance produced many inventors, as other manuscripts and, later, printed books make clear. Moreover, Leonardo published nothing. In this, of course, he was at one with many inventors who, like him, had no university education and so none of the expertise of the literary man; they could only retain what they devised if they kept it to themselves, at least until actually commissioned to construct something to which they could apply their own design. Nevertheless, Leonardo's sketches show a keen sense of observation and a lively mind. His now-famous helicopter may have been his own idea, but it could have been derived from stories about Chinese helicopter tops or 'bamboo dragonflies', and what is true of this is true of the many devices of which he made such beautiful sketches. On the other hand his studies of bird flight, which were part of an investigation into the whole question of flying, were not to be bettered for nearly another five centuries, and were certainly original.

Yet whatever he may have invented or investigated, Leonardo not only failed to publish, he did not even exploit his ideas. Essentially he was a philosopher whose desire was to contemplate the world, to revel in its wonders, not to turn them to practical account. But though he had this philosophical turn of mind, he was handicapped by being unable to read Latin. Again, he did not possess the basic training to enable him to develop a science of mechanics. Nevertheless he did gain a reputation for his mechanical acumen and in 1482 obtained the post of inspector of fortifications to the Duke of Milan.

Mechanical things were not, of course, Leonardo's sole scientific interest. He was intrigued, as any artist would be, by light, though his interest was more scientific than that of his acquaintances and led him to enquire deeply into questions of the reflection and refraction of light, especially through the eye. But then he took a slightly unusual approach to art; to him 'Painting compels the mind of the painter to transform itself into the very mind of Nature, to become an interpreter between Nature and the art. It explains the causes of Nature's manifestations as compelled by its laws . . .', and it was these laws that intrigued him. That is not to say that Leonardo was not a great painter; those of his works that remain show how great he was and his many sketches underline his extraordinary genius as an artist. But his interest was wider than could be satisfied by drawing or painting alone, even though, because of his lack of formal education, he had first to look at the world through an artist's eyes.

Leonardo's wide interests led him to the dissecting room. As an artist he would have needed to become familiar with human and some animal anatomy, but it is typical of the man that he himself should have dissected some thirty bodies. This was no light task; he had none of the modern conveniences to make the work less unpleasant and in addition he had to keep stopping so that he could draw what he saw. To do it all must have taken great determination. Yet though it doubtless brought him satisfaction it was, in a sense, all in vain. He published none of his drawings and made no attempt to impart what he had learned; indeed his anatomical notes and drawings remained largely unknown until our own times.

Illustrations page 297

Leonardo da Vinci was certainly a man of science, as we can now see with the benefit of hindsight, but in his own day this side of him was unknown, except perhaps to a few intimates. To the world at large he was a mechanician and an artist; his science was learned of only long after his death. This was his own choice for, in spite of his lack of formal education, the inventions of printing and engraving were to hand and he could have used them to make known his work if he had wanted to. But he was content to contemplate and study for his own satisfaction alone; to have sufficient for his needs seems to have been his sole material ambition. In this he was quite different from his younger contemporary Albrecht Dürer.

Albrecht Dürer

Born in Nuremberg in 1471, and so Leonardo's junior by twenty years, Dürer was also convinced that the 'new' art of the Renaissance must be based on science, though he laid stress on mathematics because its logic and precision were so helpful in questions of proportion and graphical construction. It was on matters such as these that he concentrated his attention, as well as on the aesthetics and philosophy of art. He combined these studies in his work on perspective, the illusion of three dimensions in painting or drawing. The study of perspective united art and science for many artists, and Dürer designed a machine to help him in its study. Dürer's concern with technical questions led him to write three books, one on fortifications, the second a *Treatise on Proportion* that was to become famous, and the third *A Treatise on Mensuration . . .* which gave the mathematics required for understanding the second. This mathematical text, published in 1525, was the first mathematics book to appear in German, if we except a simple basic arithmetic for builders published almost forty years earlier.

In his own day Dürer's reputation was based as much on his mathematics and his optics (his aesthetic rules in the *Treatise on Proportion* are based on sound optical principles) as on his paintings and drawings. The latter made him money, for he engraved much of his own work and made it a commercial success, but it is only in comparatively recent times that his reputation has become confined to his art alone; in the centuries that directly followed the Renaissance it was his technical writings that exerted a wide influence.

Biology in the Renaissance

Botany

One of the pictures for which Dürer is remembered now is his drawing of grasses, notable not only for its artistic execution and technique but also for its painstaking study of the grasses themselves. Yet this is only one of a considerable number of botanical and zoological pictures that he made, most of them from direct observation but a few, like his drawing of a rhinoceros, derived from sketches sent by friends. The precision and accuracy of Dürer's biological drawings made from first-hand evidence were following a true Renaissance practice. In no way did he depict things as they ought to be according to ancient authorities, however revered, but he depicted them as they were actually seen to be. Yet to do this was nothing new in Dürer's time; Leonardo, and Dürer's friend Botticelli, whose plant detail in his *Primavera*, for example, was also scientifically accurate, were both doing just the same thing. It was all part of that new scientific revolution within the Renaissance that was beginning to set observation and the precise recording of the results at a premium. Nor was Italy the sole centre of this movement to depict things realistically, for nowhere was it more evident than in early sixteenth-century Germany.

Illustration page 296

Illustration page 293

There were three notable figures who guided the German effort, Otto Brunfels of Mainz, Jerome Bock (also known as Hieronymus Tragus) of Heidelsheim and Leonhard Fuchs of Wemding, often collectively known as the 'German Fathers of Botany'. Brunfels was born about 1489 and trained at the university in his home city; he then entered a Carthusian monastery in Strasbourg, though later, in 1521, he fled the monastery and forsook the Roman Catholic faith for Lutheranism. After a period as a Lutheran pastor he returned to Strasbourg, opened his own school and married. Brunfels showed an interest in medicine and prepared one of the earliest medical bibliographies, but his fame now rests primarily on his *Living Illustrations of Plants*, the first volume of which came out in 1530, the same year as the bibliography. A second volume followed in 1531 and the third in 1536, all in Latin, though Brunfels also produced a German edition. The book drew heavily on Dioscorides and other early botanical writers, and sometimes gave a number of names for the same plant, while Brunfels himself displayed no knowledge about the geographical distribution of the plants he described. Too great a reverence was paid to past authority, Brunfels actually apologising for the fact that the book contained some plants unknown to the ancient writers. But in spite of all these criticisms the volumes are important because of their accurate delineations of 238 plants, drawn by the artist Hans Weiditz, who himself also cut the majority of the woodblocks used for printing them. The illustrations varied in size, some being full folio in extent, and all displayed a rare but happy combination of aesthetic charm and scientific accuracy. They showed the structure of each plant well and, in addition, how each plant appeared in the typical habitat in which a botanist would find it in the wild. The drawings were, in fact, so faithful that it is possible to see that some specimens had begun to wilt by the time they came to be illustrated in detail.

Botticelli and Dürer may have made accurate drawings from life of a few selected plants, but what was novel about Brunfels's work was its scope and its method; if Weiditz deserves much of the credit, Brunfels's conception and his overseeing of it were also noteworthy. Certainly, after its publication botany in the Western world was never the same; a new truly scientific era had dawned.

The birthplace of Jerome Bock is still a subject for debate but there is no doubt that he spent most of his adult life in the Saar. Brought up a Roman Catholic he also turned to Lutheranism, and was for a time personal physician to the Landgrave of Nassau, for it seems he had received formal medical training as well as his theological grounding. However it is for his botany, not his theology or medicine that he is remembered, and in that field for one work only, his *Neu Kreütterbuch* (New Plant Book). This first appeared in 1539, and marked another new departure in botany. Written in the vernacular, it made it clear not only that Bock had a flair for observing plants and writing about them, but also that he took what was then a novel approach to the subject. Concerned with the locality in which they were found, he gave a brief life-history for every plant, and, what was novel, he tried to establish links between different kinds of plants. Admittedly he used some characteristics that we should not use, such as taste, to try to establish a classification, but it was an important step in the science of botany all the same. Bock also attempted to correct some of the more fabulous and erroneous ideas that were popularly held about plants though, of course, he was not always successful. Moreover, the *Neu Kreütterbuch* was not without its errors

Illustration page 295

(for instance his suggestion that orchids were the offspring of thrushes and blackbirds), but these failings do not detract from the book's importance. However, the first edition appeared without illustrations and did not arouse much interest, attention being centred on Brunfels's volumes and, in 1542, to Leonhard Fuchs's illustrated herbal compendiary. Not until 1546, when an illustrated edition of Bock's work came out with woodcuts by the Strasbourg artist David Kandel, did the situation change and notice begin to be taken of this valuable achievement.

Leonhard Fuchs, the third of the German Fathers of Botany, was also a physician by profession and a Lutheran by persuasion, ending his career as professor of medicine at the University of Tübingen, where he exerted considerable influence. He made attempts to reform medicine by emphasizing the importance of going back to original Greek writings and ignoring the medieval commentators, an attitude owing something to the Protestant tradition of trying to go back to the Bible and ignoring the medieval traditions of the Church. But it is for his botany that Fuchs is remembered today, and this because of his pharmaceutical herbal, *The Natural History of Plants*. It was heavily illustrated, the artists Heinrich Füllmaurer and Albrecht Meyer preparing the original drawings, and the engraver Rudolph Speckle making the woodcuts and, a nice touch, all three have their pictures in the book, which also contains a full-length portrait of Fuchs. Fuchs's *Natural History* is the most famous of the early printed herbals, and though the Latin text is based on Dioscorides, the author laid out his text alphabetically so that there should be no difficulty in finding any plant. The book contains descriptions of about 400 plants native to Germany and some 100 from foreign countries. Though Fuchs made no attempt at classification he did try to establish a system of botanical nomenclature, and for this alone the book would have been a notable contribution to the science of botany. Leonhard Fuchs is still, of course, commemorated in gardens because the ornamental shrub the *fuchsia* is named after him.

Illustrations page 294

The sixteenth century saw a number of other botanists at work besides the German Fathers. Valerius Cordus of Oberhessen (1515–44), who had travelled widely to see the plants described by the ancients, prepared his own *Natural History of Plants*, a book in which he looked at plants from a botanical as well as a medical point of view. But what great contributions he might have made were cut short by his untimely death at the age of 29, and it was only through the dedication of the Swiss botanist and zoologist Konrad Gesner that it was published posthumously in 1561. Gesner himself produced an *Opera botanica* (Botanical Works) that contained nearly 1,500 plates which he drew himself, but its publication, though begun while he was still alive, was only completed after his death. In it Gesner shows that he had completely grasped the importance of plant structure as a key to the classification of vegetable life.

A considerable number of other botanical books came out during the sixteenth century, following the lead given by the Germans, and in Antwerp these were sponsored by the famous printer and publisher Christopher Plantin, who played a notable part in making first-quality scientific material available. At the height of his career he ran presses in Antwerp and Leiden and in the year 1575 alone produced no less than 83 editions of books ranging in quantity from 800 to 2,500 copies. In botany perhaps the most significant was the *New Note-book of Plants* by the Frenchman Matthias de l'Obel (or Lobel) a collection of details on some 1,200 to 1,300 plants that he himself had collected, which he classified mainly by leaf, using the divisions of single leafed seeds (monocotyledons) and two-leafed seeds (dicotyledons). There was also Charles de Lécluse, perhaps better known by his Latinized name Clusius. Trained as a lawyer, he did not become interested in botany until his mid-twenties, and then spent his time translating some contemporary Latin botanical texts into French, including a famous herbal by Rembert Dodoens, the first to be published in Flemish (1554), and a book which was the basis at least of the *Herball* or *Generall Historie of Plantes* (1597) prepared by the English barber-surgeon John Gerard. But Lécluse's great contribution to botany came in 1593 when, at the age of sixty-seven, he was appointed to a professorship at the University of Leiden, which he held until his death sixteen years later. For it was at Leiden that he made a botanical garden, the first to be seen in Europe, and at seventy-five years of age published his *Account of Rare Plants* which contained upwards of 600 new species; it was Lécluse, too, who wrote the first monograph on fungi.

Zoology

As in botany so in zoology, naturalists in the sixteenth century began to look at the animal kingdom with new eyes. The most notable of the zoologists were Pierre Belon, Guillaume Rondelet and Konrad Gesner, sometimes referred to collectively as the 'Encyclopaedic Naturalists'. Pierre Belon (1517–64), born near Cerons in France, came from an obscure family. He trained as an apothecary, studied botany under Valerius Cordus, and made three visits to England, carrying out zoological dissections at Oxford on at least one of these occasions. He became a protégé of the bishop of Le Mans and later obtained a licence to practise medicine, his winning manners earning him a place in court circles; he was granted a state pension by the French king, Charles IX. However, his life was comparatively short for at the age of forty-nine he was murdered by unknown assailants in the Bois de Boulogne. A keen traveller in the Middle East, who kept extensive notes on the flora and fauna he observed, Belon not only wrote a monograph on cone-bearing trees but also published three important books, *The Natural History of Strange Marine Fish* (1551), *On Aquatic Life* (1553), and *The History and Nature of Birds* (1555). The first contains a classification of fish and those cetaceans (dolphins, porpoises and whales) which Belon had dissected. He recognized that the two milk glands of the female cetaceans were mammalian in type, and that he was dealing with mammals that breathed air even though they lived in the water. Nevertheless, he still classed them under fishes, as he did the hippopotamus. But the most significant of his books and the one that brought him the greatest fame was his last, the one on birds. In this he corrected some current errors on the subject of avian anatomy, but his most significant contribution was his careful and detailed comparison of the skeletons of birds and men. This was a pioneering study and his work in this field has earned him the title of 'Father of Comparative Anatomy'.

Guillaume Rondelet (1507–66), the second of the Encyclopaedists, was the son of a drug and spice merchant. He went to the University of Paris to study humanities, but not finding the subject to his taste, transferred to the medical faculty. Rondelet became a noted physician and travelled widely, though in 1551, when he was in his mid-forties, he returned to Montpelier to become professor of anatomy and medicine and, later, chancellor of the university. Yet though he was a popular medical lecturer and anatomist, he is remembered now for his great interest in marine biology and his famous *Book of Marine Fish . . .* which was published in Lyons in 1554–55. Though written in Latin it came out in translation in 1558 as *The Complete History of Fish*, a more correct title, since the book deals with freshwater as well as marine zoology in all its variety; indeed it even includes a detailed description of beavers. Rondelet described the various parts of aquatic animals, their means of digestion, respiration and reproduction, and tried to relate function to environment. In some ways he harked back to Aristotle, whose work on aquatic life was so excellent, but Rondelet took things much further: he was the first to describe the swim bladder in freshwater fishes, noting too that it was present in some marine forms; he discovered the ear of the dolphin and compared the creature both with the pig and with man. He also gave a detailed description of the sea urchin, and his drawing of it is the earliest surviving picture of the dissection of an invertebrate. A great part of his book is, however, encyclopaedic in form, describing over 300 aquatic creatures, almost all of which are illustrated.

Illustration page 294

Rondelet sternly attacked those who would not accept the evidence of observation and experience but who wished to be guided only by ancient authority. In his later years he became a staunch Protestant, though he seems, by and large, to have been a somewhat Rabelasian character, fond of food, especially confectionary, and a keen musician. Not so the third Encyclopaedist, Konrad Gesner (1516–65), the godson and protégé of the Swiss Protestant reformer Huldreich Zwingli. Gesner read theology, became a noted Greek and Hebrew scholar and in 1537, at the age of twenty-one, became the first occupant of the chair of Greek at the newly founded Lausanne Academy. However, he also became attracted to medicine and before his appointment studied the subject at Bourges, then Paris and, later still after he had vacated the chair at Lausanne, at Basel, where he received a doctorate.

Throughout his subsequent life, he seems to have carried on his two widely diverse interests, ancient languages and the life sciences.

We have already seen something of his botanical work, but Gesner was also attracted to zoology and it is for his five-volume *History of Animals* that he is perhaps best remembered. Although begun in 1551, its publication was not achieved until 1587, twenty-two years after his death. It was a massive work of more than 4,500 pages and received immediate acclaim; indeed it was still being enthused over by the famous zoologist Georges Cuvier, two centuries later. In this encyclopaedic book Gesner attempted to regroup animals in a new classification, as he had tried in his botanical work when he had classified plants as flowering and non-flowering, or according to the way the sap was supplied. Gesner's reclassification was not successful, but the book's immensity underlined the need for thorough re-thinking of the whole question of classification of animal species.

Illustration page 295

The sixteenth century also produced two other outstanding zoologists, if not quite of the stature of Belon, Rondelet or Gesner. One was Ulisse Aldrovandi, born at Bologna in 1522, a typical Renaissance figure. The son of a notary, he studied mathematics and Latin, interspersing his studies with periods when he ran away from home in his eagerness to see new things and new countries. At Rome he met Rondelet, and was fired with enthusiasm for natural history, though on his return to Bologna he practised medicine and taught logic at the University of Bologna. However, his now intense interest in natural history became so predominant that in 1561 he became the university's first professor of the subject, and it was at Bologna that he established the first botanic garden in Italy. He studied the development of the egg, and these studies in embryology influenced the great embryologist, Volcher Coiter. But Aldrovandi's most significant works were a three-volume treatise on birds in 1600 and one on insects in 1603, as well as lesser writings on quadrupeds, serpents, 'dragons' and monsters, on trees and on minerals. Although not a great discoverer, Aldrovandi was a notable teacher and greatly encouraged the study of zoology in Italy.

The other outstanding zoologist was the Londoner, Thomas Moufet, who took a great interest in insects and composed a *Theatre of Insects* which, at the time and for many years after, was the best book on the subject, though it was not published until 1638, thirty years after his death. Moufet had collected specimens of insects on his travels in all the Western European countries and the book is copiously illustrated, though the text shows a naturalist who is very nearly overwhelmed by his material, so many are the facts he has noted and tried to cram in.

Medical Science

During medieval times medicine made little progress; the only way in which it developed was towards a closer association with astrology. However, things began slowly to change after Muslim medical texts had begun to filter through to the West in the twelfth century. The first noticeable effect began in Italy at the University of Bologna, where in 1312 Mondino de' Luzzi completed his *Anatomia Mundini*, a book designed to be read aloud while dissection of human bodies was carried out.

There is still argument in scholarly circles as to whether Mondino carried out the dissections himself, or whether he left this to a 'demonstrator', but that is by the way. The university boasted what was probably Europe's leading law school and, of necessity, post-mortems were carried out there. It seems that the practice was extended, probably by Mondino himself, as an adjunct to medical study. This was a vital step forward, of which Renaissance medicine was to make full use, as it did too of Mondino's anatomical terminology which was derived not from Greek but from Arabic, and much of which is still with us today.

However, though Mondino introduced the dissection of cadavers and new anatomical terms, his outlook was still Galenic, as was all medieval anatomy; not until the sixteenth century was this to be swept away. The man chiefly responsible was the great Renaissance anatomist Andreas Vesalius, who was born in Brussels on 31 December 1514, the son of an apothecary to the Holy Roman Emperor Charles V. After studying at Louvain, Vesalius moved to the medical school of the University of Paris, where he paid particular attention to anatomy, which was still, of course, Galenic in outlook. However, war between France and the Holy Roman Empire forced him to return to Louvain, and here he reintroduced the practice of anatomical dissection, which had lapsed due to neglect. But in the early sixteenth century, the most famous medical school in western Europe was neither in Louvain nor Paris but at Padua, and it was there that Vesalius went in the autumn of 1537, enrolling as a graduate, having received his bachelor's degree in Louvain. But his knowledge was such that after only two days of examinations he was granted the degree of doctor, with honours, and on the day following accepted the post of lecturer on surgery and anatomy. Clearly Vesalius was an exceptional young man.

As soon as he started his work at Padua, Vesalius made an innovation; he carried out all dissections himself, using four very large anatomical diagrams to help his students. One of these was stolen and published so, to prevent further plagiarism, he published the remainder himself, together with three drawings of the human skeleton by the Dutch artist Jan Stephen, a pupil of Titian. The anatomical drawings still promoted Galenic views, as did two texts he published about this time, but this was to be expected. Vesalius made full use of his position at Padua to dissect as often as he could, using the bodies of animals as well as humans, and over the next five years he discovered certain discrepancies between what Galen said and what he himself found by dissection. In 1539 he was helped by Marcantonio Contarini, the judge of the Paduan Criminal Court, who made the

Illustration page 298

bodies of executed criminals available to him, occasionally delaying executions to suit the anatomist's convenience; thus it came about that there was a plentiful supply of bodies, and Vesalius was able to make repeated and comparative dissections, all of which only confirmed that Galen was not completely reliable. By 1539 Vesalius was sufficiently sure of his facts to challenge Galen openly, both in Padua and then in Bologna, where he demonstrated how Galen's anatomical descriptions fitted the body of an ape but not a man.

After this, Vesalius settled down to publish his new ideas in detail, and over the next two years began to write up his results. He spared neither time nor expense, hiring the best draughtsmen and the finest Venetian block-cutters to prepare drawings for the press. He chose the printer Johannes Oporinus of Basel, and went there to oversee the printing himself. In 1543 his work, *The Fabric of the Human Body*, appeared, a magnificent example of all that was best in Renaissance book production. It contains 17 full-page drawings, together with a great number of text figures, and these are among the finest woodcuts ever made. The text, which runs to some 600 closely printed pages, is divided into seven sections or 'books'. The first deals with bones and joints, and includes pictures of the skulls of five different human races; these are the first illustrations of comparative ethnography. Book II concerns the muscles, and is the section most famous for illustrations, and Book III concerns the heart and blood vessels. Although Vesalius still accepted Galen's view that blood was formed in the liver and travelled across the septum of the heart, he obviously had his doubts, for he remarks 'we wonder at the art of the Creator which causes blood to pass from right to left ventricle through invisible pores.' In the second edition twelve years later he is even more specific, claiming that nothing could pass between the ventricles. Book IV is an account of the nervous system, in which he concludes (correctly) that the nerves are not hollow as Galen had suggested. Book V concerns the abdominal organs, and Book VI the organs of the thorax. This last contains a description of the heart, and Vesalius notes how like a muscle it is. The final section, Book VII, describes the brain, and shows parts never before described. It was here that Vesalius made it clear that human beings had no *rete mirabile* (wonderful net) of fine arteries at the base of the brain, although Galen and others had described them; they were present only in the brains of hoofed animals.

Illustration page 299

For Galen's account of the heart and the blood vessels, see pages 248–249.

The Fabric of the Human Body has been called one of the greatest scientific books ever to be written; certainly it was a milestone in our understanding of the human body and the book which marked the death knell of Galenism. At the time it appeared it came in for a great deal of criticism in academic circles, but by then Vesalius had already decided to become a practising doctor and, since there was a long history of imperial service in his family, he applied to the Holy Roman Emperor; he was at once appointed as one of the court physicians. This was unfortunate for science, since much of Vesalius's time was now devoted to 'the Gallic disease, gastro-intestinal disorders, and chronic ailments . . .' as he bluntly put it. But once in the imperial service, he could not resign, and there he had to remain for the next thirteen years until the Emperor finally abdicated. While in his service, Vesalius had sometimes to act as a military surgeon, but whenever near a medical school, he would take part in post-mortems and try to do some research if time allowed. Over the years he developed detailed improvements in some surgical techniques, wrote some complementary books to the *Fabric* and answered criticisms, his second edition having some of the contents rearranged. But essentially Vesalius is remembered for one book, because the *Fabric* marked a new era in anatomy, both in its practice and its outlook; it took time to exert its full effect, of course, but by the beginning of the seventeenth century Vesalian anatomy was accepted almost everywhere by academics as well as by medical practitioners.

Vesalius stands like a colossus over the medicine of the sixteenth century, but his stature must not be allowed to obscure the very real achievements of other medical scientists. When he left Padua he was succeeded by Matteo Colombo and, later still, by Gabriele Fallopio, who was a competent physician but, it seems, a disastrous surgeon. Appointed the chair of pharmacy in Ferrara, Fallopio moved to Pisa in 1549 to take the chair of anatomy, but here he was unjustly accused of human vivisection. He had, however dissected bodies of lions in the Medici zoo and disproved Aristotle's statement that their bones were solid and without marrow. Fortunately the accusations did not prevent his being offered Vesalius's old chair at Padua when Colombo retired. Fallopio wrote a commentary on Vesalius's *Fabric* in which he sought to correct certain errors, though in doing so he was at pains to make clear his respect for 'the divine Vesalius'. However, his commentary, *Anatomical Observations* (1561) is notable not so much for its comments on the *Fabric* but for the original discoveries that it contains about the human generative system and the growth of the foetus; this was Fallopio's great contribution. He studied muscles and the ear, but his most significant work lay in the development of the foetus, particularly the formation of its bones and teeth, and his discovery of what we now call the Fallopian tubes, those narrow ducts which run from the ovaries to the uterus. First discerned in Alexandria, they had long since been forgotten, and Fallopio's description was quite original.

For this discovery by Herophilos, see page 116.

A contemporary of Fallopio, and another notable contributor to Renaissance medicine, was Bartolommeo Eustachio. Born sometime between 1500 and 1510 at San Severino in Italy, he was both a medical man and a linguist, being expert in Arabic, Hebrew and Greek. A strong supporter of Galen, he was hostile to Vesalius and his first two books, *The Examination of Bones* and *The Movement of the Head*, both written in 1561, are an attack on ideas in the *Fabric*. Yet in spite of this unilateral hatred – Vesalius never returned it – Eustachio did some sound scientific work. In 1562–63 he produced a remarkable series of works, *The Structure of the Kidney*, *The Organ of Hearing*, and others on veins and teeth. All were of a high standard, each had something new to say, and in *The Organ of Hearing* he announced his discovery of the tube between the middle ear and the upper part of the pharynx, still known as the Eustachean tube. He also carried out some original work on the larynx and on speech.

In 1552 Eustachio enlisted the help of Pier Matteo Pini, a relative who was an artist, and with him prepared a series of forty-seven anatomical illustrations on copper plate. These were to accompany a book, *Anatomical Dissensions and Controversies*, which was, however, never published, though some of the plates were used in a smaller anatomical work to portray the kidneys and various veins, as well as the Eustachian valve. At his death in 1574, the rest of the plates were missing, and did not come to light until 150 years later, when they were published. They show careful observation, and although not the first copperplate anatomical drawings, they are nevertheless a tribute to Eustachio's flair for anatomy.

The anatomists so far mentioned either came from Italy or worked there, but in 1534 at Groningen in the Netherlands there was born a man who was to become the first systematic writer on comparative anatomy since Aristotle. Vocher Coiter (1534–76), the son of a jurist, received a good education and showed such aptitude for dissection and Galenic medicine that the city voted him a stipend for five years' study in foreign universities. He went abroad as a pupil of notable figures such as Fuchs and Aldrovandi, and himself taught for a time at Bologna and Perugia, but his Protestantism seems to have caused him trouble and for a time he was imprisoned, first in Rome and then in Bologna. By 1566 Coiter was back in Germany and in 1569 he became physician to the city of Nuremberg, remaining there until his death in 1576. A strong supporter of Vesalius, Coiter is primarily remembered today for his detailed studies in comparative anatomy. He discussed almost the whole series of vertebrates – amphibians, birds, reptiles and mammals – emphasizing their differences. His work involved vivisection – he recorded observations made on the living hearts of cats, reptiles, frogs and fishes – and his studies of the skeletons of various species of birds were unique. Illustrated with detailed drawings, Coiter's contributions underline the broad outlook of so many of the Renaissance medical men. Their medicine brought them into continued contact with herbal remedies and led them to study botany; their curiosity about the human body caused them to examine all living creatures they could lay their hands on, so it is not surprising that so many of what we should today call biologists were those who had medical training and even a medical practice. Specialization was not a characteristic of Renaissance man.

One other medically inclined, scientifically orientated Renaissance scholar who must be mentioned was the Spaniard Michael Servetus, (Miguel Serveto). Born about 1511, the son of a nobleman, he worked for a Lyons firm of publishers as corrector and editor, preparing two

Opposite above left A 14th-century 'T-O' map of the world. The 'O' is formed by the ocean. The top half of the map is filled by Asia, Jerusalem, the Red Sea and Egypt; in the bottom left is Europe, with Italy and Spain indicated, whereas Africa is bottom right. Biblioteca Nazionale Marciana, Venice.

Opposte above right Gerhard Mercator (left) and his collaborator Jesse Hondius. They compiled the first 'atlas' (1595) and independently devised the first projection that could be used with ease by sailors; their introduction of copper engraving allowed a high degree of detail on these maps and charts. British Library, London.

Opposite below left A reconstruction of the 14th-century turret clock from Dover Castle, incorporating verge-and-foliot escapement. The development of intricate mechanisms for marking time stimulated both scientific and technological advances during the Renaissance. Science Museum, London.

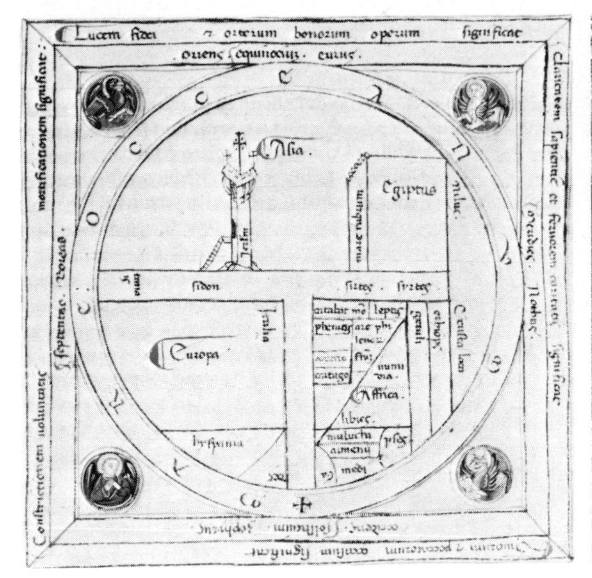

Above John Dee (1527–1608), the English mathematician, alchemist and astrologer inspired by Hermetism, and typical of the Renaissance scholars who pursued their occult studies with scientific rigour. Ashmolean Museum, Oxford.

Mare Indicum
Oceanus Indicus
TROPICUS

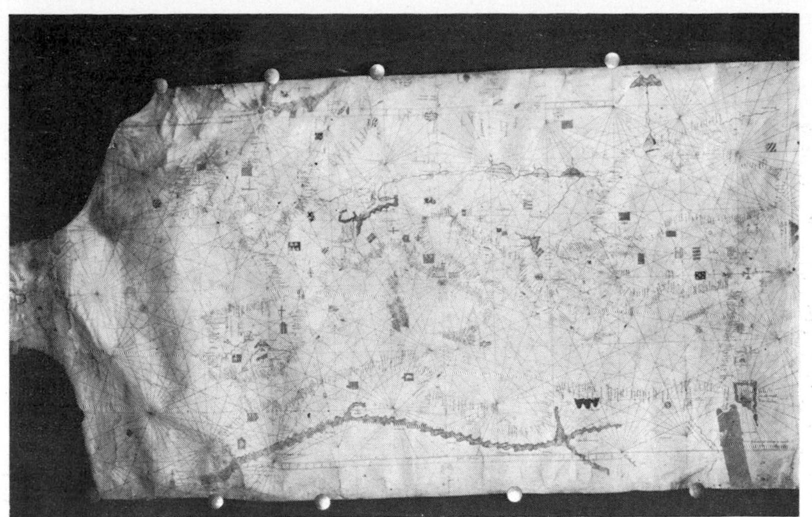

Left A 'portolan' chart of the 14th century, mapping the coasts and ports of the Mediterranean and Atlantic seaboards of Europe, derived from the experience of sailors. British Library, London.

Opposite The earliest known terrestrial globe, following Ptolemaic ideas and made by Martin Behaim in 1492, the same year as Columbus discovered the New World. This illustration shows the East African coast and Indian Ocean, as it was envisaged six years before Vasco de Gama rounded the Cape of Good Hope. Germanisches Nationalmuseum, Nuremberg.

Left A German printing press of 1522; the spread of printing in the previous 70 years hastened the creation of a community of scientists by making scientific ideas and illustrations available to all who could gain access to books.

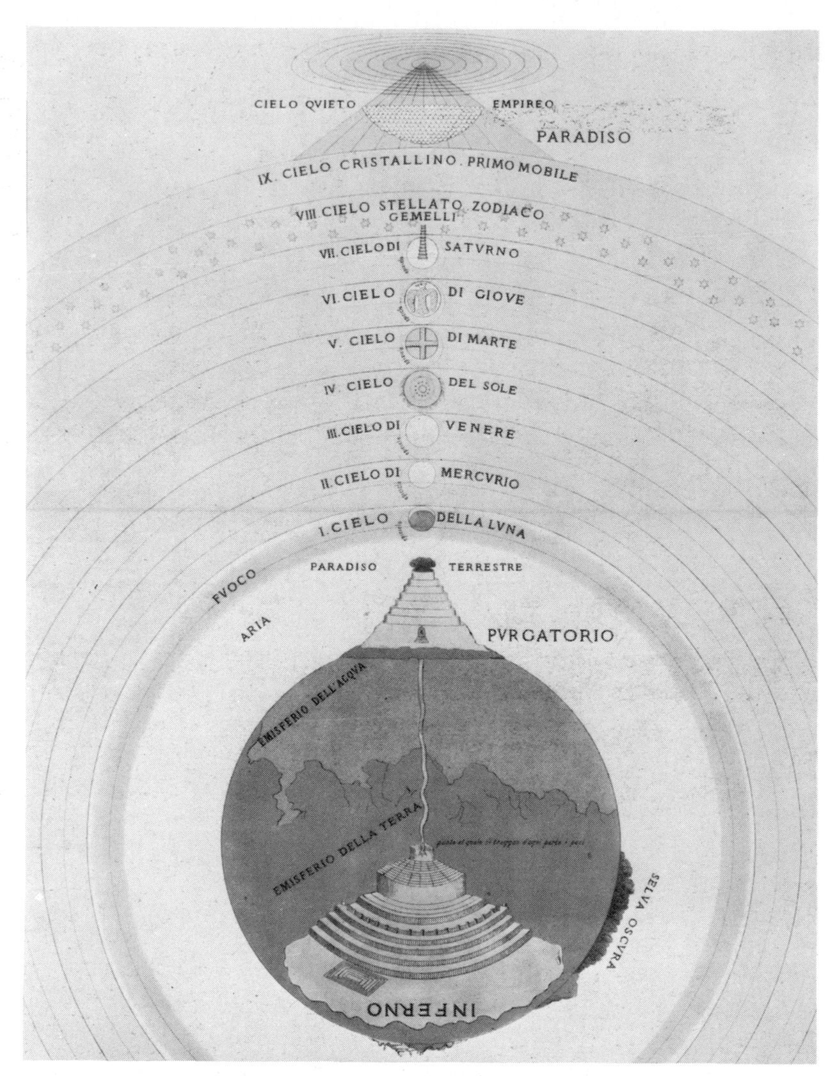

Right The universe as conceived by Dante around the year 1300, combining the Christian vision of Heaven, Purgatory and Hell with the Aristotelian concept of the spheres. From a study by Michelangelo Caetani (1872).

Right 'The Butcher's Shop' by Annibale Carracci (1560–1609), a powerful example of secularism, realism and anatomical interest in the art of the Renaissance. Christchurch, Oxford.

Above Hermes Trismegistus standing under an armillary sphere talking with astronomers, in a fresco from the Borgia Apartments in the Vatican, painted by Pinturicchio (c. 1454–1513).

Above left Detail from 'Primavera' by Sandro Botticelli (c. 1444–1510), a painting fusing Christian and pagan imagery on Hermetic and Neoplatonic principles. Galleria degli Uffizi, Florence.

Left The caravel used by Columbus according to a contemporary tapestry. The art of navigation on the high seas prompted a new independent scientific outlook among many Renaissance philosophers.

LEONHARTVS FVCHSIVS
ÆTATIS SVÆ ANNO XLI.

PICTORES OPERIS,
Heinricus Füllmaurer.
Albertus Meyer.

SCVLPTOR
Vitus Rodolph. Speckle.

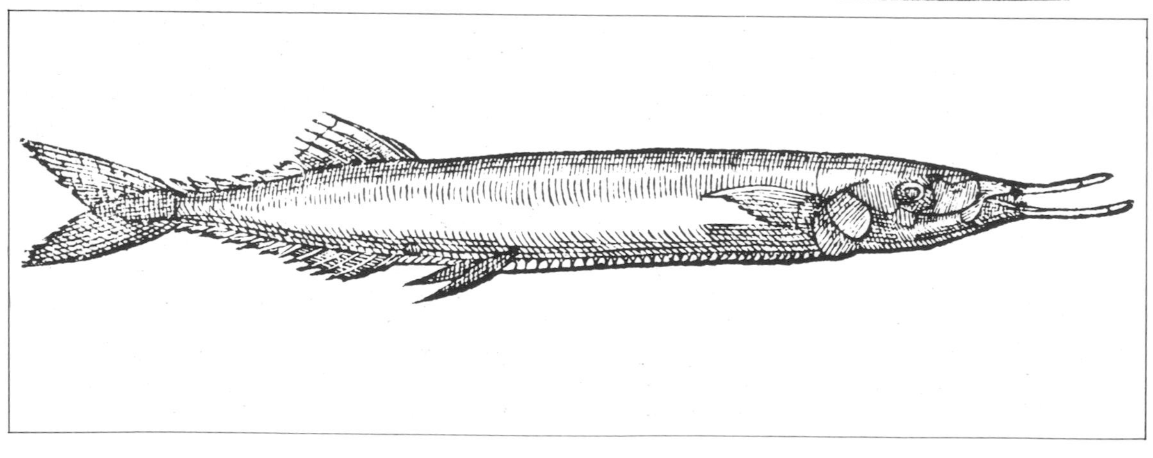

Left Diagramatic drawing of a mine from Bauer (Agricola) *De Re Metallica* (1556)

Opposite The collaborators on *De Historia Stirpium* (1542), an epoch-making botanical study: Leonhard Fuchs, author (*far-left*), Heinrich Füllmaurer and Albrecht Meyer, artists, (*above*) and Rudolph Speckle, engraver (*centre*).

Opposite below The pipefish, from Rondelet's *Book of Marine Fish . . .* (1553)

Below left A woodcut of a camel from Konrad Gesner *History of Animals*, 1587.

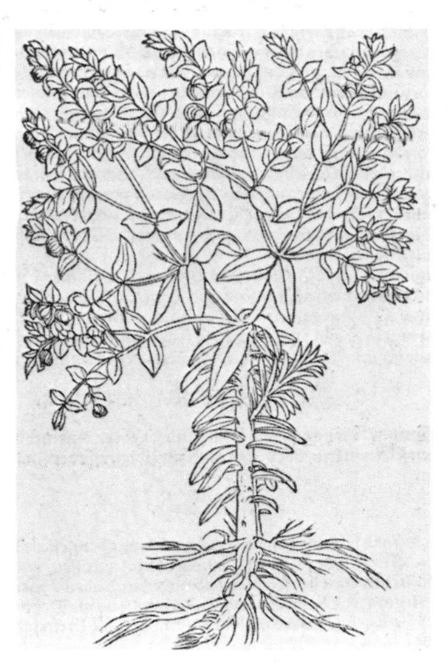

Below A woodcut of the plant *Euphorbia* (a type of spurge) from *Neue Kreütterbuch*, by Jerome Bock, 1551. This plant was considered efficacious for depilation and toothache.

Left Frontispiece of *Ars Magna, Lucis et Umbrae*, by Athanasius Kircher, 1671. Kircher contributed studies of biology, optics, telescopy and microscopy but this illustration reveals the fusion of these interests with Hermetic symbolism.

Opposite Painting of grasses (1503) by Albrecht Dürer, with an attention to realistic detail unknown in medieval art. Graphische Sammlung Albertina, Vienna.

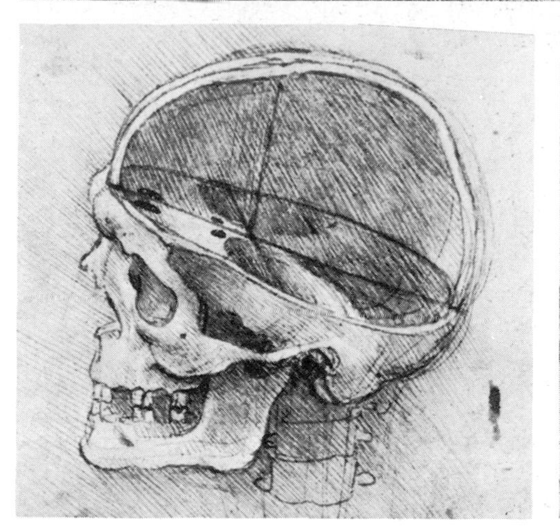

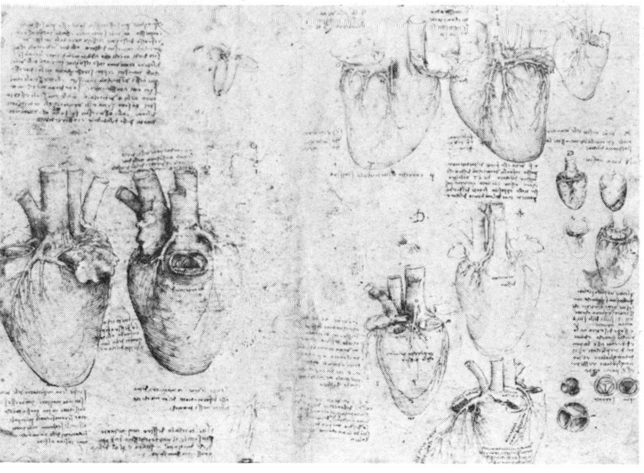

Below and below left Two anatomical sketches by Leonardo da Vinci (1452–1519), showing the detailed knowledge gleaned from repeated careful dissection of corpses. In these studies he advanced anatomy far beyond its previous limits, but his work was mainly lost for several centuries. Royal Library, Windsor Castle.

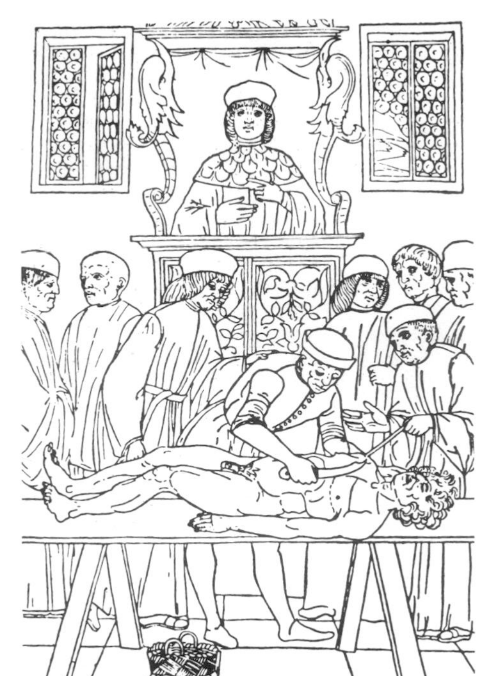

A
A
B
D
G
F
C
L
L

ANDREAE VESALII
BRVXELLENSIS, SCHOLAE
medicorum Patauinæ professoris, de
Humani corporis fabrica
Libri septem.
CVM CAESAREAE
Maiest. Galliarum Regis, ac Senatus Veneti gra
tia & privilegio, ut in diplomatis eorundem continetur.

Left The superficial musculature, from Vesalius, *Fabric of the Human Body*.

Opposite above left An anatomy lesson at Padua in the 15th century; the professor lectures while the assistant demonstrates what is described. From Johannes de Ketham, *Fasciocolo di Medicina* 1493. By comparison (*opposite below left*), on the title page of Andreas Vesalius, *Fabric of the Human Body*, 1543, the professor himself is carrying out the dissection and arguing from the evidence before him, not from preconceived ideas.

Opposite above right An assaying laboratory depicted in Bauer's *De Re Metallica* (1556), the first western practical and technical manual on mineralogy.

Opposite below right The 'Vase of Hermes' illustrated in a Hermetic alchemical treatise *Splendour of the Sun*. This 'vase' was a vessel favoured for the mystical qualities of its shape for certain alchemical reactions in the sixteenth century. British Library, London.

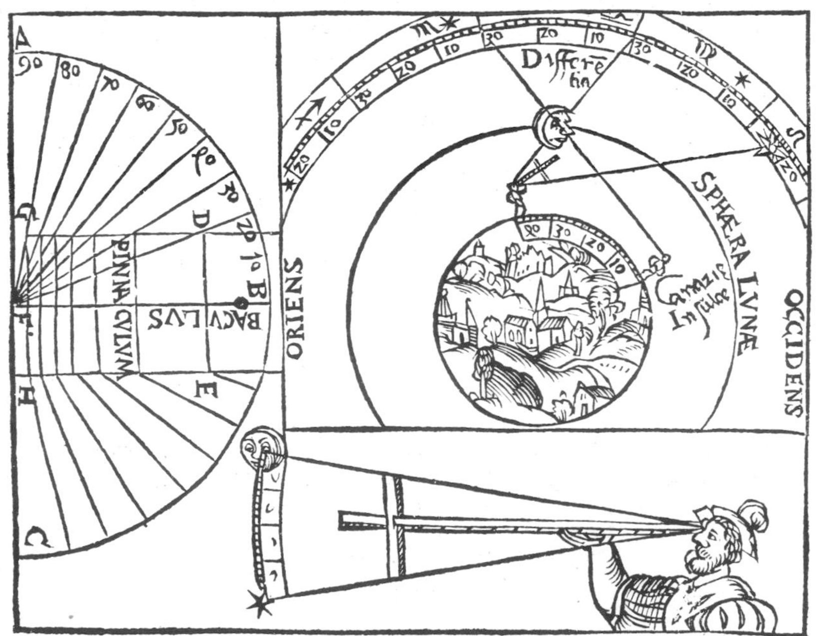
ORIENS
OCCIDENS
SPHÆRA LVNÆ

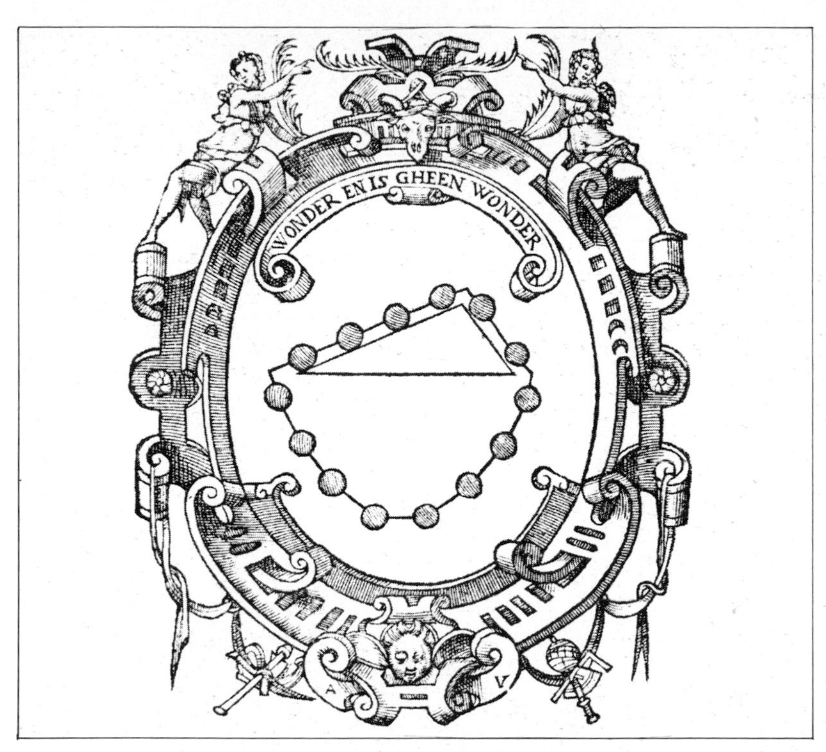

Left The title-page of Simon Stevin's *Principles of Statics* (1586) showing the 'clootcrans' and carrying the inscription 'What appears a wonder is not a wonder'.

Opposite above Painting of an alchemist (1661) by Adriaen van Ostade. This subject was by the mid-17th century treated as an image of human folly, as scientific advances in other fields had rendered the principles of alchemy obsolete. National Gallery, London.

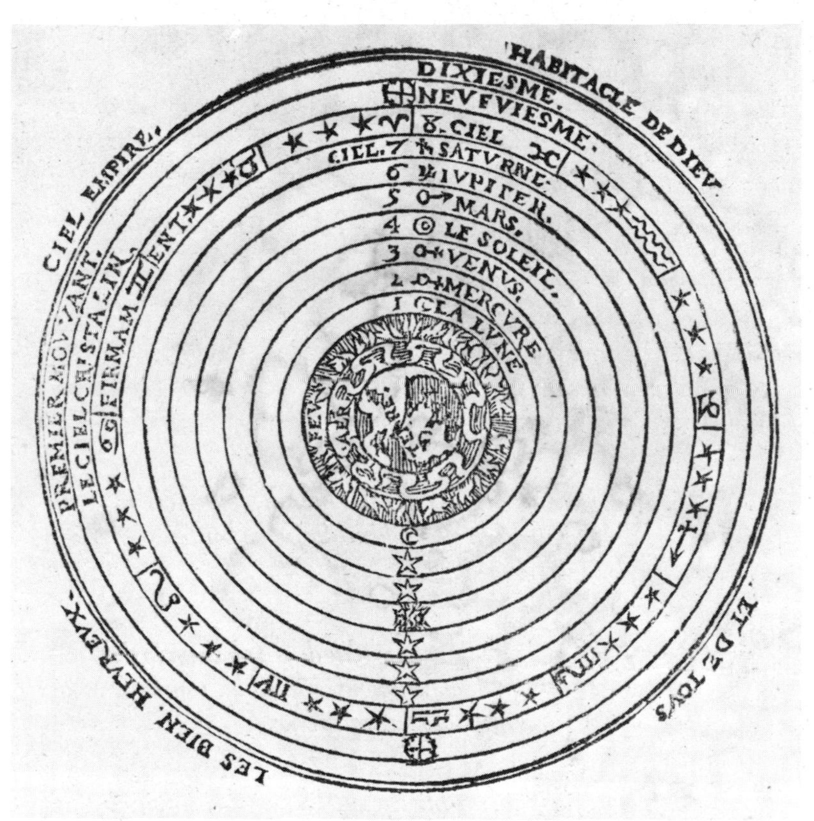

Left The geocentric universe as depicted by Peter Apian, in *Cosmographia*.

Opposite below left Portrait of von Hohenheim, usually known as Paracelsus (1493–1541).

Opposite below right The use of a cross-staff to measure the altitudes of celestial objects, from Peter Apian, *Cosmographia* (1539). With the astrolabe, the cross-staff was invaluable for the calculation of latitude.

Chaos
Fire:
Ayre:
NATURAL
MAGICK:
in XX Bookes
by
IOHN BAPTIST PORTA
a Neopolitane:
R Gaywood fecit Lond:
1658
Art:
Nature:
Earth:
I. BAPT: PORTA.
Water:

Left Nicholas Copernicus (1473–1543) with an armillary sphere.

Opposite The title-page of Giambattista della Porta's *Natural Magic* (1558) in its English edition of 1658. This book was a medley of scientific insights and others of less serious repute; the 16th-century concept of 'natural philosophy' allowed speculation on a wider range of subjects than did the later concept of 'science'.

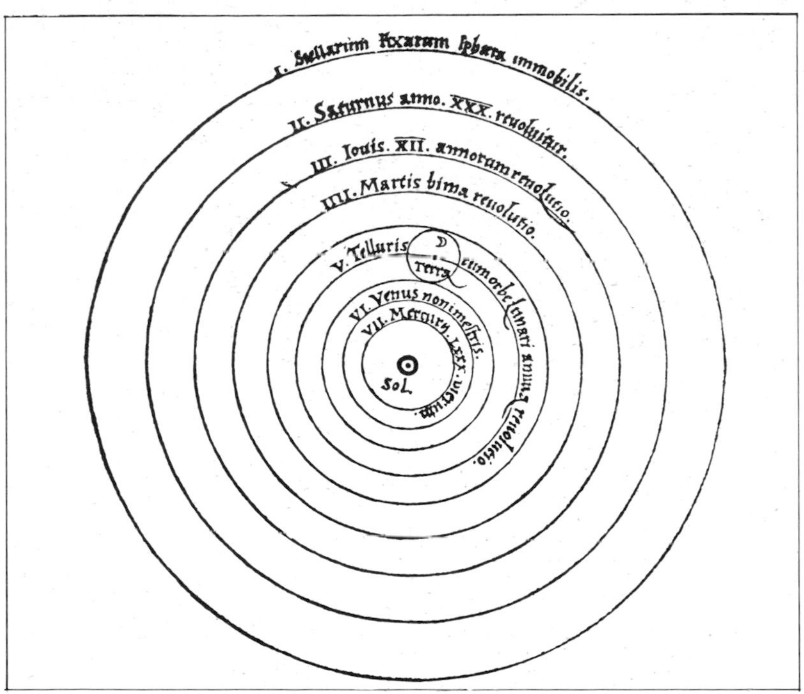

Left Diagram of Copernicus's system of the universe with the Earth and planets revolving around the Sun; from *De Revolutionibus*, (1543).

A perfit description of the Cælestiall Orbes,

according to the most auncient doctrine of the Pythagoreans. &c.

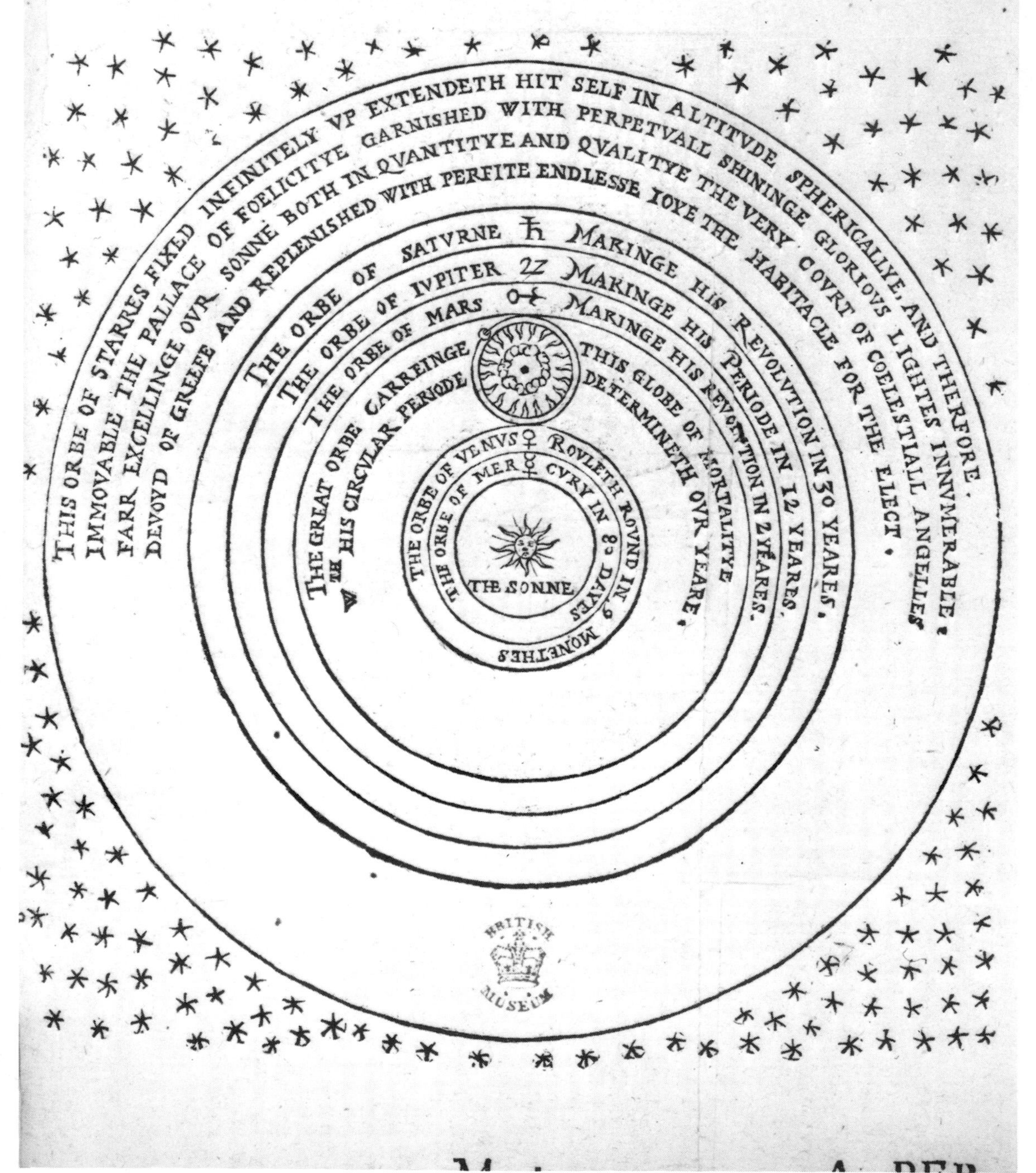

competent editions of Ptolemy's *Geography* and three editions of the Bible. It was during this work that he came into contact with many medical books and met the medical author Symphorien Champier, who advised him to go to Paris to study medicine. This Servetus did, and he soon gained an enviable reputation, becoming a member of a distinguished medical circle and gaining a reputation second only to Vesalius as a dissector.

In January 1553 Servetus published his *Restitution of Christianity*, a book which contained his valuable contribution to medicine, as we shall see in a moment. What his contemporaries saw, however, was a book with heretical views about the Trinity, and it led to his final downfall. In preparing the text, he had been in correspondence with Calvin who, however, had refused to have anything further to do with him after receiving a draft of the *Restitution*; once the book was published Calvin denounced Servetus. Servetus was arrested and imprisoned but managed to escape, though most copies of his book had by then been burned. After remaining in hiding for four months in France, he decided to move to Italy but, unhappily, chose a route that took him through Geneva. Here he was recognized and in 1553 he was burned at the stake, still inveighing against the Trinity.

The outstanding medical discovery that Servetus made was of the lesser circulation of the blood. He was the first in the West to discover it, for although it had already been noted in Islam in the mid-thirteenth century, it seems that he knew nothing of this; indeed his clearly independent discovery was motivated by theology and introduced to explain more satisfactorily the dissemination of the divine spirit through the body. Moreover, he also stated that 'vital spirit is then transfused from the left ventricle of the heart into the arteries of the whole body' – a view that comes very close indeed to a description of the greater circulation, which William Harvey was to discover in the next century. There appears to be no question of Harvey knowing of Servetus's ideas – only three copies of the *Restitution* survived so it was hardly common currency – and there is, of course, the question of how real a discovery Servetus himself had made. Was it merely part of a theological argument or was it based on medical experience? He certainly knew of the impermeability of the septum of the heart, but then Vesalius had more than hinted at that, as we have seen, and there is no evidence that Servetus performed any experiments to support his contentions, as others were later to do. Nor did his ideas become well known; they exerted no influence and stimulated no research. In brief, they mouldered in obscurity and but for their recovery after Servetus' death and, more especially after the significance of Harvey's famous discovery became clear, they would not even find a mention today. However, if Servetus had been less addicted to his anti-Trinitarianism, and Calvin less fanatical about heresies, perhaps these novel views might have entered medicine a half century earlier than they were in fact to do.

For this discovery by al-Nafis, see page 237.

Opposite Drawing of an infinite universe from Leonard and Thomas Digges's *Prognostication Everlasting* (1576). This work was of great importance in opposing the old idea of a sphere of stars, and the theory was taken up by Giordano Bruno. The possible use of a telescope to discover new stars may have helped in the formulation of this theory.

Chemistry

The chemistry of the Renaissance followed a number of parallel courses; on the one hand the traditional pursuit of alchemy continued with its two-fold aims: the tramsmutation of base metals into gold, and the discovery of an elixir which would confer eternal life and cure all bodily ills. A second, and, as it was to turn out, very important development, was medical or iatro-chemistry. This arose from the idea of an elixir, and concerned itself with the use of chemicals to supplement, or even supplant, herbal remedies. Thirdly, there was the growth of practical chemistry as a result of developments in mining and metallurgy during the Renaissance, of the production of gunpowder and of the growth of distillation. The advent of printing was to affect all three, though it played the greatest part in the last. Mining and metallurgy also led to some closer study of mineralogy and a start in the West of an appreciation of the Earth's crust which, in the seventeenth century, was to develop into the beginnings of the science of geology.

For Chinese alchemy, see pages 135–136 and 174–175; for alchemy in India, see pages 193–194; in Islam, pages 237–239.

Renaissance alchemy quickly became coloured by Hermetism. The Magus, with his esoteric knowledge of the macrocosm and the microcosm, could and did operate on the basis of the four old basic elements, though the care and precision with which he worked, as befitted the purity of vision which he believed Hermetism brought, led him on occasions to success where others before him had failed. An interesting development, epitomized in the 'Vase of Hermes', was the interest taken in alchemical apparatus. This arose not just from convenience – finding the most practical apparatus for a specific set of chemical reactions – but also because the alchemist began to favour the idea that the shapes of the vessels exerted their own mystical influence on the transformations and reactions which occurred. For example, the bird-like vessel for distillation – the pelican – was well-known, but its development, the double pelican, was mystically connected with the process of 'conjunction' whereby chemicals involved in a reaction received the 'form' of the resulting compound, rather analogous to a lump of wax taking on the form of the seal impressed on it. But in Renaissance times the alchemical reaction vessel *par excellence* was the Vase of Hermes or 'Philosopher's Egg'. Its ovoid shape resembled ancient creation symbols, and in many alchemical manuscripts and printed books it was often depicted enclosing a serpent, put there to symbolize the material of the Philosopher's Stone or, to use its other name, the Elixir of Life. But, symbolism apart, what was important in all this was the gradual development of more effective and elaborate chemical apparatus.

Illustrations pages 261, 298

Another advance in the field of proto-chemistry was the arrival of the technical treatise. Without the advent of printing such works would probably never have appeared, but once available they brought a down-to-earth approach to laboratory work. These were practical books, like the famous *Buch zu Distillieren* (Book of Distillation), printed and published in Brunswick in 1519, the first descriptive text on the subject to appear in the West. The difference between such a text and a book on alchemy was that whereas the alchemical text would purposely be written in an obscure way, drawing heavily on symbolism, the practical book aimed, above all, at clear explanation. A brief extract from the two kinds of text will make the point. An alchemical description of sublimation – the process whereby crystals of a substance are deposited in the cool upper part of a vessel containing heated solid material in its lower part – would liken the action to, say, the flight of swans or other birds;

For the invention of modern chemical nomenclature, see page 389.

Dissolve the Fixt, and make the Fixed fly,
The Flying fix, and then live happily.

But in the *Buch zu Distillieren*, describing the preparation of a medicine a 'good water' for the palsy, using a rosenhut (a form of air condenser), the author writes with a commendable clarity:

Take of parsley seed, three ounces; of green wormwood two good handfulls; of distilled wine, three ounces. Pound together, and distil either in an alembic or an ordinary Rosenhut as shown here.

The appearance of the straightforward technical text was a great step forward from the obscurity of the alchemist and the Magus, who wanted to keep their secrets away from the gaze of the vulgar. When they did write, they wrote only in allusive terms that could be fathomed, if fathomed at all, by another adept. But the technical author had a different aim; his purpose was to inform his reader, adept or not. Another famous and important book of this kind was *Pirotechnia*, a treatise on chemistry, distillation, the manufacture of gunpowder, metallurgy, and the casting of everything from type fonts and medallions to large statues and guns. The author of this comprehensive work, which made known trade processes hitherto kept as closely guarded secrets, was Vannoccio Biringuccio (1480–1540). The son of an architect, Biringuccio had first travelled through Italy and Germany looking at metallurgical workshops, and then settled down to running an iron mine and forge at Boccheggiano and, later, a mint for the Petrucci family. After their fall, he ran various foundries, cast cannon, built fortifications and designed buildings, ending his career as head of the papal foundry and director of papal munitions at Rome, where he died. Obviously, he was a man of wide practical experience, and his *Pirotechnia*, published posthumously within a year of his death, exerted a wide influence, being the first book to be published in the West on arts and manufactures requiring the use of heat in their production. Part of its influence though, was due to the great clarity with which Biringuccio was able to express himself; no other contemporary writer was his equal in this respect.

Illustration page 298

Two other famous sixteenth-century metallurgical and mineralogical books were Lazarus Ercker's *Description of Leading Ore Processing and Mining Methods* (1574) and Georg Bauer's *De re metallica*

(*Metallurgy*) (1555). Ercker's book was particularly useful for its description of the methods used for assaying precious metals, though its text does not match Biringuccio's for clarity, while though Bauer's book plagiarises some Biringuccio, it had important things to say on matters of mining. Ercker held a number of assaying posts in the mineral-rich area of Saxony, moving to Prague in 1567 as 'control tester' of coins, and his book contains, as might be expected, some useful analytical chemistry. Georg Bauer, usually known as Georgius Agricola, was more of a scholar than either Biringuccio or Ercker. Born in 1494 in Glauchau, Germany, he entered Leipzig University and obtained his first degree at the age of twenty-one. Thereafter he lectured there in elementary Greek, until in 1519 he moved to nearby Zwickau, there to organize a new Greek school. Zwickau was a centre of the Reformation, and although Bauer thought a reformation was necessary, he disliked its revolutionary aspect, and by 1523 he wanted no more to do with Greek or the reformers. He therefore returned to Leipzig, this time to study medicine. It was around this time that he visited Italy, spending three years helping to edit, for the Aldina press, editions of Hippocrates and Galen, and he also became interested in politics and economics. His subsequent travels took him through the mining districts of Carinthia, Styria and the Tyrol, and in 1527 he was appointed town apothecary and physician at what is now Jachymov in Czechoslovakia, then one of Europe's most important mining centres. Here he came face to face with the occupational diseases that beset the mining community and he studied the use of minerals and the products of smelting in their treatment. This was a new departure, because Galen had repudiated the use of mineral drugs, so Bauer began to write up his experiences, describing the principles of the geology and mineralogy of mining, and its techniques, as well as the occupational diseases and their treatment. Eventually he moved back to Germany to the quieter town of Chemnitz in order to give himself more time for his writing. When the Black Death swept through Saxony he worked night and day to alleviate suffering, but this exhausted him; an appointment as historiographer to the court of Saxony did not bring relief. He found himself forced to report honestly – and damagingly – on the claims of the ruling house, and then he became obsessed by what he saw as imminent war and a permanent break between Catholics and Protestants; the war, it is true, was avoided by the Peace of Augsburg in September 1555, but this only confirmed the permanence of the split in Western Christendom. Bauer died in the following November, tired and disheartened. Not until four months later did his extensive *Metallurgy* appear. Based almost entirely on his own observations, it was a seminal book, containing the seeds of later developments in medical drugs and the foundations for a more modern approach to geology and fossils; with its 292 beautiful woodcuts it was to be the standard work for the next two centuries.

Illustration page 295

Bauer (Agricola) was a man of science, cautious in his opinions, careful that wherever possible everything in his *Metallurgy* should be based on his own, not inconsiderable, personal experience. He was in many senses the antithesis of another German Renaissance scholar, whose interests lay in the same fields, the extrovert Theophrastus Philippus Aureolus Bombastus von Hohenheim (1493–1541). Usually known by his nickname Paracelsus, assumed when in his thirties and deriving either from a latinization of 'von Hohenheim', or meaning 'surpassing Celsus' (the Roman medical eclectic of the first century AD), or, possibly, even referring to his authorship of 'para- (doxical)' books that overturned the traditional outlook, he was a colourful character. He was born in 1493 or 1494 at Einsiedeln, Switzerland, into a medical family. His earliest education – particularly instruction in botany, mining, metallurgy and general science – came from his father; then his tutors were several bishops and the abbot of Sponheim, a notable exponent of the occult. In his early twenties Paracelsus also did practical work in the local mines or, more probably, the mining laboratories. This was a somewhat unusual training, though it seems he also travelled in Italy and spent some time at the University of Ferrara, presumably reading medicine, for soon after we find him acting as a military surgeon, first in Venice and then elsewhere.

Illustration page 300

It appears that for all his professional medical standing, Paracelsus's father was illegitimate, and his mother a bondswoman of a Benedictine abbey, and these facts seem to have rankled with him, for throughout his life Paracelsus seems to have had a love-hate relationship with authority; he was always an angry man who had the unhappy knack of alienating even his friends and patrons. He condemned traditional science and medicine, and sought to learn new cures from the peasant population, spending much time drinking with them in low taverns, a habit which brought him much expertise on wines. He treated their ailments free, balancing this by exorbitant fees to the rich.

Paracelsus's career was a mixture of success and failure. He set up a successful medical practice in Salzburg, gaining a great reputation by saving the life of the notable humanist publisher Johannes Froben; this, together with his sound medical advice to Erasmus, won him the post of municipal physician and the professorship of medicine at Basel in 1527. The academic authorities were unhappy, though, because Paracelsus refused to submit qualifying documents or take the required oath and even wrote a document denigrating Galen and promising a new syllabus; only the sponsorship of Froben and other powerful reformers ensured his appointment. But by the next year Froben was dead, and Paracelsus was forced to leave Basel, having in the interim publicly burned a copy of Ibn Sina's *Canon*, insisted on lecturing in German and admitting barber-surgeons to his classes, and brought things to a head with a lawsuit against a magistrate for non-payment of fees. After this Paracelsus wandered about from one city to another, spending at the most two years in any one place, by which time he had usually managed to long outstay his welcome. It was during his wanderings that he donned peasants' clothing and studied miners' diseases. He died at Salzburg in 1541.

A man of immense physical stamina, Paracelsus would challenge peasants to drinking bouts and, after winning, spend much of the remainder of the night dictating coherently to an amanuensis (except for moments when he would wield a sword, shouting wildly, and terrorising everyone present). Then he would spend the following day in his laboratory or at his practice. In spite of this wild life he was able to introduce a number of innovations into medicine, most especially the recognition of silicosis and tuberculosis as occupational diseases among miners, the realization that syphilis could be congenital, and that there was a connection between goitre and cretinism, but his most important contribution was his new theory of disease. Paracelsus repudiated the ancient belief that disease was due to an imbalance or disturbance of humours, and emphasized external causes, especially the invasion of the body by some 'poison'. This led him to new forms of treatment, often applying homeopathic principles and the concept of 'signatures', whereby a herbal remedy would be chosen because the plant's colour and shape resembled the affected organ. He also tried to separate drugs into their specific components and advocated the use of mineral substances as specific drugs in their own right. All this led him to develop chemical techniques and ideas which were to be of considerable use to those who practised iatro-chemistry after him.

Some of Paracelsus's techniques were practical, like his method of preparing concentrated alcohol by freezing out its watery component (a method already used by the Chinese), and his way of preparing nitric acid. He also introduced new and non-poisonous metals for medicinal use, and he was the first to attempt to construct a complete system of chemistry. This he tried to achieve by introducing three basic 'principles' – salt, sulphur and mercury. These did not replace the four elements (earth, air, fire and water) of ancient times, because they were not thought of as chemical substances; Paracelsian mercury was not the chemical substance mercury but its principle of behaviour, and his sulphur and salt were similar. Thus the principle of salt is present in every substance and if there in sufficient degree it causes the substance to be solid; likewise the sulphur principle is responsible for the inflammable nature of a material, and the mercury principle for its vaporous or liquid state. This was an ingenious and important idea which drew attention to certain universal principles of behaviour in diverse chemical substances.

For this technique in ancient China, see page 176.

Paracelsus was also a Hermetist, and this often coloured his medicine and his chemistry; for instance he gave a hierarchy of values to his principles, mercury being classed as indicative of the highest spiritual state, salt the lowest. He also had a theory that knowledge comes not from books in the old scholastic sense but from a study of Nature, because the latter enabled man to discern those invisible spiritual forces that cause bodies to behave in the way that they are observed to do. These forces, he thought, achieved results through 'knowledge', which is situated not in the observer but in the object itself. Man can acquire this knowledge only by union with the object, a meeting of the spirit of man with the spirit of the thing being observed; this is a union of two astral entities. This was, of course, a typical magical Hermetic-like view, but interestingly in Paracelsus' case led him to argue that the insane should be treated humanely. His argument here was that man's reason and knowledge are astral, and since the insane do not possess such astral knowledge, they are purer and nearer to God. But, though we should consider it an enlightened view, it was in no way a scientific approach to insanity.

A strong contrast to Paracelsus and those who shared his opinions was Andreas Libavius, a German chemist who worked in the latter half of the sixteenth century. Born at Halle in Saxony in 1560, the son of a poor linen weaver, he nevertheless managed to attend the gymnasium or high school in Halle, and then to enter first the University of Wittenberg and then that of Jena. Libavius seems to have then run two careers simultaneously. At Jena he eventually received a doctorate and the title of poet laureate, and after only a short time teaching, the university appointed him to the chair of history and poetry, though at the time he was away in Basel reading medicine. He then accepted a post as municipal physician at Rothenberg and, later, as inspector of schools, and did not finally settle down until 1607 when he was appointed rector of the high school at Coburg. A university in all but name, this was a Lutheran foundation and so never received the appropriate charter from the Holy Roman Emperor. Libavius himself was an orthodox Lutheran, strongly opposed to the Roman Catholics and, towards the end of his life, equally strongly anti-Calvinist. A man of immense industry and an overweening sense of self-confidence, he underestimated others; this led him to conflicts, even with his friends, and drove him to waste the greater part of his life in rather pointless polemics.

Libavius's range of interests was very wide but his main contributions as far as we are concerned were his voluminous writings on alchemy; these form an almost complete compendium of the chemical knowledge of his own day. They comprised a series of lectures in chemistry written in the form of letters to well-known physicians, and included a definition of what we term inorganic chemistry as 'the study of minerals', and an explanation of some obscure alchemical terms and concepts, as well as a critique of Paracelsus. But Libavius' most important work was his *Alchemia*, which appeared together with a slighter book on metals in 1597 and then in 1606 in an edition which incorporated both together. This second edition was well illustrated and is now considered to be the most beautiful of all the chemical texts published in the seventeenth century. In it Libavius divided chemistry into two parts – one concerned with apparatus and laboratory procedures, the other with the analysis of metals, minerals and mineral waters. His explanations and, particularly, his practical recipes are extremely clear and in strong contrast to the bombastic style which was so characteristic of Paracelsus.

With his own laboratory at home Libavius was able to carry out all kinds of tests, and the practical experience he gained made him a strong supporter of the use of chemical drugs. He therefore consolidated some of Paracelsus's work and opinions, and in his *Alchemia* gave details for preparing all kinds of medical substances, some including arsenic. For the whole of the seventeenth century Libavius's book had wide currency and did much to promote iatro-chemistry.

Before leaving the alchemy and proto-chemistry of the sixteenth century, there is one further figure who must be mentioned, the Frenchman Bernard Palissy. Born at La Capelle Biron in 1510, he was first trained in the manufacture of stained glass, but developed a process for enamelling earthenware which brought him fame and a reasonable fortune. Converted to Protestantism and one of the first Huguenots, he was imprisoned for his religion but released after protests by his patron, Anne de Montmorency; thereafter he decorated the new Tuileries Palace and established himself in Paris where, surprisingly, he began to give lectures on some aspects of natural history. Although without formal training in the subject, he managed to attract many of the most learned men in the capital, but in 1588, when religious persecution again erupted, he was imprisoned, and died in the Bastille at the age of eighty.

Palissy took a great interest in minerals and in geology, a subject on which he had some original ideas. He was one of the few men of his time to claim that the water in rivers came only from rainfall, arguing with great cogency against other views; he wrote on artesian wells and presented plans for constructing fountains to provide a domestic water supply. He discussed fossils and, like Xenophon in the early fourth century BC, believed them to be remains of plants and animals, though Palissy went further, connecting some of them with living forms, and claiming that some were extinct species. But for us, at this stage, his most interesting work was his experimental chemistry on minerals. He concluded that all minerals whose crystals had true geometrical shapes must have crystallized in water, and he classified salts in a way that showed a considerable understanding of their nature. He also demonstrated that drinkable gold, a well-known remedy of the time, was not really drinkable and certainly not beneficial; he also showed that mithridate – a chemical cure composed of some 300 ingredients – was not only useless but harmful. He denounced alchemists, saying that there was no hope at all of transmuting metals, and despaired of farmers who, he claimed, let manure lie in their farmyards until the useful salts in it were washed away. In some ways he was an incipient agricultural chemist though, of course, the state of chemistry in the sixteenth century was hardly conducive to pursuing such a subject scientifically.

For the development of these ideas in the 17th and 18th centuries, see pages 389–392.

Like that of Agricola and Paracelsus, Palissy's chemical knowledge had been learned by practical experience – in his case from work on stained glass windows and in developing pottery glazes. It was this practical side of chemistry, contributed by men of wide experience, which was so valuable a contribution of the sixteenth century. Together with the piecemeal esoteric knowledge of chemical reactions gained by the alchemist, it was in the next two centuries to bring about the arrival of the truly scientific discipline of chemistry.

Illustration page 300

Physics

The development of physics during the Renaissance is somewhat disappointing. Certainly there was some growth in understanding terrestrial magnetism, a little work was carried out in optics, and a small increase took place in understanding some questions concerned with mechanics, but, even so, progress was slow. It is true that the study of mechanics gathered momentum towards the end of the century, but this was due mainly to Galileo, and his work did not really manifest itself until the seventeenth century. However, it would be wrong to give the impression that mechanics remained in the same state throughout Renaissance times, though progress was due mainly to the efforts of one man, the Flemish engineer Simon Stevin.

Stevin was born in 1548 at Bruges, the illegitimate son of wealthy citizens. Little is known about his early life, though it is certain he worked in a local government post connected with finance in Bruges and Antwerp, and that he travelled to Norway, Poland and Prussia when in his twenties. On his return he established himself in the northern Netherlands, an emerging independent power that was already shaking off Spanish domination and the Spanish Inquisition, though it still needed a large army to consolidate its defences and ensure that it was free to develop its shipping and its trade, before it could reach full status as the Dutch nation. In 1581 Stevin was studying at Leiden and two years later matriculated as a student at this recently founded university; this was rather late in life to become a student – he was then thirty-five years of age – but he did so well that he was soon teaching mathematics there. In 1593, on the recommendation of the ruler, Prince Maurice of Nassau, who had been one of his pupils, Stevin was appointed 'castrametator' to the Dutch armies. This powerful post involved Stevin not only in military engineering and the preparation of defences, but also led eventually to his becoming a financial administrator and a member of the country's governing council. He initiated the practice of separating the ruler's personal accounts from those of the State and introduced the Italian method of double-entry book-keeping.

Stevin worked in both Leiden and the Hague, where he died in 1620, and it was in the northern Netherlands that he was stimulated to make his contributions to mathematics and mechanics. Stevin's mathematics, which was concerned with the important problem of algebraic symbolism, we shall come to later, but his mechanics were no less valuable and contained in three books, all of which were written in Dutch and appeared at Leiden under the Plantin imprint in 1586. The chief of these was his *Principles of Statics*, though the other two, the *Applications of Statics* and the *Principles of Hydrostatics* were also important.

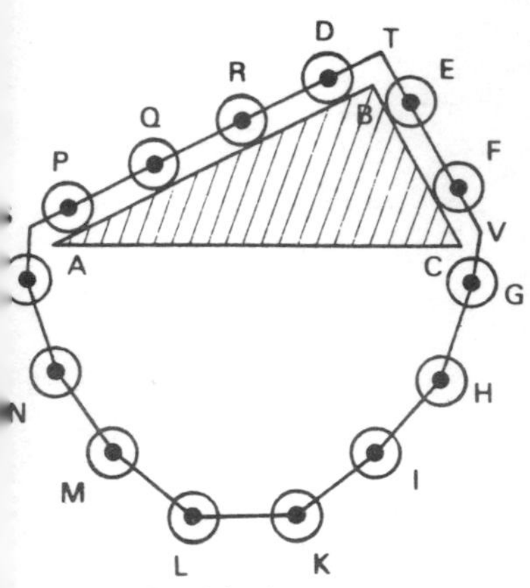

Simon Stevin's clootcrans or string of spheres.

In his *Principles of Statics*, Stevin continued the work begun eighteen centuries before by Archimedes. He discussed the theory of leverage, the centre of gravity of bodies and, above all, the theorem of the behaviour of bodies on an inclined plane. It was in dealing with the last that Stevin enunciated his most important discovery, the law of the inclined plane. He demonstrated this law by a *clootcrans* or string of spheres. As his original illustration shows, he used two inclined planes, one twice the length of the other, with a string of spheres hung around them. As the string will not rotate by itself, the hanging spheres (the spheres G, H, ... to O) may be neglected. It then becomes clear that the static string of spheres, P,Q,R, and D on the left must have an equal pull to those two spheres E and F. In other words the pull downwards towards the centre of the Earth is inversely proportional to the length of the inclined plane; or, the shorter the length of the plane, the stronger the force. Should the triangle be turned so that one of these planes is vertical, then we get the simpler situation where the weight of the spheres in the vertical plane must stand to the weight of those on the other plane (which is now horizontal) in the same proportion as does the height of the vertical plane to the length of the horizontal one. This was a profound discovery, and was reached by Stevin on the basis of careful argument and deduction; he performed what may be called a 'thought experiment', the results of which are the foundation of what we now call the 'parallelogram of forces', so fundamental in problems of statics. When he had reached his elegantly simple conclusion, he remarked (and wrote above his diagram) *Wonder en is gheen wonder*, (What appears a wonder is not a wonder).

Illustration page 301

Stevin's *Principles of Hydrostatics* was the first systematic book on the subject since Archimedes, and gave a simple and comprehensive account of displacement when a body is immersed in water. It also included an explanation of the 'hydrostatic paradox', the law which states that the force exerted by a liquid on the bottom of the vessel containing it depends only on the area of the surface under pressure and the height of the liquid above it, and has nothing to do with the shape of the vessel. His explanation included a description of a simple but effective experiment to demonstrate the principle.

For the discovery of the principle of displacement by Archimedes, see page 112.

In addition Stevin published some short treatises setting out the practical applications of some mechanical principles, while his military work led him to write also on fortifications and field encampments, on sluices and locks, and on windpower and windmills, a subject in which he made some detailed engineering improvements. His interests were indeed wide, for he also wrote on astronomy, supporting the new views of Copernicus, on civic life (his book included a guide for citizens at times of civil disorders) and on musical scales. This last contained the first description in the West of 'equal temperament' tuning, though as Stevin's publication did not appear until just before 1620, it is not clear whether this was an independent discovery. At all events, his system was similar to that used by the Chinese prince Chu Tsai-Yu (Zhu Zai-You) in 1585, and it is not impossible that Stevin knew of this work, for it seems he certainly knew of the Chinese invention of the wind-driven sailing carriage which had been reported in Europe in 1585.

For the Chinese musical scale, see page 172.

The Dutch were a seafaring people and it is not surprising that Stevin should also have interested himself in questions of navigation. Some of these were mathematical, concerned with the course on which a ship should sail to avoid getting lost. Here Stevin advocated sailing on a loxodrome or rhumb line, which always keeps at the same angle to every meridian, but the feat was beyond the mariners of his day though his idea was of use later. Trans-oceanic voyages were fraught with difficulties at this time and, as we shall see (Chapter 8), strong efforts were to be made in the seventeenth century to try to solve the serious problem of determining longitude at sea. Stevin wrote on this problem, again giving very clear explanations, especially of a system in which it was suggested that the magnetic compass could act as a guide to longitude. This was to be done by measuring the difference between true and magnetic north, since it was known that this difference was not the same at different points on the Earth's surface. It was a method that was later proved to be invalid because it was discovered (in the seventeenth century) that the variation was not constant but changed with time.

Illustration page 360

Mention of Stevin's work on navigation and the use of the magnet brings us to the subject of magnetism itself which, once the discovery of the magnetic compass had at last filtered through from China, began to arouse interest and comment. The first to write about the subject in the West was Peter Peregrinus, in the thirteenth century. He seems to have been a member of the army of Charles of Anjou, king of Sicily, probably acting as a military engineer. Little is known about Peregrinus himself except that he was a strong supporter of Roger Bacon; he is remembered today for his two-part treatise *Letter on the Magnet of Peter Peregrinus of Maricourt to Syergus of Foucaucourt, Soldier*. Who Syergus was is unknown. The letter is dated 1269, and besides describing the lodestone and its characteristics, defines the magnet's north-seeking property and contains the first use of the term magnetic pole. Peregrinus also explained how a magnet, when broken in two, becomes two magnets. The *Letter* contains an attempt to apply magnetic force to generate perpetual motion, but much more important is the mention of magnetic declination (the fact that the magnet points to magnetic north, not true geographic north).

For Chinese studies of magnetism, see page 173.

Nothing else was published on the subject after Peregrinus's impressive *Letter* until the Renaissance, and then fairly late, for the first book to appear did not come out until 1581. This was a *A New Attractive* . . . by the English instrument-maker Robert Norman. It contained not only 'A Short Discourse of the Magnet or Lodestone' but also a 'Newe Discovered Secret' of its behaviour, which was the phenomenon which now goes under the name of magnetic 'dip' and is concerned with the magnetic needle's inclination to the horizontal, an inclination which differs from one place to another. Its original discoverer was another instrument-maker, Georg Hartmann, who in 1544 had described the phenomenon in a letter to his patron, Duke Albert of Prussia, but his letter had never been published and remained unknown until the nineteenth century. Norman's description of dip was almost certainly due to his own independent discovery, and there is no doubt that he was the first to make it known to the world at large.

The most significant work on magnetism during the Renaissance was certainly that done by William Gilbert of Colchester, who was born in 1540 and read medicine at Cambridge. After this he may have travelled abroad, but the only certain thing we know about him is that by the mid-1570s he was in medical practice in London. Soon he was to become one of the most prominent physicians in London, and in 1600 was appointed physician to Queen Elizabeth I and, after her death, to King James I. Gilbert's claim to fame rests on his book *De Magnete* which was published in London in 1600. The book makes it clear that he had studied magnetism fairly deeply, perhaps in the years following his time at Cambridge, for in it he discusses the whole subject in detail, considering not only the magnetic compass but also the behaviour of the magnet itself and its powers of attraction and repulsion. It was the most comprehensive treatise on the subject since the time of Peter Peregrinus – indeed the most comprehensive treatise ever written – and was widely acclaimed both in England and on the continent of Europe; it was reprinted in 1628 and again in 1633. Gilbert also wrote about his cosmological ideas but these were not published until almost half a century after his death in 1603; it is *De Magnete* which has earned him a place in the history of science.

Illustration page 366

Gilbert clearly distinguished between the attraction exerted by amber – electrostatic attraction (though he did not call it this) – and magnetic attraction. He explained how every magnet has an 'invisible orb of virtue' around it and how it affects any iron in its vicinity. Then, after discussing other properties, Gilbert went on to show how a small spherically shaped magnet could imitate the Earth and, with the use of small magnets, demonstrated the behaviour of what today we should call the Earth's magnetic field. He accounted for magnetic declination and dip, as well as the north-seeking and south-seeking properties of the magnetic compass, by assuming that the Earth itself is a giant magnet. This was certainly a novel idea, but one Gilbert felt was supported by experimental evidence, and it was to bear fruit in later centuries, particularly our own, in which the subject of terrestrial magnetism has developed so notably.

For the development of Gilbert's work in the 18th century, see page 381.

As far as Gilbert's cosmological ideas were concerned these have often been said to be of no significance, and in the light of what had happened by the time they were published this is a valid enough criticism. Some appeared in *De Magnete*, however, despite the fact that they had been partly superseded. Nevertheless his claim that the Earth rotated daily on its axis was one that must have exerted some influence, particularly as he pointed out how such a view would fit in well with his idea of a magnetic Earth. Gilbert also denied the existence of solid celestial spheres to carry the planets in their orbital motions but this, as we shall see, was already becoming an untenable hypothesis among astronomers; what was perhaps more important was his obvious support for the new Sun-centred theory of Copernicus.

The other branch of physics to receive some attention during the Renaissance was optics. Spectacles to aid poor sight had been introduced towards the end of the thirteenth century in Italy, and were quite widely used all over Europe by the sixteenth, perhaps because of the rise in literacy stimulated by the arrival of printed books. Indeed, by this time they had become regarded as symbols of learning and even of sanctity, sometimes appearing on paintings depicting the Evangelists and, in one instance, the infant Christ. This concern with spectacles was bound to stimulate an interest in the optics of lenses, an interest which would also have been encouraged by translations of al-Haytham.

Illustration page 263

The two earliest writings on optics to be printed were a book by the thirteenth-century writer John Pecham, who had tried to reconcile the differing views of ancient authorities; and another thirteenth-century text, the *Perspectiva*, of Witelo, a Polish cleric; Witelo had drawn heavily on ideas inherited from Hero of Alexandria as well as other early writers, but he differed from Pecham, Grosseteste and Roger Bacon in wholeheartedly accepting al-Haytham's belief that there was no emission of visual rays from the eye. Witelo also accepted al-Haytham's teaching that the liquid humour in the centre of the eye is the light-sensitive part of the organ, thus again breaking with medieval tradition. In his book he discussed refraction and though his figures, supposedly obtained by experiment, were in fact not reached in this way, his ideas stimulated much interest when they appeared in print. Also influential was al-Haytham's own work, a printed edition of which came out in 1572. These last two – the works of Witelo and al-Haytham – were to trigger off much optical study, though even before their appearance something was already being done in the Renaissance by those who had access to manuscript sources. One of these was Francesco Maurolico.

For al-Haytham's description of the working of the eye, see page 228; for Grosseteste and Bacon, see pages 253–257.

Born in 1494 at Messina of Greek parents, Maurolico, often known as Maurolycus, was primarily a mathematician who spent his time in the civil service of the Holy Roman Emperor Charles V. He was one of the leading teachers of mathematics of his time and a prolific author, writing a history of Sicily, a book on astronomy and a host of translations of, and commentaries on, ancient authors; however it is for his work on optics that he is remembered today. His *Photismi de lumine* (Light concerning light) was completed by 1567, and was concerned with the casting of shadows, reflection, the formation of the rainbow, the structure of the human eye, the various kinds of spectacles available and the way they corrected defects in vision. It also contained methods of measuring light intensity and even included a description of radiant heat. In brief, the *Photismi* was a *vade mecum* of optical knowledge in the sixteenth century – 'the best optical book of the Renaissance' it has been called – but it contained no new discoveries and could exert little influence since it was not published until 1611, half a century after it was written, and by then the telescope had been invented and the book's impact was inevitably less.

The two other optical texts of the sixteenth century were by Giambattista della Porta (1535–1615). A colourful character, della Porta spent most of his life in Naples. He is now remembered mainly for his *Natural Magic,* a book which came out first in 1558, to be followed by a second much enlarged edition in 1589; it is on the second edition that his reputation is primarily based. Della Porta seems to have been self-taught and to have had a predilection for science. In 1585, at the age of fifty, he became a lay brother of the Jesuits, taking part in their charitable works in Naples, and seems to have been a devoted participant in the Roman Catholic Counter-Reformation. Nevertheless in 1580, before he joined the Jesuits, he was examined by the Inquisition, and between 1592 and 1598 all his publications were banned. Why this ban was imposed and why he was questioned in the first place is not clear, though it may conceivably have had to do with his interest in the academies of late Renaissance Italy. Naples possessed a number of 'academies' – societies composed of those having some common interest – but in 1547 they were all closed because it was suspected that they were hotbeds of political intrigue. However, they reopened in 1552 and one of them, the Altomare, was the outstanding literary centre; some of Porta's friends were members, but he set up his own organization, the Accademia dei Segreti (Academy of Secrets) or, more fully, Academy of the Secrets of Nature; it met in his own house. Precisely when he established this academy is uncertain, though it was certainly before 1580; its title alone must have caused the Inquisition some concern, and later they seem to have closed it down.

Illustration page 302

Porta's *Natural Magic* was a book about the secrets of nature, compiled from notes he had started collecting when he was fifteen; the specific title *Natural Magic* was chosen because, 'Magick is nothing else but the survey of the whole course of Nature'; the Hermetic tradition and more orthodox Neoplatonist beliefs taught a natural orderliness in things and Porta used a title that made the most of this. As far as contents were concerned, he cast his net wide and without much discretion, for the book is a quite astonishing hotch-potch of trivia cheek-by-jowl with matters of genuine scientific interest. Thus we have formulae for invisible inks and one for 'making a woman full of red pimples', together with serious discussions of husbandry, the lodestone, methods for hardening steel, and experiments in statics and pneumatics. The section devoted to optics is typical of the rest of the book; there are instructions for making a mirror to make one's face 'seem like an Ass, Dog or Sow' and 'Other merry sports with plain looking glasses' and yet, a few pages later, a serious discussion of the uses of concave mirrors. The production of a virtual image is described – 'an Image seen hanging altogether in the Air' – and then lenses are discussed, including spectacles for seeing at a distance, though Porta's description leaves something to be desired, perhaps because it is 'not a matter to be published too easily' and he still desires some secrecy. Nevertheless, Porta makes it clear that by using combinations of concave and convex glasses things 'both afar off and near hand' may be more clearly seen. From these remarks it would seem that he had experimented with lens combinations and may have hit upon the principles of both the telescope and the compound microscope. This supposition is strengthened by the fact that in 1593 he published a book on refraction *(De Refractione)* in which he again reported experiments using various combinations of convex and concave lenses. This second book also contains the first description of a camera obscura fitted with a lens. The obvious question, then, is whether Porta did invent the telescope. Doubt has been expressed because he keeps referring to 'greater clarity' as if he was seeking nothing more than sharper images, yet in *Natural Magic* he does also speak of seeing things afar off, and there is some other curious evidence, that can better be discussed in the next chapter, which make it seem likely that Porta at least knew the principle of the device. This is what one might expect of such a Renaissance experimenter. Moreover, he even wrote an unpublished work *De telescopiis* though unfortunately this does not clinch the case; it seems that he got so far and then stopped. Perhaps his problem was partly the very poor optical quality of the images that he obtained with combinations of lenses – a problem that was later to plague Galileo – and the fact that he also pursued the question of using a concave mirror as a telescopic device: a successful reflecting telescope was not to come until about a century later.

For earlier descriptions of lenses used in a similar way, see pages 254–256.

Mathematics

Mathematics flowered during the Renaissance and was therefore available to help give precision and quantitative results in those branches of science to which it could be applied. During the early stages of the Scientific Revolution this meant the use of mathematics to help clarify the behaviour of the Moon and planets as they moved across the sky and to help in solving some of the basic problems of mechanics. The great development of applying mathematical techniques to scientific questions was to reach a peak in the seventeenth and eighteenth centuries, but the foundations for this mathematization of science were laid in the sixteenth century.

In Renaissance times the first applications of mathematics lay in commerce and in the arts. The translation of classical and Arabian authors had brought everyday calculation to the fore and made the full text of Euclid readily available, to begin with in manuscript and then in print. The first printed editions appeared in Italy late in the fifteenth century, with Erhard Ratdolt's handsome Venice edition of 1482 (the first printing to have gold colouring) being specially notable. Editions of Euclid stimulated the new art of perspective with which artists were experimenting. This work was already under way in the fourteenth and fifteenth centuries in Florence as well as in Germany and the Low Countires, though it is the Florentine development which is best documented. This new art led to some geometrical investigations, first it seems by Filippo Brunelleschi (1377?–1446), though it is the fifteenth-century painter Paolo Uccello to whom the first book on the subject is credited. No copy of Ucello's book has survived; however, his artistic work makes clear his familiarity with the subject. A book that did survive was Piero della Francesca's *De prospectiva pingendi* (Perspective in Painting) which was written between 1474 and 1482, an important treatise which introduced the idea of the vanishing point. The subject was taken further by Leonardo da Vinci, and in Germany Albrecht Dürer independently developed the technique of linear perspective.

Illustration page 85

The advent of Euclid also affected cartographers, the most notable being the sixteenth-century Flemish map-maker Gerhard Mercator, whose cylindrical projection of 1569 was another application of geometry, in this case designed to aid navigators since loxodromes came out as straight lines. Again, it was Mercator's tutor, the Dutch geographer and mathematician Reiner Gemma Frisius, born in Dokkum in the Netherlands in 1508, who in 1533 proposed and illustrated the principles of triangulation, the surveying procedure by means of which places may be accurately located even though inaccessible or inconvenient for a surveyor to reach. Gemma Frisius was also the first to suggest the use of portable timekeepers for determining longitude at sea, a method that was excellent theoretically but could not satisfactorily be put into practice until the very late eighteenth century, because only then did a suitable timepiece become available.

Illustration page 289

A development of geometry which had been initiated by the Greeks and extended by mathematicians in Islam was the method of calculation using ratios of the sides of triangles and the angles between, which is the technique we now call trigonometry. The fifteenth-century development was primarily due to Georg Peuerbach and Johann Müller (Regiomontanus). Peuerbach was born at Peuerbach in Austria in 1423 and died at Vienna in 1461, but in his short life he made his mark both in trigonometry and in astronomy. Nothing is known of Peuerbach's early years, though it is certain he went to the University of Vienna and sometime during 1448 and 1453 travelled through Germany, France and Italy; he may have met the philosopher and astronomer Nicholas of Cusa while he was in Rome. In 1456 he became court astrologer to the Hungarian king Ladislaus V, and on the latter's death court astrologer to the Holy Roman Emperor. Being court astrologer did not occupy all his time, and he lectured in humanities at the university; Müller, or Regiomontanus, was his student. We shall come across Peuerbach's astronomy later, but his importance in trigonometry is that he wrote a short book explaining the way to calculate sines and chords of angles. This work later appeared in print together with tables of sines by Müller and an additional table giving sines for small fractions (minutes of arc) of angles. Similar tables of even greater precision were prepared in the next century by Georg von Lauchen (Rheticus). However they were only completed posthumously in 1596 under the direction of a pupil, Valentine Otho. A revised edition came out in Frankfurt in 1613, edited by Bartholomaeus Pitiscus, to whom we owe the word 'trigonometry'. The preparation and publication of all these tables may sound prosaic enough to us today, yet not only do they involve an immense amount of painstaking labour, but their wide availability in printed editions in the sixteenth century was instrumental in promoting the use of trigonometrical methods, which were of such great significance to the progress of astronomy, their chief field of application.

The other field of mathematics in which there was considerable progress was algebra. The arrival of mathematical works from Islam stimulated some developments, and in 1202 Leonardo Fibonacci of Pisa (sometimes known as Leonardo da Pisa) wrote the first Latin treatise on the subject, the *Liber Abaci* (Book of Computation). In this there is a long section on algebra and algebraic methods, and the book introduced the use of Indian-Arabic numerals, including the zero, to western readers. Fibonacci, remembered now for the Fibonacci numbers (a sequence obtained by adding each number to its predecessor, thus: 1, 1, 2, 3, 5, 8, 13, 21 . . .) also wrote a *Practical Geometry* which became a standard text, but after his time there was astonishingly little progress. Not until the fifteenth century was the next substantial step taken. Then, in 1484, Nicolas Chuquet produced his *Three-part science of numbers* which contained the germ of logarithms, a very clear understanding of equations and their roots, and a recognition of negative as well as positive roots. But Chuquet's book was never published and could exert little influence.

For the origin of zero, see page 149; for the development of Hindu-Arabic numerals, see page 192.

The mathematician of this period who did stimulate his contemporaries was Luca Pacioli. Born in Sansepolcro in Tuscany about 1445, Pacioli became a Franciscan monk and travelled round Italy teaching mathematics; he returned home at the age of forty-five, and in 1494 completed his *Summa de arithmetica, geometria, proportioni et proportionalita,* which was printed at Venice under his personal supervision. This was not so much an original book as a comprehensive one, borrowing, as Pacioli readily acknowledged, material from Euclid onwards to Fibonacci, but it was widely circulated and studied closely by sixteenth-century mathematicians – and book-keepers – who often admitted the debt they owed it.

It was in the sixteenth century that algebra really developed, and this was due not least to a strange situation in Italy. Between the years 1472 and 1500 no less than 214 mathematical books had been published to feed the increasing demand for mathematics by banking houses, merchants, workshops, public administrators, astrologers and scholars. This stimulated an interest in mathematics itself and created amongst the more academically minded a sense of competition which resulted in public challenges and debates on mathematical subjects. But this public interest had a drawback; it tended to inhibit the publication of results. If a mathematician discovered how to solve a particular mathematical problem, he faced two alternatives; either he could publish his results and receive honour but no financial gain, or he could keep it a secret and exploit it in public challenges. In the latter case he would also gain fame and money (since each contestant deposited a sum, which the winner collected), though he ran the risk of his discovery being published by one of his rivals.

The disputations were often concerned with the thorny problem of solving cubic equations (equations where x^3 appeared) and the first to achieve success in the sixteenth century was Scipione del Fero, a professor of mathematics at Bologna for thirty years, who solved such equations provided they did not contain the term x^2. He bequeathed his solution to a pupil, Antonio Fiore, and later it was acquired by his son-in-law, but neither published it. However, the problem was also solved independently and in a different way in a challenge thrown out by Fiore, the solver being Niccolo Tartaglia. Tartaglia – a surname he adopted from a boyhood nickname given to him because a wound in the mouth caused him to stammer *(tartagliare)* – was born in Brescia about 1500 and died in Venice in 1577, having spent the greater part of his life in military engineering, surveying and gunnery. In geometry he was a pioneer in computing the volume of a tetrahedron from the lengths of its sides, and he devised methods of inscribing circles in a triangle; in other branches of mathematics he solved various problems and published what is now called Pascal's Triangle, though others before him knew of it in a slightly different form. But his enduring fame rests on his solutions for cubic equations, for he solved not only the Scipione type, but also those where x^3 and x^2 (but not x) also appeared. He declined to publish his results but passed them on to Girolamo Cardano under an oath of secrecy. Cardano, however, did finally publish the results, though giving Tartaglia full credit, in 1545 in his book *Ars Magna* (The Great Art), the first great treatise on algebra to appear in the West.

For Pascal's Triangle in China, see page 153.

Cardano (or Cardan), born in Pavia in 1501, attended the university there in 1520 to read medicine; he completed his degree at Pisa six years later, and went into practice. In 1534 he became a teacher of mathematics as well as a doctor and, after a period as a physician of considerable repute, he accepted the chair of medicine at Pavia though, embarrassed by the dissipation of his younger son and the execution of the older for the murder of his wife, he later moved to the chair at Bologna. Imprisoned by the Inquisition in 1570, mainly for the horoscopes he had cast, he recanted and succeeded in obtaining a lifelong annuity from the Pope. Cardano wrote more than 200 works on religion, music, philosophy, physics, medicine and mathematics, but the fame of this unstable character rests primarily on his contributions to mathematics. He wrote on numerical calculation, but his major work was the *Ars Magna* in which he systematically presented many new ideas in algebra. He gave a rule for solving third degree equations without the x^2 term, he discussed the number of roots of still higher equations, and pointed out that equations with more than one root had more than one solution; in brief he originated the theory of algebraic equations. Cardano was also the first to use 'imaginary numbers' (ordinary numbers multiplied by the imaginary quantity, the square root of minus one) which are of great use in the kind of problems he was solving. He also had a passion for games and this led him to write on mathematical probabilities, though his book on this only came out posthumously.

Cardano was interested in all kinds of science; mechanics intrigued him and the gimbals suspension is credited to him – the so-called Cardan suspension – though this was in fact a Chinese invention. He realized that a projectile moves in a parabolic path not in two separate straight lines as had hitherto been thought and he declared (correctly) that perpetual motion was impossible, though this did not prevent others from pursuing this particular fantasy. He also made some minor advances in hydrostatics, and thought that a vacuum was possible, attributing its effects to rarefaction of air. In the breadth of his interests and his manifold contributions to knowledge, Cardano was a typical product of the Italian Renaissance.

A third sixteenth-century Italian mathematician to investigate cubic equations and, like Cardano, to concern himself with equations of still higher degree, was Rafael Bombelli. Born in Bologna in 1526, he spent the major part of his life as an architect and hydraulic engineer, but, stimulated by the mathematical climate in the Italy of his day, he decided to write a treatise on algebra. Cardano's *Ars Magna* had come out in 1545, an argument between Tartaglia and Cardano about breaking the bond of secrecy had erupted in 1546, and Bombelli decided that the very next year he would settle down to his task. He worked at the book for the next three years, and produced a thorough-going systematic treatise which gave a comprehensive account of the subject at the middle of the sixteenth century. He drew heavily on the work of the Alexandrian mathematician Diophantus – indeed it was Bombelli's *Algebra* that popularized Diophantus's work in the West – but he also contributed something original to the solution of cubic equations, dealing with the type where x^3, x^2 and x all appear together. The other great contribution his book made to algebra was in the symbols used to denote the powers of unknown quantities.

The symbols used in algebra are important. As we shall see in the next chapter, poor symbolism could, and did, delay progress, because symbols can act as an aid to thinking if they are explicit. Bombelli was not the only mathematician to appreciate this. In Germany some symbolism had come into use in mercantile arithmetic – the plus (+) sign was an abbreviation of the Latin 'et' – and the equals sign (=) was introduced in 1537 by the English mathematician Robert Recorde, while Leonardo Fibonacci, Nicolas Chuquet and Luca Pacioli all used some symbolism in their books. Simon Stevin also introduced new notation, in his case for polynomials (terms containing more than one algebraic expression), and advocated and used decimal fractions; he also wrote in favour of decimal coinage but it was not until the time of the French revolution that this was to be put into practice. Another contributor was Michael Stifel, a theologian as well as a mathematician, who was born in eastern Germany in 1487 and died in Jena eighty years later. It was Stifel who introduced letters for the unknown, and repeated these where powers were concerned. We use letters for unknowns today (x, y, z, for example); Stifel chose different letters from these, but letters all the same. The other sixteenth-century contributor to this mathematical shorthand was François Viète (Franciscus Vieta) and born in 1540 at Fontenay-le-Comte in the Vendee. Trained in the law like his father, Viète held senior posts in the government, and proved himself adept at decoding military ciphers. He was an 'amateur' mathematician, but this did not prevent him making innovations in the subject which all appeared in a series of books that came out during the last quarter of the sixteenth century. However, his literary style was obscure and his books had only a small circulation; his genius was not really appreciated until it was explained by commentators from 1602 onwards, most notably by the Dutchman Frans van Schooten who published *The Mathematical Works of Viète* at Leiden in 1646 in a Latin edition. What Viète achieved was that he used decimal fractions of degrees in his mathematical tables instead of the usual sexigesimal ones (minutes and seconds of arc), he gave graphical methods for solving cubic and biquadratic equations (those with x^4), and trigonometry for some others of still higher degree. Above all he used symbols not only for the quantities met with in algebra, but also for the operations (multiplication, division, etc.) performed on them. His symbols for quantities were also an innovation, since he used vowels for unknowns and consonants for given numbers. Powers were indicated by the appropriate word *(cubus, quadratus,* etc.) and all this brought about great clarity, so mathematicians were able to appreciate without difficulty what the equations indicated. Viète also applied similar methods to trigonometry, and it was he who showed how the various trigonometrical ratios – sine, cosine, tangent, etc. – could be transformed into one another. His work, and the growing use of symbolism among other sixteenth-century mathematicians, brought some simplification to algebra and trigonometry and prepared them for the far more extensive use they were to get among mathematicians in the seventeenth and subsequent centuries.

Astronomy

Almost certainly the most significant of all changes in science that came about in the Renaissance were those connected with ideas of the universe. The view which finally prevailed as the sixteenth century closed represented a revolution in outlook and one which not only affected astronomy but was also to have the most profound repercussions on philosophy and on religion. But this was not to start until the second half of the sixteenth century; before then astronomy underwent a slow process of refinement and consolidation, with little in the way of new ideas, though in the fifteenth century Nicholas of Cusa (Nikolaus von Cusa, or Cusanus), made some mildly revolutionary suggestions.

Born at Kues (Cusa) in the Moselle region of Germany about 1401, Nicholas was the son of a poor fisherman. It seems likely that he received his education away from home at Deventer in the Netherlands, where he attended a school kept by the Brethren of the Common Life, which was a community which stressed penance, meditation and the inner life, living a simple communal life with an absence of ritual. Hieronymus Bosch, the artist, was a prominent member. They aimed to educate a Christian elite and were promoters of the new learning of the Italian Renaissance, though not of its humanistic outlook. However, Nicholas also went to the university of Padua when he was sixteen – a usual age for entry at this time – and here he met humanist educators and came into contact with Renaissance science, though his formal subject for study was canon law. Ordained as a priest, he took a great interest in Church matters and particularly in a request from the Christians in Byzantium, who were then under severe pressure from the Ottoman Turks, for a reunion of Western and Eastern Christendom.

Nicholas was essentially a philosopher, but he decided to put science at the service of philosophy and this led him to some interesting conclusions. He found formal Aristotelian logic inadequate because he thought it was suited only to finite notions and he wanted to consider ideas which included the infinitely large, which was itself contained in the one absolute maximum that was God. In his discussion of the universe, these views led him to reject any cosmic centre point for the motion of the heavens and repudiate the idea either that the Earth was at the centre of all things or that it was stationary. In fact he believed the Earth moved, though not in an orbit, but with an apparent motion. It was Nicholas, too, who seriously suggested that the Earth was not the only place in the universe which supported life. His arguments in support of his views were philosophical and expressed in theological language, but they were nevertheless influential.

The imaginative concepts of Nicholas of Cusa were naturally not adopted by the majority of astronomers, who continued to pursue the Ptolemaic-Aristotelian scheme of the universe as set out in original Greek texts and refined by Arabian commentators. His younger

For the Greek concept of the universe as expounded by Eudoxos, see pages 100–101; and by Ptolemy, see pages 122–123.

contemporary Georg Peuerbach, whose contributions to mathematics have already been mentioned, worked to refine still further the astronomy of the *Almagest,* preparing tables of eclipses and of the motion of the Sun and writing a school textbook on Ptolemaic astronomy which was often to be found bound with a late medieval book on the same subject by John of Holywood (Yorkshire), better known by his latinized name of Sacrobosco ('holy wood'). The fact that the books were seen to complement one another underlines the traditional attitude of Peuerbach's work.

Illustration page 87

Illustration page 262

The prosaic nature of Peuerbach's work does not mean that it was unimportant. Astronomy is nothing if not a precision science, whose progress depends on detailed measurement and very careful calculation. Only by these means could inadequacies in Ptolemy's theories of planetary motion be detected and modifications to them checked. However, Peuerbach's death at the age of thirty-eight suddenly terminated his promising career, and would have prevented the completion of his full-scale review of Ptolemaic astronomy had it not been for his pupil Johannes Müller (Regiomontanus). Born in 1436 and so rather younger than Peuerbach, Regiomontanus nevertheless did so well at the University at Vienna that he soon found himself on the teaching staff as a colleague of Peuerbach's and the two men became close friends. It was the arrival in Vienna of Cardinal Bessarion, the papal legate, that set them off on a thorough study of the *Almagest*, for Bessarion was heading a campaign to bring ancient Greek authors to the attention of the intellectual community. Peuerbach set about making an abridgement of Ptolemy's great book, having a copy of a Latin translation by Gerard of Cremona (twelfth century) to guide him, and on his deathbed he obtained a promise from Regiomontanus that he would complete the work. This Regiomontanus did, travelling to Rome with Bessarion and completing what he called the *Epitome* some time before 1463, though it was not published until 1496, twenty years after his death. The book was much more than a précis of the *Almagest* for it added later observations, revised the computations in the original and also provided some critical comments on Ptolemy's text, one of which revealed that his theory of the Moon's motion required the Moon's apparent diameter to change more than it is observed to do. Thus it pointed out the need for a critical revision of some aspects of the *Almagest*, and so added to the evidence available for current astronomical research.

Regiomontanus remained in Bessarion's entourage, lecturing and preparing commentaries to accompany other early astronomical texts, as well as writing on some of the technical problems in computing the future positions of the planets. Later he went to Hungary at the invitation of the king, Matthias I Corvus, to collate Greek manuscripts and then, keen to publish his tables and other works which commercial publishers rejected because of the cost involved in preparing the special diagrams, he moved in 1471 to Nuremberg. Here, established in a house with his own observatory and printing press by a wealthy businessman, Bernhard Walther, Regiomontanus set himself up as the first publisher of astronomical and mathematical works whose accuracy could be utterly relied upon, for such works that had been issued by commercial publishers were full of errors. Helped by Walther, Regiomontanus also made observations of the bright comet which appeared in January 1482, observations that were accurate enough to permit its identification, two centuries later, as one of the appearances of Halley's comet.

The criticism of Ptolemy's lunar work in the *Epitome* helped bring home the realization that all was not well with Ptolemaic theory, and so played a part in preparing the way for the great revolution in sixteenth-century astronomy that is inseparably linked with the name Copernicus (1473–1543). Born at Torun in Poland in 1473, Niklas Koppernigk (Copernicus) was the nephew of Lucas Waczenrode, who took him under his protection. As newly elected bishop of Ermland (a bishopric in the western part of East Prussia which had been ceded to Poland in 1466), Waczenrode saw to Niklas's education with a view to his obtaining a canonry at the cathedral at Frauenburg, since this would make him financially independent. In 1491 Niklas entered Cracow University, where he studied classics and mathematics and developed an interest in astronomy, but when he had completed his studies there was no vacancy for him at Frauenburg – such appointments in the bishop's gift only came up at specified intervals – so he went on to Bologna, ostensibly to study canon law, although he stilll pursued his interest in astronomy. In 1501 he was at last elected to a canonry but granted a further period away, this time to study medicine at the famous medical school at Padua; here he remained for four years except for a short break at the University of Ferrara, where he obtained a doctorate in canon law.

Illustration page 303

In 1505 Copernicus returned to Frauenburg, well qualified academically for any task the cathedral chapter might require, and as things turned out, every side of his extensive training came to be used. His medical services were called on by the chapter and by the poor, his mathematical abilities when he propounded a scheme for reform of the currency, and his canon law in his administration of the diocese. Copernicus, was a man of considerable resolution; in 1520 when the Teutonic Knights invaded the area, Copernicus commanded the chapter's castle at Allenstein on the outer borders of the diocese, and he held the town until an armistice was arranged the next year. So resolute a man might be expected to take bold steps when it came to his astronomy.

Early in 1513 Copernicus had bought 800 stones and a barrel of lime from the chapter's workshops with which he had built a roofless tower for an observatory; from here he made a number of observations. The problem in astronomy that he was investigating was the motions of the celestial bodies; on the basis of the geometrical theories of planetary motion given in the *Almagest* it had been possible to compute the future positions of these bodies, but with the observations amassed over the centuries, it had become evident that some refinements in Ptolemy's original theory were desirable. Various sets of tables of future positions had been drawn up, and in every case they had given more exact positions, though each set also contained errors; theory did not fit precisely with observation, as Peuerbach and Regiomontanus had shown. But there was another consideration as far as Copernicus was concerned. He was most dissatisfied with Ptolemy's invention and use of the equant. This offset the centre of motion from the centre of the Earth, and thus introduced an uneven motion which, Copernicus believed, conflicted with the 'rule of absolute motion' whereby everything should move round the centre of the universe at an unvarying rate. He therefore sought another explanation.

For the views of Philolaos on the motion of the Earth, see page 77.

Illustration page 303

Copernicus knew some Greek philosophers had suggested that the Earth moved, and it seemed to him that a more correct view, incorporating true absolute motion, might be arrived at if the Sun were placed at the centre of the universe, and the Earth thought of as a planet, orbiting the Sun as the other planets do – Copernicus's motives in adopting this revolutionary view were primarily scientific, though it is legitimate to wonder whether possibly Hermetism may have been a factor – perhaps not a very significant one, but a factor all the same. In his book *De Revolutionibus orbium coelestium*, setting out his ideas, on the same page as that on which the heliocentric idea of the universe appears, Copernicus wrote:

At rest, however, in the middle of everything is the Sun. For in this most beautiful temple, who would place this lamp in another or better position than that from which it can light up the whole thing at the same time? For the Sun is not inappropriately called by some people the lantern of the universe, its mind by others, and its ruler by others. [Hermes] the Thrice Great labels it a visible god, and Sophocles's Electra, the all-seeing.

Certainly Copernicus was mustering every authority he could to support his case – a typical method to persuade an intellectual readership that still held some reverence for the wisdom of the past – for he himself was worried about the reception of his ideas. Interestingly this seems not so much to have been because a moving Earth was contrary to the Bible, but because Copernicus was afraid of ridicule. After all what evidence was there for the Earth's motion? There seemed to be none at all; direct evidence was lacking, for if the Earth moved, as Copernicus suggested, then there would be an annual shift in the apparent positions of the stars but no such shift was observed. Copernicus was well aware of this, and in his treatise countered the argument by the explanation that the sphere of the stars (in which he still believed) was too far off for this shift to be detectable. He could not, however, answer the question of why God should leave a vast gap between the planets and the stars. However, Copernicus was right; detection of such 'annual parallax' needed techniques unknown in the sixteenth century, and such observations could not be made for 300 years, though a proof of the Earth's motion was, in fact, obtained in the eighteenth. But there were other objections – the same or similar to those made in ancient times – that a moving Earth would be at the mercy of perpetual gales and tidal waves, or that such motion would shake the Earth to bits. Again Copernicus anticipated these and tried to counter them, but their refutation really demanded a whole new physics of moving bodies, and, again, the theory for this lay in the future. Copernicus himself used no new ideas of motion; he merely gave the Earth perpetual motion in a circle about the Sun, but this was no more than that possessed by every orbiting planet. Falling bodies still fell towards the centre of the Earth because this was their natural place, as Aristotle had pointed out, though he did not discuss the question of why they did not fall into the Sun instead, since it, and not the Earth, was now situated at the centre of the universe.

For Aristotle's use of these arguments, see page 104.

In many ways, then, Copernicus had good cause for caution, so he first circulated his ideas privately among friends in a *Little Commentary*. This was well received, and even provided the material for a lecture to the Pope, Clement VII, and some of his cardinals in the Vatican gardens by Johan Widmanstadt, the papal secretary; later, Copernicus was to be urged to publish his views, Cardinal Nicholas von Schönberg (whose secretary Widmanstadt had become) writing to him to this effect. Schönberg's letter was in fact published at the beginning of *De Revolutionibus*, but although Copernicus obviously thought highly of it, he needed more than a letter from a prelate to stir him to action. In the spring of 1539 the stimulus finally arrived in the person of Georg Joachim (Rheticus), professor of mathematics at the University of Wittenberg.

Together Copernicus and Rheticus studied the new theory, and within a few months Rheticus knew enough to write his own short pamphlet, *Narratio prima (First Account)*, (1540). This dealt only with the movement of the Earth and was to be followed by other accounts, but at last the ice was broken and Copernicus himself took the plunge, preparing his own full account of the theory. This he sent by way of Bishop Giese to Rheticus who had agreed to see the work through the press. In the event Rheticus, who had chosen his Nuremberg publisher friend John Petrejus because he was keen to bring out the work, had to leave the city before the book was finished so he could take up a new post at Leipzig. Rheticus therefore handed the technical supervision to a local Lutheran cleric, Andreas Osiander, and this had unexpected consequences. Osiander held the view that had been stated by Jean Buridan two centuries earlier – 'For astronomers, it is enough to assume a way of saving the phenomena, whether it is really so or not' – and since Copernicus was by then a sick man and a long way from Nuremberg, Osiander took it on himself to express this opinion in an unsigned preface. He did not tell the author. Thus it was that readers of *De revolutionibus orbium coelestium* (The Revolutions of the Heavenly Spheres) learned that what they were to read was not a real picture of the universe, but 'a calculus consistent with the observations'. Many thought this was Copernicus's opinion, though it seems that some imagined it had only been inserted by the author to forestall religious opposition. Perhaps this was what lay behind Osiander's decision, for when the theory first became known, and before *The Revolutions* was published in 1543, Luther had immediately expressed strong disapproval – 'The fool will turn the whole science of Astronomy upside down. But, as Holy Writ declares, it was the Sun not the Earth which Joshua commanded to stand still' – and after the book came out the reformer Melanchthon wrote a short physics text inveighing against the theory. But Copernicus was by then dead; a copy of the printed text is supposed to have reached him on his deathbed.

In many ways Luther was right. The Copernican theory did turn astronomy upside down, but in a sense that was long overdue, as events in the seventeenth century were to show. But Luther's disquiet underlined another and more serious problem than the question of literal interpretation of the Scriptures, and this was the dethronement of man and the Earth from the centre of the universe to a place of no special significance. No longer was man situated in a place befitting his unique nature as an image of God, at the centre of all things, but banished to a mere planet among the other planets. In due time this was to have the most profound repercussions on man's view of himself, and his place in creation.

If the Protestants on the continent of Europe objected to *The Revolutions*, no such criticism appeared in England where, by and large, there was either a general welcome for the theory or people kept an open mind. The mathematician and physician Robert Recorde in his *Castle of Knowledge* of 1556, which was written as a dialogue, admittedly makes his pupil refer to the Copernican theory as a 'vaine phantasie', but immediately counters this with the teacher's reply 'You are too yonge to be a good judge in so great a matter . . . you were best to condemne no thing you do not well understand'. The famous John Dee, alchemist, Hermetist, proponent of the occult and at the same time erudite mathematician, and philosopher and astrologer to Queen Elizabeth I, adopted the theory though, typical of his reserve in keeping such matters to himself, he wrote no publication in its support. On the other hand his pupil Thomas Digges did make the theory known, and widely, when in 1576 he produced a fresh edition of his father, Leonard Digges's, very popular perpetual almanac *Prognostication Everlasting*. In this Digges not only described the new theory but also gave a diagram of the universe which placed the Sun at the centre and the planets, including the Earth, in orbit round it. But this was not all, for the stars were shown as extending infinitely into space, not fixed to a celestial sphere as Copernicus had believed. This was a great leap forward; it removed one of the objections to

Illustration page 289

the new heliocentric theory and brought the idea of an infinite universe before the public, for the diagram appeared in six subsequent editions issued between 1578 and 1605. Why Digges adopted an infinite universe is still a matter for conjecture, but it will be better if we defer the question until we meet him again in connection with the invention of the telescope.

Illustration page 304

For the invention of the telescope, see page 341.

Copernicanism may have received a warm welcome in England, but how was it received in Roman Catholic countries? Strange to relate, there were no immediate objections. The dedication of the book to Pope Paul III, and the fact that Johan Widmanstadt's advocacy had raised no murmurs of dissent, doubtless quietened any incipient opposition, and perhaps Osiander's preface also allayed any suspicions of a real overthrow of geocentric ideas. However, the situation was to change by the second decade of the seventeenth century, due in no little degree to the open advocacy of Copernicanism by the turbulent and arrogant Giordano Bruno (1548–1600).

Bruno was immersed in Hermetism, and it was this that motivated his desire to see the Church return to ideas reflecting the religion of the ancient Egyptians. For Bruno was little concerned with the idea of Hermes as a Gentile prophet, the view that sanctified the Hermetic writings for Christians, but believed so strongly in the magical aspects of the teaching that he wanted to return to a full Egyptian outlook which, nevertheless, he believed could be incorporated somehow into the reformed Catholic Church. With these views in his mind, it is not surprising that he should have fallen under suspicion of heresy, and proceedings were instituted against him, though the charge was of holding Arian views (the opinion that Christ was not God incarnate but a created being). At all events Bruno fled from Italy, and wandered through France, Germany and England, though he went first to Geneva where he again found trouble and acquired an intense distaste for Calvinism.

After a period in France between 1579 and 1583 when he lectured on the art of memory in Toulouse and Paris, Bruno moved to England with a royal letter recommending him to the care of the French ambassador, Michel de Mauvissière. Here he stayed for two years, which were to be among the most productive in his life, for under the ambassador's protection he was able to publish some very provocative works. Within a few months of his arrival Bruno visited Oxford in the entourage of the Polish prince Albert Laski, and then returned to give some lectures. During these he spent most of his time quoting astral magic from Ficino's translation of the Hermetic writings and with these he associated the Copernican theory. Oxford, which was then still a bastion of Aristotelian teaching, did not take kindly to Bruno's hectoring, and later he wrote abusively about the university and, incidentally, about the Reformed Church in Elizabethan England as well. For a time after this he dared not leave the protection of the embassy.

While a guest of de Mauvissière, Bruno wrote a number of books, all against established philosophical opinions; some were brilliant and elegant in style but all propounded his own particular interpretation of Hermetic teaching. If this were all he had done, his life might by now have been to a large extent forgotten, but he coupled with his views those of Copernicus, whom he praised because of the heliocentric theory but criticized because he had not appreciated the Hermetic implications of his new idea of a heliocentric universe. Moreover, Bruno was a bold speculator. He adopted Thomas Digges's concept of an infinite universe of stars, each one like the Sun, and Nicholas of Cusa's idea of life elsewhere in the universe; these views he promoted too, but all under the broad umbrella of Hermetism. Bruno's universe was, like the Chinese universe, a living organism, though Bruno's view was motivated by Hermetic magic, not by a desire to rationalize the entire natural world in an organic fashion.

In due course the Catholic League in France grew in power, the liberal de Mauvissière was recalled, and Bruno followed. But soon Paris became too dangerous for him and he fled to Germany. For a time Bruno was at peace at Wittenberg, but in the end his alchemy and his Hermetism were too much for his hosts and he moved, first to Prague and thence to Helmstedt and Frankfurt. An invitation from the nobleman Zuan Mocenigo to teach his art of memory, for which he was renowned, then took him to Venice. Bruno, who was never one to doubt his own cause or his abilities, entered Italy without a qualm. He even took with him the manuscript of a book which he intended to dedicate to the Pope. But, alas, things did not go as he intended; Mocenigo informed on him and he was imprisoned by the Inquisition. After a long trial Bruno recanted his heresies and threw himself on the mercy of the inquisitors, but even so he was sent to Rome for another trial and there his case dragged on for eight years. In the end, after wavering, he refused to recant and in 1600 was burned at the stake for, it seems (the original records are lost), denying the divinity of Christ and diabolical magical practices. This second charge may well have been what sent him to the stake, for it is not unlikely that Bruno was the promoter of a magico-religious movement of some kind; Hermetic perhaps, or possibly connected with the origins of Freemasonry or with that other strange hotch-potch of magic and religion, the brotherhood of the Rosicrucians.

As far as can be made out there was no specific mention of Bruno's advocacy of Copernicanism as a reason for his condemnation. Yet, clearly, any views he promoted in his polemical writings would be suspect, not least by the Dominicans of which order Bruno had once been a member; it is interesting that it was a Dominican friar who was to launch the attack against Galileo's advocacy of the Copernican theory fourteen years later. That was when the first great battle between science and religion was launched, and it was the scientist Galileo, not the magician Bruno, who was the first martyr of science. If Bruno is to be remembered, it is because it was he who turned the whole Roman Catholic Church against the greatest scientific hypothesis of the new scientific revolution, and managed to do so without himself making any substantial contribution to science.

The Copernican theory was a typical product of Renaissance speculation and perhaps its culminating point. It demonstrated how, being prepared to throw out preconceived ideas and accepted doctrines, it was possible to come to a new synthesis, to formulate a totally new view of Nature. Copernicus's reorientation, like that promoted by Vesalius, whose *Fabric of the Human Body* appeared in the same year as *The Revolutions*, changed man's view of himself. It changed also the way he was to pursue his science. No longer was he to set authority above observation; instead he was to forge ahead on his own, testing each new hypothesis against the touchstone of experiment. It was a technique that was to produce some astounding results, as the subsequent centuries were to show.

Chapter Eight

The Seventeenth and Eighteenth Centuries

We come now to a period in which modern science was finally launched and set out on its unprecedented voyage of conquest. From the beginning of the seventeenth century to the end of the eighteenth, the general outlook on the natural world altered in a way that would have astounded Copernicus. The revolution he had started developed so rapidly and so broadly that not only astronomy was transformed but physics also. When this had happened, the break with the last vestiges of the Aristotelian universe was complete. Mathematics became an increasingly essential tool of the physical sciences; results were expressed in numbers and qualitative assessments were rejected. There was also a considerable development in the design and manufacture of scientific instruments, for if the natural world was to be more closely and more precisely investigated then specialized equipment was called for. The design of what were in fact a new generation of precision instruments started in the latter part of the sixteenth century with the work of Tycho Brahe (1546–1601).

The Precision Universe

Tyge Brahe (Tycho was a form he adopted later) was born in 1546 at Skåne, then in Denmark but now part of Sweden, son of a privy counsellor. Because of a family agreement, the boy was brought up by his paternal uncle Jörgen Brahe and his wife, who were childless; his early education was with a private tutor, but at the age of thirteen Tycho entered the Lutheran University of Copenhagen, and it was here that he developed an interest in science. But his uncle disapproved; law was the appropriate study for a young man in his social position, and he was sent off to Leipzig for legal training. In consequence Tycho had to pursue his science and his growing predilection for astronomy in secret until free to study science openly on his uncle's death in 1565. He moved from Leipzig to Wittenberg, then to Rostock (where, incidentally, he lost half his nose in a duel and made a metal substitute – a fact evident in every portrait) and Augsburg. Tycho built a giant wooden quadrant at Augsburg, for in 1564 he had observed a close approach in the sky of Jupiter and Saturn and noticed how incorrect even the best astronomical tables were in predicting this event. By then he was convinced of the need for a new standard of precision in astronomy.

Illustration page 353

Tycho's determination to make more precise measurements was heightened in 1572, when a brilliant 'new star' (a supernova) suddenly blazed forth in the constellation of Cassiopeia. When he analyzed his own measurements and those of other European observers, including Thomas Digges, he discovered that it lay further away than the Moon. This was a crucial observation. It meant a complete break with Aristotelian tradition, which taught that such an object must lie in the sub-lunary sphere because the heavens were changeless. When the Danish King, Frederick II, gave him the island of Hven (now Ven) in the Danish Sound two years later, Tycho set to work without delay to establish an observatory dedicated to accurate measurement. Some of the instruments were later mounted outside in sunken pits to avoid the worst effects of wind disturbance while observing. The instruments themselves were made of metal and built with a degree of precision that was wholly new. They consisted for the most part of sextants and quadrants of large size, each mounted on very sturdy supports. They were carefully checked against each other, and a note was made of the inherent errors of each instrument. This was something entirely novel. It was based on the realization that no measuring instrument, however carefully made, could be perfect; some error was certain. Tycho realized, a fact appreciated by scientists ever since, that a small inherent consistent error does not matter provided one knows what it is and makes due allowance for it. To understand this and take action to nullify its effects was a vital innovation.

Illustration page 353

The instruments were as big as was consistent with easy handling, so that the scales marked on them might be as large as possible and, therefore, the divisions on them marked to ever smaller fractions of a degree. In addition, Tycho built a large mural quadrant on a wall which ran precisely north-south. The radius of the quadrant was about 1.8 metres (6 feet) and the gradations on its scale were actually marked down to arc minutes (one arc minute is $^1/_{60}$ degree). But here Tycho introduced yet another important novelty. Each arc-minute division was itself engraved with a diagonal line composed of small points. These 'transversals' enabled the observer to read accurately in between the arc minute divisions; the dots were at ten arc-second intervals (1 arc second is $^1/_{60}$ of an arc minute, or $^1/_{3600}$ of a degree), and an observer could estimate readings in between the dots; thus the accuracy of the mural quadrant was five arc seconds (0.0014 degrees). The accuracy of Tycho's other instruments was not equal to this, though he regularly observed to within fractions of an arc minute. This was a degree of accuracy never before approached in the whole history of astronomical observing. It was to have almost immediate consequences of immense significance.

In 1577, almost twelve months after Tycho had started observing on Hven, a large bright comet with a very long tail appeared in the evening skies. It captured the popular imagination and was the subject of a vast number of pamphlets, most of which considered it a portent of disaster. This was on the basis of early Greek views, particularly on the opinion of Aristotle that comets were hot dry exhalations; as such they had a drying effect on the air and generated conditions ripe for diseases and epidemics. In addition, there were plenty of astrological implications, so a bright comet was viewed with awe and wonder. Tycho's scientific approach bypassed these considerations – though it seems likely that he was not averse to Hermetic magic, and he certainly cast horoscopes – and he made observations of his own as well as collecting the observations of several other astronomers such as Thomas Digges. These proved conclusively that the comet lay further away than the Moon, which showed it to be a truly celestial object, not a meteorological phenomenon such as a 'dry exhalation'. Moreover Tycho noticed how the tail always pointed away from the Sun, a circumstance he put down to sunlight streaming through the comet's head; in consequence the tail could not be formed out of 'dry fatness' as Aristotle had thought. But most important of all, the observations proved that the comet moved right through the supposed celestial spheres, and so they could have no physical reality – they were a figment of the Greek imagination. This was yet another break, and a fundamental one, with Aristotelian tradition.

On various fundamental points, therefore, Tycho disagreed with the Aristotelian universe. Above all, the heavens were not changeless – his observations of the new star of 1572 and the comet of 1577 both proved that. But if he did not agree with Aristotle on many points, Tycho was still unwilling to accept the Copernican view of the cosmos. His Protestantism revolted against the violence it did to the Scriptures, and so he formulated his own cosmology. In this the Earth remained fixed at the centre of the universe, with the Moon and Sun in orbit around it, though Tycho conceded that the planets could be permitted to orbit the Sun. It was a compromise, and although almost forgotten by the close of the seventeenth century, it had a great vogue for a time in Protestant countries.

Illustration page 353

Unfortunately Tycho was arrogant, haughty with the royal family and neglectful of the welfare of his tenants, and when King Frederick died in 1588 funds for the upkeep of the observatory were no longer forthcoming. This and family problems made him decide to emigrate, and by the middle of 1597 he had left Hven. Two years later he had finally settled in Prague under the patronage of the Holy Roman Emperor, Rudolph II, and so it was to Rudolph that he dedicated his book *Mechanics of the New Astronomy* – the great descriptive treatise on his instruments and their methods of use. The Emperor gave him a pension and the castle of Benatky, some 35 km (22 miles) north-east of Prague, though it seems Tycho actually lived in Prague until his death in October 1601.

Illustration page 354

Among the mourners at Tycho's extraordinarily elaborate funeral was one of his assistants, Johannes Kepler (1571–1630). Son of a mercenary and twenty-five years Tycho's junior, Kepler was originally intended for the Lutheran church, and had gone to Tübingen to read theology. While there he had become interested in astronomy, converted to the Copernican theory and, in addition, showed such outstanding mathematical ability that when the post of mathematics teacher at a well-known Lutheran school at Graz became vacant he was persuaded to leave theology and take the post. This was to prove the turning point in his career.

Kepler practised astrology as well as astronomy, and when he arrived in Graz he issued a predictive calendar which, by good fortune, turned out to be correct in its predictions of the weather and peasant uprisings. His local reputation became almost legendary, even though other calendars followed with predictions that were not so striking. The attitude Kepler took to astrology was interesting, for he rejected most of the rules by which astrologers worked; what he did strongly believe in was the harmony of the universe and the existence of a sympathetic correspondence between the cosmos and the individual – hence his astrological bias. On the other hand he always referred to astrology as the 'foolish little daughter of astronomy', and was later to write that if astrologers sometimes do tell the truth, it ought to be attributed to luck. Nevertheless none of this prevented him from casting horoscopes, especially when asked to do so by the rich.

Kepler's theological leanings had implanted in him a very firm belief in the divine design of the universe and this, in its turn, led him to what he believed to be a great discovery. He published it in 1597 under the title *A Precursor to Cosmographical Treatises containing the Mystery of the Universe* – usually known as *Mystery of the Universe* for short – and it was clearly intended to be the first of a series of astronomical treatises. These Kepler did write, though not all were as mystically inclined as this one; some were purely scientific in tone. The discovery Kepler made, and which delighted him so much was that, taking the Copernican cosmos with the Sun at the centre, he could fit into the spaces between the spheres that carried the six planets in their orbits the five regular polyhedrons of Euclidean geometry. Thus between the spheres of Saturn and Jupiter a cube fitted exactly, in between the spheres of Jupiter and Mars, a tetrahedron, and so on. Since there are only five regular polyhedra, Kepler believed he had found the clue to the universe, the reason why there were only six planets and the reason why they are spaced out in the way astronomers then believed them to be. It was a mystico-mathematical argument and underlines one aspect of Kepler's nature, an aspect that was to be put to good use again for astronomy many years later.

Illustration page 354

Tycho Brahe was much impressed by Kepler's book, but not because he believed the young man had discovered the secret of the universe, for by then his own observations had made it clear that the celestial spheres were a fiction. What mainly impressed Tycho was Kepler's mathematical ability. Here, clearly, was a man who could take his planetary observations and use them to extract the true motions of the planets. At this time religious persecution took a hand, forcing Kepler and his family out of Graz and, finally, to Tycho's side at Prague. Kepler settled there in 1600, but the very next year Tycho died; Kepler was then appointed 'imperial mathematician' in his stead.

Tycho had been secretive over his planetary observations, but on his deathbed had urged Kepler to use them to prepare a new set of tables of planetary motion – the *Rudolphine Tables* – which he believed would validate his own compromise planetary theory and prove the Copernican scheme untenable. At the time Kepler was working on observations of Mars, a fortunate circumstance because the eccentric orbit of Mars – about which neither Tycho nor Kepler knew anything at the time – was well suited to showing up the vital differences between the theories. The observations themselves were not only more precise than any ever made before, they possessed another advantage. Tycho had observed continuously instead of adopting the customary procedure of observing only at astronomically or astrologically significant times, such as conjunctions and oppositions. The result of this was that the observations covered the movements of the planets right across the sky, and so were far more complete than any previously made.

Kepler's work on the Mars observations took him years; the amount of calculation involved was immense and there were no mechanical aids to help him. What is more, a brilliant supernova appeared in the autumn of 1604 and this captured his interest for a time, resulting in a book, *The New Star*, which appeared in 1606. Nevertheless, as his computations slowly proceeded it became clear that Tycho's planetary theory was not acceptable – it just did not fit the evidence of his own observations – but then nor did the Copernican theory either. Mars did not behave according to either hypothesis. It was at this stage that we see the other side to Kepler's character, for in spite of his preconceptions of celestial harmony, he was willing to go wherever his observations led him: a totally scientific attitude. His fundamental belief may have been that this could only lead to further discoveries of divine design in the cosmos, but at least he was willing to be guided by results.

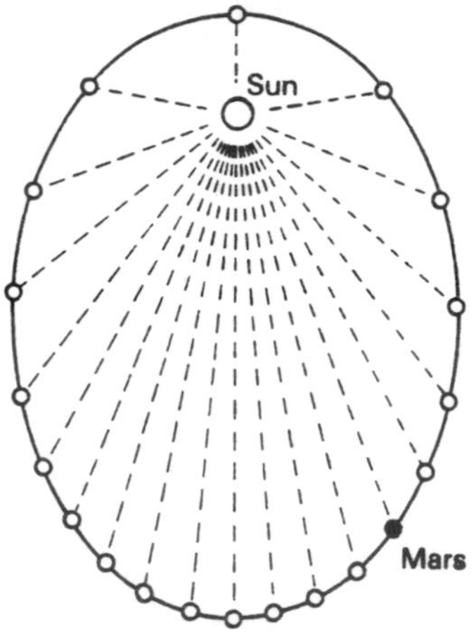

The outcome of this investigation was published in 1609 as *The New Astronomy*, and new it certainly was. For what Kepler had found broke with tradition, with all that the Greeks and every subsequent astronomer had taken for granted. He conclusively showed not only that Mars orbited the Sun but also, and much more significantly, that it did so in an ellipse. Gone was the circular orbit of the Greeks and gone, too, was the belief in uniform planetary motion, for the planet Mars varied its orbital velocity as it moved along its elliptical path, moving faster when near to the Sun (perihelion) and slower when far away (aphelion). Admittedly, in 1609 Kepler could only prove this for Mars, but when he had time to examine the rest of Brahe's observations, it was evident that the other planets behaved in the same way. Between 1619 and 1621 he published a three-part description in his *Epitome of Copernican Astronomy*.

Kepler's revolutionary view of the orbit of Mars; the dotted lines and positions marked on the ellipse indicate equal areas, and equal intervals of time on the orbit.

The laws of planetary motion had now to be revised: instead of uniform circular motion about the Earth it was now clear that the planets moved in ellipses about the Sun and at varying speeds, the speed changing so that a line drawn between the Sun and the planet always swept out an equal area of the space inside the ellipse in the same period of time. These were fundamentally new ideas for astronomy. The mystical side of Kepler still, however, sought for some underlying regularity, some evidence of divine design behind this novel system of planetary motion, and it was while writing the *Epitome* that he also busied himself seeking such a principle. His perseverance was rewarded and in 1618, a year before his purely astronomical *Epitome* was finished, he published his *Harmony of the World*. This, to him, was his crowning achievement. He had discovered a relationship between the velocities of thc planets in their elipitcal orbits and musical harmony; he was able to relate the greatest and least velocities of each planet to the musical scale. This was Kepler's culminating vision, the apotheosis of the 'music of the spheres' of Pythagoras and Plato. Though today no scientific value is attached to this astronomical-musical relationship, we do value an additional discovery Kepler made while he was working out the musical law. This is a relationship which shows that the ratio between the time each planet takes to complete one elliptical orbit and its average distance from the Sun is the same for them all. Its signal importance lies in the fact that if we know the orbiting times (which are relatively easy to determine) of the planets and the average distance of only one of them from the Sun, we can compute the distance of the rest. Later generations of astronomers were to make good use of this.

Illustration page 354

For 'the music of the spheres', see pages 74–77.

Kepler's time at Prague was fraught with difficulties. His salary was always badly in arrears, and in 1611 tragedy overtook him. His wife and eldest son died, Prague was the scene of a bloody revolt and his patron Rudolph abdicated. In 1612 Kepler migrated to Linz, where the *Epitome* and *Harmony* were published, and here he remarried, though he was only allowed to settle there for a time. In 1625 he was forced to move again, this time to Ulm. It was at Ulm that the *Rudolphine Tables* appeared in 1627, three years before his death.

If Kepler was pushed from pillar to post because of wars and uprisings involving persecution, religion was also to impinge on the work of his great Italian contemporary, Galileo Galilei. Born in Pisa in 1564, and so some eight years older than Kepler, Galileo was the son of the independently minded composer and musicologist Vincenzo Galilei. He was brought up in a household that valued the arts and gave a typical Renaissance welcome to new ideas; indeed Galileo once said that he might well have chosen a career as an artist, while his brother, Michelangelo, did become a professional musician. The family moved to Florence – Vincenzo Galilei was a member of a Florentine family prominent in medicine and public affairs – and Galileo was sent to the famous Jesuit monastery school at Vallom-

Illustration page 355

brosa some 20 km (12 miles) away. In 1578, at the tender age of fourteen, he became a Jesuit novice, but his father at once removed him from the school. Three years later we find Galileo enrolled as a medical student at Pisa University. But medicine was not to Galileo's taste; his predilection was for mathematics and it was while he was a medical student that Galileo discovered the isochronism of the pendulum by using his pulse to time the swings of a chandelier during services in church. He came down from Pisa without a degree, and then devoted himself to mathematics, mechanics and hydrostatics.

In 1588 a mathematically slanted lecture to the Florentine Academy on the geography in Dante's *Inferno* earned him much praise and the help of Guidobaldo del Monte, through whose influence he obtained the chair of mathematics at Pisa. Galileo was now thirty-five years of age and, becoming increasingly critical of Aristotelian teaching about motion, he wrote a small tract, *Motion*, in which he demolished Aristotle's distinction between two different kinds of motion – forced and natural; to Galileo they were both essentially the same. This important step he was to elaborate later on. It was while a professor at Pisa that he is said to have dropped different weights from the Leaning Tower of Pisa, though not in front of the entire university as one story has it. At all events, Galileo certainly investigated the motion of falling bodies, proving, contrary to Aristotelian theory, that however light or heavy they took precisely the same time to fall to the ground. By rolling balls down inclined planes he was also able to discuss the motion of bodies along a surface and came close to, though he did not reach, what was later to be called Newton's First Law of Motion. These discoveries were all notable in themselves, but they had an additional significance because to achieve them Galileo used mathematical techniques when analyzing his results. In no other way could he have come to these conclusions. His powerful mathematical approach was, in fact, so effective that it was to become the hallmark of the new physics which was to develop throughout the seventeeth and eighteenth centuries; that is the justification for calling him the Father of Mathematical Physics.

For Aristotle's theory of motion, see pages 105–106; for late medieval criticisms of this theory, see pages 265–268.

For Descartes's extension of Galileo's ideas, see page 344.

The chair at Pisa was poorly paid and when his father died in 1591 Galileo had to assume responsibility for the whole family. He needed to move to obtain a better-paid appointment and, with Guidobaldo's help, he obtained the chair at Padua where the academic atmosphere was much more to his liking; here there was freedom of academic thought, for the university came under the powerful Venetian government and brooked no interference from outside. At Padua Galileo lectured and did more research on motion, explaining how a projectile moves in a curved (parabolic) path and not in straight lines as demanded by Aristotelian laws of motion. He also set up a workshop in his house for the manufacture of his calculating device, the 'Geometrical and Military Compass', as a way of supplementing his income which, though an improvement on Pisa, was still not adequate for his needs.

It was while he was at Padua that Galileo became concerned with the telescope. In the spring of 1609 one appeared in Venice and was offered to the Doge at a high price. Galileo received a report on it and learned that it had two lenses, one at each end of a tube. This was a sufficient clue, because Galileo's knowledge of 'perspective' as he called it (i.e. his knowledge of optics) was enough for him to design his own device for seeing at a distance, and to do so very quickly. Almost overnight he designed a telescope to magnify three times and soon one with a magnifying power of ten. Yet to make an efficient telescope was not just a matter of arranging suitable types of lenses in combination; it involved using specially prepared lenses, because the customary spectacle lenses were not sufficiently powerful to do more than exhibit the principle, as Galileo was quick to realize. Eventually he was able to make an instrument that magnified as much as thirty times.

Illustration page 354

But Galileo did not invent the telescope. The example in Venice had come from the Netherlands and it was here, at the beginning of October 1608, that Hans Lipperhey (often spelled Lippershey), a spectacles-maker from Middelburg, had submitted a patent claim for an instrument for seeing at a distance (the name 'telescope' was not coined until 1611 in Italy). However Lipperhey was not alone in claiming discovery; within fifteen days a second application for a patent had been received by the States General (the parliament of the Netherlands), and later on claims to the invention were made by another Middelburg optician. The story of the discovery is also complicated partly by the fact that della Porta also claimed the invention – and certainly an Italian telescope of 1590 is mentioned in contemporary accounts – and partly by some very strong claims from the English, notably Leonard and Thomas Digges. The actual inventor is, therefore, still a subject for debate.

For the earlier use of lenses, by Grosseteste and Bacon, see pages 254–256; and by della Porta, page 319.

Galileo's importance in the early history of the telescope is that he made immense scientific use of it; by 1610 he had published his *Sidereal Messenger*, a book in which he detailed some vitally important astronomical observations he had made. It may well be that he was not the first to use the telescope astronomically; Thomas Digges's conception of an infinite universe of stars, each like the Sun, may have been prompted by looking at the heavens through a telescope, while there is incontrovertible evidence that Digges's friend Thomas Harriot was mapping the Moon with a telescope within a couple of months of Galileo's telescopic observations. But Harriot did not publish, and it was Galileo's *Sidereal Messenger* that drew the attention of the Western scientific world to what the telescope could do. In the book Galileo explained how he had observed many more stars than were visible to the unaided eye – and hence many more than were known to the ancients – an argument against the 'perfection' of ancient science so dear to his Aristotelian antagonists. He also saw that the Moon had mountains – he even measured their heights from the lengths of the shadows they cast – he discovered that the Milky

For Thomas Digges's vision of the infinite universe, see page 331.

Illustrations pages 354, 356

Way was composed of myriads of separate stars and, above all, that the planet Jupiter was accompanied in its orbit by four small orbiting moons. This was indeed important because it showed that an orbiting planet could carry its own satellites with it, and effectively countered the argument that if the Earth moved as Copernicus supposed, the Moon would be left behind. Later in 1610, Galileo was to notice that Venus presents phases like the Moon and he observed, too, the strange nature of the planet Saturn, which looked to him as if it were a triple planet. Sometime between 1610 and 1611 he also observed sunspots with the telescope.

To Galileo all these observations amounted to strong supporting evidence for the Copernican theory. However his desire to move back to Florence and his success in doing so – during 1610 he became philosopher and mathematician to young Cosimo de Medici, the Grand Duke of Tuscany – transferred him from the protection of Venice to the more Aristotelian-orientated atmosphere of a state where the Church had a powerful influence. In addition, he was viewed with suspicion, his court position arousing jealousy which Galileo did little to disarm. He visited Rome and was feted for his telescopic observations and elected to the Academy of the Lincei, but back in Florence there were those who declined to look through his telescope and see for themselves. There, intellectual conservatism prevailed and academics were totally unwilling to let an instrument which they did not understand give them cause even to consider rejection of the Aristotelian universe, let alone the whole of Aristotelian physics and mechanics as Galileo wanted. It was in Florence too, at the church of Santa Maria Novella, that in 1614 the Dominican, Tommaso Caccini, preached his sermon against Copernicanism and against 'mathematicians', though it was probably those who practised cabalism and numerology – the Hermetic 'mathesis' – that he had in mind. The Dominicans had by then set their face against the Copernican theory, and Galileo was advised to tread warily. He went again to Rome, but found it impossible to convince the authorities of the significance of his case; indeed the Church's leading theologian, Cardinal Robert Bellarmine, admonished him because a moving Earth went against the Scriptures.

Galileo returned to Florence, and for a time kept quiet, but in 1623 he launched one of his scathing attacks on a book about the nature of comets. Here he used biting wit to ridicule and demolish his opponent – a technique of which he was master – but his book, *The Assayer*, was not all polemic. In it Galileo set out his views on scientific reality and on the new scientific method; he explained his doctrine of primary qualities (which were those that could be measured) and secondary qualities (which were not measurable, i.e. qualities like odour and taste). He also explained how to define a problem with the help of preliminary experiments and, from the results, to form a theory, which could then be used to 'predict' consequences that could be observationally tested. It was in *The Assayer*, too, that he made his famous remark, 'The Book of Nature is . . . . . . written in mathematical characters'.

The Assayer was dedicated to Maffeo Barberini, the new pope and a friend and protector of Galileo who received it with enthusiasm. The Jesuits, one of whose members had, unknown to Galileo, written the original book on comets, were not pleased however. What Galileo hoped, in view of the success in the Vatican of the *Assayer*, was that Barberini would revoke the decree condemning Copernicanism and he visited Rome to beg him to do so. Although unsuccessful, Galileo did receive permission to write about the whole question of the two views of the universe, the Ptolemaic-Aristotelian and the Copernican, though he was warned not to come to any definite conclusion. Elated with his partial success, Galileo returned yet again to Florence and settled down to write his *Dialogue on The Two Chief World Systems – Ptolemaic and Copernican*. Published in 1632 it was hailed all over Europe as a literary and scientific masterpiece. Yet in Italy it caused a storm to break about Galileo's ears.

Illustration page 356

The Church found the book to be too biased towards Copernicanism, and the prescribed argument Galileo had been told to put at the end of the book written in such a way as to appear pointless. In 1633 Galileo was arraigned before the Inquisition, but in view of his age – he was sixty-nine – he was treated indulgently. Nevertheless, forced to recant, he was sentenced to house arrest, but fortunately his mind was still active and, in spite of the rebuff, he worked on his last book, *Discourses Concerning Two New Sciences . . .*, a text which discusses his final conclusions concerning mechanics. It was a valuable book which, however, had to be published in Protestant Leiden; it came out in 1638. He spent his last years applying the pendulum to the regulation of clockwork, though it was to be the Dutch scientist Christiaan Huygens who actually put this into practice in 1656 and thereby ushered in an era of precision in mechanical timekeeping. Galileo died in 1642, the same year in which Isaac Newton was born.

Illustration page 355

The view of the universe adopted by Galileo was based on observation, experiment and a liberal application of mathematics. A somewhat different attitude was taken by his younger contemporary René Descartes (1596–1650) who was born in La Haye (now La Haye Descartes) in Vienne, France. Descartes was essentially a mathematical philosopher who set about formulating a new philosophical approach to the universe that broke completely with the old medieval scholastic outlook. The delicate son of a lawyer, in 1604 Descartes was sent to the new Jesuit school at La Flèche where his intelligence was remarked upon, though his own reaction to the teaching he received there left him dissatisfied, disappointed and full of doubts. In 1616 he took a degree in law at Poitiers University but two years later when the Thirty Years' War began, he became involved in military affairs, though he probably did no fighting. During this period he embarked on some mathematical research and also decided to set about constructing a general scheme of knowledge. To complete this work he

Illustration page 356

found he needed more peace and quiet than military service would allow so he resigned his commission and, in 1628, he moved to the Netherlands where he stayed for the next twenty-one years. In 1649 he moved to Sweden at the invitation of the young Queen Christina, a passionate patron of the arts and an ardent collector of learned men for her court, but here his health deteriorated; he developed pneumonia and died early in 1650.

We are concerned here not with Descartes's philosophical writing but with his contributions to mathematics and his ideas about the universe. Descartes was continually concerned with what was certain and what was uncertain and, after much consideration of the question, came firmly to the conclusion that the one thing he could be certain about was his own existence as a sentient being; he expressed this in the phrase 'Cogito ergo sum' ('I think, therefore I am'). With this basis, he then argued that a good God would not deceive his creatures, and proceeded from here to construct his picture of the universe. He stated his principle in two works, the *Discourse on Method* of 1637 and the book with which we are principally concerned, the *Principles of Philosophy* of 1644. Written in three parts, the first of which restates his philosophical doctrines, the second and third then explain his interpretation of the cosmos. He argued that since we cannot think of any limit to the extent of space, the universe must be infinite. On the other hand, we can think of dividing matter *ad lib*, and so he rejected the existence of atoms. Descartes also rejected the existence of a vacuum and considered space a 'plenum', that is, filled with matter, all of the same kind, and all in motion. He believed that God always conserves the same amount of matter and motion – the first statement of the important law of the conservation of momentum (momentum = quantity of matter × its velocity) – and that if it suffers no collision a body will always move in the same direction with the same speed indefinitely, the first formal statement of what has come to be known as the law of inertia. Galileo had come close to this, though he had not included celestial motions because he considered them to be a special case; as far as he was concerned celestial motions were naturally circular.

In part three of the *Principles* Descartes showed how, with one further assumption, the universe we observe must arise. The additional assumption was that at Creation, God took matter, divided it up arbitrarily and set it in motion. Since there cannot be a vacuum, one particle must always be replaced by another, and this led Descartes to his doctrine of vortices. Although all matter is the same, its particles vary in size; the largest make earthy material, those of medium size, air, and the smallest fire. These particles collect into vortices. At their centres the fire particles congregate because they move the fastest; thus at the centre of each vortex a star will be formed. Such stars rotate and so exert an outward pressure, just as a stone does when whirled round at the end of a string. Every star is therefore pushing on every other star. However, the stars have a tendency to skin over with gross matter; when this happens they display sunspots to begin with, and finally cease to shine at all. Outward pressure now ceases and the vortex collapses. The skinned-over star now wanders into another vortex to become a planet, unless it is very massive when it wanders from one vortex to another and so becomes what we call a comet. Planetary satellites were old planets, formed long ago.

Descartes's system was an intellectual *tour de force* and made an immense impression when it appeared. Clearly, it supported the Copernican theory and helped to make that theory acceptable in many academic circles where it might otherwise have remained in limbo. At Cambridge University it was welcomed and taught to students, of whom Isaac Newton was one. Descartes's theory of a plenum of vortices and his ideas on the evolution of planets, comets and stars were wrong, though their disproof lay in the future. Nevertheless, his discussion of planetary motion was interesting and underlined the fact that, during the seventeenth century, physical principles were being used to try to explain the motions which Kepler had so brilliantly investigated. Not everyone was convinced that his elliptical orbits were correct, for they differed only slightly from circles, and some wanted to retain the old circular motion. But elliptical or circular, the crucial question was to explain why the planets move in their closed orbits. Kepler had made an attempt to explain it all by postulating a magnetic attraction, but this had not been satisfactory.

In 1666, Giovanni Borelli, born in 1608 at Naples and who in his youth had met Galileo, published his *Theory of the Medician Stars*. This was a book about the four large satellites of Jupiter, which Galileo had christened the 'Medician Stars' when he was trying to obtain Cosimo de Medici's patronage; its significance lay in Borelli's discussion of the physics of their motion, and more especially of the wider question of the motion of planets round the Sun. On the latter, Borelli reviewed the various ideas of the time – that they were moved by invisible 'spokes' emanating from the rotating Sun, or that they were floating in some kind of solar atmosphere – but rejected them for lack of evidence. As a mathematician – Borelli held Galileo's old mathematical chair at Pisa – he concluded that they moved due to a combination of three forces; one a 'natural instinct' pulling them towards the Sun, the second a tangential or sideways force coming from the Sun's rays (he derived this from some ideas of Kepler), and the third a tendency of a body to recede outwards from the Sun (the kind of force that operates when a stone is whirled round on a string, i.e. a centrifugal force). In short, Borelli believed that though the planets were subject to various forces, they moved in stable orbits because they were in equilibrium. This was a notable step forward, though it was only to be brought to its full development by Isaac Newton twenty years later.

An attempt to examine the matter further was made by Robert Hooke, a scientist of great brilliance who, however, seems to have been incapable of following up his often very original ideas. Hooke realized that the Sun must exert a strong pull on the planets, though the precise mathematical expression of the force – Hooke favoured the idea that it was magnetic – eluded him. It eluded other contemporaries too; the young astronomer Jeremiah Horrocks of Liverpool, the architect and astronomer Christopher Wren and the astronomer Edmond Halley. All felt sure that the Sun exerted an attractive force; they even suspected that it might be an 'inverse square' force (i.e. it diminished as the square of the distance; a planet twice as far off as another would be pulled with one quarter the force, one three times more distant with only one ninth of the force, and so on.) They thought, too, that the path such a body should follow would be an ellipse, but none could prove it mathematically. That even Halley, no mean mathematician himself, failed is perhaps not surprising for the solution to the question required the invention of a radically new mathematical technique. This need arose because as a planet moved in an elliptical orbit, its distance from the Sun would be continually altering, with the result that the attractive force would constantly change. A mathematics was needed which could specifically deal with changing quantities and the independent invention of such a technique – the 'calculus' – was one of Newton's achievements.

Illustration page 358

Illustration page 356

Isaac Newton was born at Woolsthorpe in Lincolnshire on Christmas Day 1642. A premature infant who came into the world two months after the death of his father, he was so small that he would have 'fitted into a quart mug'. The Newton family were yeomen farmers, and it was taken for granted that Isaac would continue in the tradition, but he had no aptitude for farming. He was interested in mechanical contrivances and in the natural world, and his mother was finally prevailed upon by the headmaster of the King's School in Grantham, and by an uncle, to let the boy be prepared for entry to the university. In consequence, in 1661 Newton went up to Trinity College, Cambridge as a subsizar, a poor scholar who would have to earn his keep by doing menial tasks for the fellows of the college.

Newton showed no particular promise in his early years at Cambridge, though we know he read, among other works, books on optics and mathematics, as well as Descartes's *Principles of Philosophy* and Galileo's *Dialogue on the Two Great World Systems* but not, it appears, the later *Discourses on Two New Sciences*. It was Isaac Barrow who seems to have been the first to recognize Newton's abilities, and this in spite of the fact that on examining him in 1664, he found the young scholar deficient in his knowledge of Euclidean geometry. Barrow himself was a theologian and mathematician who, in 1663, had been appointed to a new chair of mathematics endowed by Henry Lucas, an old graduate of the university and once its member of parliament. Barrow encouraged Newton who, in 1665, took his first degree, though without distinction. Normally Newton would now have prepared for his MA degree, but 1665 was an abnormal year. In December 1664 bubonic infection – the Great Plague – had broken out in London, and in Cambridge the university authorities began to be troubled by the fact that the plague had arrived in the city. As no effective medical treatment was available, they closed the university and Newton returned to Woolsthorpe. Here, but for a brief trip back to Cambridge, he stayed until 1667.

During these plague years 'I was', Newton later wrote, 'in the prime of my age for invention, and minded mathematics and philosophy [i.e. science] more than at any time since', and it was certainly at this time that he performed optical experiments and laid the foundations for his theory of light, as well as working out the groundwork of his ideas on gravitation and planetary motion. In the garden at Woolsthorpe there was an apple tree, and the story that the fall of an apple gave him a vital clue in the solution of the planetary problem may possibly be true. On his return to Cambridge, Newton became a junior fellow of Trinity College and in 1668 a senior fellow. With his MA also granted in 1668, he was now well established as an academic, especially when, in the year following, Isaac Barrow resigned his Lucasian chair in favour of his young pupil.

The falling apple had posed in Newton's mind the question of whether the force exerted by the Earth in making the apple fall was the same force that made the Moon 'fall' towards the Earth, and so pull it in to an elliptical orbit round the Earth. Calculations showed him that it did, but much more work remained to be done. It was perhaps not until 1684 that, after an exchange of letters with Robert Hooke, Newton was fully in command of all the dynamical principles involved and was able to settle the question. At all events, 1684 proved to be important for another reason, for it was then that Edmond Halley visited Newton to discuss the planetary question. Wren had offered to buy a book for either Hooke or Halley if they could solve the problem but since neither could do so, Halley decided to enquire of Newton when he next visited Cambridge. To his surprise Newton replied that the force between Sun and planets, resulting in an elliptical orbit, did operate according to an inverse square law and that he had proved it; he could not, however, find his proof, but promised to write it out and send it to Halley. In due course he did, Halley receiving what was virtually a small treatise on the subject. Fortunately he at once realized its immense significance and pressed Newton to write a full book on the whole question of motion, telling him that the Royal Society – Britain's equivalent to the Italian academies – would publish it. In the event, the Royal Society's finances were not up to the task and Halley, whose father had been a prosperous merchant, paid for the publication out of his own pocket, as well as taking on the responsibility of seeing the work through the press. However, while Newton was busy writing the book, Hooke grumbled that he should be given credit for discovering the inverse-square law of attraction and that it was he who gave Newton the basic clue. This was an exaggeration and Newton would have nothing to do with it; indeed for a time he refused to write another word, and it took all Halley's diplomacy and tact to get Newton to take up his pen again. At last, in 1687, the book was not only complete but printed, appearing under the title of *Philosophiae Naturalis Principia Mathematica* (The Mathematical Principles of Natural Philosophy), usually known as the *Principia* for short.

The *Principia* was a masterpiece; it has been called the greatest scientific book of all time and even if this is an overestimate, it was certainly a work of supreme importance. Its impact was immense. But perhaps this was only to be expected for, between the covers of one volume, Newton had rewritten the whole science of moving bodies with a mathematical precision it had never before possessed. He completed what the late medieval physicists had begun and Galileo had tried to bring to fruition; his three 'laws of motion' forming the basis of all further work. Newton had also solved the two-thousand-year-old problem of astronomy, the movement of the planets in space. With a mathematical analysis that was astonishing in its completeness, he showed how an inverse-square law was bound to result in motion in an ellipse, and force the planets to obey those laws which Kepler had deduced so painstakingly from Tycho's observations.

Illustrations page 358

Yet these two astonishing achievements were not all the *Principia* contained. The attractive force acting from the Sun was not magnetism but gravitation, and Newton took the bold step of identifying this attraction, which operated in space, with the attraction of the Earth for the Moon and for every other body on its surface. In other words, he proposed the powerful concept of universal gravitation, whereby every body in the universe attracts every other body, thus bringing the whole universe under one basic law. No longer was there one set of laws of behaviour for celestial bodies and another for terrestrial ones: physics was universal.

It need hardly be said that there were some objections to Newton's theory, most notably from the French who were still wedded to the ideas of Descartes and wanted to know the precise nature of gravity. This was a question Newton could not answer; when asked whether it was some property inherent in physical objects, he replied 'pray, do not ascribe that notion to me'. All he could say was that gravity was a force that operated according to an inverse-square relationship; that it alone was sufficient to explain all celestial motions, and he was prepared to go no further. The French criticism that this was dangerously near to a medieval attitude and so was unscientific was not without substance. Indeed, it was not until the eighteenth century, when Voltaire wrote his *Elements of Isaac Newton's Philosophy* (1738), that a general exposition of Newtonian ideas was available in French, and this book, together with the efforts of the mathematician and physicist Pierre-Louis de Maupertuis, changed the general attitude in France.

Newton's *Principia* so completely closed a period in astronomy that those who followed tended rather to consolidate the achievements he had made. Halley, for instance, applied the mathematics of Newtonian motion and gravitation to the question of comets. He saw that the appearances of comets which had been seen in 1531 and 1607 seemed to tie in with the bright comet of 1682; he assumed these were one and the same body, and computed its return for December 1758, using the law of universal gravitation to calculate the delay which Jupiter would cause. It did, in fact, reappear as predicted, though a few days late (the effect of Jupiter was a little more than Halley had expected) and 'Halley's Comet' marked the first triumphant vindication of Newtonian theory. Later, other mathematical astronomers were to apply the theory to ascertain in detail the mutual effects or 'perturbations' of the planets on one another, and it was in this field that a number of mathematicians made great advances, most notably the Swiss Leonhard Euler and Alexis Clairaut, Jean d'Alembert, Joseph Lagrange and Pierre Laplace in France.

For records of Halley's comet in China, see pages 161–162.

For Newton's design for a reflecting telescope, see page 377.

Edmond Halley did, it is true, try to extend astronomy in other ways. For example he noticed that the stars were not fixed in position as had been universally believed; his own observations compared with those made in Greek times showed such discrepancies that there was clearly a factor at work that had hitherto been overlooked. Correctly, Halley identified a real shift in star positions as the cause. Again, he discussed what later came to be called 'nebulae' – bright cloudy patches in the starry sky visible only with a telescope – and also the question of why, if there is an infinite universe of stars, the sky is nevertheless dark at night (a question which, in our own day, has come to be called 'Olbers' Paradox'). Halley's pioneering efforts in stellar astronomy were not taken up by other astronomers at the time; not until the second half of the eighteenth century was stellar astronomy pursued, and it is really a subject that only came into its own in the nineteenth and twentieth centuries. Most were content to look at the solar system and glory in the precise applications of universal gravitation.

The *Principia* was the culmination point of planetary astronomy but, of course, it would not have been possible without the accurate measurements of Tycho Brahe and his successors. When the telescope came it brought with it a still higher standard of precision observing, though not all astronomers recognized its potential in this respect. Indeed Hooke, who was adamant about the telescope's superiority but was a tactless man, almost caused an international incident when he severely criticized Johannes Hevel of Danzig because he still measured celestial positions using 'open' not telescopic sights. Hevel, or Hevelius as he is better known, was a rich brewer in Danzig (now Gdańsk) and a noted telescopic observer of the Moon and of planets and comets, yet he still believed he could obtain sufficiently precise results without a telescope. Only the intervention of the Royal Society and the diplomatic visit of 'the pleasant Mr Halley' to Danzig, where he observed with Hevelius's open sights and was full of praise for their accuracy, defused the situation. But Hooke was right, and at

Illustration page 358

Greenwich and Paris where ever more accurate positional observations were being made, the telescope was in regular use.

These two national observatories, the first of their kind in the West, had been set up at about the same time, the Paris Observatory being completed in 1671 and Greenwich in 1676. The French observatory was directed by a dynamic Italian, Giovanni Cassini, whose son and two grandsons succeeded him as directors, and Greenwich by the rather taciturn John Flamsteed, the first Astronomer Royal, a man who was a careful and indefatigable observer. It was Flamsteed who supplied Newton with some observational evidence on the Moon's movements to help in his *Principia* and in Newton's later work on lunar theory, but because he was loath to publish his observations until they had been subject to minute time-consuming corrections, the two fell out. As president of the Royal Society, Newton caused Flamsteed's observations, which had been deposited with him under seal, to be published, with Halley acting as editor. By then, Halley and Flamsteed had long ceased to be friends, ostensibly because of Halley's carousing with Peter the Great of Russia (who had come to London both to study British shipbuilding and also to seek first-hand knowledge of the new scientific movement), though the cause was more probably due to Halley's criticism of some of Flamsteed's observational results.

Illustration page 359

The establishment of both the Paris and Greenwich observations was not undertaken for reasons of pure research, but for making such celestial observations as would enable mariners to find longitude at sea. The idea was to use the Moon's movement across the backcloth of the stars as a giant lunar clock from which to determine the time 'back home'; such home time could then be compared with 'local time' obtained at sea by observing the Sun, the difference giving the longitude. For the method to work satisfactorily very precise celestial observations were needed and in spite of the hard work of Flamsteed, Halley (his successor as Astronomer Royal) and those who followed him, it was not until 1767, the time of the fifth Astronomer Royal, Nevil Maskelyne, that the requisite tables for mariners were satisfactory enough to be published. But by that time the English clockmaker John Harrison had made the first successful chronometer for use at sea. This was a more direct method, obviously – one had only to look at the chronometer for 'home' time – though for safety both methods remained in use until very recent times.

Illustrations page 360

The national observatories represented only one aspect of the desire for greater precision in astronomical measurement. During the eighteenth century two particularly notable achievements of this kind were made; one concerned the distance of the Sun, the other the movement of the Earth. In the light of Kepler's law relating the distances of the planets and their orbital periods, the measurement of the distance Earth to Sun was fundamental because, once this was found, all the rest of the planetary distances within the solar system could readily be calculated. It was Halley who promoted the idea that the most satisfactory way of measuring this basic distance, this 'astronomical unit' as it came to be called, was to time transits of the planet Venus across the Sun's disc from different places on the Earth's surface. Such timings would provide the distance Earth to Venus and from this the astronomical unit could be computed. The method was both elegant and precise; the one trouble was that such transits are rare, occurring in a pair eight years apart followed by a gap of some hundred years or more before the next pair. Nevertheless, though Halley died in 1742 and the next transits were not due until 1761 and 1769, his reputation and advocacy had been such that international expeditions were mounted and results obtained which were far and away superior in accuracy to anything obtained before, giving a distance of 152,855,000 km (95 million miles). A later and improved analysis gave 149,637,000 km (93 million miles), close indeed to the present day figure of 149,597,870 km (92.96 million miles).

The measurement of the movement of the Earth was a byproduct of an unsuccessful attempt to measure the distances of the nearer stars. As soon as the telescope was recognized as a device which could provide great precision, various people, including Robert Hooke and John Flamsteed in England and Giovanni Cassini in France, tried to make measurements to detect the apparent shift of nearer stars against the background of more distant ones. All failed; the angles they were measuring were, we now know, still too small for their equipment to detect. In 1725 James Bradley, the professor of astronomy at Oxford and Samuel Molyneux, a keen amateur astronomer, began making their own attempt using more sophisticated techniques. They too failed, though they did succeed in measuring a shift which, in the end, turned out to be caused by a combination of the time taken by starlight to travel down the telescope tube and the shift sideways of the tube due to the orbital motion of the Earth. Thus at last, in 1729, there was definite proof that the Earth really does move in space, an observational vindication of the basic Copernican concept. With this, and with Newton's theory of universal gravitation, the gigantic revolution in man's concept of the universe, begun almost two centuries before, was now complete. But discoveries could still be made as was shown by the career of William Herschel.

Illustration page 361

Herschel was born in Hanover in 1738 into a family of musicians, his father being bandmaster in the Hanoverian Foot Guards. The family atmosphere encouraged learning, and so William was free to indulge in a wide range of interests. At the age of fifteen he entered the band of the Foot Guards, but army life, even the less militant life of a bandsman, was not Herschel's forte. In 1757 he emigrated with an elder brother to England, a country which he had visited the year before with the regiment. Here he followed his profession as a musician, first in London and later in the provinces, where there were more musical opportunities. At the end of 1766 he moved to Bath, which was a fashionable centre and gave great promise for a musical career. Herschel did well and made enough money to go back to

Hanover for a holiday; on his return he brought with him his sister Caroline, who hoped for a career as a singer but in the event was to help her brother in his astronomy and to become a competent astronomer in her own right. Herschel's astronomical interests arose fortuitously; he read a book *Harmonics, or the Philosophy of Musical Sounds* by Robert Smith, Plumian professor of astronomy and experimental philosophy at Cambridge, and was so taken with it that he immediately read the same author's *Compleat system of opticks*. This contained descriptions of what could be seen through a telescope and fired Herschel's imagination. He read other astronomical books and began to observe, but he was so disappointed by the quality of the telescopes he could hire that he resolved to make his own instruments. It turned out that Herschel had a flair for fashioning mirrors, and the reflecting telescopes he began to build were of superior quality to any available from professional instrument-makers.

As soon as he had suitable equipment, and quite unaware of the general dynamical bias of professional astronomy, Herschel began to catalogue the heavens, noting down everything he could see in a very methodical way. In 1781, during one of these surveys, he noticed 'a curious either Nebulous Star or perhaps a Comet' (many comets when first observed telescopically have no tail, so the suggestion was no wild guess), and he watched its movements among the background of more distant stars. Later analysis of his observations and those of others showed the object to be a planet, the first planet to be discovered in historic times, the rest of the planets having been known from time immemorial. The discovery was widely acclaimed, though Herschel's name for the planet – he chose Georgium Sidus (George's Star) after George III of England – did not commend itself internationally, and the less politically slanted classical name of Uranus was chosen.

Herschel was now famous and it was not long before he received a royal pension and was free to devote himself entirely to astronomy, an interest which had gradually been superseding his musical obligations. Herschel now moved to Slough to be nearer Windsor – his pension entailed his showing the heavens to members of the royal family – and it was while there that he persuaded the king to finance the construction of a giant telescope and provide a pension for his sister to act as assistant, the first British government support to be given for pure scientific research. The instrument was a reflector with an aperture of no less than 1.2 metres (48 inches), but it was cumbersome – in the 1780s large-scale mechanical engineering had not yet developed – so that in the event Herschel did most of his observing with a smaller telescope having a mirror of 47 cm (18½ inches) diameter and only half as long as the larger telescope. Nevertheless the '40 foot' did help Herschel in some aspects of his work and certainly added lustre to his reputation, for it was far and away the largest telescope in the world. Herschel was the first to appreciate the advantages of larger-aperture telescopes for penetrating deep into space because of the dimmer objects they are able to render visible,

Illustration page 361

Left A contemporary sketch of Tycho Brahe (1546–1601).

Far left A brass quadrant movable in azimuth and altitude, designed by Tycho and built at his observatory at Uraniborg. From Tycho Brahe, *Mechanics of the New Astronomy* (1598).

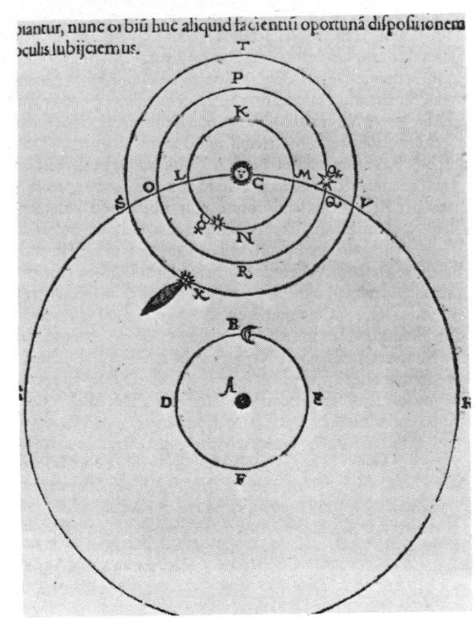

Left Tycho's system of the universe, a compromise between the heliocentric and geocentric theories. Interestingly, it shows a comet as a truly celestial body. From Tycho's *On a recent phenomenon in the Earth's atmosphere* (1588).

Above Portrait of Johannes Kepler (1571–1630).

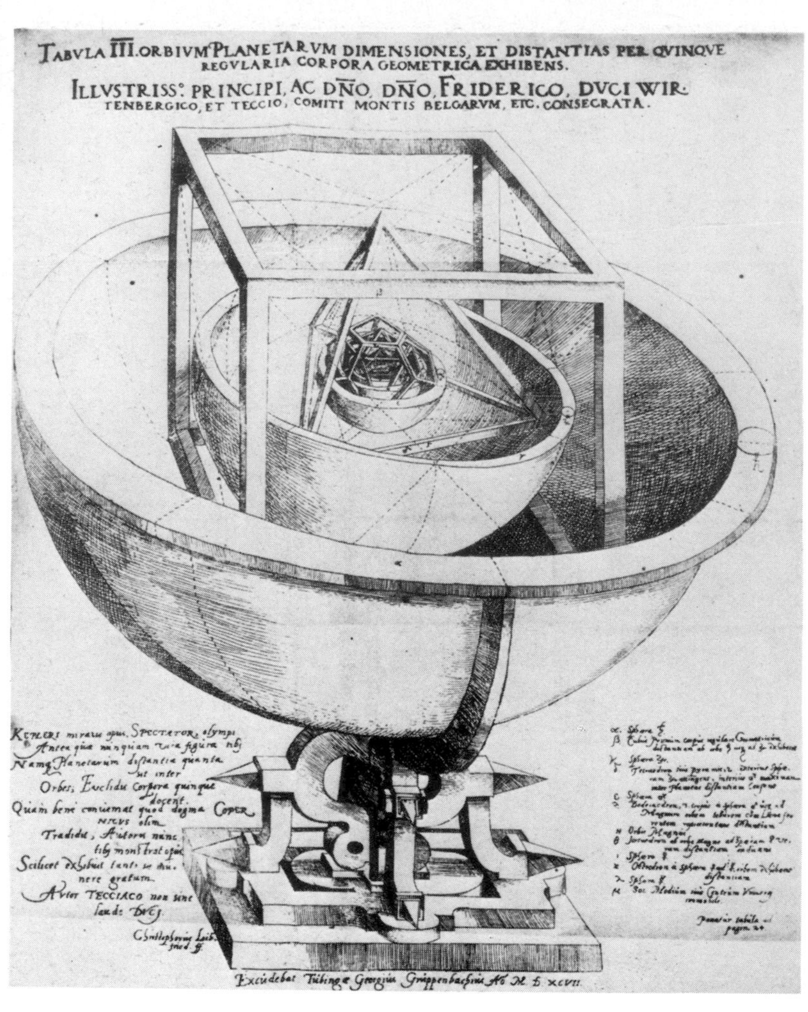

Right Kepler's heliocentric theory based the five regular geometrical solids, from which the orbits of the five planets are derived. The discrepancies between this attractive pattern and Tycho's detailed observations proved a stimulus to Kepler's later ideas. This appeared originally in his *Mystery of the Universe* (1597).

Opposite above left Portrait of Galileo Galilei (1564–1642). Biblioteca Marucelliana, Florence.

Opposite below Sketches of the Moon as seen through the telescope and showing craters and mountains by Galileo from *Sidereal Messenger* (1610).

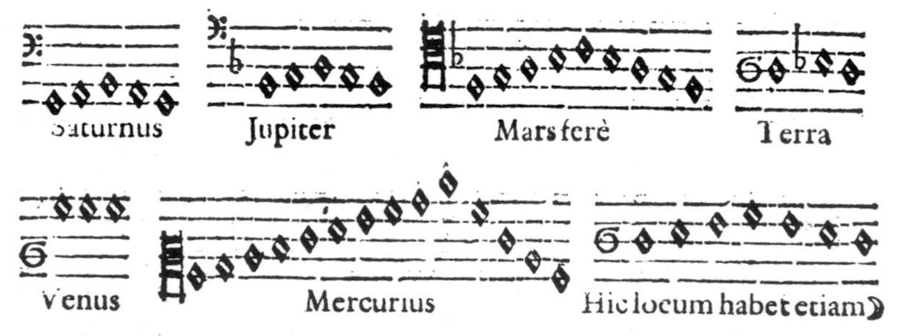

Right The 'divine' musical scales of the planets which Kepler calculated from the greatest and the slowest speed of each planet in its elliptical orbit. From his *Harmony of the World* (1619).

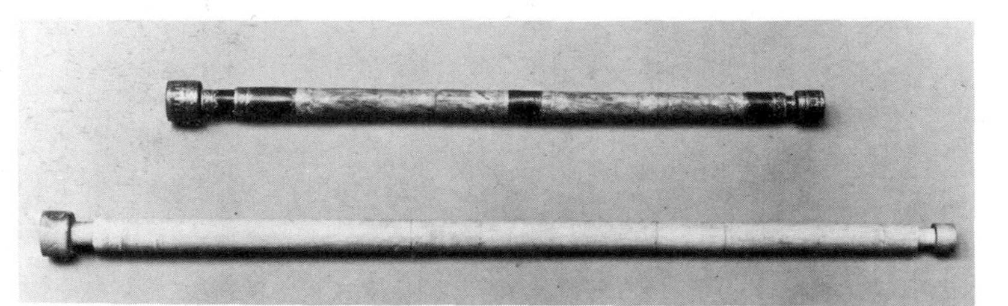

Right Replicas of two telescopes built by Galileo. Science Museum, London.

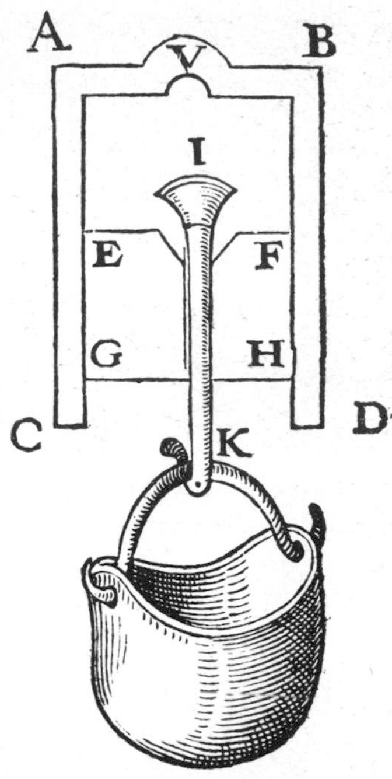

Above Experiment to demonstrate the existence of a vacuum, from Galileo's *Discourses on Two New Sciences . . .* (1638). This work summed up his studies in mechanics, a field that he revolutionized by his clear perception of the connections between force and motion, and by his experimental approach.

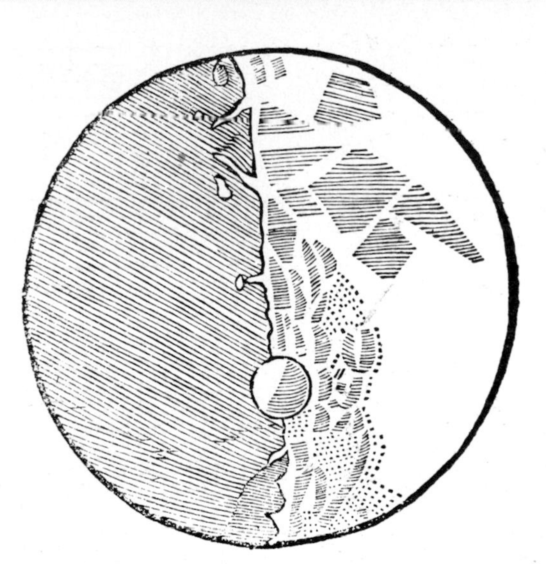

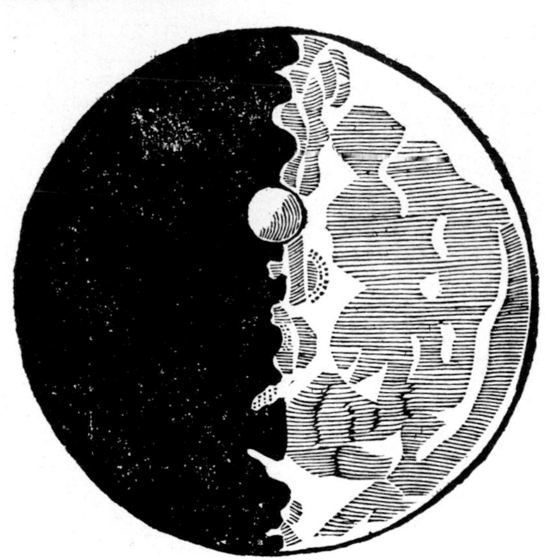

Right Galileo's method of measuring the heights of lunar mountains, from a letter he wrote in 1611 to Christoforo Grienberger.

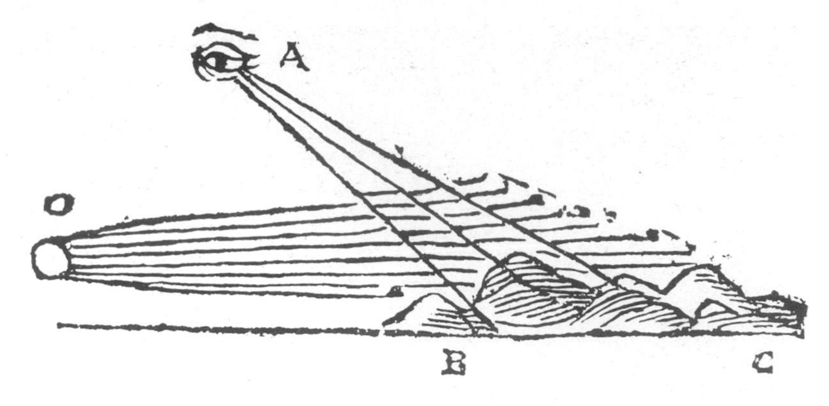

Right Galileo's drawings of his observations of Jupiter's four chief satellites, reproduced in his *Sidereal Messenger*, 1610.

Opposite Dedication page of Galileo's *Dialogue on the Two Chief World Systems* . . . in the edition of 1642 published in Leiden. On the left is Aristotle, in the centre Ptolemy, and Copernicus is on the right.

22 OBSERVAT. SIDEREAE

que talis poſitio. Media Stella oriẽtali quam pro-

Ori. ** * Occ.

xima min. tantum ſec. 20. elongabatur ab illa, & a linea recta per extremas, & Iouem producta paululum verſus auſtrum declinabat.

Die 18. hora 0. min. 20. ab occaſu, talis fuit aſpectus. Erat Stella orientalis maior occidenta-

Ori. * * Occ.

li, & a Ioue diſtans min. pr. 8. Occidentalis vero a Ioue aberat min. 10.

Die 19. hora noctis ſecunda talis fuit Stellarũ coordinatio: erant nempe ſecundum rectam li-

Ori. * * * Occ.

neam ad vnguem tres cum Ioue Stellæ: Orientalis vna a Ioue diſtans min. pr. 6. inter Iouem, & primam ſequentẽ occidentalem, mediabat min. 5. interſtitium: hæc autem ab occidentaliori aberat min. 4. Anceps eram tunc, nunquid inter orientalem Stellam, & Iouem Stellula mediaret, verum Ioui quam proxima, adeo vt illum fere tangeret; At hora quinta hanc manifeſte vidi medium iam inter Iouem, & orientalem Stellam locum exquiſite occupantem, ita vt talis fuerit

Ori. * • * * Occ.

configuratio. Stella inſuper nouiſſime conſpecta admodum exigua fuit; veruntamen hora ſexta reliquis magnitudine fere fuit æqualis.

Die 20. hora 1. min. 15. conſtitutio conſimilis viſa eſt. Aderant tres Stellulæ adeo exiguæ, vt vix

RECENS HABITAE 23

Ori. * * * Occ.

percipi poſſent; a Ioue, & inter ſe non magis diſtabant minuto vno: incertus eram, nunquid ex occidente duæ an tres adeſſent Stellulæ. Circa horam ſextam hoc pacto erant diſpoſitæ. Orien-

Ori. * ** Occ.

talis enim a Ioue duplo magis aberat quam antea, nempe min. 2. media occidentalis a Ioue diſtabat min. 0. ſec. 40 ab occidentaliori vero min. 0. ſec. 20. Tandem hora ſeptima tres ex occidente viſæ fuerunt Stellulæ. Ioui proxima aberat ab eo min.

Ori. * * * * Occ.

0. ſec. 20. inter hanc & occidentaliorem interuallum erat minutorum ſecundorum 40. inter has vero alia ſpectabatur paululum ad meridiem deflectens; ab occidentaliori non pluribus decem ſecundis remota.

Die 21. hora 0. m. 30. aderant ex oriente Stellulæ tres, æqualiter inter ſe, & a Ioue diſtantes; in-

Ori. *** * Occ.

terſtitia vero ſecundũ exiſtimationem 50. ſecundorum minutorum fuere, aderat quoque Stella ex occidente a Ioue diſtans min. pr. 4. Orientalis Ioui proxima erat omniũ minima, reliquæ vero aliquãto maiores, atq; inter ſe proxime æquales.

Die 22. hora 2. conſimilis fuit Stellarum diſpoſitio. A Stella orientali ad Iouem minu-

Ori. * ** * Occ.

torum primorum 5. fuit interuallum a Ioue

Right René Descartes, a portrait painted by Frans Hals in 1649. Musée du Louvre, Paris.

Far right Sir Isaac Newton (1642–1727) at the age of 59, painted by Godfrey Kneller. National Portrait Gallery, London.

DIALOGVS
DE SYSTEMATE MVNDI,
Auctore
GALILÆO GALILÆI LYNCEO.
SERENISSIMO
FERDINANDO II. HETRVR. MAGNO-DVCI
dicatus.
ARISTOT.
PTOLEMEVS
N. COPERNICVS

PHILOSOPHIÆ
NATURALIS
PRINCIPIA
MATHEMATICA.

Autore *J S. NEWTON*, *Trin. Coll. Cantab. Soc.* Matheseos Professore *Lucasiano*, & Societatis Regalis Sodali.

IMPRIMATUR.
S. PEPYS, *Reg. Soc.* PRÆSES.
Julii 5. 1686.

LONDINI,
Jussu *Societatis Regiæ* ac Typis *Josephi Streater*. Prostat apud plures Bibliopolas. *Anno* MDCLXXXVII.

Above Title page of the first edition (1687) of Newton's *Principia*.

Right Edmond Halley (1656–1742), a portrait by R. Phillips sometime before 1721. Halley's questions on planetary motion prompted Newton to publish his inverse square law. National Portrait Galley, London.

Right Johannes Hevelius (1611–87) observing from a rooftop with one of his long-focus telescopes. From his *Selenographia*, 1647, a study of the surface of the Moon.

Below Diagram of the paths of projectiles ejected from the Earth at different velocities illustrating how the theory of gravitation could explain planetary orbits, from Newton's *A Treatise on the System of the World*, 1728.

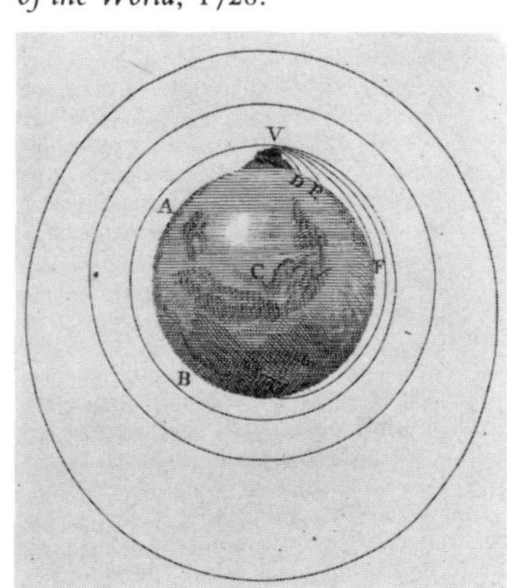

Above Sextant with a telescope for accurate sighting and installed at the Royal Observatory at Greenwich in 1676 by John Flamsteed (1646–1719), the first English Astronomer Royal.

Left Equatorially mounted telescope made in 1791 by the instrument maker Jesse Ramsden for Sir George Shuckburgh.

Right John Harrison's first marine timekeeper, 1751, National Maritime Museum, London.

Right John Harrison's fourth and fifth marine timekeepers, which won him the prize offered by the British Admiralty for a clock to keep time at sea and accurate enough to permit longitude to be determined. Harrison's life was devoted to the problem of the marine timekeeper: a good example of the dovetailing of mercantile, strictly scientific and technological interests at this period. National Maritime Museum, London.

Left William Herschel, often regarded as the greatest observational astronomer in history, from a portrait painted by L. F. Abbott in 1785. National Portrait Gallery, London.

Below Engraving of Herschel's giant reflector built in the 1780s, with an aperture of 1.2 m (48 in) and a focal length of no less than 12 m (40 ft.)

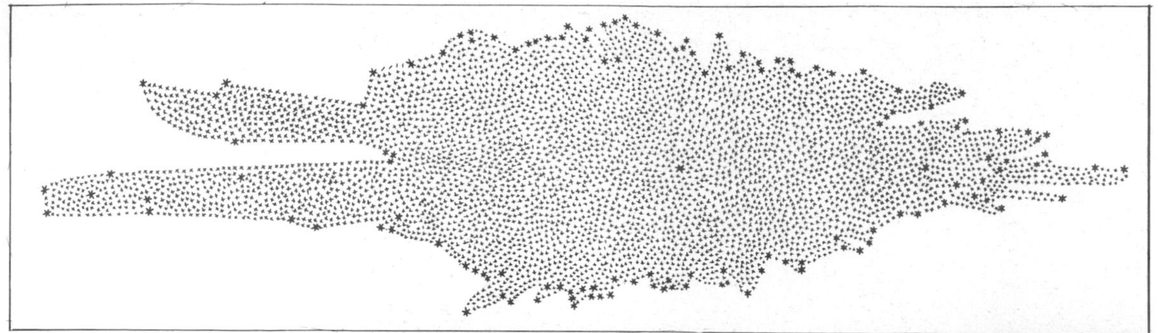

Above Herschel's idea of the system of stars; the larger star near the centre is supposed to be the Sun. From his paper 'On the Construction of the Heavens' published in 1785.

Right Visit by Louis XIV of France to the Académie des Sciences, founded in 1666. The Paris Observatory can be seen through the window. The scientific equipment demonstrates the varied types of science pursued in the early academies.

Far left Gottfried Wilhelm Leibniz (1646–1716), the philosopher, psychologist and mathematician who developed the calculus. Engraving by A. Scheits in 1695.

Left Blaise Pascal (1623–62), who, in addition to his mathematical and religious writings, demonstrated variations in atmospheric pressure.

Left Frontispiece to Thomas Sprat, *The History of the Royal Society of London for Improving Natural Knowledge*, 1667. On the left of the bust of the founder, Charles II, is the Society's first president, and on the right Francis Bacon (1561–1626) whose writings stimulated the Society's formation.

Right Manuscript of about 1672 containing a sketch of Newton's experiment to show not only that white light (arriving at S, left) is dispersed into its separate colours by a prism, but also that these colours can be recombined using a second prism which has been inverted. Royal Society, London.

Right Woodcut of cannon boring. The heat produced by this work led to the realization that heat could be continuously generated, and stimulated research into the subject in the 17th century. From Vannoccio Biringuccio, *De la Pirotechnia*, 1540.

Opposite Precision in measuring the specific heat of a body was much helped by a carefully designed device such as this calorimeter of about 1800 by Antoine Lavoisier and Pierre Simon de Laplace.

Right James Watt's (1736–1819) design of a steam engine of 1769. This application of the theoretical study of heat to practical machinery was of vital historical importance.

The Calorimeter of Lavoisier and La Place.

Right A battery of Leiden Jars being charged by an electrical machine of rotating glass plates, designed by John Cuthbertson in the 1780s.

Right A globe-shaped magnet with lumps of iron to represent mountains, used by William Gilbert to demonstrate the north-seeking properties of the magnetic needle. Gilbert's work laid the foundations of the studies in magnetism of the 17th and 18th centuries. From his *De Magnete*, 1600.

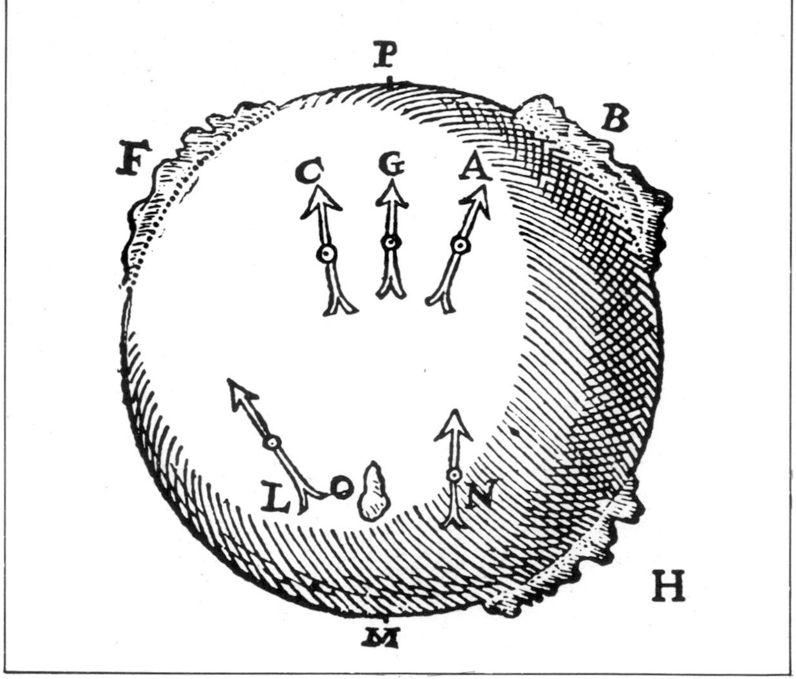

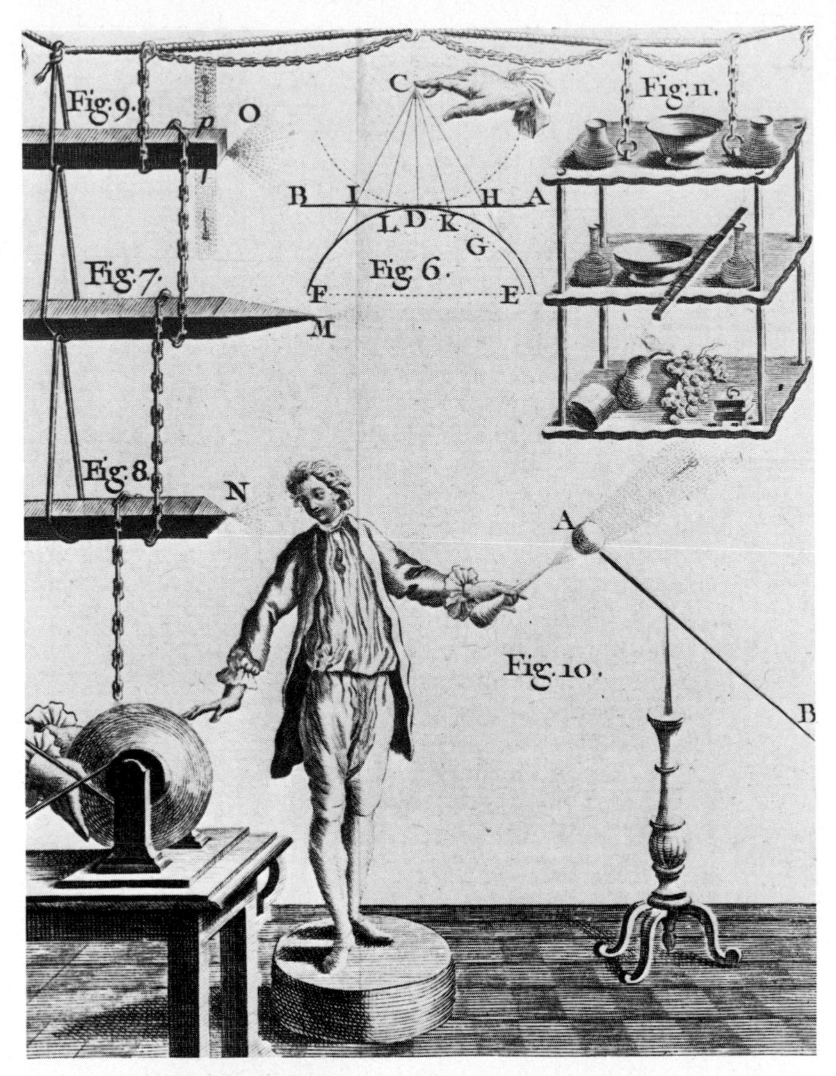

Left Various effects produced by charges of static electricity generated by a rotating glass globe (left), elegantly illustrated by J. A. Nollet in 1749.

Left Franz Anton Mesmer's (1734–1815) 'tub', a vat filled with dilute sulphuric acid round which patients sit holding the protruding iron bars, to be cured by 'animal magnetism'.

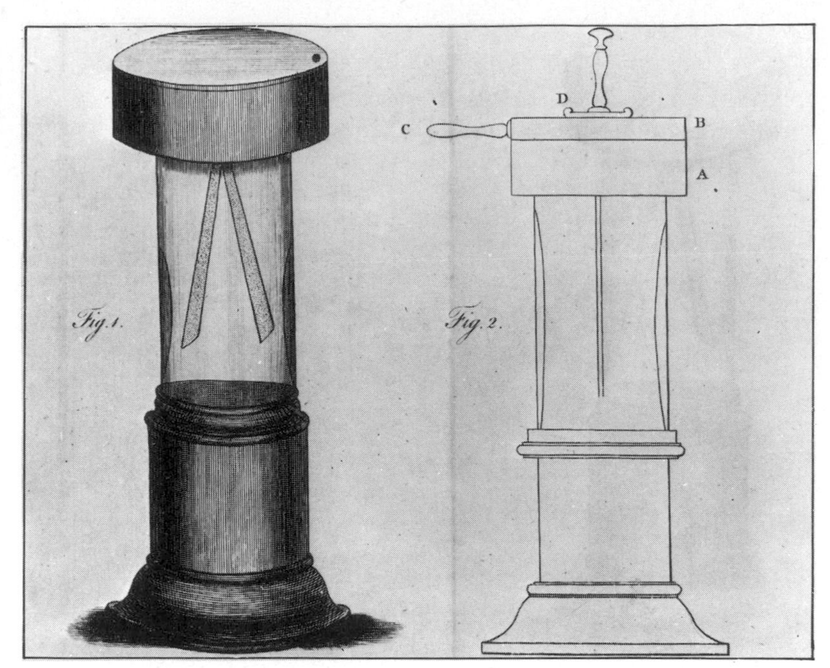

Right An electrometer for measuring electrostatic charge by observing how strongly two pieces of gold leaf repel one another when charged; designed in 1787.

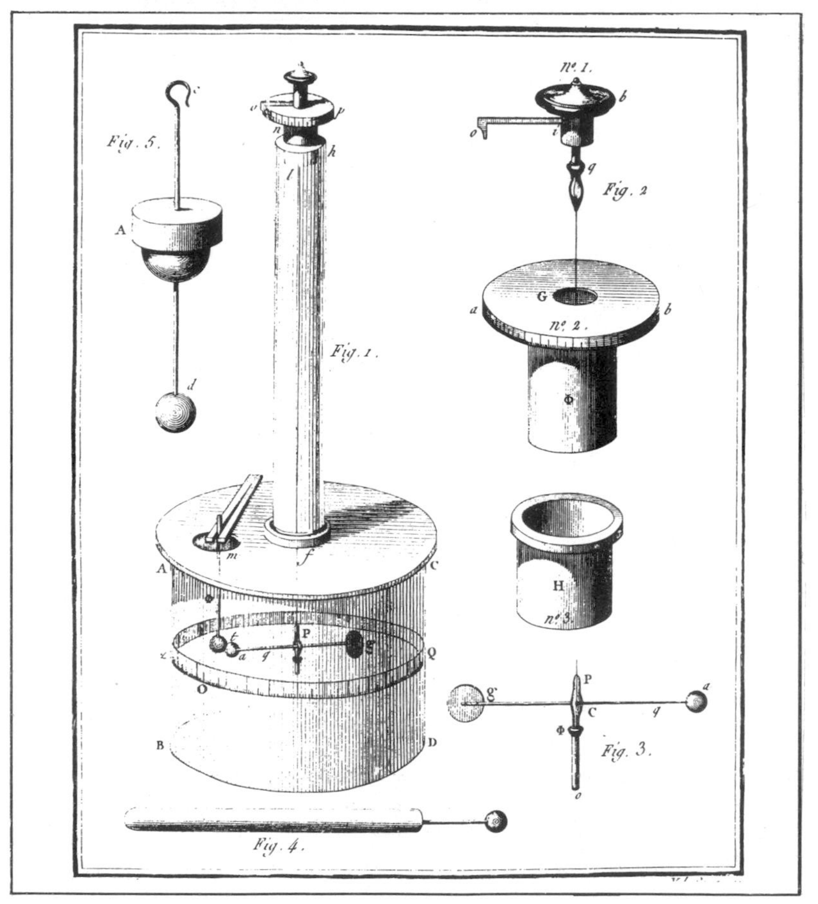

Right Charles Coulomb's torsion balance, from his *Memoirs on Electricity and Magnetism*, 1785–89. This device enabled electrical and magnetic charges to be measured accurately.

and this attitude was to affect every later generation of astronomers.

Herschel's observing programme was prodigious; he was a man of great physical stamina and immense perseverance and observed sometimes under the most uncomfortable conditions. He surveyed the sky continuously in the thirty years between 1778 and 1808, and produced a host of important results. A hunt for a shift of nearer stars against the background of more distant ones, a 'parallactic' shift to allow stellar distances to be detected failed, but it brought Herschel to the realization that some stars formed binary systems, the stars orbiting round one another. Detailed studies of them showed that they orbited according to Newton's gravitational laws and so gave proof of the existence of gravity out in the depths of interstellar space.

Herschel's mapping of the heavens and particularly his 'gages' of stars whereby he counted the stars in various areas of the sky led him to two important conclusions. On the one hand, a simplified statistical analysis of his results led him to produce evidence that the Sun was moving in space and determine the point in the sky to which it seemed to be moving. The result, which agrees by and large with modern estimates, was somewhat fortuitous since Herschel had comparatively few stars on which to base his calculations (only those which seemed to have shifted their positions over the years could be used), but his very attack on a problem of this kind opened up new possibilities. Herschel's other important conclusion was that the stars were not distributed evenly in space but arranged in a large elongated shape with the Sun and its planets close to the centre. This was a view which was extremely novel and confirmation of it lay in the future, since the equipment available to other astronomers was not powerful enough to check Herschel's results.

Illustration page 362

Herschel's sky surveys also revealed a great number of hazy patches of light or nebulae (Latin *nebula*, cloud). He was not the first to observe these, the French astronomer Charles Messier had catalogued 104 of them though his purpose was merely to record them because of their nuisance value; he wanted to avoid confusing them with comets in the early stages of their approach to the Sun. But it is a measure of Herschel's thoroughness that he catalogued not a hundred but some 2,500 examples in the northern skies. His observations of them raised a basic question, however. When Herschel moved to ever larger aperture telescopes he saw that some of these hazy patches could be resolved into separate stars; the question then arose as to whether any were really clouds at all and whether, perhaps, still larger instruments would enable all nebulae to be seen as separate stars. Herschel's own experience with the giant 40-foot left the question still open.

For later investigations into the nebulae, see pages 475 and 523.

Developments in Mathematics

Newton's achievements in the *Principia* were possible because of his development of the calculus, although the book itself is written in geometrical terms. This is because geometrical methods were then the ones with which mathematicians were most familiar, and the book was enough of an intellectual hurdle without introducing the stumbling block of an entirely new mathematical technique. Yet without his calculus, his technique of 'fluxions', as he called it, Newton could never have reached the results he did. Newton was not alone in his development of a technique of this kind; the German mathematician Gottfried Leibniz had invented an equivalent method and, indeed, Newton and he had a long wrangle over plagiarism, though we now know that both reached similar conclusions quite independently. Nor did their ideas come completely out of the blue; in a sense they were only following a development to which a number of mathematicians had contributed.

Illustration page 363

One of these was Kepler who, when he married a second time, laid down a cellar of wine and decided to compute the volume of his barrels. This he did by imagining each barrel to be composed of a very large number of very thin circular sections, whose areas he calculated and added together; intrigued by the method he used it for other shapes and published a small tract about it. This procedure of using thin sections, of adding infinitesimal increments which changed their value, was a fundamental concept in the Leibniz-Newton technique. Again, Bonaventura Cavalieri of Bologna, a contemporary of Kepler, proposed a 'method of indivisibles' whereby a plane figure could be divided into an infinite number of lines. This was useful for computing the areas underneath curves, and helped to lead to what was for a long time known as the 'integral calculus', though it was strongly criticized by some mathematicians because a line is defined by having length but no breadth and therefore no amount of lines could fill a figure or a curved area. Nevertheless Giles Roberval in Paris wrote on the subject and used the technique to find the areas under various curves with such success that the Cavalieri method became more widely accepted. Other work on drawing tangents to curves by Isaac Barrow and a study of curves themselves by John Wallis also helped to expand these ideas.

For the use of the same idea by Eudoxos, see page 100.

The development of the calculus as a tool of wide applicability to all kinds of problems, using algebraic methods to operate it, was due to the 'analytical geometry' (algebraic geometry) of René Descartes and Pierre Fermat. In his *Géométrie*, published first as an appendix to his *Discourse on Method* (1637), Descartes gave a method of denoting points and lines by numbers, or algebraically by letters. The method is similar to that used in giving map references by numbers counted from an 'origin' and based on the idea of a grid of perpendicular lines across a map. Descartes found that using the technique a line or a curve could be expressed in the form of an algebraic equation. His younger contemporary Pierre Fermat, who had independently experimented with applying algebra to geometrical problems, had also arrived at the idea of representing a curve by an equation. He then decided to pursue this technique further and was the first to discover how to solve questions of maxima and minima; that is, finding the greatest and least values which can be taken by some variable entity, and expressing these by an equation. This, too, was to have considerable influence on the development of the more general methods of the calculus, which now incorporated problems of maxima and minima, areas under curves, infinitesimal quantities and the rate of change of variables of all kinds.

The development of the powerful technique of the calculus by Leibniz and Newton, first demonstrated so effectively by Newton's solution to the whole problem of planetary motion, has an interesting postscript. In order to indicate the fact that he was considering the rate of change of a variable – the quantity x, say – Newton wrote the x with a dot over it, thus $\dot{x}$. On the other hand Leibniz would write the same thing as dx/dt. Now the point at issue is that $\dot{x}$ does not indicate what is happening, whereas dx does. The pair of letters dx means an infinitesimal increment of x, similarly the pair dt indicates an infinitesimal increment of time, so that dx/dt does in fact make it clear that what is happening is that infinitesimal increments of x are being considered with respect to infinitesimal increments of time t; in other words we have a rate of change of x with time. The difference may seem marginal, and in a sense it is, but when one comes to more complex equations the advantages of a more descriptive style become increasingly significant. The British, with their great respect for Newton – he was buried in Westminister Abbey and his tomb bears Alexander Pope's epitaph:

Nature and Nature's laws lay hid in night,
God said let Newton be and all was light.

– kept to his notation, but European mathematicians adopted the Leibniz method. What is so significant is that there was little development of the calculus in Britain but much in Europe, where in the eighteenth century Leonhard Euler, Joseph Lagrange and Pierre Laplace were at work. Indeed, the situation became so serious that in the second decade of the nineteenth century, John Herschel (son of the famous astronomer William Herschel) formed a society at the University of Cambridge to get the Leibniz notation adopted in Britain.

Powerful though it was, the calculus was not the only mathematical innovation of the seventeenth and eighteenth centuries. There were, for instance, the introduction of logarithms and the development of a mathematics to deal with probabilities. Logarithms are in one sense a sixteenth-century development, since the men concerned, the Scotsman John Napier and the Englishman Henry Briggs, were born respectively in 1550 and 1561. Nevertheless, Napier's method of multiplying numbers by a process of addition (i.e. by adding their 'logarithms') and finding roots, by dividing a logarithm by the number of the root, did not appear until 1614 (*Description of the Wonderful Logarithmic Canon*); moreover Briggs's later development of the system with Napier and his promotion of it in his *Logarithmic Arithmetic*

(1624), meant that this valuable aid to speedy calculation was not available to mathematicians until the second and third decades of the seventeenth centuries. Nevertheless, it is a measure of the effectiveness of this system that it was used for most arithmetical processes (multiplication, division, calculating powers and roots) and for trigonometry right up to present times. It has only been ousted by the introduction of the pocket electronic calculator.

Illustration page 363

It was Blaise Pascal (1623–62), resident most of his life in Paris, who did much to lay the foundations of a mathematics to deal with probabilities. A man of wide talents, he also made some contributions to the physics of fluids and of the vacuum, and became famed as a religious philosopher. Indeed, it was his preoccupation with religion, which was later to occupy all his time, that brought him into contact with discussions on chance and uncertain events in individual and community life. Pierre Fermat had already considered the question of calculating what throws of a dice were likely to produce, but Pascal extended the work, developing what has since become known as the 'calculus of probabilities', a powerful technique which can be widely applied to solving scientific problems as well as calculating the outcome of games of chance. It was Pascal, too, who made an extensive study of the binomial theorem and Pascal's Triangle.

For Pascal's Triangle in Chinese mathematics, see page 153.

Another original contribution to mathematics made in the seventeenth century was that of the architect Girard Desargues who, in 1639, laid the foundations of 'projective geometry' (the study of geometrical figures when drawn from various viewpoints). It was, however, a subject that remained undeveloped except by a few successors, Pascal among them; not until the nineteenth century did it come into its own. And, of course, both seventeenth- and eighteenth-century mathematics had the advantage of contributions from the Bernoulli family whose lives and works spanned the period between the 1650s and 1780s. Their work was concerned with the development of the Leibnizian calculus and its application to a wide range of astronomical and engineering problems. Together with their compatriot Leonhard Euler (1707–82), who finally eased the calculus away from geometry and treated trigonometry as a part of algebra, Pierre Laplace (1749–1827), whose famous *Celestial Mechanics* (1799–1825) was a compendium of the application of gravitation to astronomy, and Joseph Lagrange (1736–1813), they introduced an advanced form of algebra. This 'higher analysis', using not only the calculus but also techniques of dealing with periodic variations such as are met with in astronomy, and probability methods, was to provide the scientist of the nineteenth century with a host of powerful techniques. In brief, they consolidated what Galileo and Newton had begun and set the seal on the mathematization of every branch of physical science. It was also a technique that, in the hands of pure mathematicians, was to generate new mathematical ideas that, nevertheless, were later to be found useful in describing the real world.

Scientific Academies

A new factor that emerged during the seventeenth and eighteenth centuries was the growing government interest in science. The foundation of the observatories at Greenwich and Paris was one aspect of this, though it was a rather utilitarian aspect since its aim was to assist navigation in the interests of overseas trade. However, there was another facet and that was the establishment in some countries of scientific academies. In late sixteenth-century Italy, as we saw (Chapter 7), there were a number of learned academies in Naples, one at least of which was devoted to science, and there was the Academy of the Lincei in Florence. But the most famous was the Florentine Accademia del Cimento (Academy of Experiments), established in 1657 by two of Galileo's pupils, Vincenzo Viviani and Evangelista Torricelli. Helped by two of the Medici family who took an active interest in it, its members carried out a wide range of experiments in biology and physics, but when one of the Medici became a cardinal in 1667, the academy closed. Nevertheless it existed long enough to publish a useful account of its work.

In England things occurred in a rather different way. In London in 1596 Sir Thomas Gresham, a wealthy merchant and financial advisor to Queen Elizabeth I, instituted Gresham College in the City of London. This was a bold move; at the time the only English universities were Oxford and Cambridge, and the vice-chancellor and senate of the latter had written to Gresham pointing out that while they thought London, Oxford or Cambridge were the only suitable places for such a college, they hoped that as one 'bred in Cambridge' he would 'not choose Oxford'. Gresham's idea was that his college should have seven resident professors who were to give public lectures in English as well as in Latin, and some of these were to deal with practical scientific subjects not then on any university curriculum; perhaps this was why he chose London rather than Oxford or Cambridge. Then, twenty-four years later, when Gresham College had become a flourishing concern, Francis Bacon, the Lord Chancellor, advocated the new experimental science in his *Novum Organum* (New Instrument). His purpose was more the 'relief of man's estate' than the acquisition of pure knowledge, and he suggested that what should be done was to collect as large a body of facts as possible about each subject to be investigated; these facts should be classified and then tabulated to show whether any particular property or quality was present, absent or present 'in different degrees, more or less'. An examination of the results would show the nature of the phenomenon under study. But, of course, no one ever discovered anything in this way; it has been likened to 'research by administration' and was clearly a method devised by someone whose practical knowledge of experimental science was negligible; it was all logic with no imagination. But Bacon's importance lay in the stimulation his ideas aroused, in his vision of the betterment of human conditions through the application of science, and in his suggestion for the establishment of an academy of scientists which appeared in 1627 in his *New Atlantis*.

The upshot of Bacon's ideas and the establishment of Gresham College was that those interested in science did begin to meet, first in London, then, during the Civil War, in Oxford as well and, on the restoration of the monarchy in 1660, they gathered again in London to set up a scientific society. Given royal patronage, it was known as 'The Royal Society of London for Improving Natural Knowledge', though this soon became abbreviated to 'The Royal Society'. Still Britain's premier scientific society it, too, like the Academia Cimento, began not only to hold scientific discussions but also to publish scientific discoveries, though in the case of the Royal Society its *Philosophical Transactions* were in the form of a scientific journal; this began publication in 1665 as a private venture of the Society's secretary, William Oldenburg, though it was later published by the Society, as it still is today. The *Phil. Trans.* and the Society itself were to encourage the new experimental science throughout the length and breadth of the United Kingdom.

Illustration page 363

Like the Royal Society, the Académie des Sciences in France had its beginnings in the informal meetings of scientific men. They included Descartes and Pascal and used first to meet at the 'cell' of the Jesuit scientist Marin Mersenne in the Convent of the Annunciation in Paris; later they gathered in the large houses owned by senior members of the government. Finally, the author and administrator Charles Perrault suggested to Jean-Baptiste Colbert, the minister of finance to Louis XIV, that a regular academy be established. Originally to have embraced literature and history as well as science, at its first meeting just before Christmas 1666 it turned out to be devoted exclusively to science. The members of this Académie des Sciences received a royal pension as well as financial help with their research, and met twice a week in a room of the royal library.

Illustration page 362

Seventeenth-century Germany was not without its own scientific academies. The earliest was a short-lived society in Rostock founded in the 1620s, though the German equivalent of the Royal Society or Académie des Sciences did not appear until 1700. This was the Academy of Sciences in Berlin and, like its English and French equivalents, it materially encouraged science and the publication of scientific results.

Physics

Optics

Physical science received great impetus during the seventeenth and eighteenth centuries and advanced considerably in certain fields, most notably in optics, in investigations into the nature of a vacuum, in studies of heat and of electricity and magnetism. As far as optics is concerned, the invention of the telescope has already been mentioned, but nothing has so far been said about that other combination of lenses, the microscope. It is now generally thought that the microscope was developed after the telescope, but yet again its origins are

For the invention of the telescope, see page 341.

obscure. The magnifying glass had, of course, been known at least as early as the thirteenth century, but it was not until the seventeenth that it reached its peak in the tiny pebble-like lenses ground and polished by the Dutch microscopist Anton van Leeuwenhoek.

Illustration page 409

Yet, good though van Leeuwenhoek's simple microscopes were, the possibilities of really high magnifications could only be reached by using compound microscopes, containing at least two lenses. Galileo is said to have experimented with them, and in 1665 Robert Hooke published a wonderful descriptive text, the *Micrographia*, which included not only a description of his microscope but also detailed descriptions of objects he had examined. However, the compound microscope was bedevilled by the optical fault of 'chromatic aberration', which meant that every image was surrounded by coloured fringes that made detail difficult to observe. Not until the early nineteenth century was this defect to be fully overcome. Nevertheless the microscope was to prove its use at once, as will become evident when we begin to discuss the biological sciences.

Illustrations pages 406, 407

For the use of the microscope in biology, see pages 392–394.

Scientifically the most important advances in optics during the seventeenth and eighteenth centuries were those concerned with the refraction of light, and with theories about the nature of light itself. That lenses refracted light, bending it from its original path to a new one, had long been known, but the precise relationship between the angle at which the incident beam met glass or water, and the angle through which it was bent or turned was not clear. However, in about 1621 Willebrord Snel (Snellius) of Leiden discovered the relationship, known, because it uses the trigonometrical sines of the angles, as the 'sine law', though it was left to others, notably Christiaan Huygens and René Descartes, to publish this important result.

If Snel's law expressed simply but exactly what happened when light travelled from one transparent medium (air, water, glass, etc.) into another, the reason why it did so was far more difficult to determine. With his all-embracing philosophy of the natural world, one might expect Descartes to have some theory about light and why it underwent refraction, and he certainly did. He thought light was a force caused by vibrations of the particles of which he believed all bodies to be composed. Descartes also suggested that although light travelled infinitely fast through his transparent medium which permeated all space, it travelled more slowly through a material such as water and slower still through air, because air was less dense and so would transmit the vibrations less efficiently. Such a view we now know to be incorrect; the denser the medium the slower light passes through it, but Descartes was not the only one to interpret the idea from the point of view of the contiguous vibrations of particles within bodies and reach the wrong answer. Robert Hooke and Christiaan Huygens came to the same conclusion.

Christiaan Huygens, who was born in the Hague in 1629 of a prominent Dutch family, lived in the Netherlands all his life except for fifteen years, when he spent most of his time in Paris as a member of the Académie des Sciences. An able mathematician and brilliant physicist, he and his brother Constantijn ground and polished lenses, built their own microscopes and telescopes, and solved Galileo's problem of Saturn during the winter of 1665–66 when they discovered the planet's ring system and, incidentally, a satellite of the planet as well. Christiaan also invented the pendulum clock, studied the mathematics of oscillatory movements and of hanging chains and was a strong believer in the existence of other inhabited worlds. But without doubt his greatest contribution lay in optics and in his theory of light.

For an early claim that light did not travel at infinite speed, see page 81.

According to Huygens light is a series of shock waves pushing through an invisible substance, the aether. These waves travel very fast, but not with infinite speed as virtually everyone assumed. Huygens was led to this correct view of the velocity of light by recent work carried out by the Danish astronomer Ole Römer. In 1675 Römer had found that when the Earth and Jupiter were in certain relative positions in their orbits, eclipses of Jupiter's bright satellites appeared delayed; the delay, he discovered, was due to the extra time light took to travel across the Earth's orbit. This gave him a velocity of 193,000 km/sec (120,000 miles/sec) for the velocity of light, a little over half the correct value.

For the suggestion of this idea by al-Haytham, see page 228.

Huygens's waves had no regularity about them and were, he thought, emitted from every part of a luminous body. He also conceived the idea of 'secondary wavelets', each point at the front of one of his shock waves giving rise to other shock waves and so on. These secondary wavelets were used for explaining reflection and refraction. But since his waves were irregular, Huygens was unable to use them to explain colour, and he was forced to ignore this question. Confirmation that light was indeed a wave disturbance of some kind came from some independent experiments carried out by Francesco Grimaldi, whose results were published posthumously in 1665. Grimaldi showed how shadows were not really sharp but displayed coloured fringes at the edges, but his experiements were questioned – Newton could not successfully repeat them all – and their full significance had to wait a long time for recognition.

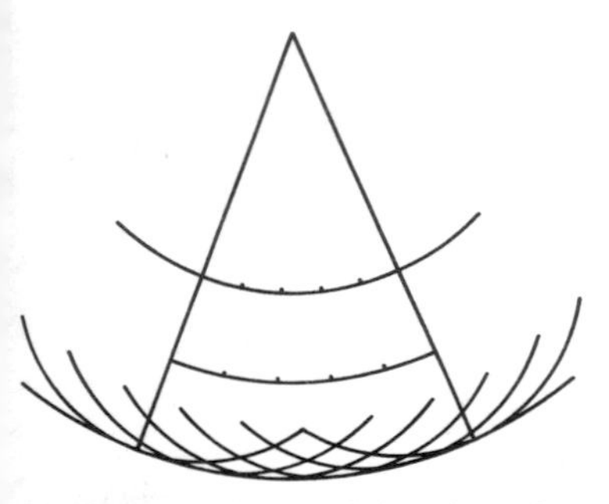

Huygen's theory of the spherical secondary wavelets of light. The light travels from the source in the form of a spherical wave, and from every point along this line secondary waves are created and so on.

Huygens's theory met with some severe criticism, especially in England. Huygens had claimed that light waves travelled slower in a denser medium, and Halley questioned where the impetus came from to make them speed up again when entering a less dense medium, a query that seemed valid enough at the time considering the type of wave envisaged by Huygens. But more significant were the criticisms of Newton, who wanted to know why, if light was propagated by pressure waves, it did not bend round objects. Light does in fact do this, as Grimaldi had shown, but Newton had assumed Grimaldi's effects were simply due to refraction. Newton also questioned why Huygens's theory could not explain the phenomenon of 'double refraction' (whereby crystals of a substance like Iceland spar (calcite) split up a beam of light into two separate components), recently discovered by Erasmus Bartolinus. This was, perhaps, not a very fair criticism because Newton's own theory could offer no explanation either.

Newton's own view was that light was a stream of 'corpuscles', and he accounted for the fact that light is partly reflected and partly refracted (the 'shop-window' effect whereby we not only see through a window from outside but also see our own reflection) by an ingenious system of 'fits of easy reflection and fits of easy refraction' which his particles underwent when passing from one transparent substance to another. Such fits were caused, he thought, by the way the corpuscles were vibrating when they shot out from the emitting source. It was also vibrations at the edges of objects that caused the corpuscles to give those coloured fringe effects Grimaldi had observed and Newton had partly confirmed in his own experiments. If this had been all there was to it, Newton's theory might not have exerted the immense influence it did, but there were other factors which played a part – notably his theory of colours and his invention of the first successful reflecting telescope. For Newton had come forward with the first really satisfactory explanation of colour; 'white light' (the kind of light we get from the Sun) was, he claimed, composed of light of all colours mixed together. This suggestion he was able to prove by a famous experiment in which he used one glass prism to split up a beam of sunlight and a second prism to combine the dispersed coloured band (the spectrum) back to white light again.

Illustration page 364

Newton's development of the reflecting telescope arose out of his experiments aimed at dispersing sunlight into its component colours, for these led him to two conclusions. In the first place he discovered that the different colours are each refracted to a different degree, blue and violet most strongly and red least strongly; that was why a prism acted to disperse white light into a spectrum, and raindrops likewise to give a rainbow. In the second place he concluded that every colour is always refracted in the same proportion. This meant that any attempt to combine different kinds of glass to make a compound lens to overcome chromatic aberration, the false colour fringes seen in a microscope or in a lens (refracting) telescope, would never be successful. In the late 1750s Newton was to be proved wrong on this last point, but in the meantime it remained unchallenged, and its very error turned out to be actually an advantage because it stimulated Newton to devise the first practical reflecting telescope (a telescope where the front lens is replaced by a curved mirror at the rear end of the tube). The reflecting telescope was not his invention; it may even have been known in the sixteenth century, while there is no doubt that the Scots mathematician James Gregory had designed such an instrument in the early 1660s, though its construction had defeated the skill available at the time. Newton severely criticized Gregory's design and his own was certainly simpler, though it had the disadvantage that the observer had to look through the side of the tube, instead of the rear as with Gregory's. Indeed its simplicity was such that Newton was able to build an instrument himself. Though it

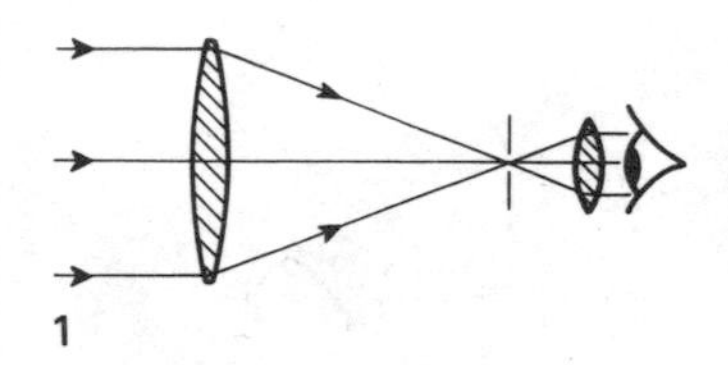

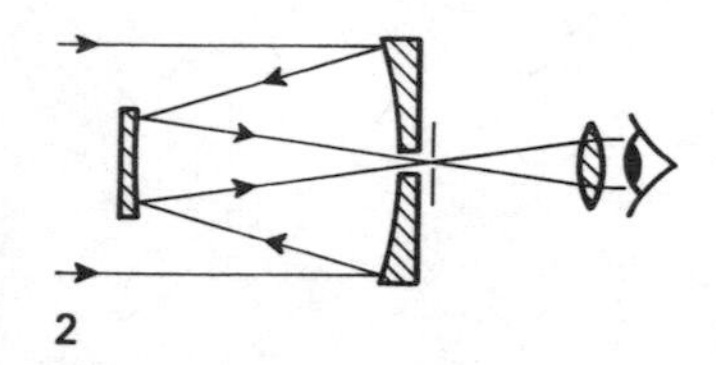

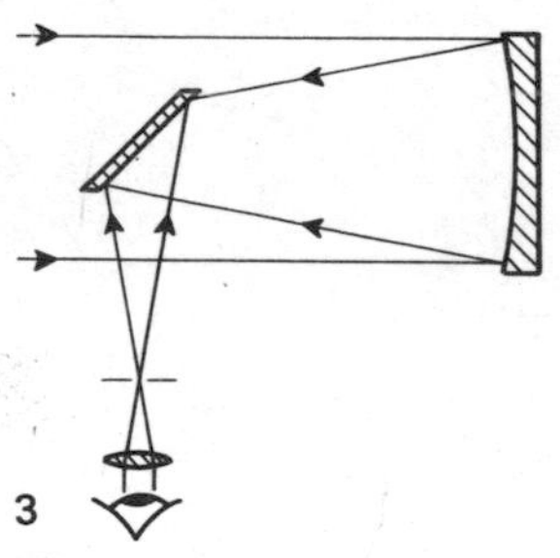

The principles of the three types of telescope. 1. The refracting telescope, as developed by Kepler; 2. the reflecting telescope designed by Gregory; and 3. the reflecting telescope of Newton, to be viewed from the side.

worked well, brought him to the notice of the Royal Society and demonstrated the power of his arguments, strangely enough it was not until towards the end of the eighteenth century, in the 1780s, that this excellent design came into its own. It was Gregory's reflector that opticians took up from the 1740s onwards; the fact that one could use such a telescope by looking through the end of the tube instead of the side, thus making alignment on a celestial object much easier, seems to have been the main reason.

Newton communicated the design of his telescope, along with his theory of light and colour, to the Royal Society in 1670, but it did not become widely known until the publication of his *Opticks* in 1704. Delay in publishing the book was primarily due to the fact that Newton had waited until 1702 when Hooke was dead before finishing it, for Hooke had his own wave-theory of light and Newton feared long and bitter arguments should the book appear during Hooke's lifetime. Nevertheless by the time 1704 arrived not only was Hooke dead but Newton's reputation was international; the *Principia* had seen to that. This acted as an additional factor that set the *Opticks*, and with it Newton's theory of light and colours, on the triumphant course it was to take for almost a century. But Newton had by then forsaken science; he was, it is true, President of the Royal Society, but he was also Master of the Mint and therefore a government administrator during his working day, while he seems to have devoted himself almost exclusively to his extraordinary theological interests in his spare time. He was now a revered figure, knighted in 1705 by Queen Anne – the first knighthood to be given for scientific research – and so respected that when he died in 1727 he was given a state funeral.

The reverence in which Newton was held degenerated into what can best be called 'Newton worship' parodied in the 1860s with the couplet

You think that Newton told a lie,
Where do *you* hope to go when you die?

Yet this reverence was real enough, though the spirit of enquiry in the eighteenth century had become so deeply ingrained as to permit questioning of some of Newton's ideas. Indeed as early as 1695 David Gregory, the Savilian professor of astronomy at Oxford, was querying whether Newton really was correct to assume that all colours were always refracted in the same proportion. By the 1730s the Swiss mathematician and physicist Euler had gone so far as to start experimenting with glass and water lens combinations to see if he could overcome chromatic aberration. Meanwhile, in England a lawyer, Chester Moor Hall, had hit upon the idea of using an object-glass (the lens at the front of the telescope) with two components, each made of a different kind of glass. Hall had argued that the two kinds of glass could be designed so that the chromatic aberration in one could be to a large extent cancelled by the other, and his idea worked.

Nevertheless, for various practical reasons, it was not for another twenty years – not until 1758 – that the manufacture of such 'achromatic' lenses became commercially practicable. Interestingly enough, Hall had argued his case for the cancellation on a mistaken analogy with the human eye; though he achieved the desired result in practice he did so for the wrong theoretical reasons.

Heat

Galileo's approach to physics was to try to devise experiments that would allow him to express his results mathematically. This affected many branches of the subject and not least studies of heat. Heat had always been something of a mystery, but by the time of the founding of the Accademia del Cimento experiments were at least opening the way to the first steps in a scientific approach to the subject by designing devices to measure temperature. The inventor of the first thermometer is unknown; it may have been Galileo, the Hermetist Robert Fludd, the Dutch physicist Cornelius Drebbel (who certainly invented a thermostat for maintaining the temperature of an oven) or the Italian physician Santorio Santorio. At all events it was some time before it was appreciated that a thermometric scale required two fixed points if it were to be scientifically useful, a 'freezing point' and a 'boiling point'; this did not come until the eighteenth century, when Ole Römer devised such an instrument in 1708, using alcohol as the liquid. Römer used a mixture of water and ice and then, it seems, a mixture of ice and ammonium chloride, to give him his zero. For the upper fixed point he took the temperature of boiling water, and he divided the range in between into 60°. Römer's work is now largely forgotten and the credit for the scale – or, rather, for a modification of it – given to the Dutchman Daniel Fahrenheit, who visited Römer in 1708 and on his return began to produce thermometers of his own. It seems that Fahrenheit had somewhat misunderstood the details of Römer's scale – for instance, he thought the upper fixed point represented blood temperature, not boiling point – but he did appreciate the need for two fixed points. He constructed his thermometers on what he believed to be Römer's methods, and he settled on 212° as the boiling point of water; zero on his scale was Römer's zero.

Illustration page 411

Two other well-known thermometric scales appeared in the eighteenth century – the Celsius or centigrade scale and the Réaumur. The first was devised by Anders Celsius and published by the Swedish Royal Society in 1742, using the boiling point of water at one extreme (0°) and its freezing point at the other (100°); it was the Swedish biologist Linnaeus who inverted the scale to give us our present 'Celsius' thermometer. Around the same time René-Antoine Réaumur was experimenting with an alcohol thermometer. He noticed that alcohol expanded by 80 parts per thousand when going from the freezing point of water to its boiling point. In 1730 he therefore produced a thermometric scale of 80°, and this was long in use in some Western European countries.

Thermometry was an attempt to measure the degree of heat, but what was heat itself? This was a question which scientists in both the sixteenth and seventeenth centuries tried to answer. By and large there were two main suggestions, one that heat was due to the vibrations of parts of a substance, the other that heat was an 'imponderable' fluid (i.e. a fluid that could be neither weighed nor measured), though the French astronomer and philosopher Pierre Gassendi did suggest that there were hot and cold particles whose presence was the cause of what is called heat and cold. Francis Bacon and Robert Hooke favoured the vibration theory, but it was the concept of heat as an imponderable substance that finally became accepted, the 'caloric' theory as the French chemists Lavoisier and Berthollet called it. Of course there were those who thought otherwise, and around the end of the eighteenth century the American-born Benjamin Thompson, later Count Rumford, pointed out that his experience boring out iron cannons for the elector of Bavaria, showed that heat could be generated by friction. This hardly seemed to fit the caloric theory and Rumford favoured vibration, but the question was not to be resolved until the middle of the nineteenth century, and even then as famous a physicist as William Thomson (Lord Kelvin) still favoured the imponderable fluid hypothesis.

Illustrations pages 364, 365

If the nature of heat was primarily a matter for speculation, the development of thermometry stimulated the desire for other quantitative work, some way of measuring heat, whatever its nature might turn out to be. The heat of mixtures – some hot water and some cold – was examined, especially by Jean-Baptiste Morin and Georg Richmann who realized that bodies lost heat to their surroundings and that any measurements of temperature in experiments should take account of this, and of the heat losses within the apparatus itself. This was useful in leading to greater precision, but the most significant eighteenth-century work came from the Scottish medical man, chemist and physicist, Joseph Black. Born in 1728, Black studied in Glasgow and then in Edinburgh, became professor of chemistry at Glasgow for a time and then in 1766 moved to take the same chair in Edinburgh; side by side with his academic appointments he ran successful medical practices and amassed a small fortune, though he himself seems to have lived frugally enough.

Black's great contribution to the study of heat was his appreciation that different bodies have differing capacities for heat. Previously it had been suggested that when bodies are all at the same temperature they all possess the same quantity of heat, but Black disagreed. He claimed that a block of iron felt hotter than an equal-sized block of wood at the same temperature because the iron possessed more heat; its capacity to store heat was greater. This led him in the 1760s to the concept of 'specific heat' – the capacity of a body to absorb heat – and he was able to give experimental methods for measuring it. Black also studied the heat required to change the physical state of a body, for example, turning ice into water or water into steam. Experiments on this topic led him to propound a second concept which he called 'latent heat', that is the heat required by a body to cause a change of state. These were both important ideas and led to a quantitative approach to what, at the beginning of the seventeenth century, had seemed a very intractable subject. They also had important practical applications, the most notable of which was James Watt's development of the separate condenser for a steam engine, an invention which revolutionized steam engines and made them an economic proposition. This was a direct result of Black's theoretical work, since for a time Watt was scientific instrument-maker to Glasgow University and had become acquainted with Black and his ideas.

Illustration page 364

Electricity

The third great development in physics during the seventeenth and eighteenth centuries was the study of electricity. Electrical effects had been known in antiquity – indeed the very word 'electricity' is derived from the Greek *elektron*, meaning amber, for it was the ability of amber when rubbed to attract small leaves that seems to have been the sole electrical fact known. Since amber acted in a way similar to lodestone, which attracted pieces of iron, there was some confusion in early times between magnetism and what we should today call 'static' electricity – the electricity derived from rubbing certain substances.

William Gilbert made some studies in static electricity, but came to the conclusion that it was a trivial subject, and so indeed it might have remained had not the 'electrostatic machine' been developed. The original invention was due to Otto von Guericke, German diplomat and engineer and sometime mayor of Magdeburg, who in the late 1640s interested himself in physics. One idea which intrigued him was the suggestion that celestial bodies reacted on one another by magnetism, and he sought to modify Gilbert's experiments with a spherical lodestone by casting a sphere with minerals to mimic what he believed to be the actual composition of the Earth. The finished sphere contained a good deal of sulphur – later on he used a sphere of pure sulphur – and by rotating it and rubbing it at the same time he found it acquired attractive powers and gave off sparks. He did not recognize these as due to static electricity – though that is what they were – he thought they were demonstrations of the powers or 'vertue' of the body. However, some thirty years later, in 1705, Francis Hawksbee in London noticed that a glow was sometimes emitted from the empty part of a barometer tube, and he correctly diagnosed this as due to the mercury rubbing against the glass.

Illustration page 366

For the work of William Gilbert, see pages 316–317.

But the study of static electricity really only got under way from the 1720s. Then Stephen Gray, stimulated by the work of von Guericke and Hawksbee, performed a series of electrical experiments using an electrostatic machine. His first discovery was that the 'electric virtue' could be transmitted great distances along a thread if it were made of suitable material, like silk, but leaked away if the material were unsuitable. Secondly, he found that a body could be electrified if another electrified body was held near to it. Meanwhile, in France, Charles-François Dufay discovered that electrified bodies could either attract or repel one another, and this led him to suggest that there were two kinds of static electricity.

By the 1740s and 1750s many variations of von Guericke's electric machine were built, using a glass globe or glass cylinder to generate the electric charges. Later these became more powerful, especially when John Cuthbertson in the 1780s designed one using two oppositely rotating glass plates, and they all provided stronger sources of static electricity and enabled more extensive experiments to be carried out. One consequence of these was the discovery in 1745 by Ewald von Kleist of a device to store electricity so that, when discharged, it would produce a big enough spark to ignite alcohol. To a great extent this was nothing more than a curiosity – von Kleist used a simple medicine glass coated with metal – but in Leiden, where Pieter van Musschenbroek was conducting electrical experiments, he found by chance a way of confirming von Kleist's result. Musschenbroek designed a glass vessel coated inside and out with metal that became known as the Leiden Jar and was the first capacitor. Subsequent experiments showed that the spark emitted was greater the thinner the glass between the metal coatings. This was significant because it simply could not be explained on the theory of two kinds of electricity, and paved the way for the theory of the American scientist Benjamin Franklin, who found himself able to account for the action of the Leiden Jar by a one-fluid theory. Nevertheless a modified two-fluid theory was to return for a time, though it found favour only among a few of the more mathematically minded experimenters.

Illustrations pages 366, 367

The charges stored by a Leiden Jar could be very large indeed, and the discharge from Jars wired together was found to be capable of killing animals. They could be used to store electricity for further study, and it was in charging up such a jar from a kite flying during a thunderstorm in 1750 that Franklin was able to show that lightning was nothing more than static electricity.

In electricity, as in other fields of physics, experimenters were keen to take a quantitative approach. During the 1760s and 1770s it became clear from experimental work by Franklin, by the chemist and physicist Joseph Priestley and the eccentric millionaire experimenter, Henry Cavendish, that the force of attraction or repulsion between electric charges varied, like gravity, according to the inverse square of the distance between them. Clearly this was a discovery of the first importance, but it needed confirming by measurements of the utmost precision. It was the Frenchman Charles Coulomb, a physicist and engineer, who managed to achieve the necessary accuracy in 1785 with his 'torsion balance'. Essentially this was a device that held an electrically charged ball or a magnet at the end of a horizontal arm suspended at the end of a thin silver wire. The whole apparatus was housed in a glass case on which a scale was engraved. Once the horizontal arm was at rest a second charged ball (or magnet) was brought to within a specified distance; this caused the arm to turn, the torsion of the silver wire resisting this turning so that the amount by which the horizontal arm rotated was proportional to the forces of attraction or repulsion. Ingenious and elegantly simple, like all the best experiments, the balance was very sensitive and allowed Coulomb to prove that not only electrical charges but also magnets attracted each other with a force that did vary precisely according to the square of the distance between them.

The last electrical developments in the eighteenth century took place in Italy at Bologna and Pavia. In 1780 at the Univeristy of Pavia, Luigi Galvani, the professor of obstetrics, began a series of careful investigations into the responses obtained from the hind legs of frogs when static electricity was applied to them. First of all he found that the legs, mounted on a glass sheet with metal foil each side, twitched when an electric charge was applied to the spinal column above them. But as he proceeded, with slight variations, to make other tests he came up with a strange and unexpected result. The frog's legs twitched even when they were fully insulated so long as the appropriate nerves were given an electrical path to the ground, and an electrical machine some distance away was made to produce a spark. Similar contractions occurred in legs fixed by a brass hook through the spinal cord to iron railings outside the laboratory. They would twitch not only during a thunderstorm but even when the sky was quiet. He got the same effect indoors.

Galvani interpreted his results as due to 'animal electricity'. He concluded that he had proved an idea that had been discussed time and again throughout the eighteenth century, that the nerves and muscles of animals contained a subtle fluid akin to the electric fluid. However, his results were not in fact due to this at all – there was no such fluid. The curious and unexpected events he had observed were due to the contact of two dissimilar metals (in his case brass and iron) in a damp environment. That this was the cause was discovered by Alessandro Volta, professor of experimental physics at Bologna, who in 1792 and 1793 published in the Royal Society's *Philosophical Transactions* his belief that the 'metals used in the experiments, being applied to the moist bodies of animals, can by themselves . . . excite and dislodge the electric fluid . . .' Argument with Galvani followed but in 1799 Volta proved his point by constructing a device – a pile of copper and zinc discs separated by discs of moist pasteboard – which actually gave out electricity. This was not only the first electric battery, it was also the first source giving a continuous flow of electricity, and its implications both as a source of electricity and in the theoretical questions it raised were immense. It raised chemical problems and was instrumental in forging a link between electricity and material substances, and so opened a new dimension of research of which the nineteenth century was to take advanatage.

Illustrations page 401

Chemistry

During the seventeenth and eighteenth centuries new developments in chemistry owed their origin to various seventeenth-century experiments on combustion and respiration using vacuum pumps. It was Galileo who, in 1638, in his *Discourses Concerning Two New Sciences*, showed that in spite of Aristotle's claim that there could not be such a thing as a vacuum, one could be generated without undue difficulty. This investigation was extended by his pupil Toricelli and by Otto von Guericke in Germany. Both studied atmospheric pressure by creating vacuums at the top of sealed tubes (thus making the first barometers), while Guericke himself went further and designed efficient vacuum pumps. With these he displayed the power of a vacuum, in a famous public demonstration of his 'Magdeburg hemispheres'; he exhausted air from two copper hemispheres placed mouth to mouth and then showed how even two opposing teams of horses could not pull them apart.

Illustration page 355

Illustration page 402

A third notable experimenter in this field was Robert Hooke (1635–1703). The sickly son of a Protestant minister, Hooke managed in 1653 to get to Christ Church, Oxford, as a chorister, and while at the university he met a number of men interested in science who were to become some of the founders of the Royal Society. Among them was Robert Boyle (1627–91), son of the first Earl of Cork, who employed the undergraduate Hooke as a laboratory assistant. Although primarily interested in chemistry, Boyle was also fascinated by the physical properties of air – he had already written a book on it – and, intrigued by von Guericke's work, he instructed Hooke to build an improved vacuum pump. Hooke had a strong mechanical flair – later he was to invent the universal joint and a spring-driven clock – and his pump was highly satisfactory. Indeed Hooke was so able an experimenter that in 1662 he was appointed 'curator of experiments' to the Royal Society and thereby enabled to carry out some research of his own in meteorology – he designed hygrometers, barometers and a wind-gauge – and in the elasticity of materials, a subject in which he is still remembered by Hooke's Law (which states that the strain is proportional to the stress within a substance). With Boyle, and using his improved vacuum pump, Hooke helped derive the relationship we know as Boyle's Law relating the pressure and volume of a gas.

Illustration page 403

Another outcome of the Boyle-Hooke co-operation was Boyle's publication in 1661 of his books *The Sceptical Chymist* and *Certain Physiological Essays*. These made clear his support for atomic-like theories of material substances and his wish to move away from Aristotle's combination of matter and form. In *The Sceptical Chymist* he emphasized a need for a new definition of elements, suggesting they could best be described as the perfectly homogeneous substances into which mixed bodies could ultimately be resolved. Boyle's injunction was timely, for by the mid-seventeenth century the old earth-air-fire-water classification had become unproductive. Using it still left chemical reactions a confusing muddle of ideas, as the work of Jan Baptista van Helmont in Brussels had made abundantly clear. A medical man and experimental scientist, van Helmont properly set great store by accurate weighing and measuring in all his chemical work – a procedure that was to pay great dividends in all subsequent experimental chemistry and was vital in any investigations into the various aspects of combustion. In the first half of the seventeenth century van Helmont had found that a chemical study of the smoke remaining after solids and fluids are burned showed that such smoke was different from air and from water vapour; it displayed characteristics of its substance of origin. For such smokes he coined a new word, 'gas', derived either from the Greek *chaos* (empty space) or the Dutch *gaesen* (to ferment or effervesce), and in subsequent experiments he showed that many different kinds of gas existed. Clearly, these required further investigation.

Hooke, as well as Boyle, took up the challenge, arguing that since the saltpetre in gunpowder burns under water, it must contain something that air does. Indeed, he extended this idea to all combustible bodies and concluded that air contains a 'solvent' that permits combustion to occur. Hooke already knew that air was necessary to maintain the 'fire of life'; and his experiments also showed that animals need air blown through the lungs to live and plants need air for growth. All this was taken further by John Mayow, a medical man and experimenter whom Hooke had met on a number of occasions. Mayow devised a method for collecting gases over water – an important experimental step forward – and by 1679 had shown that both respiration and combustion consume something from the air. He suggested that air also carried 'nitro-aerial particles' which were consumed during these processes.

Such studies were not, of course, the sole prerogative of English experimenters. The German economist Johann Becher, who was also an experimental chemist, studied the chemistry of metals and minerals, publishing his results in 1669 under the title *Subterranean Physics*. Becher reckoned that all minerals and metals were composed of three qualities, a *terra lapida* or transparent vitrifiable component (equivalent to the 'salt' of Paracelsus), a *terra mercuralis* or subtle and volatile component, and *terra pinguis*, an igneous, fatty and combustible component. As far as other substances were concerned, those that were combustible also contained *terra pinguis*. It was an ingenious explanation, especially since Becher made the point that none of these components were true chemical elements; they were only behavioural qualities.

Becher's ideas were taken up by the German physician Georg Stahl, who was a strong believer in vitalism, the idea that there is a vital force operating throughout Nature, and especially noticeable in living things. A writer of great influence, he welcomed Becher's theory and produced some new editions of *Subterranean Physics*. But Stahl's place in the history of chemistry rests on his *Fundamentals of Chemistry*, published in 1723. This book was important because it promoted two ideas. The first was a definition of chemistry; to Stahl it was a method of resolving compounds into their elements and studying their recombination, a view that provided chemical experimenters with a much clearer idea of their aims. The second, and more important, was the introduction of 'phlogiston' (from the Greek *phlogistos*, burnt) to replace Becher's *terra mercurialis*, and the extension of this to all combustile materials. The importance of the 'fire principle', phlogiston, is that it acted as a great unifying concept in chemistry; it correlated a wide variety of facts, being applicable not only to combustion, but also to respiration and calcination (the roasting of metals to a high temperature, but without fusing), thus bringing some deeper understanding to all kinds of reactions. For the next thirty or forty years chemists pursued the theory with vigour, until it became so entrenched in chemical thinking that a great intellectual effort had to be made to break away from it.

After the advent of the phlogiston theory, the study of combustion and gases was pursued because chemists were still puzzled by the nature of air, and more significant research was carried out, especially by Joseph Black. At the time he took his MD degree in the 1750s, there was a widespread idea among physicians that limewater (a weak solution of calcium hydroxide in water) was efficacious in dissolving 'stones' in the bladder. Two Edinburgh professors agreed on this but violently disagreed on the chemical processes involved, and Black decided to investigate them for the subject of his thesis. However he did not wish to become involved in the conflict and ended up examining other similar materials to see whether he could find a more powerful substitute for limewater. The upshot of this was that he examined the white powder *magnesia alba* and found that when strongly heated it proved to have unexpected qualities. When calcined only a fraction of its weight was lost, so Black concluded that what had happened was that air had been driven off. This seemed confirmed by the fact that while *magnesia alba* gave off air bubbles when combined with acid, in its calcined form (the calyx) it did not. Moreover by the use of an alkali, Black was able to turn the calyx back into an amount of *magnesia alba* equal in weight to the original. Now it was known that a crust formed on limewater, but Black discovered that this did not happen when the liquid was kept in a stoppered bottle. Thus in both cases – with the calyx of *magnesia alba* and with limewater – it seemed that something in the air was combining with them. Black was therefore led to the important conclusion that air was not one single substance but more than one. This was a crucial step forward.

Black called the combining component 'fixed air' since it became fixed in various substances. Further investigations led him to the discovery that fixed air was produced during respiration, in combustion and in fermentation. So from 1756, when Black published his results, chemists were aware that air was composed of more than one chemical substance. The problem next to be solved was to determine what substances these were. Henry Cavendish carried out some research and in 1766 presented the Royal Society with his results in a scientific contribution he called 'On Factious Airs'. It was a study of the physical properties of Black's fixed air and of what he called 'inflammable air', which seemed to be the same as Stahl's phlogiston. He found that an acid acting on a metal released inflammable air and concluded that this came from the metal itself. Thus, according to his results, there seemed to be three kinds of air – air, fixed air and inflammable air – though the general opinion was that these three were merely air altered in its properties. As yet no one had followed up van Helmont's idea of different gases.

The next steps to solve the problem were taken by Joseph Priestley, an English Presbyterian preacher and teacher, and a man with very radical political views, who strongly supported the ideals of the French Revolution. Deeply interested in chemistry and physics, and author of a successful book *The History and Present State of Electricity . . .* (1767), Priestley decided he too must experiment on air. An accurate and critical observer, in 1772 he was ready to publish an account in the *Philosophical Transactions* of his first experiments. Using his own improved pneumatic trough (this was a container partly immersed in mercury) for collecting his 'airs', he described how he had prepared different varieties – 'nitrous air', 'phlogisticated air', 'acid air' and seven others. Other experiments showed him that the volume of air decreases by a fifth during respiration, and in 1772 he had also found that combining his 'nitrous air' (our nitrous oxide) and ordinary air gave a product which was reddish in colour and diminished in volume. He had carried out this experiment by exploding the gases with a spark inside a sealed vessel; a powerful technique that allowed quantitative experiments to be made. Priestley also discovered that plants 'restored' air breathed by a mouse or altered by the presence of a burning candle.

Illustration page 404

However, Priestley's most important discovery came in April 1774 when he obtained a colourless air by heating red oxide of mercury using sunlight concentrated through a large burning glass – an efficient method of obtaining great heat in the laboratory. The air he obtained surprised him; in it a candle burned brightly yet the air was itself not soluble in water. He called it 'dephlogisticated air'. In the autumn Priestley travelled through northern Europe, and in Paris discussed his results with the French chemist Lavoisier; this was to prove a crucial meeting. On his return to England, Priestley worked as a minister of religion, but this did not prevent him continuing his experiments. In 1781 he used a spark to explode a mixture of inflammable air and dephlogisticated air in a bottle, and noticed that 'dew' (i.e. water) resulted: he commented only that 'common air deposits its moisture when phlogisticated'. Cavendish later repeated this experiment and found the dew to be 'pure water', concluding that 'dephlogisticated air is in reality nothing but dephlogisticated water'. Clearly, a link was being forged between water and the components of air, but the overall picture was not yet clear, though there was no doubt about the experimental evidence. Indeed, Priestley's discovery of dephlogisticated air had been anticipated in 1772 by the Swedish pharmacist Carl Scheele, whose results led him to claim that air was of two kinds, one supporting combustion – 'fire air' – and one preventing it. Scheele's results were not published, however, until 1777, with an English translation three years later.

The man who finally solved the problem, though he had to make the great intellectual effort of casting aside the phlogiston theory before he could do so, was Antoine-Laurent Lavoisier. Born in 1743, he was a well-educated Parisian government servant with a genius for science. That he should have met with death by the guillotine in 1794 during the Reign of Terror in the French Revolution was certainly a great loss to the scientific community – the mathematician Lagrange remarked, 'It took them only an instant to cut off that head, and a hundred years may not produce another like it' – but, unhappily, he had become a member of the unpopular Ferme Générale which collected taxes on behalf of the government.

Lavoisier began his scientific career in the 1760s in geology, to which he brought as much precision measurement as possible – a trend that was to characterize all his work – then he became concerned with the Paris water supply. This led him to chemical experiments and to disproving the idea that water changed to earth if heated long enough, an erroneous view due to the fact that when impure water is boiled away, a solid residue remains. Again it was careful quantitiative experiment that led Lavoisier to his results. Not until the early 1770s did he begin to study combustion, but as soon as he did he was able to show that when a metal was calcined it did not take up 'fire particles', that is it did not absorb 'fixed air'. It was at this juncture that Priestley visited Paris and told Lavoisier of his idea of dephlogisticated air.

After Priestley's visit, Lavoisier made a number of other crucial experiments. By the late 1770s he was himself convinced that air was a compound of some kind, containing an eminently combustible part and an unbreathable part. Studies of calcination alone and then in the presence of charcoal had led him to conclude that Black's 'fixed air' was some kind of charcoal compound. He therefore put all these results together and reached the first of those conclusions that were to overthrow old chemical ideas and usher in modern chemistry. In 1779 he claimed that the combustible part of air was a constituent of all acids and he named it the 'acid principle' or '*principe oxygine*' (derived from the Greek *oxus*, acid). At this stage, however, he had not openly discarded the phlogiston theory; he merely said that his view gave another explanation of the theory.

Illustrations page 405

By now Lavoisier could explain a host of chemical processes, but was still puzzled by inflammable air. However, it was at this time that Priestley and Cavendish were using sparks to produce dew and

Cavendish concluded that the dew was pure water. News of Cavendish's identification reached Lavoisier in Paris, and of Cavendish's explanation, which still assumed that all gases contained water and used the phlogiston theory to account for it. However Lavoisier performed other experiments, not only producing water, but also investigating the precise action of acids on metals and the production, during these last experiments, of 'inflammable air'. The outcome was that he recognized that water breaks down under certain conditions, giving his *principe oxygine* on the one hand and a water principle – *principe hydrogen* (Greek *hydor*, water) on the other. Once this recognition had taken place a lot of chemical reactions could now be more effectively explained than before, and this was especially so in the case of reactions involving acids on metals. Lavoisier showed how oxygen and hydrogen each played a part, and was able to formulate a whole new chemistry without invoking the 'fire element' phlogiston.

Based on careful analysis and backed by meticulous measurement, it was a serious challenge to the old phlogiston theory and over the years completely supplanted it. However, its advantages were evident at once to numerous chemists and in 1784, a year after Lavoisier announced his results, Joseph Black was teaching it to his students in Edinburgh, while in 1785 Claude Berthollet, a well-known chemist, embraced the new ideas. Indeed, during the next two years, Berthollet, Lavoisier, and the chemists Guyton de Morveau and Antoine de Fourcroy set about reorganizing the whole of chemical nomenclature in the light of the new theory, giving each substance a name that described its chemical composition, and defining their elements very carefully, following the principles set out by Robert Boyle more than a century earlier. Thus 'oil of vitriol' now became sulphuric acid, 'aqua fortis' nitric acid, and so on; it was a system very similar to the one we use today, though it appeared first in 1787 under the title *Method of Chemical Nomenclature*. Two years later Lavoisier produced his famous *Elementary Treatise on Chemistry* which in its clarity and comprehensiveness popularized the new ideas. The modern era of chemistry had at last dawned.

For the development of chemical notation, see page 439.

Geology

It was in the seventeenth and eighteenth centuries that geology moved from the realms of opinion and speculation to the dignity of a science. Palaeontology – the study of fossil plants and rocks – had at last begun to receive careful study. Robert Hooke described fossil wood and commented on petrified bodies in 1665, suggesting that these might result from natural processes over a period of time while similar views were put forward by Niels Stensen (often known as Nicolaus Steno) in Denmark only five years later. Others, like the biologist John Ray and the botanist John Woodward, agreed with these opinions, though it was still early days for a completely new understanding; in the seventeenth century there was still a strong belief in

For Chinese awareness of fossil remains, see page 168; for Palissy's observations see page 312.

the West in the original formation of species at the Creation, and in the Flood as a historical event.

It was, however, in ideas about the Earth, its interior, its crust and its age that there was a more substantial advance. In 1617 in his *Principles of Philosophy* Descartes made some original suggestions about the constitution of the Earth. He suggested that the Earth was once molten, like the Sun, but had now cooled and condensed except in its central regions, where he pictured a metallic and a second even denser layer covering a central core that was still incandescent. Just below the surface, he suggested, lay both subterranean water and a layer of mud and sand, and it was this layer that some later writers drew on to account for the Flood. Variations of Descartes's idea came in the years that followed; the Jesuit writer Athanasius Kircher published his *Subterranean World* in 1665 in which he also pictured a fiery centre, though Kircher's had cavities filled either with fire or water as an explanation of volcanic eruptions. Gottfried Leibniz also accepted the idea that the Earth was once incandescent, and interestingly enough suggested that igneous rocks formed the original crust and that sedimentary rocks were laid down later.

The age of the Earth was another intriguing question that came up for discussion. In the 1650s James Ussher, the archbishop of Armagh in Ireland, concluded from scriptural and ancient historical sources (by totalling the ages of Old Testament patriarchs) that the Creation of the World had occurred in 4004 BC, and well into the nineteenth century this figure was still printed in the marginalia of Protestant Bibles. Yet there were some who tried to tackle this question from a scientific rather than a historical point of view, though the first serious attempt did not take place until the early eighteenth century, when Edmond Halley presented a paper on the subject to the Royal Society in 1715. He suggested that the salinity of the oceans might offer a clue to the Earth's age, basing his argument on the assumption that, due to evaporation, the saltiness of every ocean should increase with time. He was willing to accept that 4004 BC might mark the beginning of man's existence, but he thought the Earth was much older than that. Later on in the eighteenth century studies of possible changes in animal species, especially by the Comte de Buffon and the Chevalier de Lamarck, led to suggestions for a much longer time scale, of the order of 50,000 years or more.

In the nineteenth century the formation of geological features was to prove a valuable source of evidence for a long time scale, but in the seventeenth and eighteenth centuries evidence was still a little too uncertain for geologists to venture much of an opinion. All the same, it was in these two centuries that the necessary foundations of physical geology were laid. This began in 1671 with Niels Stensen's *The Prodromus to a Dissertation Concerning Solids Naturally Contained Within Solids,* a book in which he outlined the principles of modern physical geology and showed how layers of rock could collapse to form mountains and other features. He also suggested that all rock layers

Illustration page 405

or strata must originally have been laid down horizontally on a solid surface and that their folding and breakage occurred later; Stensen also claimed that if remains of marine animals were found, this meant that the sea must once have covered the area. Hooke also discussed changes in topography, though he concentrated only on the effects likely to be caused by earthquakes.

Following Stensen's work, more detailed studies of rock strata were made, and by the middle of the eighteenth century a considerable amount of evidence had been amassed, enough for Johann Lehmann, a German medical man much interested in mining and metallurgy, as Bauer and Paracelsus had been, to produce a *History of Stratified Rocks* (1756). This work was followed by others, especially by G.C. Füchsel who drew one of the first geological maps to illustrate stratigraphic descriptions, while in 1759 Giovanni Arduino, an Italian mining expert, divided stratified rocks into 'four general and successive orders', which he named primary, secondary, tertiary and quaternary – terms not unfamiliar today. From his study of fossils Arduino also concluded that '. . . as many ages have elapsed during the elevation of the Alps, as there are races of organic fossil bodies embedded within the strata'.

In 1775 at the mining town of Freiberg in Saxony, Abraham Werner founded a geological school that was to achieve an immense reputation, while his own research was to add materially to late eighteenth-century knowledge of minerals and their connection with geology. In clarifying them, he rightly pointed out that the external characteristics of mineral crystals were not only important in identifying them but also acted as a clue to their composition, though he did not ignore the importance of chemical analysis as well. But Werner's greatest contribution lay in his ideas about the origin of the Earth's crust – indeed he is sometimes known as the father of historical geology – and it is certain that he did much to turn geology into a science and a sound academic discipline. His theory was based on two beliefs: that the Earth was once enveloped by a universal ocean, and that all the important rocks which compose the crust were either formed as sediments or were precipitated from the enveloping ocean. Possibly his most important conclusion was that the time scale of the Earth's history was very long indeed; he wrote about waters covering the Earth 'perhaps 1,000,000 years ago . . .' and then went on to show his appreciation of so vast a period by saying 'in contrast to which written history is only a point in time.' His full theory traced the gradual changes that he conceived to have occurred over this long geological time. There were, he thought, five periods of formation; a very primitive period when rocks were precipitated from the still and calm ocean, and which was devoid of life. Then came a transition period, followed by a period of storms and the development of life; some of the old rocks were then broken down and a great inundation followed. Next there were volcanic and alluvial periods. His theory also allowed of local variations, and though we are now well aware of its faults, it was an immensely effective synthesis of stratigraphic knowledge which, coming at the end of the eighteenth century and the beginning of the nineteenth, exerted considerable influence on later thinking.

Important, too, for later developments was the physical geology of James Hutton. Hutton, a Scottish medical man who developed a penchant for geology, produced in 1795 his own *Theory of the Earth*. Like Werner, Hutton was concerned with the processes at work, but his approach was different. He took no notice of the general belief in a universal flood, nor did he invoke any catastrophes to account for the way he found the Earth's crust to be; instead Hutton suggested a continuous process of change. The forces operating on the crust were, he believed, cyclic; first, there was the action of erosion which transported material that gradually accumulated, next came a period of consolidation, then an expansion due to heating, followed by elevation of material, after which erosion would again come into play. In all this Hutton was correct, though the reasons he gave for consolidation, expansion and elevation were not. Nevertheless his book was important, for in producing a theory that accounted for everything on a basis of gradual cyclic change he introduced a totally new outlook into geology. In the nineteenth century this was to prove of the greatest significance.

Biological science

Microscopy

The invention of the microscope brought a new dimension into the biological sciences, and although its results were not as spectacular as when the telescope was first applied to astronomy, they were startling enough, all the same. To begin with, in the seventeenth century, a number of the 'new philosophers' took up the new instrument and applied it to biological study; most notable among them were Robert Hooke, Jan Swammerdam, Marcello Malpighi, Nehemiah Grew and Anton van Leeuwenhoek. Hooke's research was given in his *Micrographia* which, published in London in 1665, was the first great work to be devoted to microscopy. Illustrated with 37 beautiful copperplate engravings, only three of which were not concerned with 'Physiological Descriptions, of some Minute Bodies made by Magnifying Glasses . . .', the book is a tribute to Hooke's artistic talents as well as his care as an observer, and it exerted a considerable influence for the rest of the century because it made clear what the microscope could do for biological science. Well described as a 'banquet of observations', it contained the first modern biological use of the word 'cell' (though Hooke used it for cork not for 'living' cells), and it initiated the study of insect anatomy by showing in some detail the multiple eye of the fly and giving a description of a bee sting. The book also made known Hooke's views on light and colours.

Illustrations pages 406, 407

For these, see page 375.

More detailed and systematic was Swammerdam's microscopical research. Born in Amsterdam in 1637, the son of an apothecary,

Swammerdam studied medicine at Leiden, though there is no evidence that he set up in medical practice. He did, however, conduct some medical research, though here his work lay totally within the general outlook of his day; he struck out on no important path of his own. Today Swammerdam is remembered for his work on insects; he was the first to perform dissection under the microscope and wrote on the anatomical details of bees, wasps, ants, gnats, the dragonfly and the mayfly, and made comparative studies of them. He claimed that insects were no less perfect than the higher animals – a view totally in opposition to Aristotle and his scale of nature – and showed that the development of a winged insect was essentially a matter of growth and a change of form. Swammerdam also attempted a classification of insects, dividing them into those which developed directly without transformation, those which gradually acquired wings, those which did so under larval skins, and those which went through a pupal stage. Yet although most of his time was spent on insects, Swammerdam did manage to carry out some other biological research; he discovered the red corpuscles in the blood of frogs, whose methods of reproduction he studied closely, and he also found that if a nerve were cut and the end stimulated the appropriate muscle might contract, but that in doing so the muscle itself did not increase in volume. This was significant; it meant that no fluid had passed into the muscle from the nerve, contrary to the then universal belief which took the nerves to be hollow tubes. Swammerdam's microscopical research showed this was not so.

For Aristotle's 'ladder of souls', see page 107; for a Chinese alternative, see page 134.

Swammerdam's work was only partially published during his lifetime. His *General History of Insects* came out in 1669, and his *Ephemeral Life* – a book on the mayfly – in 1675, five years before his death. Yet it was not until almost fifty years later that the bulk of his research appeared when the Dutch scientist Hermann Boerhaave published it between 1737 and 1738 under the title *Bible of Nature*. Such posthumous publication was not, however, the fate of the other three great microscopists of the period. Marcello Malpighi, who was born in Bologna in 1628, was, like Swammerdam, a trained medical man and he used the microscope both in medical research and in studies he made in embryology, where he examined in particular the development of the chick. Malpighi observed all kinds of tissues, discovered tiny 'capillary' tubes in which he found blood flowed only in the direction arteries to veins, and in his research on the chick his microscope showed him very early stages of development never previously observed. Nehemiah Grew's microscopic work was carried out on plants, and his studies demonstrated that there were sexual differences between some plants, a view hinted at previously but never confirmed. He recognized that stamens, which carry the pollen, are the male organs; he described in detail their action in plant reproduction.

Illustrations page 408

Of all the pioneering microscopists, there is no doubt that the greatest was Anton van Leeuwenhoek, whose work spanned the latter half of the seventeenth century and the early decades of the eighteenth.

Born in 1632 in Delft, he spent most of his life there working for the municipality in a variety of responsible positions. Microscopy was a spare-time interest. Leeuwenhoek constructed his own microscopes, developing them from the magnifiers used in the cloth trade (his first father-in-law was a cloth merchant), his lens-grinding technique improving with each instrument. At the university museum at Utrecht, one of his single 'pebble-lens' microscopes still exists; it has the astonishing magnifying power of 270 and with it detail can be detected down to 1½ thousandths of a millimetre, yet it seems he made microscopes more powerful even than this. They were all tiny instruments, held close to the eye, with the specimen to be examined fitted in a holder, also very close to the lens, but their freedom from gross chromatic aberration enabled Leeuwenhoek to advance knowledge very considerably.

Illustrations page 409

His interests were wide. Like Malpighi he studied the blood, confirming his results, but also noting the existence of red corpuscles in the blood of a tadpole, and then in his own blood. Later investigation showed him that they were round in the blood of man and mammals, but oval in fish and birds, and he was even able to detect the small nucleus present in the oval ones. Leeuwenhoek also ventured into embryology, observing spermatozoa, making the first drawings of them and noting the differences between those of a man and those of a dog.

Next he spent time studying protozoa, those minute single-cell animals found in fresh water, in the oceans and in moist soil, noting their development and observing, for instance, how one hydra gave birth to others by budding. Leeuwenhoek also devoted some attention to insects, observed the life-cycle of the flea from the larval stage onwards, studied aphids and the way they reproduced their young, and he also found that cochineal, much used as a dye, was composed of the bodies of insects. It was Leeuwenhoek too who carried out the first microscopic examination of yeast, discovered the sheath surrounding nerve fibres, and, in 1683, after examining scrapings taken from between the teeth, noticed the presence of minute 'rods' which he later referred to, rather delightfully, as the 'flora of the mouth'. These were the first observations ever to be made of bacteria. There is no doubt, then, that Leeuwenhoek was a microscopist of extraordinary ability, but his technique was very much his own and he had no successors. Not until the achromatic compound microscope was developed in the 1830s did microscopy again make any notable progress.

Illustration page 408

For developments in the study of bacteria in the 19th century, see pages 431–346.

Botany, Zoology and Medicine

Three particularly notable developments took place in these fields during the seventeenth and eighteenth centuries: one was the application of physics to botany and then to animals, including man; the second was the hunt for a new and improved classification of plants and animals, a study which brought some of its investigators near to a theory of evolution; the third and last was the development of an understanding of human and animal anatomy. All three were interlinked in one way or another.

The pioneer in applying the laws of Newtonian physics to the vegetable kingdom was Stephen Hales (1677-1761). He was the natural successor of Grew and Malpighi, who had sought parallels between the behaviour of plants and animals, since both scientists subscribed to the belief that flora and fauna were both 'contrivances of the same wisdom', both works of the same God. Hales was an English clergyman who carried out his research in his spare time and did not publish his discoveries until he was in his fifties, though when they did appear their importance was immediately recognized. They came out in his book *Vegetable Staticks* which was published in 1727, though he had by then given the Royal Society an account of some of this work. The book concerned his physical experiments on sap pressure, on the passage of water through a plant and investigations on 'perspiration' from the leaves. It also contained Hales's suggestion that it is through their leaves that plants '. . . draw some part of their nourishment from the air', and also that 'one great use of leaves is . . . to perform in some measure the same office for the support of vegetable life, that the lungs of animals do, for the support of animal life'. Yet he was prepared to go further than this; after describing experiments showing the precise way leaves expand as they grow, he remarked 'May not light also, by freely entering the expanded surfaces of the leaves, and flowers, contribute much to the ennobling principles of Vegetables?' Hales's work was unique; his only predecessor was van Helmont who had planted a tree in a pot, weighed it, watered it (measuring the weight of water used) and also weighed the tree from time to time, a series of experiments that led him to conclude that it was water alone that nourished plant life. Hales's work was more productive and *Vegetable Staticks* was really the pioneering study of plant physiology. It was, however, not followed up until the nineteenth century.

Illustration page 411

As far as human and animal anatomy was concerned, this was revolutionized at the early part of the seventeenth century, as was the whole of subsequent medical science, by William Harvey's discovery of the primary circulation of the blood. Harvey (1578-1657) went up to Cambridge when he was fifteen to study arts and medicine, leaving in 1599 to go to Padua to complete his medical training. There he studied under Fabricius and may, perhaps, have encountered Galileo. Harvey returned to England in 1602, set up a successful medical practice in London, and numbered Francis Bacon among his patients. In 1609 he was appointed physician to the famous Saint Bartholomew's Hospital, and in 1615 Lumleian lecturer in anatomy at the Royal College of Physicians. In 1616, the very year Shakespeare died, he began a course of lectures to the Royal College in which he sketched his theory of the circulation, though his book *On the Movement of the Heart and Blood* did not appear until 1628. In 1618 he was appointed one of the physicians extraordinary to James I, in 1632 physician to Charles I, and after the King's surrender he lived in retirement in London, having had his manuscripts, drawings and anatomical collection vandalized during the Civil War.

Harvey's discovery of the circulation was due to a long series of observations. In the first place he noticed how the heart hardened when it contracted, leading him to regard it as a muscle. Next he put down the dilation of the arteries when the heart contracted as due to the heart forcing in blood (previously thought an independent action on the part of the arteries themselves), and his observations of the slower heart-beats of animals led him to realize that the heart does not beat as a whole. He noticed the action of the valves between the heart's upper and lower chambers, and then realized that the flow of blood is continuously in one direction only. He now tackled the problem quantitatively, noting that though the heart only holds 2 ounces (60 gm) of blood, since it beats on average 72 times per minute, the left ventricle must pump 2 × 72 × 60 ounces of blood every hour into the main artery or aorta. This amounted to 8640 ounces or 540 lbs (250 kg) per hour, or three times the weight of an average man. Harvey was clear that the veins just could not supply this amount of blood in the time. He then asked himself whether, perhaps, the blood moved in a circle and, finally, repeated an experiment of his old professor Fabricius showing that the valves in the veins only permit the blood to travel one way. This, he felt, clinched his theory.

Illustrations page 410

Harvey's discovery was a brilliant piece of reasoning based on observation, but at the time he published his results, he still did not know how the ends of the arteries are linked up with those of the veins; it was Malpighi with his microscope who discovered this important link. Later, other anatomists made other related discoveries which helped to bring an understanding of the organs concerned with blood supply. Harvey's discovery was also applied, of course, to animals as well as humans. Marco Severino, professor of anatomy and surgery at Naples, used it in his *Democritean Animal Anatomy* of 1645, the first really comprehensive work on comparative anatomy, where he recognized similarities in structure between man, apes and mammals in general. This attitude was continued, for instance by Edward Tyson, who was the first anatomist in England to carry out anatomical dissection from a comparative point of view; the evidence he provided was useful, even though he concluded in his *Anatomy of the Male Pygmy* (1699) – useful for its description of the cranial nerves – that the pygmy was an immature chimpanzee.

In the eighteenth century a mechanical approach to Harveian circulation was taken by Stephen Hales who discussed blood pressure in man and animals, but the most outstanding figure of the century to apply the new knowledge in the field of anatomy, comparative and otherwise, was the Scottish surgeon John Hunter (1728-83). The founder of a more scientific attitude to surgery – before his time it had been something of a matter of trial and error – and one of the greatest practising surgeons of all time, he devised new surgical methods, among them a novel technique for dealing with local swelling in an artery – up to his time a method from the second century AD was still in use. But John Hunter's interests were very broad, and spread far outside the bounds of normal surgical practice, for he took a great interest in every type of living creature. He wrote on the electric organs of certain fishes, on the structure of whales, on air sacs in birds and bees, and studied the temperatures of a wide range of animals. And over the years he amassed a vast collection of some 14,000 animal specimens which, in 1795, twelve years after his death, was bought for the nation and passed to what became the Royal College of Surgeons for safe keeping. Hunter still exerted great influence not only in surgery, but also in the later development of the zoological sciences. Before, however, going on to mention the main trends of eighteenth-century zoology, there is one other aspect of medicine in that century which must be mentioned, and that is the practice of inoculation.

Inoculation in Western Europe had its beginnings in the advocacy of the technique by Lady Mary Wortley Montagu, wife of the British ambassador to Turkey. While at Istanbul, Lady Mary had learned of the practice and had decided to have her six-year-old son inoculated against the smallpox, and on her return to England in 1718, began a campaign to promote inoculation. She achieved some success; in 1721 the Prince of Wales (later George II) issued instructions that a few condemned criminals should be inoculated and, when this trial succeeded, the two young princesses Amelia and Caroline were treated, so that they both 'had the distemper favourably'. Yet in spite of the successes of this method of giving a patient a mild dose of fever to ward off a worse one later, inoculation was only practised spasmodically and was not examined closely until the 1780s. Then Edward Jenner, one of John Hunter's pupils, who had gone into medical practice in Berkeley, in Gloucestershire, decided to investigate the subject. In 1768 a Suffolk physician, Robert Sutton, had devised a slightly improved technique that had revived the desire for inoculation, and Jenner found himself asked to carry it out. In doing so he discovered that some patients seemed to be completely resistant to the disease, and on inquiry learned that they had contracted cowpox from the teats of cows during milking. He also discovered that it was widely believed among milkmaids and milkmen that contracting cowpox prevented one getting smallpox, though local medical opinion was uncertain. Nevertheless, a careful investigation led Jenner to the conclusion that the local belief was valid and in May 1796 he inoculated an eight-year-old boy with matter taken from a milkmaid's cowpox pustule. In July the boy was inoculated with smallpox but without effect. In 1796 Jenner published at his own expense a 75-page book *An Inquiry into the Causes and Effects of the Variolae Vaccine,* describing twenty-three detailed cases. The effect of the book was enormous; the practice of vaccination spread with astonishing speed, and when Jenner then found that the lymph taken from smallpox pustules could be dried and stored, the vaccine began to be sent over long distances. He named the matter producing cowpox a 'virus'.

The last great development in seventeenth- and eighteenth-century biological science was the question of classification and change of species. Classification was important because it coloured the naturalist's outlook on what it was he observed, and if the question of change or development of species was to be discussed, then the definition and classification of species was obviously of paramount importance. In the seventeenth century, the naturalist John Ray, who was born in Black Notley, Essex in 1627, and had for a time lectured at Cambridge, tried to devise a new classification system. Forced to leave Cambridge because he would not sign the English Church's Act of Uniformity, he was supported in his later work by his young Cambridge contemporary, the marine biologist, Francis Willughby.

In 1660 Ray produced a *Catalogue of Plants Growing in the Neighbourhood of Cambridge* and later, between 1686 and 1704, the year before his death, he produced his monumental *General Account of Plants*. It contained descriptions of some 17,000 items which were classified according to fruit, flower and leaf, and particular attention was paid to whether they were monocotyledons or dicotyledons. But Ray did not confine himself to plants; his contact with Willughby stimulated him to consider zoology and he wrote on animals (*Synopsis on Quadruped Animals and Serpents*, 1693), on insects (*Account of Insects*, 1710) and on birds and fish (*Synopsis on Birds and Fish*, 1713). Like most of his contemporaries Ray considered there was only one correct way of classifying animals, because it should echo the way they had been divinely designed according to the surroundings in which God had planned they should live. Nevertheless he was a careful and accurate observer and was well aware that not all animals were ideally adapted to their surroundings, so he suggested there must be a 'plastic force' in nature operating to change them.

Ray was never certain in his own mind that he had achieved the correct classification. He had divided animals into those 'with blood' (a group that included all we should now call vertebrates – mammals, birds, fishes) and those 'without blood' (the invertebrates, such as insects and shellfish). On this basis he had made certain subdivisions – invertebrates were classed according to size, vertebrates by the structure of the heart, differentiating those with a pair of chambers from those like fish with simpler hearts. Yet ingenious though Ray's methods were, it was not these that came to be adopted by biologists; they preferred the system devised by the Swedish naturalist Carl von Linné, now better known as Linnaeus (1707–78). He was to produce three classification systems, one covering plants, another animals and a third minerals. His father had little money and Linnaeus had to be very frugal while studying at the universities of Lund and Uppsala. At the latter he spent much time in the botanical garden and it was at Uppsala that he came across a book by the French botanist Sebastian

Illustration page 411

Vaillant in which the sexuality of plants was mentioned; this greatly impressed him and with a good library and a botanical garden to hand, he began to attempt a major classification of plants using their sex organs as his basis. In 1732 Linnaeus made a trip to Lapland, then went to Harderwijk in Holland to obtain his degree as a doctor of medicine, after which he moved on to Leiden University where he stayed for three years. While there his literary output was astonishing; it included one of his most important works *The Genera of Plants*, which appeared in 1737 with short descriptions of 935 plant genera, though even more significant was his *System of Nature* that had come out in 1735 and was the first of his books to be published in Holland. In the latter Linnaeus set out his ideas on the way a classification should be made. It consisted only of eight large sheets of tables, but was to exert an immense influence, especially in the enlarged and improved editions that appeared as the years passed. The tenth and last edition appeared in 1758, and is notable for the fact that it was in this that Linnaeus announced his binomial or two-name system for plant and animal classification, one name giving the genus or common characteristic, the other the species. (Thus, the dog family is given the genus name *Canis*; the wolf – *Canis lupus* – is one species of this genus, and the domestic dog – *Canis familiaris* – is another). For animals he also gave the common name where appropriate. This was not only biologically sound, but also very useful, so much so that it is still current today. The method for the classification of animals followed Ray in some respects; Linnaeus too picked specific organs as classification criteria; mammals were sorted according to teeth, birds by their beaks, fishes by their fins and insects by their wings, while there were also separate classes for reptiles and 'worms'. In the end Linnaeus managed to include 5,897 species. His classification of minerals attached great importance to crystal structure but he had no real concern for chemical composition, and of all his work this had the least effect on subsequent research.

A man of great charm, popular with students and university staff alike – he spent most of his time, from 1741, to his death in 1778, at Uppsala – Linnaeus was always sure in his own mind that species were fixed. He therefore asked himself time and again whether his classification system was the 'true' one, was it the way God had really constructed the vegetable and animal kingdoms? For there is no doubt that he believed that in his science he was following God's handiwork. In the introduction to later editions of the *System of Nature* he wrote 'I saw the infinite, all-knowing and all-powerful God from behind as He went away and I grew dizzy. I followed His footsteps over nature's fields and saw everywhere an eternal wisdom and power, an inscrutable perfection'.

Not every biologist took Linnaeus's attitude, and chief among the rebels were some of those members of the Enlightenment in France. The Enlightenment is the name given to the movement which looked on the world with a new rationalism that to a greater or lesser extent

For the Jesuit Matteo Ricci in China, see page 125.

severed any connection between the natural world and God's continuing concern with it. It began in the seventeenth century when people became more aware of the religion and culture of non-Christian nations, due partly to the reports from Jesuit missionaries in Egypt, Siam and especially China where it was recognized there were 'exemplary and virtuous men' though 'no real religion at all'. The concept arose of the 'noble savage' who should be treated with respect and not as an uncouth barbarian, and with this was coupled the idea that all men had some kind of religious feeling implanted by nature: a natural religion rather than a divinely revealed one. Later on, this view was coupled with one that looked on God more as an architect and a mechanic than a heavenly Father, especially after the Newtonian synthesis of the *Principia*. He had created a universe which, once constructed, must obey the laws of natural behaviour. Miracles became unthinkable, biblical prophecy unacceptable, and Holy Writ itself was called into question. After all, if the Earth moved in space, as it obviously did, then this was something on which Scripture was in error, and if it were wrong in this, where else might it be unreliable?

In England the belief in 'natural religion' led to a movement known as Deism. In the 1620s, while the Dutch jurist Hugo Grotius was drawing up his case for natural law in international relations, Edward Herbert, diplomat, historian, metaphysical poet and philosopher, brother of the devotional poet George Herbert, published in Paris his book *On Truth (De Veritate)*. Herbert claimed that 'instructed reason' was the surest guide to truth and though he gave five religious ideas that are God-given, it is significant that revelation was not one of these. If Deism was launched with Herbert's book, its swan-song came in the 1790s with the *Age of Reason,* written by the American Thomas Paine while he was in prison under Robespierre. Deism was, however, never an organized movement. It was an outlook, and one which flourished very strongly in France because of the disillusionment and discontent there with Louis XIV's reign. One of its great advocates was Voltaire, who came across English Deism when he visited the country in 1726; already committed to an anti-Christian rationalism he returned in 1728 to write his *Philosophical Letters*, which expounded Deist ideas, and later in 1760, his *Philosophical Dictionary* which turned into what one commentator has termed a 'compendium of malice' against Christianity. It was soon after this that the political theorist and philosopher Jean Jacques Rousseau also supported Deism in his novel *Émile* (1762) and his book *The Social Contract* (also 1762).

Another analogous outlook was what has come to be called scepticism. This seems to have been set off in France in 1697 when Pierre Bayle published his *Historical and Critical Dictionary*, a volume of critical and sometimes spicy biographies aimed at demolishing the 'vices of religion'. Thus, with an Old Testament character like King David, Bayle would show how his life was morally offensive and then point out that in spite of it, the biblical account refers to the king as a man after God's heart. The book had a huge success.

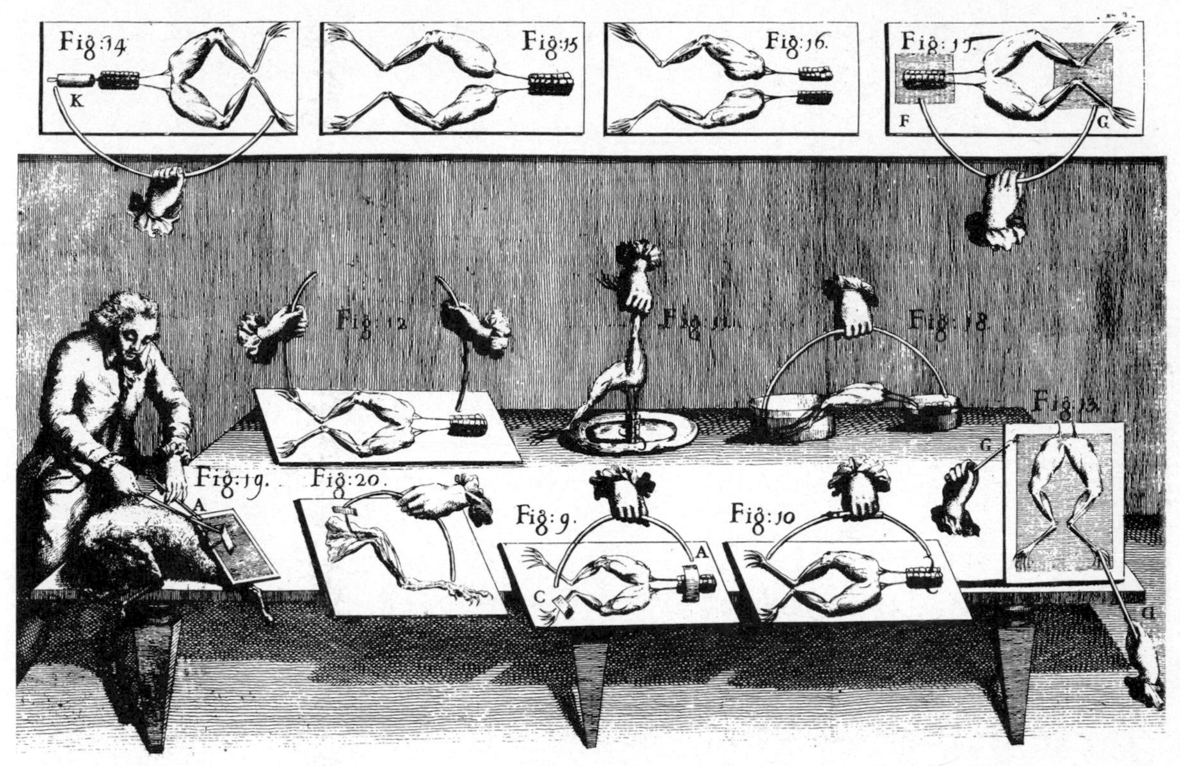

Above Experiments on 'animal electricity' as published by Luigi Galvani (1737–98) in *De viribus electricitatis in motu musculari commentarius* of 1791.

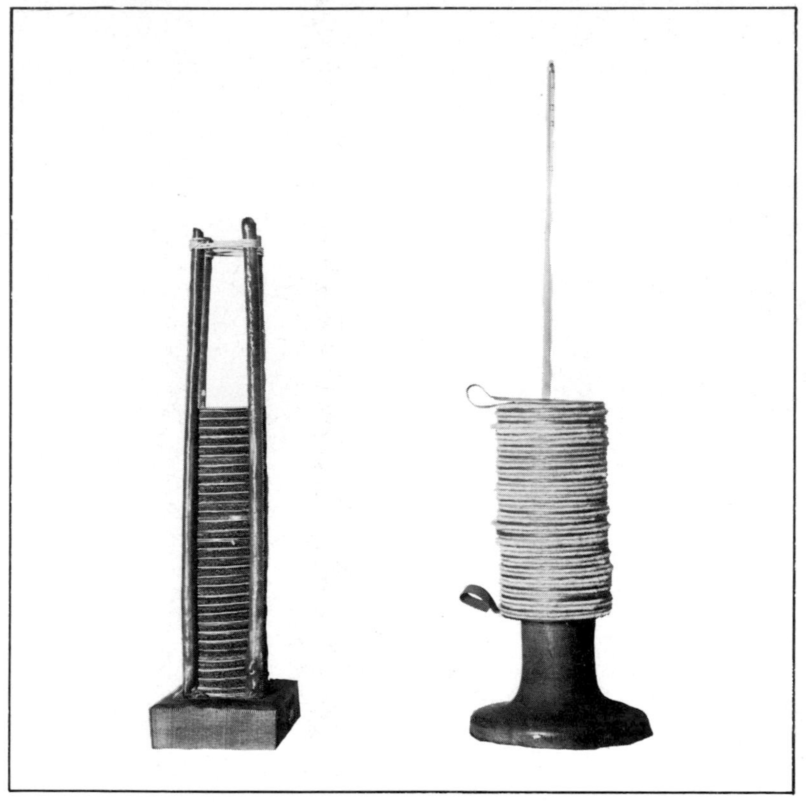

Left Two 'Voltaic piles'. The discovery by Alessandro Volta (1745–1827) that electricity could be generated continuously by these 'piles' of, alternately, copper, zinc and paste-board was an important step forward in the study of electricity. Museo Nazionale della Scienza e della Tecnica "Leonardo da Vinci", Milan.

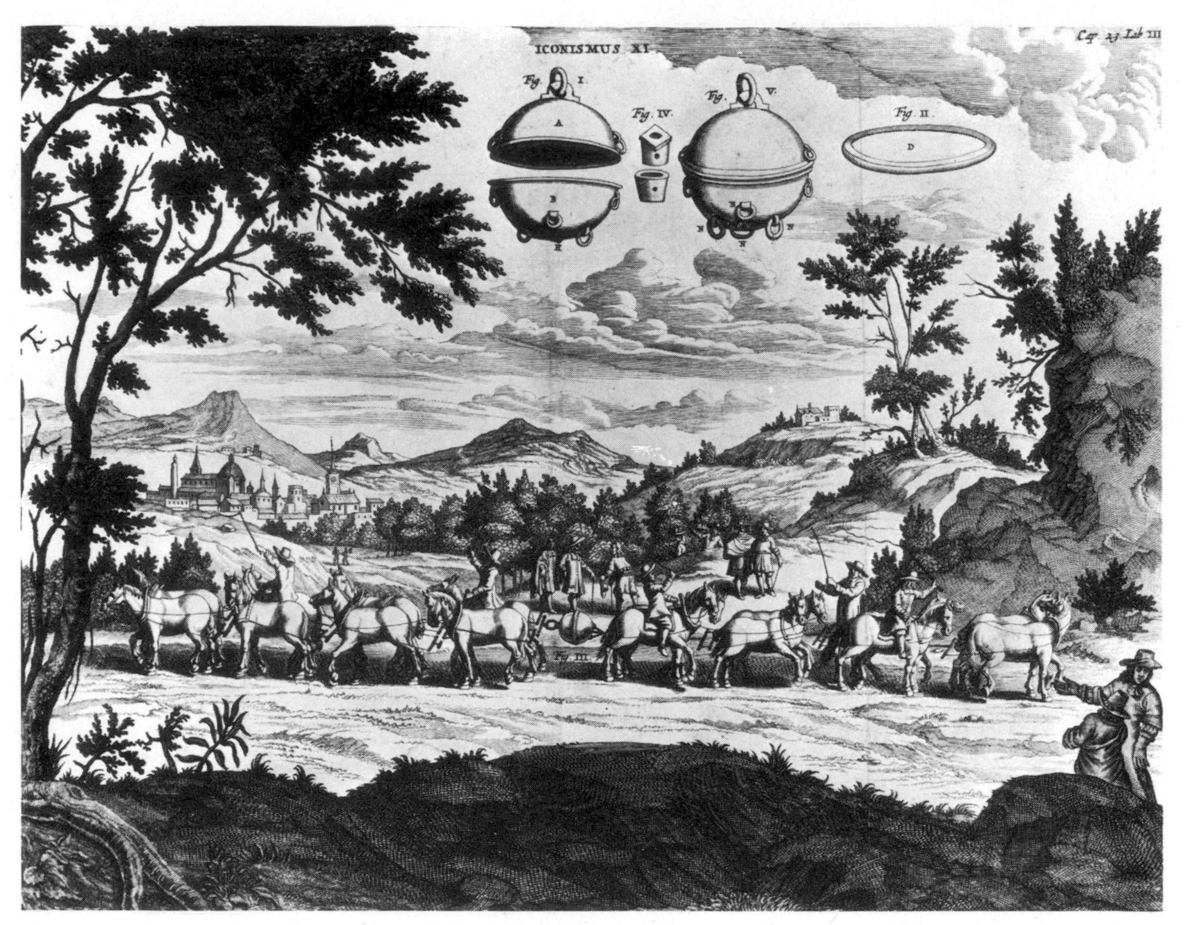

Above The experiment by Otto von Guericke to demonstrate the strength of a vacuum, using two 'Magdeburg' hemispheres of brass from which air had been exhausted. From his *New Experiments*, 1672.

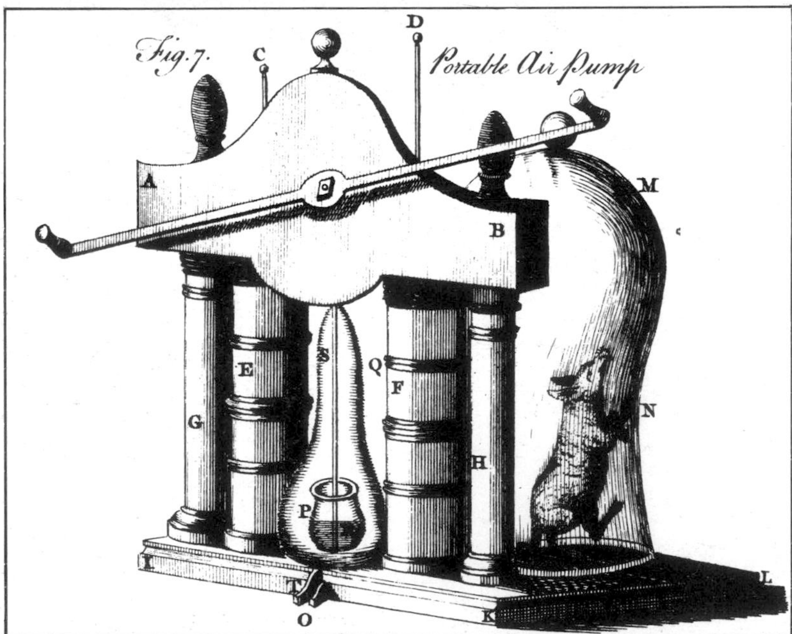

Right A mid-18th century experiment of evacuating a bell-jar by an air pump to investigate the effect of a vacuum on an animal.

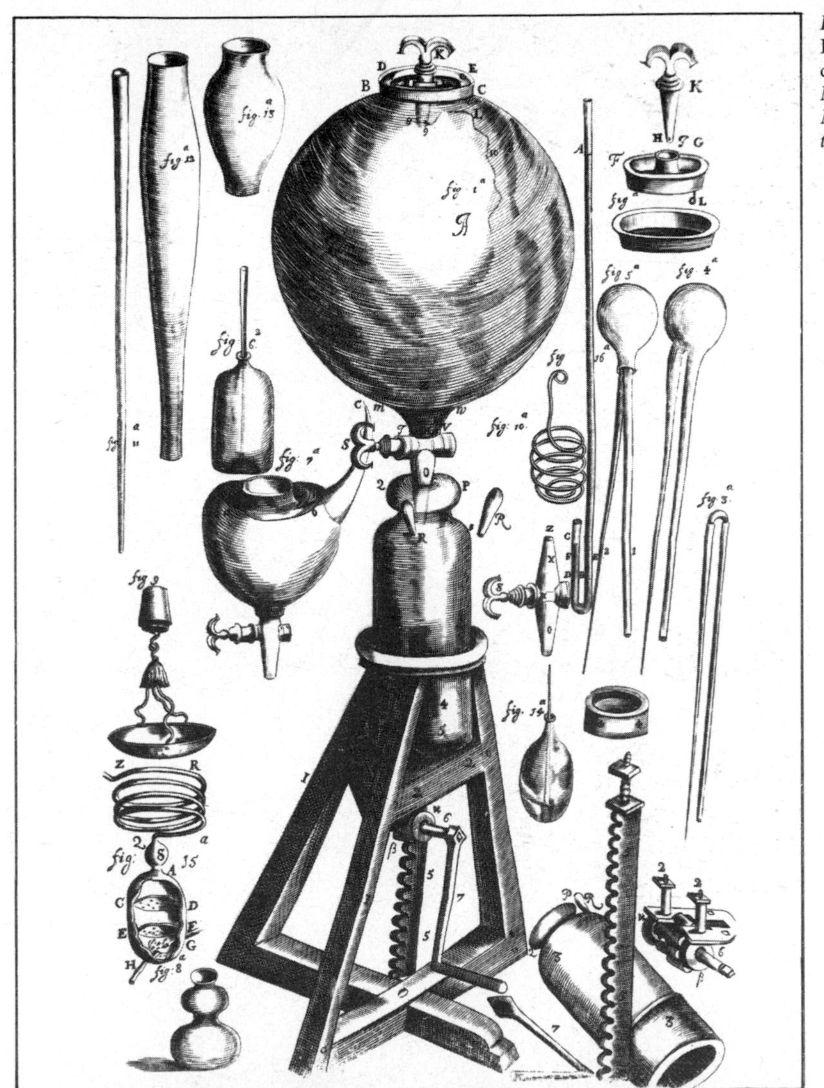

Left Air pump built by Robert Boyle and Robert Hooke for creating a vacuum. From Boyle's *New Experiments Physico-Mechanicall, Touching the Spring of the Air*, 1660.

Left A late 17th century chemical laboratory.

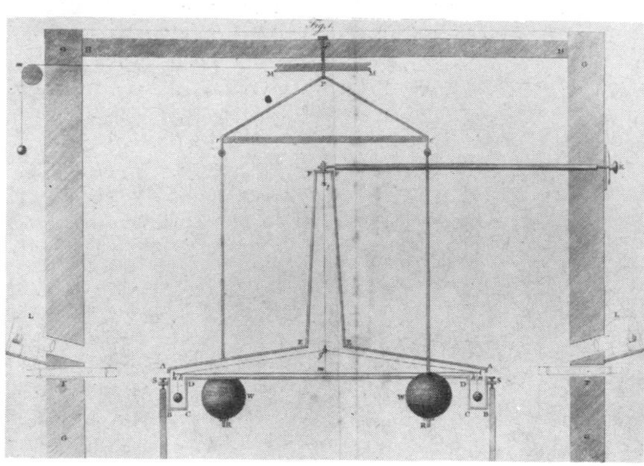

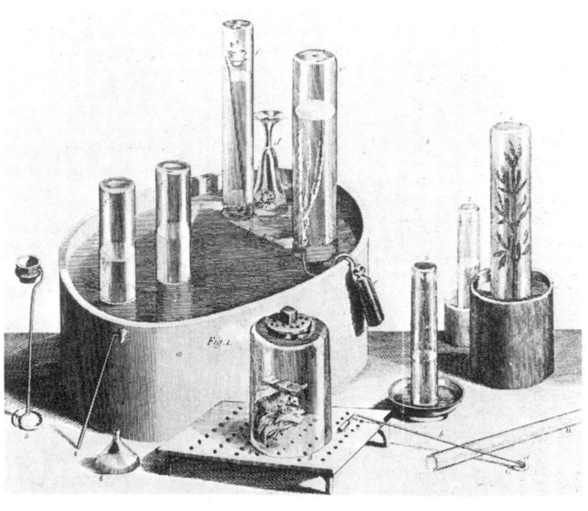

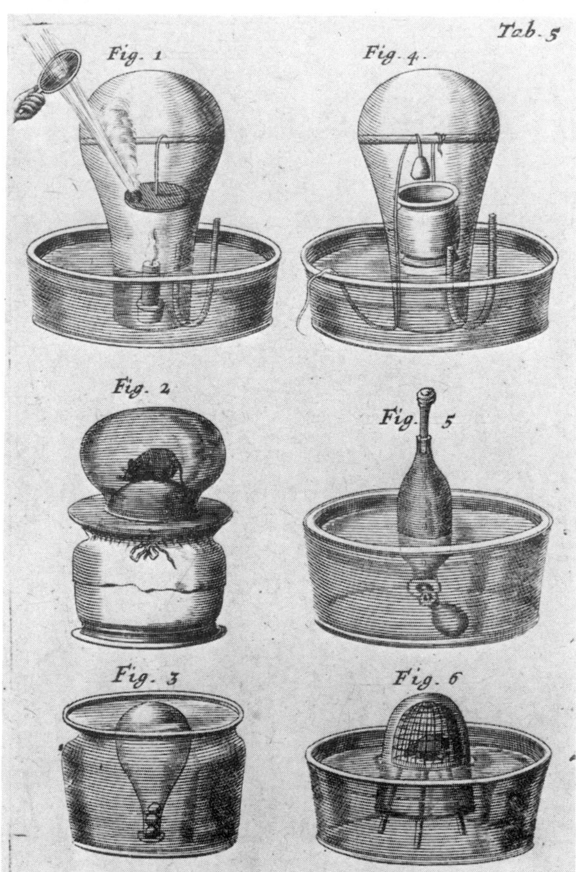
Tab. 5
Fig. 1
Fig. 4.
Fig. 2
Fig. 5
Fig. 3
Fig. 6

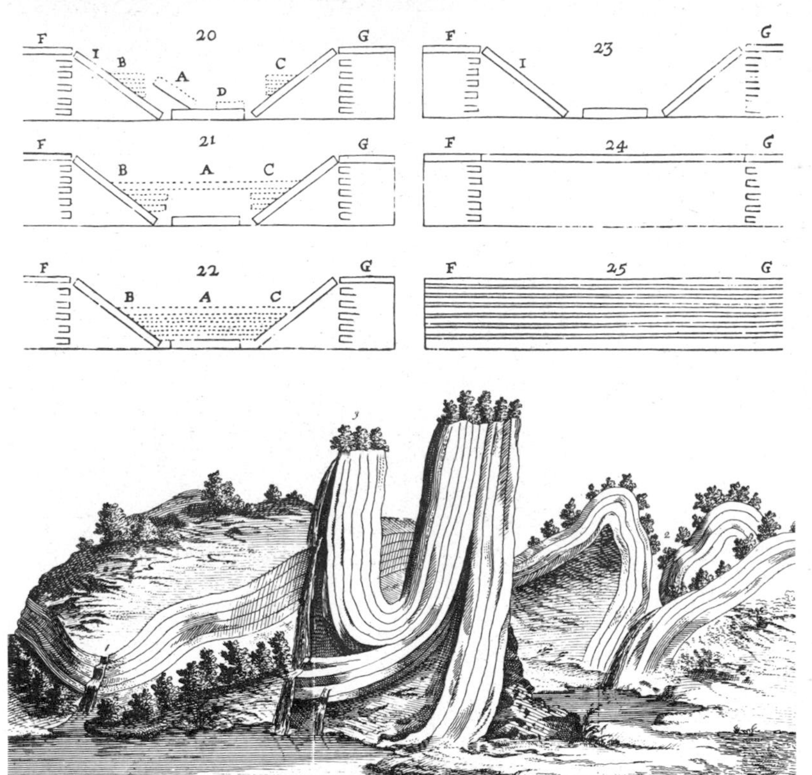

Above left Engraving of Antoine Lavoisier (1743–94). *Above*, his apparatus for the investigation of oxygen. The trough (right) contains mercury as does the retort (left). On prolonged heating some red oxide of mercury was found in retort, while the volume of air in the bell jar (right) was reduced.

Centre left Drawing from Niels Stensen's *Prodromus* . . . 1671, showing schematically how rock strata can fold.

Left The formation of surface geological features by the bending and folding of strata. From *Nature Display'd*, 1740.

Opposite centre left Henry Cavendish (1731–1810) in 1798 devised this apparatus for determining the density of the Earth.

Opposite below left Joseph Priestley's pneumatic trough for isolating 'airs'. From his *Experiments and Observations on Different Kinds of Air*, 1774–77.

Opposite below right John Mayow's apparatus for collecting gases over water. From his *Tractatus quinque medico-physico*, 1674.

Opposite top An 18th-century chemical laboratory, as shown in W. Lewis *Commercium philosophio-technicum*, 1763.

Right Fossil fish, from Peter Wolfant, *Historiae naturalis . . .* 1719.

Right Robert Hooke's microscope (bottom) together with other apparatus designed by him, including a barometer (upper left) and lens grinding machine (upper right). From his *Micrographia*, 1665.

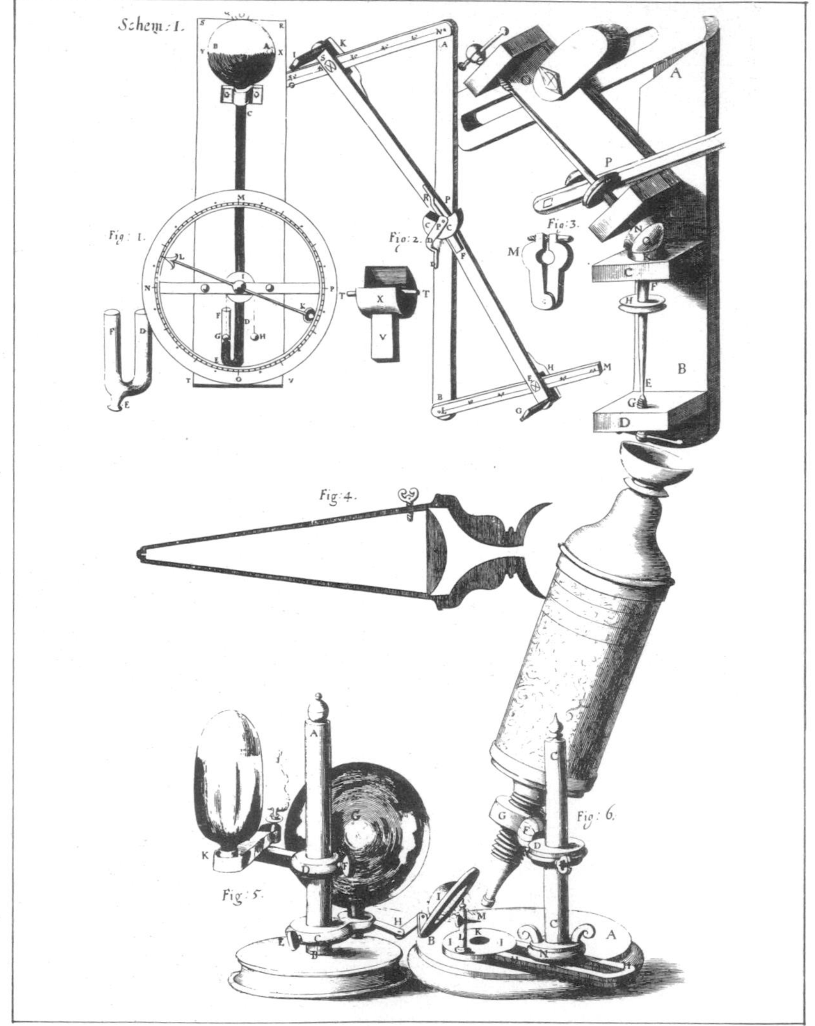

Opposite Hooke's view of a nettle seen through a microscope, together with a piece of wild oat and a drawing of his hygrometer. From the *Micrographia*.

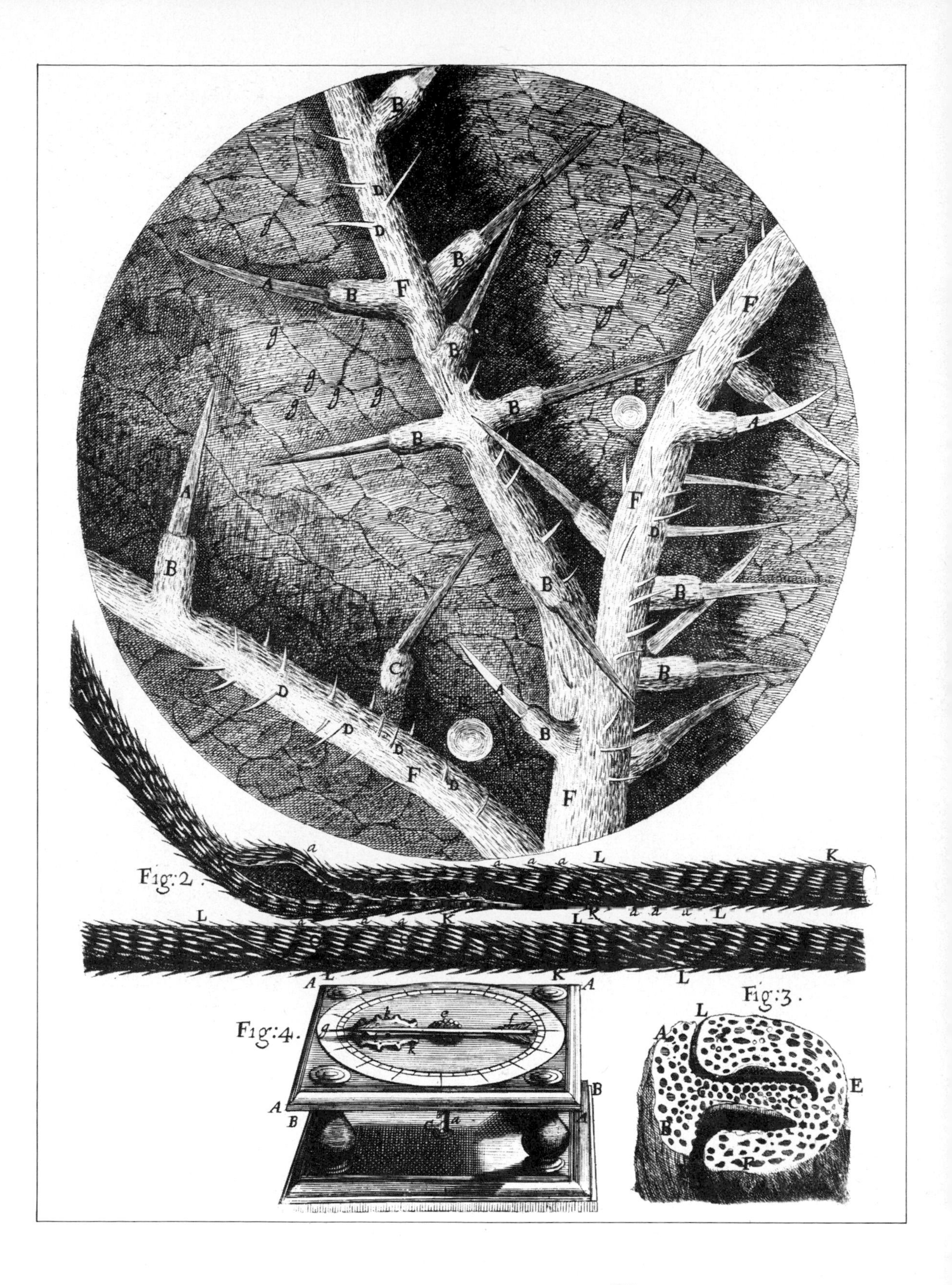
Fig:2.
Fig:3.
Fig:4.

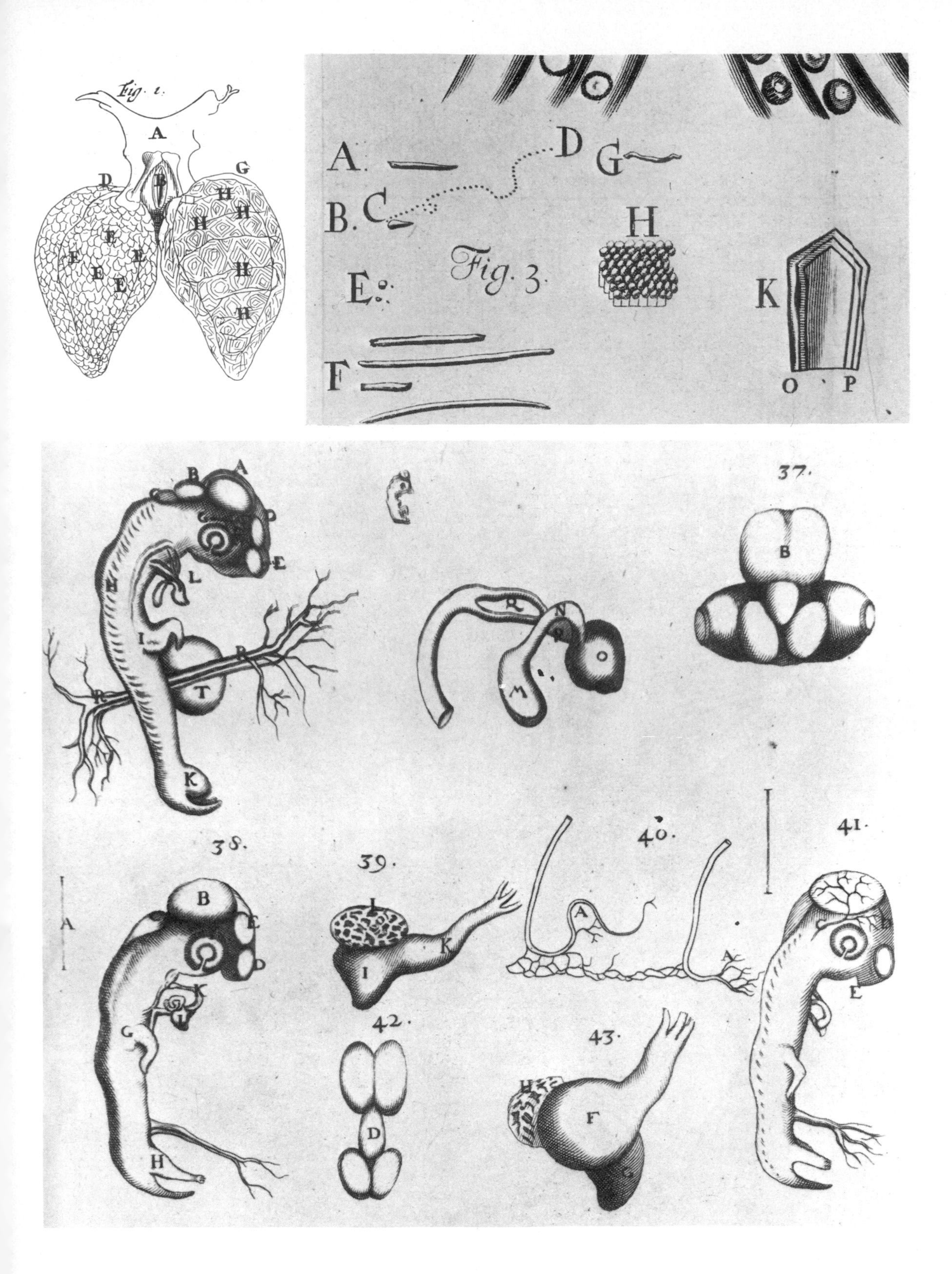
Fig. 1.
A
B
D
G
E
H
A.
D
G
B.
C
H
Fig. 3.
E
K
F
O
P
37.
B
A
B
E
L
I
T
K
N
O
M
38.
39.
40.
41.
B
E
D
K
I
G
H
L
K
I
A
42.
D
43.
F
E

Opposite above left Drawing of the blood capillaries in a frog's lungs by Marcello Malpighi (1628–94). He was the first actually to observe the circulation of blood in the capillaries. From his *Opera Omnia*, 1687.

Opposite above right The first representation of bacteria, taken by Leeuwenhoek from scraping between the teeth. Published in the Royal Society's *Philosophical Transactions*, 1684.

Opposite below Drawings by Malpighi of the development of the chick in an egg. From an appendix to his *Anatomy of Plants*, 1675.

Left Copperplate engraving of Anton van Leeuwenhoek (1632–1723).

Below Leeuwenhoek's microscope, with attachments for holding a test-tube in which a living specimen could be inserted. From his *Arcana naturae delecta*, 1722.

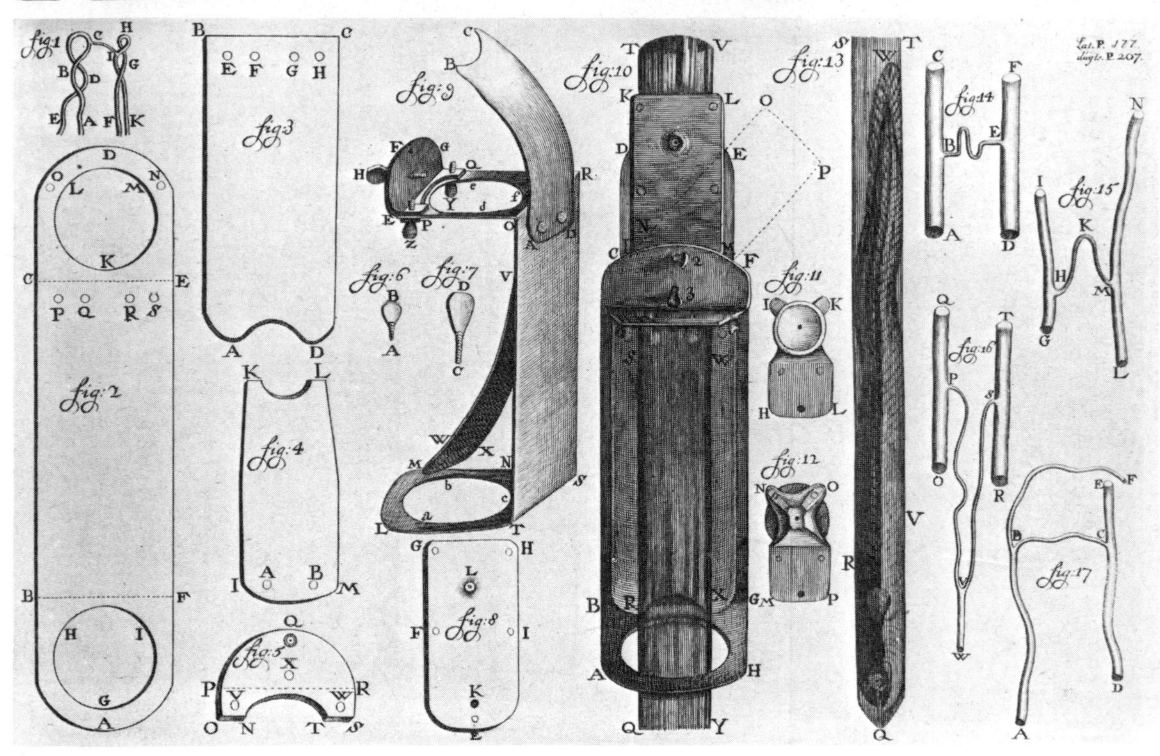

Right Portrait of William Harvey (1568–1657), painted in about 1627. National Portrait Gallery, London.

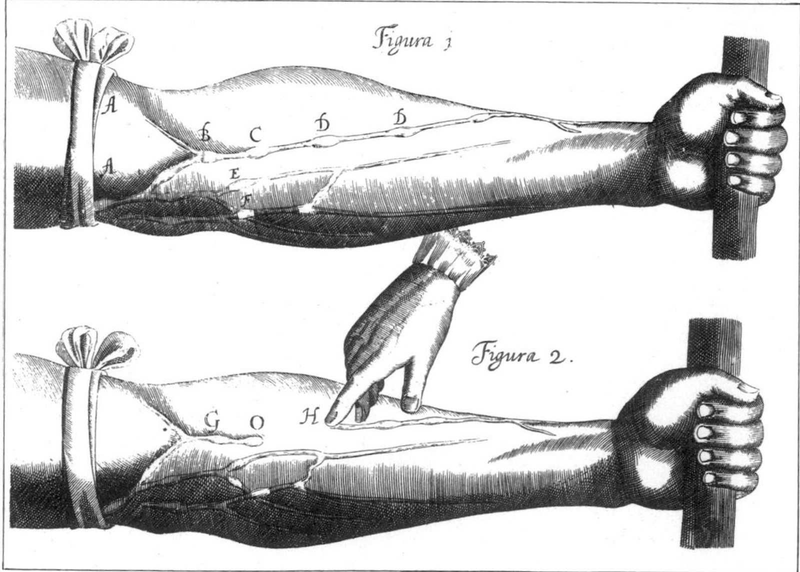

Right Harvey's demonstration of valves in the veins from his book on the circulation of the blood, *De Motu Cordis*, 1628.

Left Carl von Linné or Linnaeus (1707–78) in Lapland dress; he made a botanical visit to that country in 1732.

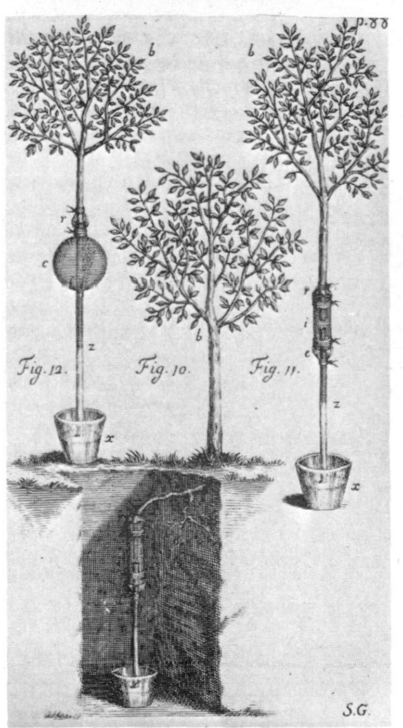

Far left Methods of investigating sap pressure and movement of sap, devised by Stephen Hales. From his *Vegetable Staticks*, 1727.

Left Santorio Santorio (1561–1636) in his 'weighing chair', in which he could eat and sleep. He used it for the first experiments in metabolism. From his *Ars de statica medicina*, 1711.

1
2
3
4

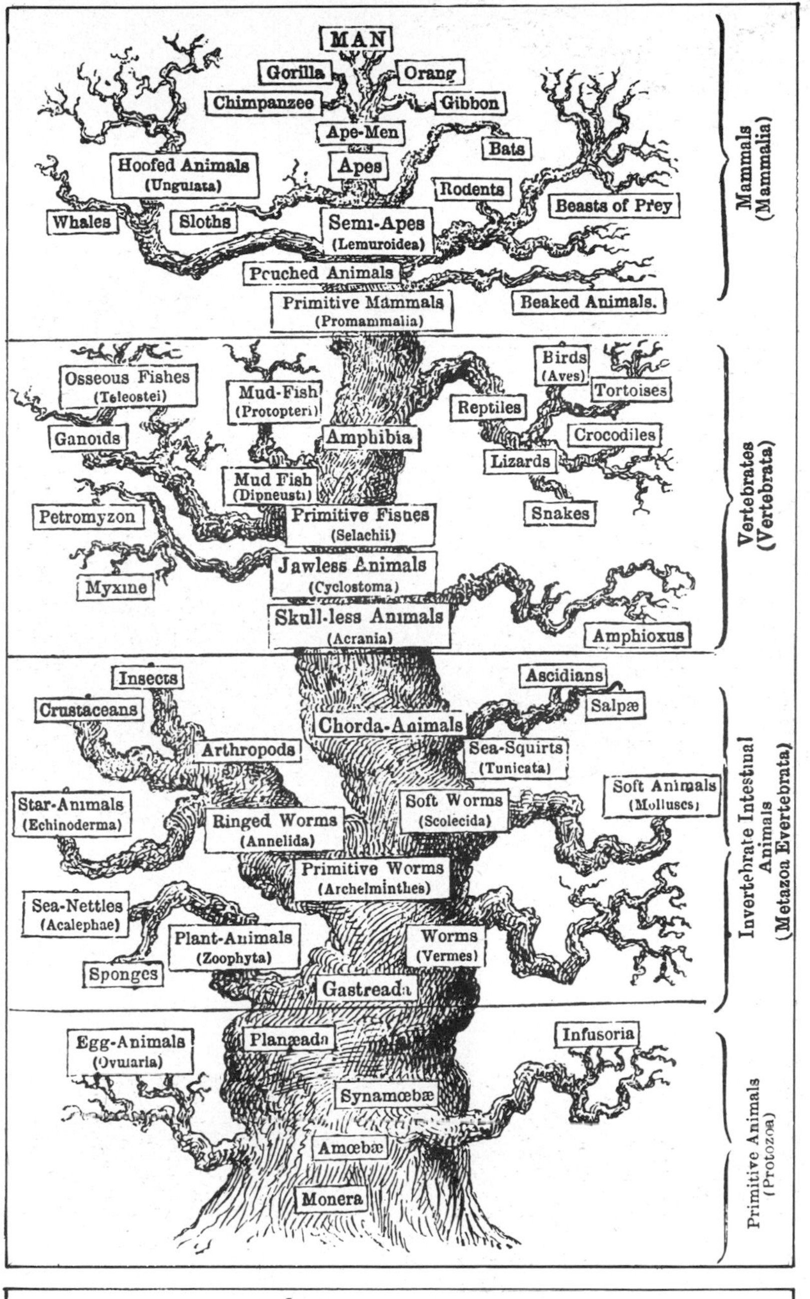

Opposite above The botanical and zoological Jardin du Roi in Paris where Jean Baptiste de Lamarck (1744–1827) worked after 1788.

Opposite below left Charles Darwin (1809–82) sketched in 1878. National Portrait Gallery, London.

Opposite below right Differences in the beak formation of finches observed in the Galapagos Islands; these differences seemed to Darwin to have evolved to fit the finches better for their different habitats. From his *Voyage of HMS Beagle*, 1839.

Left The pedigree of man as envisaged by Ernst Haeckel (1834–1919) from his *Evolution of Man*, 1879. Haeckel was the first advocate of Darwinism in Germany and was the first person to draw up such an 'evolutionary tree'.

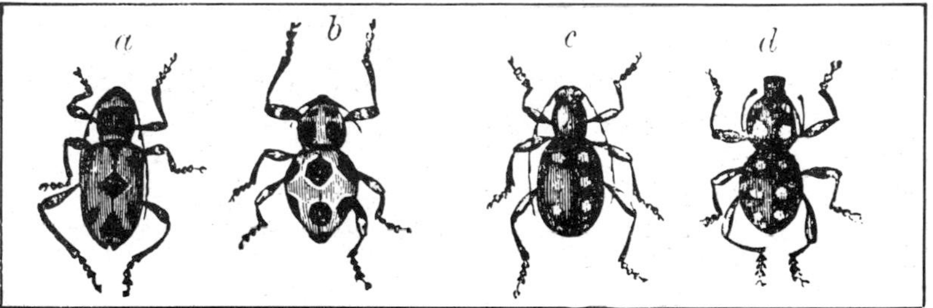

Left Protective coloration in insects caused by the evolutionary acquisition of an external appearance deterring predators, from *Darwinism*, 1889, by Alfred Russell Wallace (1823–1913). In 1858 Wallace proposed the concept of the 'survival of the fittest' to account for the origin of species, thus prompting Darwin to publish his own theories. Wallace's later work found room for religion and spiritualism within the theory of natural selection.

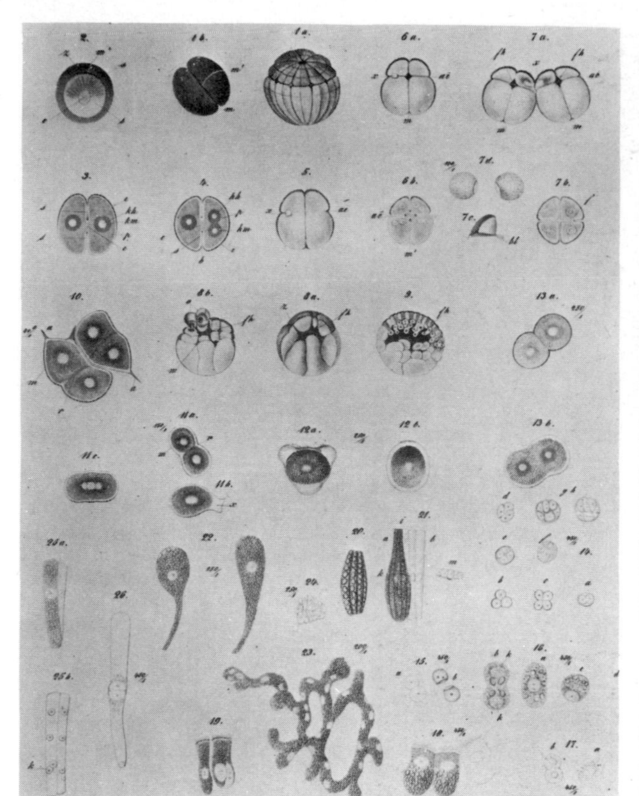

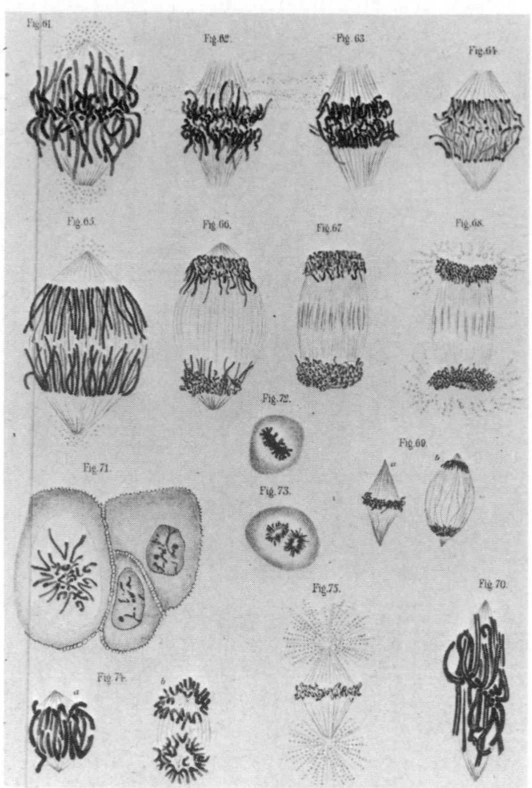

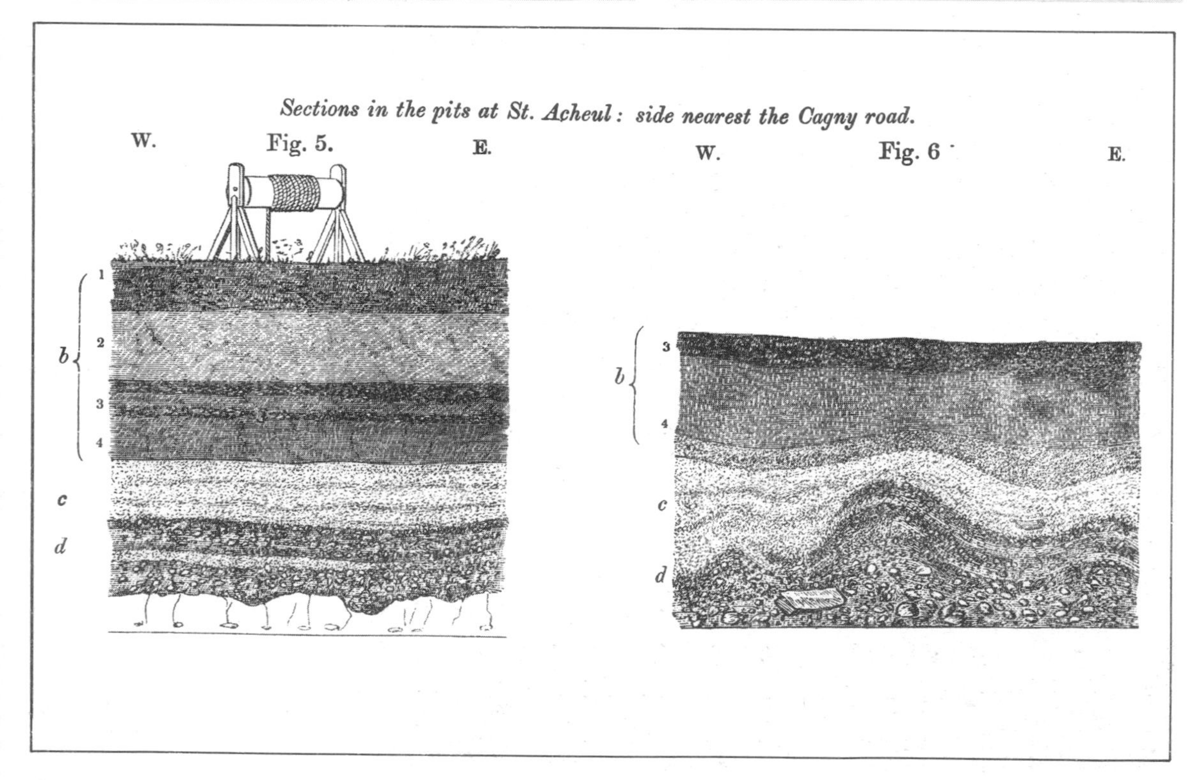

Sections in the pits at St. Acheul: side nearest the Cagny road.

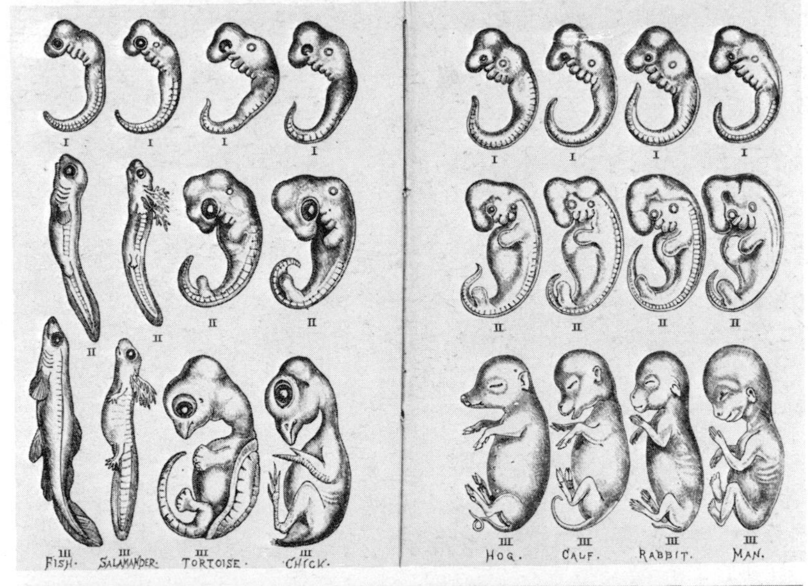

Left Comparative development of the embryos of a number of different species, from *Darwin and After Darwin*, by the British friend of Darwin, George Romanes (1848–94).

Opposite above right The behaviour of chromosomes during cell division in the lily. From Walther Fleming, *Zellsubstanz, Kern und Zelltheilung*, 1882.

Opposite above left Stages in cell development. From Robert Remak (1815–92), *Untersuchungen*, 1855.

Opposite below Cross-section of pits at St Acheul, northern France, where remains of extinct mammals and flint instruments were found in the late 1850s. This find enabled a technique to be developed of dating fossil remains by relating them to the geological strata in which they were found.

Left Louis Pasteur (1822–85) in his laboratory. Musée du Louvre, Paris.

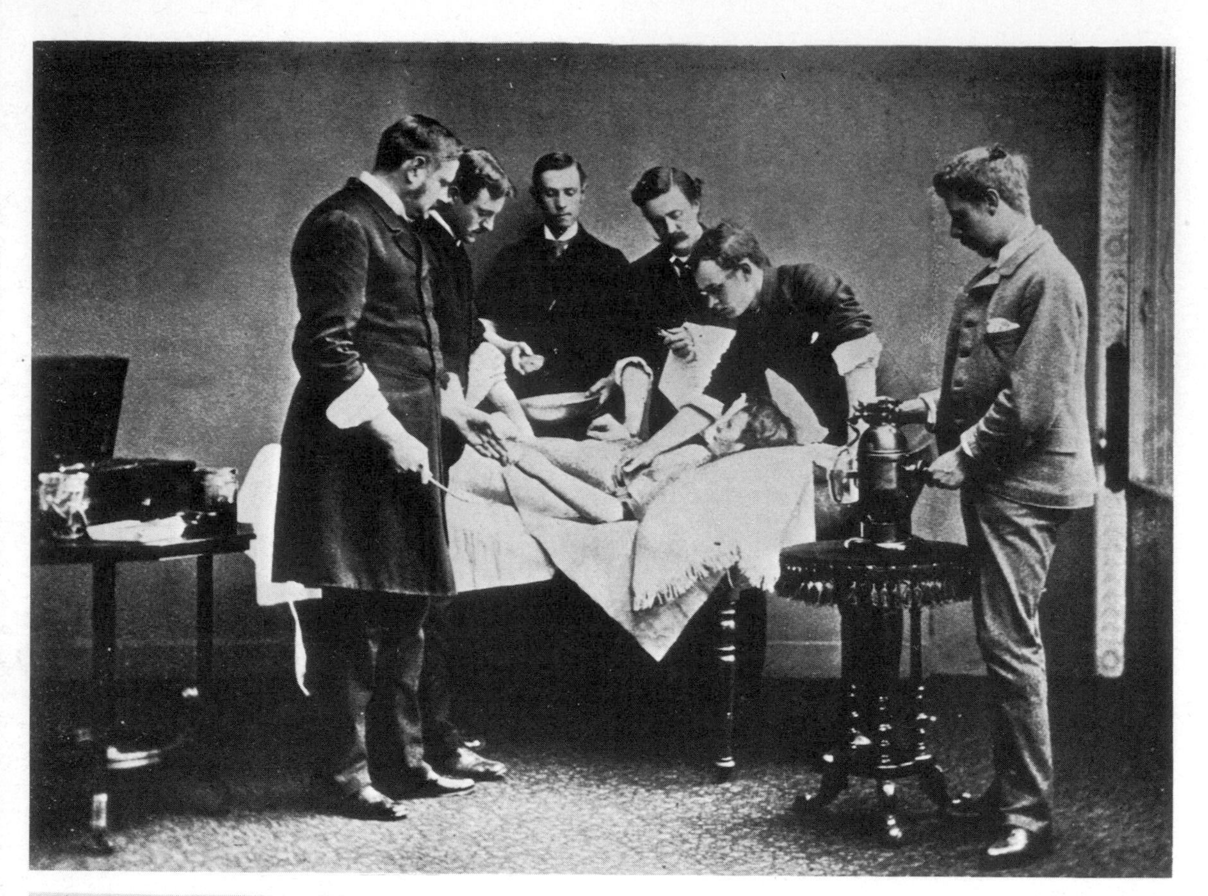

Drawn & Etch'd by J. Stephenson.
John Dalton,
D.C.L. LL.D. F.R.S. L.&E. &c &c,
President of the
Lit. & Phil. Soc. Manch.

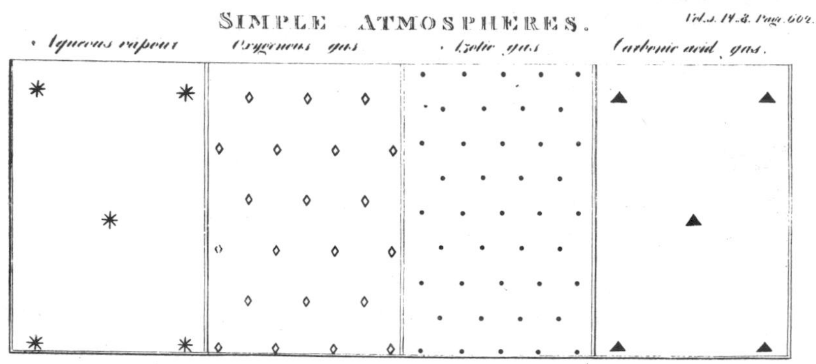

SIMPLE ATMOSPHERES.
Aqueous vapour
Oxygenous gas
Azotic gas
Carbonic acid gas.

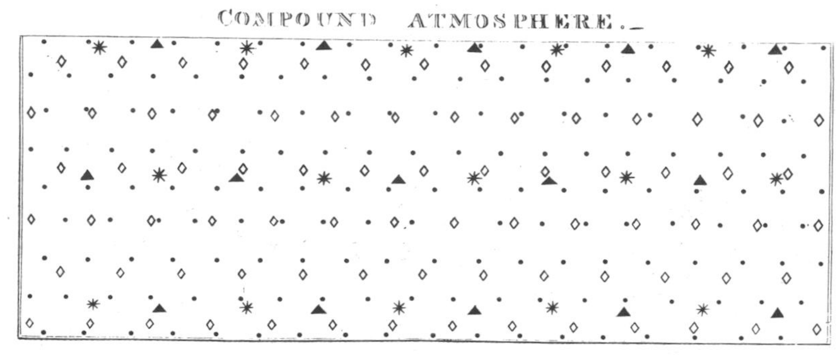

COMPOUND ATMOSPHERE.

Almost half a century earlier the philosopher John Locke had applied the methods of Galileo and Newton to the study of man in *An Essay on Human Understanding* (1690). Here Locke had set out to find the scope and limitations of man's knowledge and had suggested that all knowledge arises from experience, from sensations sorted out by reflection; there are no innate ideas, there is no divine inspiration or inherited gift of original sin and all children are born with a mentally 'clean slate'. It is, Locke said, environment that makes man what he is; improve the environment and you improve man.

The late seventeenth century and the eighteenth were therefore times of philosophical change, alterations in man's outlook about himself and an attack on what had for long been the cherished beliefs of Western Christendom. In France this went further than elsewhere, and atheism was openly proposed. The physician Julien La Mettrie wrote a book *The Man-Machine* (1747) epitomizing a totally materialistic outlook; man was reducible to nothing more than matter in motion and La Mettrie even went on to say 'We are no more committing a crime when we obey our primitive instincts than the Nile is committing a crime with its floods, or the sea with its ravages'. This attitude was taken up by the *philosophe* Denis Diderot who, with Jean d'Alembert (mathematician and permanent secretary to the French Academy) edited the famous *Encyclopaedia or Classified Dictionary of Sciences, Arts and Trades*, a compendium of the new rationalist thinking which contained severe criticisms of religion and of the Establishment. The twenty-eight volumes of the work appeared between 1751 to 1772, and a further five in 1776 and 1777. Its influence was immense.

It was in this mêlée of atheistic, sceptical and deistic views, in the revolutionary new thinking of the Enlightenment, that Linnaeus's *System of Nature* was examined by a number of influential French biologists. Not for them a path following in the footsteps of God, theirs was a journey of pure materialism. Notable among them was Pierre Maupertuis, who in the 1740s and 1750s suggested that biological explanations must be in terms of particles in motion, and that it was such particles from both male and female animals, including man, that determined what the offspring would be like. For him there was no fixity of species; species as observed were not examples of divine perfection, just the outcome of a mechanistic nature. His views were echoed by Diderot who suggested only one change, he preferred 'filaments' rather than particles due, perhaps, to the drawings of spermatozoa by Leeuwenhoek that he had seen. But perhaps the most advanced of the French biologists, whose work lay wholly in the eighteenth century, was Georges Leclerc, the Comte de Buffon. Born in Montbard, near Dijon in 1707, he was a rich man who lived in Paris but spent every summer on his estate Montbard, where he observed Nature and read widely on natural history. Gradually he amassed a vast amount of material, which he collated and published between 1749 and his death in 1788 in no less than 36 volumes, under

Opposite above An early photograph of an operation in progress, with the assistant on the right using Joseph Lister's carbolic spray invented in 1869 to prevent bacterial infection.

Opposite below left Etching of John Dalton (1766–1844), the English chemist.

Opposite below right Dalton's diagram on the mixing of separate gases in the atmosphere, 1802.

the simple title of *National History, General and Particular.* A further eight volumes appeared posthumously. This vast work contained his ideas on the history of the Earth, in which he accepted fossils as evidence not only of animal life but also of extinct species, though its main importance lay in his idea of evolution or change of species with time. Indeed, he even went so far as to suggest that there might be one common ancestor for man, apes and quadrupeds. Buffon also pointed out that not only men but some animals possessed organs that seemed to serve no use at all, and these he felt must indicate that some kind of change had taken place.

With Buffon we come very close to a theory of evolution, but not close enough to allow one to be formulated in any detail. More work was needed; this was to be done by Lamarck and others, and finally, of course, by Charles Darwin. But an account of this lies more properly in a discussion of nineteenth-century science.

Chapter Nine

Science in the Nineteenth Century

The nineteenth century was to see great developments in every branch of science. New specialist scientific societies, supplementing the established scientific academies, were symptomatic of the growing degree of specialization that increased knowledge and more elaborate techinques were making necessary. Moreover, science began to put on a still more public face as its practical consequences began to be evident in everyday life. Probably the most notable was the development of steam power which, as we have seen, was due to James Watt applying the theoretical work of Joseph Black at Glasgow, while at the end of the nineteenth century the new technique of electrical engineering was to be generated from the pioneering work of Michael Faraday. And it was during the nineteenth century, in 1840 at Glasgow, that the word 'scientist' was coined, appropriately enough by the British Association for the Advancement of Science, which had been founded nine years earlier to provide an annual forum where scientists could meet to discuss their work openly in a way that could be generally understood.

Illustrations pages 364, 453

The founding of the British Association was symptomatic of the growing realization of scientists of the significance of their work and their need to communicate it to the public. In the United States the American Association for the Advancement of Science was set up in 1848, and elsewhere the need for some popularization was appreciated, while it was in the United States that popular scientific periodicals devoted mainly, though not exclusively, to applied science, first appeared; soon, though, they were to be found in other countries. In one sense this was not new; in the eighteenth century some more literary periodicals had included contributions on scientific topics, it was just that in the nineteenth century this movement developed at an accelerated pace and became more specialized. One of its effects was to increase an already wide interest in science. Certainly popular and instructive scientific lectures and popular scientific books were not new; they had been a feature of late eighteenth century life, but now, from the mid-1830s onwards, the pace quickened. For their part the public did not fail to respond, sometimes with unexpected enthusiasm. Thus when Charles Darwin's *Origin of Species* appeared late in November 1859, the entire printing was sold out on the first day of publication. But, of course, this was a subject that the reading public saw must have the most profound philosophical and religious implications besides its significance for the scientific community.

Evolution and the Age of the Earth

One thing that became clear from eighteenth-century biology was that there were those who were beginning to question the whole principle of the fixity of species. Buffon, in particular, had felt that there might be some form of common descent, at least among mammals, but how could one become sure? One way to try to get closer to the answer seemed to be to study comparative anatomy and see if it really were possible to derive one animal from another by modification. Among those who discussed this was the German philosopher Immanuel Kant, who thought that perhaps all animals might one day be traced back to some early primitive type of creature. But if this were so, then there was another question that had to be asnwered, and that was to find what it was that caused a change. Why was a child like its parents, what was it that caused an animal to propagate its species, and what could it possibly be that might cause a species to change with time? There was no very clear answer, but by the end of the eighteenth century and the beginning of the nineteenth at least the problem was well and truly aired by two very different men, Erasmus Darwin and Lamarck.

Erasmus Darwin, physician, poet and naturalist, was born near Nottingham in 1731, youngest son of a retired barrister of independent means. He went up to Cambridge University in 1755 where he studied classics, mathematics and medicine. He also spent some time at the medical school in Edinburgh and attended some of John Hunter's anatomy lectures in London, but it was from Cambridge that he obtained his medical degree. In 1756 Darwin started in medical practice in Nottingham, and then later moved to Lichfield where he gained a great reputation as a physician. His final years were spent at Derby.

It was while at Lichfield that he came to know the Birmingham manufacturer Matthew Boulton, at whose Soho Works James Watt developed the steam engine which, between 1776 and 1800, was to have an immense impact as an efficient and economical power source for pumping in mines and elsewhere and in driving all kinds of machinery. Boulton was a man with wide scientific interests, and with Erasmus Darwin and a Dr William Small established a society for discussing scientific topics. This met on nights of full Moon, because only on such nights could its members find their way along unlit roads, and so became known as the Lunar Society. Watt and Joseph Priestley were also members, as well as about eight others, of what was a rather radically minded society, which soon became well-known; scientists visiting the area were glad to accept invitations to attend. It was a rich and brilliant company which even gave financial support to Priestley so that he could continue his scientific research.

Illustration page 364

For Priestley, see page 387.

The outlook of the members of the Lunar Society echoed the views of the age, in which the idea of progress had begun to grow. The air was full of scientific developments and their growing applications, and Erasmus Darwin was caught up in this just as much as his friends; it was doubtless a factor that exerted some influence when he began to formulate ideas about the evolution of animal species, ideas which he published in two volumes as *Zoonomia or the Laws of Organic Life* (1794 and 1796). In this bold and forthright book, Darwin was seeking some explanation of 'animate activity', some answer to the question of what was the essential difference between something that was living and something that was dead. He concluded that living things possessed an 'irritability', a spirit of animation, and in considering their early development as embryos, drew a parallel between their embryonic changes and possible earlier evolution. For Erasmus Darwin rejected the idea of pre-formation, the concept that the whole animal was formed at the start in the female and had only to grow in size to reach its true development; instead he subscribed to the view that life began with a filament from the male and that the female's job was to provide the nourishment for subsequent growth.

To Darwin the capabilities of an animal were a consequence of its material organization, its structure giving the clue to its function. Following on from this – and it is the crucial point – he believed evolution was a similar process; change of structure would come because of changes in the environment and an organism's response to that change; lust, hunger and a desire for security would be the forces that motivated this response. Any changes would, he thought, be passed on to the offspring. In brief Erasmus Darwin believed in the inheritance of acquired characteristics, and with this belief he produced what was indeed an emergent theory of evolution, though of course it still left many questions unanswered.

Jean-Baptiste de Monet, chevalier de Lamarck (1744–1829) became interested in botany when, compelled by illness to leave military service, he hoped to take up medicine. However, his finances did not run to this and he had to work in a Paris bank. Indeed, throughout his tragic life – he married three or (possibly) four times, but death overtook each wife – Lamarck lived on the poverty line; his personal effects, books and biological collections had to be auctioned to pay for his funeral and he left his large family unprovided for. It included one son who was deaf and another who was insane. Added to all this, he went blind eleven years before his death in 1829 and was frequently the object of ridicule for his speculations. Yet Lamarck did, nevertheless, receive some recognition from the scientific community in France, owing to the efforts of Buffon, who had been impressed with the younger man's *French Flora* and arranged for government publication; it appeared in 1779. This was the fruit of Lamarck's interest in medical botany and provided an improved method for plant identification. By 1788 Buffon had also arranged a botanical post for Lamarck in the King's Garden (later to become known as the Botanical Garden) in Paris, though rearrangements there five years later

Illustration page 412

made it necessary for him to change to zoological work; however, he seems to have quite welcomed the change.

Lamarck's best-known works are his *Zoological Philosophy* of 1809, and the seven-volume *Natural History of Invertebrate Animals* which came out between 1815 and 1822. They demonstrate his abilities to classify animals, for he separated spiders and crustaceans (lobsters, crabs, shrimps, wood-lice etc.) from insects and began to sort out Linnaeus's very broad group of 'worms', by separating those that had really worm-like forms. And it was Lamarck who introduced the valuable classification of vertebrates and invertebrates. But he did more than classify, for he came to the firm conclusion that there was a 'natural sequence' for all living creatures. He appreciated that this sequence was incomplete, that there were gaps in our knowledge, but he thought further research would fill them in. As far as classifications were concerned, Lamarck appreciated their usefulness, but became convinced they were artificial, and he was driven to the conclusion that all living things, both animals and plants, really formed a continuous series.

From an evolutionary point of view Lamarck was of great importance, because he thought that since classification into species was an artificial process, the idea of fixity of speicies was similarly quite without foundation. He gave examples of selective breeding to support his argument that species change, and concluded that the cause of such changes was external conditions. In the case of selective breeding the change came about because of man; in the wild it occurred because of environmental changes. He also attempted to explain how changes would occur, and was led to a law of use and disuse; changes in the environment led to extra demands on certain organs which thereby became more exercised and so more developed. Such development would be passed on to the offspring. He cited the giraffe as an example of this process. So, too, any lack of development was caused by disuse. Thus Lamarck's evolutionary ideas were to be explained by the inheritance of acquired characteristics.

When Lamarck died in 1829 his theories were largely ignored, though in England his work did receive some consideration in the generation before Charles Darwin. In France the official eulogy on Lamarck prepared for the Académie des Sciences was unsympathetic, being written by Georges Cuvier (1769–1832), whose influence was so immense that he had earned the nickname 'dictator of biology'. The delicate son of a Swiss Protestant officer in the French army, Cuvier had attended the University of Stuttgart where he studied administration, judicial matters and economic science which, then, included some botany. It was here that he met Karl Kielmeyer, one of the founders of the German *Naturphilosphie* (nature philosophy) school whose teaching was based on the ideas of the philosopher Immanuel Kant and the literary giant and scientist Johann von Goethe who, incidentally, invented the now widely used term 'morphology' to describe the study of form and structure in biology. The essence of *Naturphilosophie* was that it reflected the action of mind on nature; man was the summit of nature and the nature-philosophers tried to derive the diversity of living things from a belief in the unity of matter and basic principles. This could, they believed, lead to the complex organisms which culminated in man.

Cuvier was influenced by this philosophy, and especially by Goethe's discussion on morphology, but it was in France that he settled, not in Stuttgart. A brilliant lecturer, he soon gained a great reputation by his anatomical descriptions of animals. He also studied fossils and laid the foundations of what has since become known as palaeontology, for his basic conception of the 'principle of correlation of parts' proved to be very useful. The principle is essentially one that relates the separate parts of an animal to the animal as a whole; for instance, the presence of feathers means a particular development of the forelimb (because this is used as a wing) and this, in its turn, is related to the breast bone, and so on. When only parts of a fossilized animal are found such a principle can be of great use in leading to the reconstruction of the original animal.

Cuvier's study of fossils did not lead him to any Lamarckian view. Though he realized that there had been a succession of animal populations, and that a great number of species had existed that were now extinct, his method of accounting for all this was to suppose that the Earth had experienced a series of catastrophes; of these the Flood was the only one of which there were historical records. He expressly denied the existence of fossil men and believed that after each catastrophe the remaining species repopulated the Earth. What of the appearance of new species? These, he thought, were not really new but came from parts of the world not yet adequately explored.

For the early Chinese awareness of the significance of fossils, see page 168; for the 17th- and 18th-century views, see pages 389–390.

Cuvier was a firm believer in the fixity of species and he produced an immense amount of detail in support of his case, in spite of his administrative and consultative work for the government, first under Napoleon and then under the restored Bourbon monarchy. Of all that he published his most significant work was his vast book *The Animal Kingdom, Distributed According to its Organization* (1817), the first important advance in classification since Linnaeus. In it Cuvier argued against arranging animals in any single scale of nature, such as an evolutionist like Lamarck thought might be possible. Instead he insisted on a fourfold classification – vertebrates, molluscs (slugs, snails, oysters, etc.), articulated creatures (insects, spiders, lobsters, etc.), and radiata: the last was a heterogeneous collection, though the other three were what seemed to be natural groupings. These groups were further broken down into a more detailed classification.

Cuvier's views stimulated a positive interest in comparative anatomy and in palaeontology that lasted throughout the rest of the nineteenth century. Of those who followed him, a typical example was Richard Owen, who was also much taken with the *Naturphilosophie* outlook. Born in Lancaster in 1804, apprenticed to surgeons first in Lancaster and then in London, Owen also attended some anatomical lectures in Edinburgh and finally, in 1826, qualified as a member of the Royal College of Surgeons. He succeeded William Clift as keeper of the Hunterian Collection, and in 1856 he was appointed superintendent of the natural history departments of the British Museum in Bloomsbury, London, departments for which Owen managed to get a new museum building (now the Natural History Museum at South Kensington, London) in 1871.

Owen's scientific work lay mainly in palaeontology, though he did dissect a great number of rare animals in order to identify species in the Hunterian Collection and produced a useful account of this research. His palaeontological interests led him to reconstruct many prehistoric animals, and also to undertake an immense investigation of mammal teeth, an important study since teeth are the hardest bones in the body, and so the most likely parts to be found in fossilized form. However, Owen suffered from jealousy and an intolerance of competition; when Charles Darwin's *Origin of Species* appeared in 1859 he feared eclipse of his pre-eminent position, and though Darwin was a friend of twenty years standing, set out to discredit him. Owen wrote a long and anonymous review in the *Edinburgh Review* on which Darwin commented 'It is extremely malignant, clever and . . . requires much study to appreciate all the bitter spite of many of the remarks against me . . . He misquotes some passages, altering words within inverted commas . . .'

Illustration page 412

Charles Robert Darwin (1809–82) the son of a doctor, was almost of an age with Owen. His grandfather was Erasmus Darwin, who, by a second marriage, was also grandfather of Francis Galton, whom we shall meet later. Charles Darwin was a keen collector of wax seals, franks (the equivalent of postage stamps), shells and minerals, but he made no mark at school, and when in 1825 he went to Edinburgh University to study medicine, he was a failure; the lectures he found dull, and attending operations, performed without anaesthetics, repelled him. Darwin was then sent to Cambridge to prepare for the Church, but he did little work though his discussions with John Henslow, the professor of botany, and some other distinguished scientists, gave him some confidence in himself and awoke a very real interest in natural history. In 1831 the Admiralty asked for a naturalist to accompany Captain Robert Fitzroy on a voyage in HMS *Beagle* to survey the coasts in the south of South America, and Henslow recommended Darwin. He set sail from Devonport in December to start on what was to be a brilliant scientific career.

Charles Darwin's work on the voyage was much affected by the views of Charles Lyell, a Scotsman who, while at Oxford and supposedly reading law, had become fascinated by geology and with the dating of rock strata by the fossils embedded in them. A visit to Yarmouth on the English east coast, and discussions about the slow erosion of the coastline there, together with other experiences, had convinced him that slow change seemed to be a characteristic geological phenomenon. In the event Lyell did become a lawyer, but his interests in geology still continued; on a visit to Paris he met Cuvier and other French scientists including Constant Prévost, who took him on a tour of the Paris basin and convinced him that the strata had been formed by very slight geological changes and not, as Cuvier thought, by incursions of the sea over the land. Much other research over the next five years confirmed Lyell in his views and eventually he wrote up his ideas in his monumental three-volume *Principles of Geology*, the first volume of which appeared in June 1830 and created a sensation. The second and third volumes came out in 1832 and 1833, and by then Lyell had forsaken the law and become professor of geology at King's College, London. The *Principles* contained a careful survey of current processes which, as he wrote, were altering the Earth's surface and showed how unnecessary it was to postulate vast upheavals and catastrophes to account for them. He enunciated what has come to be called 'uniformitarianism', the uniform course of nature throughout the Earth's history in changing the Earth's crust.

To support his case, Lyell used evidence drawn from the geographical distribution of plants and animals, claimed each species had grown up in a particular centre from which it spread, and showed how it had persisted for a time before becoming extinct and being replaced by another species. Thus he held that the emergence of new species was a steady process throughout geological history. Because the first volume of Lyell's book caused a furore, Darwin was advised by Henslow to take a copy 'but not to believe it'; the second volume Darwin arranged to be posted to him, and he collected it at Montevideo on the way down the east coast of South America.

The voyage on the *Beagle* lasted five years, and increased Darwin's practical knowledge by leaps and bounds. In Brazil he came across tropical forest for the first time, on the island of Tierre del Fuego he saw a race of men 'so savage, so devoid of beliefs, that they hardly seemed human', and in Chile he observed an earthquake and saw its effects in raising the level of the land. Ashore, Darwin often went on long expeditions, collecting samples. But his interest and concern with natural history did not mean that his geological interests were forgotten; at Santiago on the Cape Verde Islands, off the west coast of Africa, he found he could follow the entire island's geological history, from the lowest strata which contained crystals and were composed of igneous rock to the limestone above which contained shells of marine creatures and was obviously deposited by water that must once have covered the island. Finally, Darwin noticed that the topmost volcanic layer was comparatively recent and had baked the limestone directly below it. His observations in South America convinced him that rocks could be altered after they were laid down – a view different from that generally believed at the time – but confirmed by an earthquake at Concepción where examination after the event showed how the strata had been altered and lifted to new positions. Moreover, the coral reefs that he visited also supported his ideas about the elevation and subsidence of land. He knew that the coral polyps which build the reefs can only do so when underwater and at depths not greater than some 36 metres (120 feet), but since the tops of these reefs were now above water and all approximately the same height, Darwin concluded that they indicated subsidence of the sea floor, the corals growing upward as their bases dropped. This view, which differed from that suggested by Lyell, has since been confirmed by deep borings. Darwin discussed it with Lyell on his return before he published his geological results.

An account of Darwin's collection of specimens was published in detail as part of the official record of the voyage, but his own biological conclusions were another matter. When he set out in 1831 he had no idea of questioning the idea of the fixity of species, but as he voyaged over so wide an area various questions set him thinking. Why were there similar animals in far distant places? Why was the South African ostrich so like the South American rhea, for instance, and why was this so with other animals; why did closely parallel species appear everywhere? On the other hand he had also to answer the question which was brought forcibly home to him when he visited the Galapagos Islands in the south-east Pacific, for here he found fossil animals similar to, but not the same as, living forms, and that closely parallel animals replaced each other as one travelled over the area from south to north. He also made a close study of finches which were to be found everywhere and noticed that though varying from island to island they all had characteristics best explained by assuming they came from a common ancestor.

Illustration page 412

Gradually, when he began to think over the evidence after his return in 1836, he concluded that the simple answer must be that species did undergo a change. Yet what caused such a change? Was it due to alterations in the environment and, if it were, how did they operate to give such an effect? He considered the man-made changes evident from selective breeding and realized these were equivalent to the evolution of one species into another. Moreover, he appreciated that a species of animal will become extinct if it is not well adapted to its environment. Then, late in 1838, while he was still turning the matter over in his mind, Darwin read *An Essay on the Principle of Population* by the clergyman and political economist Thomas Malthus. In the book, which had come out in 1798, Malthus had argued that, unless checked, growth in population would outstrip any increase possible in food production. What is more, he had pointed out that, unchecked, population will double every twenty-five years. This gave Darwin the final argument he needed. He saw that the species which survived must be those best fitted for their environment, and that the forces operating were so strong that they formed gaps in the animal population by weeding out species that underwent variations that made them unsuitable for the environment. These gaps were then filled by those of the species whose variations made them better adapted. It was a case of natural selection of those most fitted to survive in changed surroundings.

EVOLUTION AND THE AGE OF THE EARTH

In July 1858 Darwin began to write *On the Origin of Species by Means of Natural Selection, or the Preservation of Favoured Races in the Struggle for Life*. In it he expressed his ideas and was able, by means of his theory of evolution by natural selection, to give an explanation for change of species. Lamarck had considered an 'inner feeling' to be the operative factor, but Darwin provided a sound physical cause. He had, in fact, done for biology what Newton had done for physical science; he applied a natural law to correlate a vast amount of diverse evidence. No longer did each species need a special act of divine creation; it's particular form was due to natural causes.

For a theory of adaptation proposed in ancient Greece, see page 82.

Not only was the *Origin of Species* the best seller of 1859, but by its detailed evidence it made the concept of evolution scientifically respectable. It also raised a storm of protest, especially from people who could not or would not follow the detailed evidence, and most vociferously from those who took a literal interpretation of the Biblical account of Creation. Fortunately, Darwin had eloquent and well-informed friends, among them Charles Lyell, Alfred Russell Wallace (who, as Darwin knew, had independently come to similar conclusions), the botanist Joseph Hooker, and, most significant of all, the anatomist and anthropologist Thomas Henry Huxley. A few scientists, of course, could not accept the theory on purely scientific grounds, among them Louis Agassiz, a Swiss paleontologist and pupil of Cuvier, and the physicist Lord Kelvin (William Thomson). Kelvin had accepted a long time scale for the age of the Earth, but calculations he made on its rate of cooling seemed to him to limit the age to 100 million years; this was too short for Darwinian evolution. We now know Kelvin's computation was in error because of factors which only become known later, but his criticism does not seem to have carried much weight at the time.

Illustration page 413

For Kelvin, see page 446; for the resolution of this point, see page 482.

Because of his great reputation, a more serious opponent was Richard Owen, and another public figure who set his face against Darwin was the bishop of Winchester, Samuel Wilberforce. Both were involved in the debate at the British Association's annual meeting in Oxford in 1860. Since great public interest had been aroused by the controversy over Darwin's book a public discussion was held, though Darwin himself was not to be there. The defence primarily rested on Huxley, while the anti-evolutionists were led by Owen and Wilberforce. Huxley soon crossed swords with Owen over technical matters, for the *Origin of Species* was an implicit rejection of Owen's classification of mammals, but the real clash came with Wilberforce. Owen had briefed Wilberforce, who had either not understood what he was told or had been inadequately instructed; at all events he made a number of scientific blunders and then capped them all by asking Huxley a loaded and impertinent question. Was Huxley's ancestry from an ape on his grandfather's or his grandmother's side? When the audience called on him to answer, Huxley carefully corrected the Bishop's scientific mistakes and then said that he would rather be related to an ape than to a man of proven ability who used his brains to pervert the truth. The audience went wild with excitement. The evolutionists had won the day and made clear that no ill-informed Church opposition was to be tolerated.

Huxley, born in 1825 and Darwin's junior by sixteen years, became his most powerful advocate – 'Darwin's bulldog' he came to be called. In due course Huxley was able to show quite clearly that the differences between man and the apes were less than those between the apes and lower primates, and all through, by meticulous anthropological work, and a careful scientific treatment of man as a zoological animal, Huxley did much to promote the evolutionary cause, both in the scientific community and outside it. It was Huxley, too, who had a great influence on getting proper scientific teaching established in England, emphasizing its importance at every level of education, and the need for schools and universities to have laboratories where students could themselves verify what they were told and try applying the scientific method.

As far as religion was concerned, Huxley was an agnostic; he regarded the Bible highly as literature and as 'the Magna Carta of the poor and oppressed, at least in those passages which support the concept of righteousness'. Darwin, for his part, gradually became an agnostic. Yet when Karl Marx wrote asking permission to dedicate the English edition of *Das Kapital* to Darwin, Darwin refused because he did not wish to be associated 'with attacks on religion'. Sadly, Darwin was now a sick man; on his return from his voyage in the *Beagle* his health had begun to deteriorate due to a trypanosome infection contracted in the Andes, and his *Origin of Species* as well as all his later works, including the *Descent of Man* (1871) and five concerned with botany, were all written while continually suffering a debilitating illness. It is no wonder, then, that he stayed at home, comforted by his family, and avoiding public appearances. Yet his reputation grew and when he died in 1882, special permission was given for him to be buried in Westminster Abbey.

Biology

Cell Theory

Pure biology made great strides during the nineteenth century; various physiological processes were examined in increasing detail on the basis of what may be termed the 'animal-machine' outlook, while studies were made of allied subjects like colouration and mimicry. But of all the advances, two were to have a most significant effect on work during the century that was to follow; these were cell theory, and the question of the spontaneous generation of life.

Hooke's first use of the term 'cell' has been noted, but cell theory as it developed in the nineteenth century was more than a question of defining material within a certain specified area; it was an attempt to tackle the basic question, 'what is an organism?' The first attacks on the problem came from France, where from around 1800 physicians combined post-mortem examinations with a clinical description

For Hooke, see page 392.

of the state of the corpse, making it possible to locate where in the body things had gone wrong. This, in its turn, led to a detailed study of the organs themselves. The pioneer in this field was Xavier Bichat who, by the time of his death in 1802, had identified no less than twenty-one types of tissue – mucous, fibrous, etc. – in organs and noted their distribution. The next step had to wait until after the 1830s and the advent of good achromatic microscopes. These were developed to the point when, around the 1880s, due primarily to the work of Ernst Abbe at the Carl Zeiss optical works, optical microscopes could distinguish particles which were no more than two-thousandths of a millimetre across, so bringing the instrument almost to its theoretical limit of definition.

Once the achromatic microscope had arrived the main ideas of a cell theory could be formulated, and this was done by two men, Matthias Schleiden and Theodor Schwann. A lawyer, who later became the arrogant but able professor of botany at Jena, Schleiden reacted against the dry systematization which then passed as botany, and in a paper about phytogenesis (plant origins) he seized on the cell as the essential unit of the living organism. Theodor Schwann was totally different in character, a quiet, simple and pious man, and his research went deeper than Schleiden's. Devoted to the investigation of the elementary structure of animal tissues, which are more difficult to observe than those of plants, he nevertheless saw that their development was similar to the development of plant tissue. Schwann also discussed eggs, including the eggs of mammals, discovered in 1828 by Karl Ernst von Baer, and came to the conclusion that eggs were essentially cells, even though many, like the hen's egg, were enormously extended. Thus Schleiden and Schwann promoted a view of common cell development and thereby stimulated a great deal of research, even though they were wrong in their belief that the cells themselves were formed by a purely chemical process.

In the 1830s *naturphilosophie* was in full swing in Germany. Here Lorenz Oken, who was concerned with finding a unit from which the immense diversity of living creatures could be generated, pounced on the cell as the answer, though his arguments in favour of it were philosophical; he spurned microscopical examination. Nevertheless, scientific study was stimulated, and in particular studies were made of individual embryonic changes that occur within eggs, using a technique of careful microscopic observation. This allowed Robert Remak and Rudolf Albert von Kolliker to show that such development took place by cell division.

Illustration page 414

The main triumphs of cell theory did not arrive, however, until the second half of the nineteenth century. First, the medical microscopist Rudolf Virchow, professor of pathological anatomy at Berlin and Germany's leading physician, began to investigate the question of cells in living tissue and in 1858 produced his important *Cellular Pathology*. In this he disproved the Schleiden-Schwann theory of cell formation and came out with the proposition that cells can only arise from pre-existing cells. Virchow then went on to show that cells '. . . are the last link in the great chain of subordinated formations that form tissues, organs, systems and the individual'. He also saw the cell as a richly detailed unit, the irreducible organized centre of activity within the body. Yet in spite of his advocacy there still remained a need to demonstrate conclusively that the cell really was the fundamental functional unit, and it was here that the French physiologist Claude Bernard played a vital part. In all his medical research – he did valuable work on the digestion, the physiology of the nerves and on toxic substances – Bernard sought to demonstrate the vital unity of all organisms. Certainly, he leaned towards the ideas of Schwann rather than those of Virchow but, all the same, the broad picture he drew of the whole organism was important because it showed clearly the relationship between cells and tissues and the interactions between cells and the body fluids surrounding them.

In spite of the contributions of Virchow and Bernard, more remained to be done, and it was at this stage that the embryologists made special contributions. They had been primarily concerned with demolishing the theory of preformation and replacing it by the doctrine of epigenesis, in which, starting with the more or less simple egg, there was a straightforward organic development resulting in increasing complexity. This new theory had been helped first by the realization that no mammalian egg would develop without sperm being present and, secondly, by the observation in 1843 of the presence of sperm within the egg. However, the action of the sperm was still not understood; the general explanation being that the presence of sperm merely triggered the development of the egg. Not until 1876 and 1877 did things change, for then Oskar Hertwig and Hermann Fol were at last able to show the presence of not one but two nuclei, one male and one female, in the fertilized egg. This transferred the unit binding successive generations of animals and plants away from the cell to its nucleus, and was a crucial discovery.

Biologists now began to hunt for something, perhaps something that could be defined chemically, which could be transferred from sperm to egg. They therefore started to make detailed microscopic examinations of the cell nucleus, though this was not easy; strong stains had to be used to make the transparent components show up in the microscope field. Nevertheless this work resulted in the discovery that during cell division thin thread-like structures were present. These, it was found, were what were transferred when cells divided. Because chromatin was the stain used to make them visible, Wilhelm von Waldeyer-Hartz gave them the name 'chromosomes'.

Illustration page 414

Spontaneous Generation of Life

From the earliest times there had been speculation about how life could start. Rotting meat bred maggots, fleas and lice were associated with moisture and filth, and it seemed that the combined action of heat, water, air and putrefaction could spontaneously beget life. Nevertheless in the seventeenth century this almost universal belief was questioned and Francesco Redi, physician at the court of Tuscany, discussed the whole matter in his *Experiments on the Generation of Insects*, 1668. With a rare insight, Redi suggested that the maggots that appeared on putrefying matter were introduced from outside in some way, and experiments with such material led him to the conclusion that the maggots and worms came from eggs. He therefore devised other experiments in which the putrefying material was covered with very fine Naples gauze and then saw that the eggs had been deposited on the gauze. This proved conclusively that the maggots were not bred by the putrefying matter, though strangely enough he did consider that grubs could appear spontaneously in fruit, vegetables and leaves. Leeuwenhoek also carried out some studies which allowed him to explode the idea that edible mussels could be bred from sand, and showed that they came from spawn.

For Leeuwenhoek, see page 394.

In spite of the research by Redi and Leeuwenhoek, the idea of spontaneous generation was too deep-rooted to die, so that the chief investigators in the next century, John Needham and Lazzaro Spallanzani, were not at all in agreement. Needham was born in London in 1713 to Roman Catholic parents, but spent most of his life in Belgium where he became the first director of the Royal Academy of Belgium. A secular priest, he argued with Voltaire over the question of miracles, but he is now remembered for two experiments on spontaneous generation that were reported in the *Philosophical Transactions* of the British Royal Society and in the later volumes of Buffon's *Natural History*. In the first of these he took an almond kernel which he placed in a tightly corked glass container full of water. In due course he saw part of the kernel rot and observed minute particles floating about; these fixed themselves to the container walls. He concluded they had come from the 'infusion' of almond and water, not from the air. His second experiment was with hot mutton gravy, and was more significant. He sealed the gravy in a phial (which had not, of course, been sterilized) and after some days found the gravy swarming with life. This led him to the view that there was a vegetative force existing in every minute point in matter and every filament that goes to make up the whole animal. This fitted in well with his more general belief that all animals consisted of organic particles which were themselves indestructible.

Lazzaro Spallanzani (1729–99), born near Modena in Italy, held professorships in Modena and Pavia, and carried out valuable research in physiology, particularly on the digestion. An expert microscopist, he published an account of his work on spontaneous generation in 1765 and his *Observations and Experiments concerning the Animalcules of Infusions* in 1776. Opposed to the ideas of Needham and Buffon, he carried out tests using glass flasks; he found 'animalcules' arrived quicker the larger the flask, and if he had an infusion that did not show them, he made a small crack in the glass and noted that they then soon appeared. He next disproved Needham's ideas by showing that when the flasks were sealed after heating for some time, no animalcules appeared. Comparative trials with flasks having necks like capillary tubes, with some sealed off and others left open, showed clearly that air was a vital factor in the development of the little creatures, which Spallanzani classed into two orders – a superior order that contained those which were destroyed by heating to boiling point for half a minute, and an inferior order which survived boiling for half an hour. The last were probably bacteria.

When the nineteenth century dawned the question was examined further by a number of investigators. The German agricultural chemist Franz Schultze, who was born at Naumburg in 1815, devised a most ingenious experiment whereby he placed a vegetable infusion in two flasks. One flask was boiled and then left open to the air, but the other was fitted into an apparatus which allowed air to be drawn into the flask through sulphuric acid. The air was changed twice a day, but after two months no 'animalcules' (micro-organisms) had appeared, though they were flourishing in the control flask. When the flask without micro-organisms was opened to the air, it too became full of life. The experiment showed that air in the flask was not, of itself, the cause of generation of life, though Schultze's experiment was criticized by others because when they repeated it they did not obtain the same results – due, it is now clear, to lack of care.

The matter was also taken up by Theodor Schwann. In 1836, the same year as Schultze's experiment, Schwann used a large globe with a little infusion. The whole was sealed off and the globe inserted in boiling water for 15 minutes; no micro-organisms appeared. Critics said he might have affected the oxygen in the air, others that the organic matter might have given off carbon dioxide during heating and that the latter gas might have inhibited growth. Schwann then made sure he removed any carbon dioxide, but still there were criticisms. He therefore redesigned the entire experiment and in 1837 performed a Schultze-like test in which the air was constantly renewed though, in Schwann's case, the air had been heated before admission to the flask. This silenced the critics and showed conclusively that it was not air itself that caused putrefaction, but something in the air which was destroyed by heat.

Yet in spite of all previous work, in 1858 the biologist Félix Pouchet still believed in spontaneous generation and presented a paper to the Académie des Sciences in which he claimed to have proved his point. The next year he published a book *Heterogeny or Treatise on Spontaneous Generation based on New Experiments*; the burden of this was the existence of a vital force which must be present, so that while no mineral salts could ever produce life, this was bound to happen where the host material had once been living. Only in the latter half of the century was the whole question finally settled due to the work of Pasteur and Tyndall.

Illustration page 415

Louis Pasteur (1822–95) was born at Dôle in eastern France in 1822, the son of a tanner. No more than an average student, he studied chemistry at the Sorbonne and then worked there for a time on crystal structure and the optical effects of crystals. Then in 1854, after holding a chair in chemistry at Strasbourg, he became professor of chemistry and dean of the newly founded faculty of sciences at Lille University. The brewing industry in Lille was having trouble with its beer going sour after fermentation and Pasteur agreed to look into the question. His studies showed him that the process was due to the presence of living organisms, and through a series of careful and logically thought-out experiments proved that the organisms were airborne. What he did was to use dripping water in a tube to draw in air from outside the laboratory, this incoming air passing through a wad of gun-cotton (cellulose nitrate). After the air had passed through, the gun-cotton was dissolved in alcohol and ether and the 'dust' examined microscopically. Minute oval bodies were seen, indistinguishable from the eggs of animalcules or plants. From this series of experiments Pasteur was able to show that the number of 'organized bodies' depended upon the movement, moisture and temperature of the air and the height above ground at which it entered. Therefore they were not chance results. He also repeated Schwann's experiments with great success, and in 1861 published his *Memoir on Organized Corpuscles which Exist in the Atmosphere.*

Pasteur next performed another series of experiments to investigate putrefaction. He used a sterile infusion containing air which had been heated, and showed that putrefaction invariably occurred if the 'gun-cotton dust' was introduced. He next proved that an infusion could be sterilized and remain sterilized in a flask which was open to the air provided the neck of the flask was drawn out to be very narrow and was bent downwards. This showed that the 'corpuscles' had weight. Lastly he demonstrated that the corpuscles were not evenly distributed through the air and compared the number in various places under different conditions.

Pouchet, who been working along similar lines, dismissed Pasteur's results when they came out and so serious a situation developed that the Académie des Sciences appointed an official committee to discuss the case; the committee found in favour of Pasteur. Yet in spite of this there were still objections, notably by Henry Bastian, the professor of pathological anatomy at University College, London. In 1872 Bastian published his *Beginnings of Life*, in which he emphatically supported the old idea that life could be produced from non-living matter. His thesis was, of course, wrong but his experiments did draw attention to the fact that some micro-organisms were more resistant than others to heat and, as Pasteur acknowledged, this was to help him in his later work.

The final rejection of the spontaneous-generation theory took time. Bastian and others refused to give in, but in the 1880s help in confirming the micro-organism theory came from an unexpected quarter, the Royal Institution in London. Founded in 1799, the Institution was designed to be 'a centre of philosophical and literary attraction' for giving instruction to the young (its Christmas lectures to 'a young auditory' are still a major feature) and to the mature, as well as to work to apply science and technology for the general good of the community. Its main endeavours lay in the physical sciences; it had no appointment for a biologist. It was the professor of natural philosophy John Tyndall who entered the spontaneous generation conflict and did so as a result of some physics research he was doing on radiant heat and gases. The gases and air used in his experiments had to be free of suspended particles; this led him to find that the presence or absence of such particles could be shown up by an intense beam of light, and that the particles, if present, could be removed by passing the gas or air through a flame. With the argument raging over spontaneous generation, he decided to examine the evidence himself, and made infusions of beef, mutton, liver and other substances, exposing some to the air and keeping other samples sealed up. The significant point, though, was that those infusions which took more than two hours to sterilize in his London roof-top laboratory could be sterilized in five minutes in a laboratory at Kew, 11 km (7 miles) away in the country. He therefore built a new roof-top laboratory some 8 metres (24 feet) from his old one on the roof of the Royal Institution building and made it as sterile as he could. He again found a five-minute sterilization time. It then became obvious that it was so unlikely spontaneous generation could occur only in one laboratory and not in the other, that airborne micro-organisms must be the cause of putrefaction.

During his experiments Tyndall found some solutions more difficult to sterilize than others, and he devised a method later known as 'Tyndallization' – heating a solution for a short time (about one minute), then pausing and heating again, followed by repeats of the whole process. The technique permitted sterilization by, say, five short heatings, of an infusion that could not be sterilized by an hour's constant boiling. It seemed to be a very strange effect until the German botanist and bacteriologist Ferdinand Cohn saw that it fitted in with what he had found to be the life-cycle of rod-like micro-organisms or 'bacilli' (from Latin *bacillus*, small rod). Cohn, who had worked with Pasteur, had discovered that bacilli grow from small round bodies or 'spores', and that while the spores could resist heat the bacilli could not. Tyndallization caught the micro-organisms at the bacillus stage, before they could reproduce.

These were important results not only because they finally demolished the concept of spontaneous generation but also for other implications. They led to a greater understanding of the way diseases are transmitted and wounds become septic, and helped Pasteur in his later research, such as work on the souring of milk, and made it clear that the 'pasteurization' of milk, when it is heated to 65°C (149°F) for half-an-hour, destroyed the specific airborne bacilli associated with it. Again when in 1865 Pasteur was asked to investigate a disease which was killing off millions of silkworms in the south of France – a serious matter, for production of cocoons had dropped by 75%

and the industry was in dire trouble – he found that the disease was a bacterial infection, and was at once able to suggest precautions against further trouble.

Encouraged by his success with silkworms, Pasteur decided to examine two diseases which were then devastating cattle and poultry, anthrax and chicken-cholera. Both were due, he found, to bacilli, and were transmitted from one animal to another, but the anthrax results had wider implication. Anthrax was a disease that could be transmitted to man, and the discovery of its cause raised the general question of what, if any, other diseases in man might be bacterial infections. As far as Pasteur was concerned, his immediate problem, once he had diagnosed bacilli as the cause, was to find how to effect a cure, and he turned to the inoculation work of Jenner as a possibility for further application. His difficulty was to obtain bacteria which would immunize yet not kill the patient, and in 1880 he came across a culture of cholera which had grown less deadly with the passage of time. This gave him the clue he wanted; by growing cultures of bacteria from animals and then filtering them, he found safe vaccines, first in 1881 against anthrax and, a year later, against the dreaded disease of rabies. But by now he was not alone in this research; others, too were carrying out similar work, and soon vaccines against tetanus and diphtheria were available. Moreover, methods were devised for separating strains of bacteria and cultivating them in the laboratory away from their animal host, techniques which in the twentieth century were to provide many other vaccines including those against poliomyelitis and typhoid.

Many medical men did not accept Pasteur's results, at least to begin with, but one who did was the surgeon Joseph Lister, a man whose father had been responsible for some of the recent technical improvements to the microscope. Born at Upton in Essex in 1827 and educated at University College, London, he worked at University College Hospital and was shocked at the number of deaths after even quite minor surgery, due to the wounds putrefying and subsequent blood poisoning. This too, he found, was the case in army hospitals during the Crimean War and at Glasgow Infirmary, where he went in 1860 as professor of surgery. Yet Lister noticed that no putrefaction occurred in internal wounds. While he was turning this over in his mind, a successful attempt to attack this problem was being made in Vienna by the physician Ignaz Semmelweis. Semmelweis had noticed that in one maternity ward of the Vienna General Hospital up to 30% of the pregnant mothers died from fever, while in another ward the death rate was ten times less. Then, while performing a post mortem on a medical friend who had died from a cut inflicted while he himself had been carrying out a post mortem, Semmelweis found his friend's body to contain diseased tissue similar to that of the pregnant mothers who had died. Knowing that those in the ward with the high death rate were treated by students who had come straight from the post-mortem room, he concluded that they carried some infection with them. Against great opposition, he instituted a strict rule that after being in the post-mortem room they must wash their hands in carbolic and water before visiting patients. In barely a month, the death rate in the lethal maternity ward dropped to only a fifth of its previous figures. Semmelweis then extended his antiseptic procedures to include all instruments used in treatment and again documented favourable results. But a proposal to create a commission to examine his results was turned down, partly for political reasons and partly due to some medical opposition; his appointment at Vienna was not renewed. He returned to Hungary where he was able to repeat his success and set up a good private practice, and finally, in 1861 published his results. Yet in spite of the details he gave, general disbelief followed; Semmelweis became alternately depressed and angry and, finally, general sepsis from a surgically infected finger led to his death in an asylum in 1865.

Once Pasteur's work on micro-organisms as a cause of putrefaction was published, Lister at least immediately saw its surgical significance and began to search for something which would kill airborne bacteria without seriously damaging body tissues. In 1867 he found a solution of carbolic acid was satisfactory and by spraying the air in operating theatres with the solution, and insisting on cleanliness, he brought about a dramatic reduction in post-operative mortality.

Illustration page 416

Lister's work on asepsis coincided with another surgical innovation in the West, the introduction of anaesthetics. In 1800 the chemist Humphry Davy had published the discovery of the anaesthetic properties of nitrous oxide, and in 1845 the American dentist Horace Wells tried the gas in this way for dental operations but it seemed unreliable. A colleague, William Morton sought for a more infallible substitute and the Boston chemist Charles Jackson suggested ether. Morton first used ether during a tooth extraction in 1846 and demonstrated its effectiveness in a tumour operation. Its use was immediately taken up at Edinburgh by the surgeon James Simpson, but the very next year he substituted chloroform which he found more satisfactory, and from then on this became the standard anaesthetic. With the use of anaesthesia and asepsis the whole practice of surgery was revolutionized, and though anaesthesia was not the direct result of nineteenth-century fundamental scientific research, the introduction of antiseptic conditions and, later, the growth of general hygiene and the purification of water supplies certainly were.

Chemistry

In the nineteenth century chemistry underwent considerable development along two main lines; the atomic theory of matter was revived but in a new and powerful form, and the subject of organic chemistry was born. Both were to have wide implications not only for chemistry itself but also for physics and biology.

The rebirth of the atomic theory was due to the Quaker teacher John Dalton (1766–1844). Born in Eaglesfield, Cumberland where his

Illustration page 416

father was a weaver, Dalton was sent to the village school which he seems to have taken over as teacher when he was twelve, though without much success. After a short time as a labourer on local smallholdings he moved at the early age of fifteen to teach at a boarding school in Kendal. The school was well equipped with books on science and some scientific equipment – this included a telescope, microscope and air pump – and it was here that Dalton seems to have taught himself science, with help from John Gough, a local naturalist. It was at this time that he began to keep a daily meteorological record. In 1792 he had progressed enough to be appointed as professor of the Nonconformist New College in Manchester, and became involved with the Manchester Literary and Philosophical Society, one of England's oldest learned societies.

Dalton's scientific work covered observations of the aurora, trade winds and the cause of rain; he discovered a law of partial pressures relating to gas and another to their expansion (the last is now known as Charles's Law since it was discovered independently and earlier – about 1787 – by Jacques Charles). Dalton also made the first systematic study of colour blindness, a defect from which he himself suffered. But of course it is his atomic theory which was his most important contribution, yet strangely enough he was led to it not by any study of chemistry but by his interest in meteorology. Knowing that Lavoisier had discovered that air was composed of at least two gases of different weight, Dalton asked himself some basic questions about the atmosphere. These were questions concerned more with the physics of the subject than with its chemistry. For instance what were the proportions on which these gases occurred? Was the water vapour in the air chemically combined with the gases or only mixed with them? Again, it was found that the amount of gas absorbed by water vapour differed with different gases; why was this so? And as far as the atmosphere itself was concerned, since its two gases – oxygen and nitrogen – differed in weight, why did not gravity separate the heavier from the lighter gas? Dalton's experience from examining the resulting pressures when gases were combined, together with some logical thinking about all the problems mentioned, led him to conclude that the gases were mixed but not chemically combined. The reason why this happened was, he thought, because heat surrounded all the gas particles (or 'atoms' as he called them), keeping them apart and thus preventing them settling down into separate groups. Dalton then extended this idea by assuming that each gas was composed of its own kind of atoms; the heavier the gas the heavier its atoms, with atoms of one kind attracting one another. The outcome was that Dalton had reached a powerful concept which explained both the state of matter as well as why one chemical substance differed from another in weight and, indeed, in chemical behaviour.

Illustration page 416

He announced his ideas first in 1803 in a paper about the absorption of gases that he gave to the Manchester Literary and Philosophical

ELEMENTS

	W^t		W^t
Hydrogen	1	Strontian	46
Azote	5	Barytes	68
Carbon	5,4	Iron	50
Oxygen	7	Zinc	56
Phosphorus	9	Copper	56
Sulphur	13	Lead	90
Magnesia	20	Silver	190
Lime	24	Gold	190
Soda	28	Platina	190
Potash	42	Mercury	167

Dalton's original list of the atomic weights of the elements.

Society, then in lectures he gave in London and Edinburgh, but the complete publication of the theory did not come out under his own name until 1808 when it appeared in his *New System of Chemical Philosophy*. Nevertheless it had been mentioned the year before by Thomas Thomson in a new edition of his *System of Chemistry*, and this helped it to become more widely known.

The essence of Dalton's theory was not just that all chemical elements were composed of their own kind of atoms but also contained the proviso that chemical reactions did no more than separate and unite these elementary particles. 'No new creation or destruction of matter is within reach of chemical agency', he wrote. This theory was more than an ingenious idea, as subsequent research was to show. It was a theory that explained the quantitative laws of chemistry. For although Dalton could not determine the weight of a particular atom since he did not know how many billions of atoms there were in a particular piece of material, he was able to measure their relative weights by assuming that the same number were always present in a given volume. This gave chemists a simple, basic explanation based on definite rules. Certainly an atomic theory was not new; the Greeks had one and Boyle had also adopted a primitive atomic idea, but it was now propounded with a degree of precision no atomic theory had ever had before. It meant that the relative weights of atoms could be found by weighing the substances involved in a chemical reaction, and brought about a revolution in precision and understanding.

For the Greek atomists, see pages 82–84; for Boyle, see pages 384–385.

As the usefulness and validity of the theory were confirmed by later research, so its circle of supporters grew. Two of the chief men who contributed to this acceptance were Gay-Lussac (1778–1850) and Amedeo Avogadro (1776–1856).

Joseph Gay-Lussac was born in Saint-Léonard near Limoges in France and entered the new École Polytechnique when he was nineteen years of age. He had gone to the École after success in a competitive entrance examination and kept up this early promise for he moved to the exclusive School of Bridges and Highways where he came under the wing of Berthollet. At the age of thirty-one he was appointed professor of chemistry at the École Polytechnique and many years later at the Jardin des Plantes. Before he held the chemistry chair at the École Polytechnique he had gone on balloon flights to study the upper air and these led him to carry out research on the combining properties of gas, research in which he was assisted by the Prussian naturalist, explorer and scientist Alexander von Humboldt (1769–1859). As part of their research, Gay-Lussac and von Humboldt passed an electric current through water – an experiment that was being repeated by others now that a constant source of electricity from the voltaic pile was available – and then, when at the École, he extended his work. In 1811 Gay-Lussac had gathered enough experimental evidence to show not only that water was composed of two parts hydrogen and one part oxygen, but also that after examining other substances, it was clear that all gases combine in volumes that are in simple relation to one another. This, as he pointed out, was just what was to be expected if Dalton's theory were correct.

Count Amedeo Avogadro, born at Turin in 1776, son of a distinguished ecclesiastical lawyer and administrator, followed his father's profession until 1806 when an interest in science and especially in Volta's novel discoveries led him to forsake law and take up mathematical physics. He is remembered now for what is still called 'Avogadro's hypothesis', which was a powerful idea that cleared up a problem unsolved by Gay-Lussac. Gay-Lussac's work did not explain why, when gases combined, they sometimes seemed to occupy less space. For instance when two volumes of hydrogen and one of oxygen combined to give steam, the steam occupied less space than the hydrogen and oxygen separately. This was not to be expected if equal volumes of gases contained equal numbers of atoms. However, what Avogadro realized was that the atoms might combine together when the gases were mixed to give groups of atoms. Thus in the case of water two volumes of pairs of hydrogen atoms combined with one volume of pairs of oxygen atoms gave two volumes of water (or putting it in symbols $2H_2 + O_2 = H_2O + H_2O$). Incidentally the word 'molecule' (Latin *molecula*, small mass) now used to describe groups of atoms was coined by Avogadro, though to him atoms were 'elementary molecules' and combinations of atoms were 'integral molecules'. But whatever the groupings were called, Avogadro's hypothesis was important not only because it explained an anomaly but also because it paved the way for measurement of atomic weights more precisely than Dalton's original theory alone could do. Avogadro became enthusiastic about this aspect of his idea but not enough facts were to hand to allow him to obtain the reliable results he wanted. Not until the late 1850s was there sufficient evidence. Stanislao Cannizzaro then reviewed all the information available and was at last able to make a series of satisfactory measures of atomic weights.

With the precision that was now part of chemistry, the new nomenclature that Lavoisier had introduced towards the end of the eighteenth century, and the atomic theory itself, it became increasingly desirable to have some form of chemical shorthand by which one could describe precisely what was happening in a reaction, rather as we have just done for Avogadro's hypothesis. The atomic theory provided an ideal opportunity for this, and Dalton himself led the way by using circular symbols to represent the elements concerned, though this had certain drawbacks, and it was the Swedish chemist Jöns Jakob Berzelius (1779–1848) who devised a really satisfactory method. His chemical shorthand was simplicity itself and is the basis of the method still in use today; the atom of each element was represented by the first letter (or letters) of its name. Thus H represented a hydrogen atom, O an oxygen atom, Zn a zinc atom, and so on. It was immensely effective, allowing one to see at a glance what happened in a particular chemical reaction.

For Lavoisier, see pages 388–389.

Berzelius did more than invent an effective chemical symbolism.

He also made a serious attempt to measure atomic weights, and though some of his values have had to be altered since his time, by and large his measures were good enough to provide other chemists with values they could rely on for the more accurate determination of the composition of different substances. He also discovered some new substances himself while he was analysing a large number of chemical compounds in the process of making his measurements of atomic weights. There were other chemical problems, too, that engaged his attention. He tried to tackle the strange question of why some substances that were chemically the same nevertheless acted in different ways though it was early days for a solution to be found and Berzelius was unsuccessful. He did however coin the name 'protein', from it from the Greek *proteios*, primitive, because although he himself was not conducting research on such substances, he did appreciate their basic chemical importance.

While Berzelius and others, notably Pierre Dulong and Alexis Petit, mainly pursued the question of atomic weights some other chemists devoted their time to investigating the chemical challenge posed by Volta's electric pile. One side of this new electrochemistry was the use of the pile in decomposing chemical solutions. We have already come across its use in disassociating water into hydrogen and oxygen, but it was used extensively with other solutions, notably by Sir Humphry Davy. Born in Penzance in 1778, Davy began studying medicine but became increasingly interested in chemistry. After an apprenticeship with an able surgeon-apothecary, at twenty years of age he was put in charge of the Pneumatic Institution in Bristol. Founded by Thomas Beddoes, a physician who had also held the chair of chemistry at Oxford, the Institution's purpose was to cure diseases, particularly consumption (pulmonary tuberculosis) by the use of gases of various kinds – hence the use of the word 'pneumatic' in the title. Davy stayed in Bristol for three years but in 1801 moved up to the Royal Institution in London where there was more scope for research, his move following on his discovery of the intoxicating and anaesthetic properties of nitrous oxide ('laughing gas'). A year after he had moved to London as an assistant he was promoted to be professor of chemistry at the Institution and here he worked with remarkable success until 1812 when he married a wealthy widow and resigned his professorship. It was not a happy marriage, but fortunately Davy had not completely relinquished his ties with the Institution and continued to do research there.

Davy is now popularly remembered for his invention in 1815 of the miner's safety lamp which arose directly out of a commission to investigate the explodable gases ('firedamp') in mines. Yet this was after he had retired from the Royal Institution where his time was occupied in pure chemical research; he discovered a number of new substances and did pioneering work on agricultural chemistry, though his most important work was his study of electrochemistry. To begin with he used the Voltaic pile to decompose all kinds of solutions, a

Illustration page 449

technique which led him to find that the poisonous gas chlorine is a true chemical element and to the important discovery that alkaline substances are all compounds – something suggested earlier but then unprovable. Moreover, Davy followed up this aspect of his research in which he had analysed alkalis like caustic potash (KOH) and soda (Na_2CO_3) by pointing out that such analyses showed Lavoisier had been wrong to claim that oxygen was the 'acid principle'. By dint of further experiments he was able to show that it was the ability of a substance in solution to release electrified hydrogen particles (what we today call 'hydrogen ions') that made it an acid; in other words hydrogen not oxygen should be considered the essential 'principle' of an acid.

Davy not only used the electric pile as a tool of research but was also intrigued by what went on chemically inside it. His 'piles' were really batteries of zinc and copper plates in water and once he began to examine their action, he noticed that no electricity was generated if the water was pure, but only when it contained a substance that would enable it to add oxygen to (i.e. to oxidize) the zinc. This made it clear that electricity could be generated by a purely chemical process, and further research using plates made of other metals in his batteries, led him to the conclusion that 'Chemical and electrical attractions are produced by the same cause'. In other words that the chemical affinity of one substance for another is similar to the electrical attraction between bodies.

Davy's reputation was immense and international. He was awarded the Napoleon Prize from the Institut de France in 1807 even though Britain and France were then at war, and between 1813 and 1815 went on a European tour, travelling through France and visiting French scientists on the way with the express permission of Napoleon. Interestingly he received no censure on his return and three years later was made a baronet; reactions today to such an international attitude to scientific research would hardly be the same, but then the situation has changed somewhat and scientific research has many more applications and implications as far as waging war is concerned. At all events, the trip through France enabled him to discuss electrochemical research with Gay-Lussac and his colleague Louis Thénard who were using a vast battery of many plates. They told him that they suspected that the liquid between the plates decomposed at a rate that depends on the electric current (i.e. on the quantity of electricity that flows) and not on the metal of which the plates were made. This puzzle was solved by Davy's young assistant Michael Faraday, who demonstrated that the electrical effect worked throughout the liquid not just at the points where it touched the plates. Faraday, to whom we owe the terms 'electrolyte' for the liquid in a battery and 'ions' for the positively electrified atoms that move through it, was next able to establish the laws of chemical and electrical action inside a battery. In 1833 he made a statement that showed great scientific insight; 'the atoms in matter', he wrote 'are in some way endowed or associated with electrical powers, to which they owe their most striking qualities, and amongst them their chemical affinity'. The significance of this we shall see later.

As the years passed chemistry had become an increasingly precise science, and by the mid-1830s it could certainly be classed as one of the 'exact' sciences. At the same time it also started to become more complex so that it is not possible here to follow its progress in great detail, though there is one new aspect that must be mentioned, the growth of 'organic chemistry'. The term 'organic' became used because the substances dealt with were those to be found in living things, though it was later realized that they were not solely associated with vegetable and animal life. The subject was a new one because in the earlier days of scientific chemistry it had been assumed that plant and animal materials were essentially different from other substances and that the same rules and ideas could not be applied to both. But by the nineteenth century the application of the same basic science to everything, animate or inanimate, was characteristic of the scientific outlook. Once the barrier was down it became clear that at least some organic substances like starch, sugar and certain oils and dyestuffs were amenable to ordinary chemical research. Moreover acids had been isolated from sour milk, and certain kinds of natural gum, and it was obvious that chemistry had a wider field to cover than had once been thought.

Though chemists in general still believed in a vital force operating within living things, chemical analysis began to show that, whether or not this was so, some living substances certainly contained ordinary chemicals. Indeed Lavoisier himself had found that all the 'organic' compounds he had examined had carbon in them and usually hydrogen as well. Then in 1828, the German chemist Friedrich Wöhler, who studied and worked with Berzelius, discovered that urea ($H_2N.CO.NH_2$) which is to be found in the urine of mammals, birds and some reptiles, as well as in milk and blood, could actually be synthesized from ammonium cyanate, a substance which had the same chemical composition, though its chemical action was quite different. This additional example of isomerism, as Berzelius called it, was seen as due to the way the atoms of the substance were arranged, and in due course this led to chemical formulae being written to display the structure of a substance as well as content. But Wöhler's discovery did more than help bring a new symbolism to chemistry, for the fact that an organic compound could be synthesized from inorganic materials established that animal compounds were indeed composed of ordinary and familiar chemical substances.

The subject of organic chemistry was next given impetus by new methods of chemical analysis devised by Justus von Liebig. Born in 1803 at Darmstadt in Germany, Liebig became interested in chemistry at an early age as his father had a laboratory for servicing a family business making medicinal chemicals and painting materials, and in due course he was apprenticed to an apothecary. He later studied chemistry in Bonn and Erlangen but though he obtained his doctorate in 1822, at the age of nineteen, Liebig realized he must go abroad for further instruction. He went to Paris and attended lectures by Gay-Lussac, Thénard and Dulong. A paper by him on the explosive acids known as fulminates (CNOH) impressed Humboldt who arranged for the young man to work with Gay-Lussac. Later Humboldt persuaded the Bavarian sovereign Ludwig I to provide Liebig with a laboratory and an academic position at the university of Giessen. Here, as 'extraordinary' professor, Liebig taught generations of students from 1824 until 1852 when he moved to Munich.

Liebig carried out much research on organic compounds, especially with Wöhler, and was a prolific author, writing well over 300 scientific papers, a book on organic chemistry, an encyclopaedia of chemical science and starting not one but two scientific journals. He also tackled questions of agricultural chemistry, but above all he was a brilliant teacher and his influence spread far and wide as his pupils moved into all kinds of chemical research; in fact, Leibig's greatest contribution may well have been the inspiration to carry out research into chemistry, and especially organic chemistry, that his teaching managed to arouse. Among these dedicated pupils was the organic chemist August Hofmann. Born in 1818 at Giessen, son of the architect who enlarged Liebig's Giessen laboratory, he studied chemistry and followed Liebig's tradition of giving instruction by teaching in the laboratory. In 1845 in his late twenties, Hofmann moved to London to take charge of the Royal College of Chemistry which had been founded by some of Liebig's English pupils and was later to be absorbed into London University. Here he taught with great success and began research into the organic chemistry of coal-tar substances which are the products of the distillation of coal and contain a wide variety of hydrocarbon compounds. He encouraged some of his pupils to carry out similar work, most notably William Perkin, the son of a builder and contractor, who became a very able chemist – he actually assisted Hofmann for a time – and built a laboratory at his home. Perkin's research centred on the relationship between quinine (derived from the bark of the cinchona tree and well known as a medicine), and a substance derived from the coal-tar product quinoline. In 1856 Perkin's research led him, quite by chance, to discover a remarkable purple dye and, being a man with a strong practical sense, he at once realized that here was a product of great commercial importance. Purple had always been a rare dye, derived from a secretion of a sea-snail (*Murex brandaris*), and its use confined to the wealthy, but now it could be synthesized at will; moreover, the synthetic purple had a far greater degree of permanence than the sea-snail product. With this synthetic purple dye Perkin went into commercial production and laid the foundations of what was to become a great chemical industry, for a whole series of beautiful synthetic dyes were discovered in the years that followed. But important scientific results also arose from this discovery.

Illustration page 449

Illustration page 449

Benzene

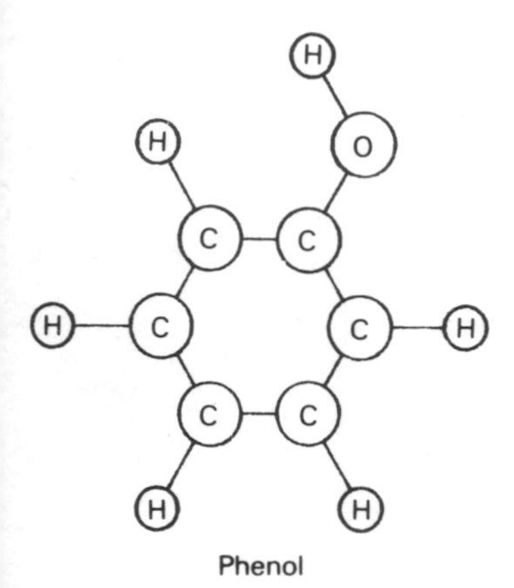

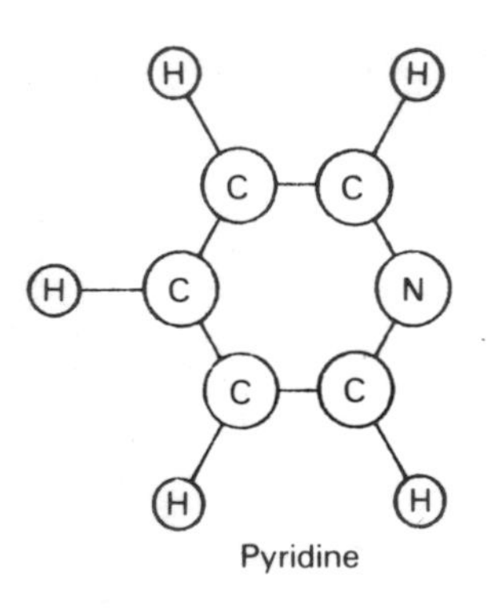

The arrangement of atoms in the benzene ring C_6H_6 (top), and its modification to form phenol (centre) by replacing one hydrogen with a hydroxyl group; and pyridine (below), in which one of the carbon atoms is replaced with nitrogen.

Hofmann had extracted benzene from coal tar, and this substance lay at the basis of all the aniline dyes, but the scientific problem they posed was not their chemical composition but the way in which the atoms were arranged within their molecules. The problem, though difficult, was nevertheless soon to be solved by August Kekulé von Stradonitz (1829–96). In the 1850s Kekulé had begun to carry out research, studying the replacement of hydrogen by other substances during chemical reactions, and this led him to the important conclusion that although chemists had been placing different substances into groups according to their chemical behaviour, it would be preferable to classify them according to the way atoms, or groups of atoms in them, were replaced during chemical reactions. In other words Kekulé wanted to use a system based on the combining power of the substances, on what was called their 'valency' (Latin *valens*, strength).

This concept, coupled with the idea that valency caused, or was at least indicative of, the way atoms were arranged, led Kekulé to make a particular study of carbon, which was present in all organic materials. He found it had a fourfold combining power (it was 'tetravalent') and realized that this would be the reason why it could form into the long chain molecules typical of organic compounds. In the case of benzene, Kekulé knew he had a substance whose molecules contained six carbon atoms, but there was a particular problem here. Benzene did not behave like a substance with an open-ended chain of carbon atoms as did so many other organic compounds. Kekulé puzzled over this for a long time and then one day in 1865 while dozing over the fire, he suddenly came to the realization that if the six carbon atoms were linked in a closed chain, all the experimental results obtained with benzene could be explained. He had discovered that basic structure, the benzene ring.

Kekulé's discoveries were important, not only because they explained many facts about organic reactions which had not previously been understood, but also because they gave chemists a basis on which to construct the atomic layout of a host of other organic molecules. In the twentieth century this was to lead to new chemical and biological insights and, commercially, to the vast petrochemical industry and the development of plastics.

Physics

During the nineteenth century physics forged ahead at an ever accelerating rate and subjects which had once been separate disciplines began to converge. All the same it will be more convenient if we begin by taking the sections – heat, electricity and light – and see the relationships as they unfold historically.

Heat

At the turn of the century when Count Rumford was favouring a vibratory theory to explain the heat generated when cannon are bored, he carried out a number of experiments to try to get some kind of quantitative result to support his view. Using a metal cylinder and inserting a blunt borer, he had two horses harnessed to rotate the cylinder 34 times a minute. In some experiments the cylinder was packed round with flannel to conserve its heat, in others Rumford immersed the cylinder itself in a barrel and discovered enough heat was generated to melt blocks of ice floating in it or to boil the water. In every case the supply of heat seemed inexhaustible and it became clear to him that heat could simply not be an imponderable fluid.

After Rumford's research, it fell to the Frenchman Sadi Carnot to take the matter farther. Elder son of one of the leading figures in the first French Republic, Carnot made a particularly penetrating analysis of machines that produce mechanical power from heat, concentrating especially on the energy and heat losses of a steam engine. It was a careful study that allowed Carnot to show that every engine, whatever kind it might be, could be divided into three constituents; each possessed a source of heat (in a steam engine this is the firebox), a working substance to carry the heat (the water and steam in a steam engine), and a receiver for the heat (in a steam engine this would be the condenser). In passing from source to receiver heat moved from a higher temperature to a lower one and work was done. From the results of this analysis Carnot claimed that in a perfect engine where no heat is lost to its surroundings and there is no loss by friction, no heat or caloric escapes (he still used the imponderable fluid theory). The mechanical work an engine does is due to the fall in temperature not to loss of caloric as such. In brief Carnot's idea was that no heat is ever created or destroyed; it is only moved from one body to another (in an engine from source to receiver). In an ideal engine, then, it should be possible – at least in theory – to move the caloric back from receiver to source and everything would be the same as it was at the beginning of the process. Put in other words, in theory the heat cycle is reversible though it is certainly never so in practice. Carnot published these results in 1824 and then continued his research but unfortunately died in a cholera epidemic in 1832 at the early age of thirty-six. His research was incomplete and, doubly unfortunate, his posthumous papers were undiscovered and unpublished for nearly half a century. For Carnot had finally come to the conclusion that 'heat is nothing else than motive power, or rather motion which has changed its form', and had begun to work out the foundations of a kinetic (vibratory) theory of heat. He also suggested that the total motive power in the universe was constant; indeed his notes enshrine most of the groundwork for what was to be called the first law of 'thermodynamics', as will become clear shortly.

The next stage in the story of the development of a modern theory of heat concerns the research of three men, James Joule, Lord Kelvin and Rudolf Clausius.

Joule (1818–89) began his working life in his father's brewery at Salford, near Manchester, but later studied under Dalton and after this began experimenting on the constitution of gases. He studied the work done by a gas when it expands and the heat generated when it is compressed – both instances of a relationship between mechanical action and heat – and in 1847 he described what has now become a famous experiment. This consisted of a paddle-wheel in water and measurements made of the work done in turning the wheel and the changes of temperature of the water bath. The experiment gave a precise determination of how much work is required to generate a given amount of heat or, as it is usually termed, the mechanical equivalent of heat.

Illustration page 451

Joule's results were taken up by William Thomson, later Lord Kelvin, who was born in Belfast, Northern Ireland, in 1824. The family moved to Glasgow when Kelvin was five years old on his father's appointment to the chair of mathematics there. Kelvin himself was trained at Glasgow and Cambridge in physics and mathematics and in 1846, at the age of twenty-two, was appointed professor of physics at Glasgow, due not a little to his father's influence. Here Kelvin remained for the next half century. A brilliant mathematician and physicist, he also had great practical and financial flair; he became involved with the laying of the first Atlantic telegraph cable in 1866 and made great improvements in the design and accuracy of the magnetic compass. For this work he was first knighted and then raised to the peerage, taking the name Kelvin after the river on whose banks the University of Glasgow lies.

Kelvin tried to find the appropriate mathematical laws to express Joule's work, and at once realized that to formulate them he needed some absolute scale of temperature rather than an arbitrary one based only on the freezing and boiling points of water, or of some other convenient liquid. Kelvin turned back to Carnot's published work on the heat cycle, and realized that the mechanical work done by a perfect frictionless engine depends only on the amount of heat or caloric, (at this stage Kelvin was still thinking of heat as an imponderable fluid), and the temperatures of heat source and receiver. In consequence he tried to find a temperature scale in which the unit of heat and the mechanical work developed will always be the same in whatever part of the scale the temperature difference happens to lie; in other words so that they are the same whatever the actual temperatures involved. Such a scale would be independent of the working substance – water, alcohol, mercury or whatever – or on the body in which the heat exchange is occurring. To achieve this Kelvin had to be clear about the precise relationship between the heat generated in Joule's experiment and the caloric Carnot had talked about.

It is at this point that the German physicist Rudolf Clausius enters the picture. Born in Köslin in Prussia in 1822, son of a pastor and schoolmaster, Clausius studied at Berlin University and in 1850 made his name with a famous paper on the theory of heat. Reviving Carnot's published work on the steam engine, Clausius realized that Carnot had been mistaken when he thought of an engine doing work solely because its caloric dropped in temperature. Clausius agreed that caloric could not be destroyed but claimed it could be converted into something else; in an engine, for example, it was converted into mechanical work. This led him to formulate two laws for expressing a relationship between the flow of heat and mechanical work – the first two laws of thermodynamics. The first stated that in any closed system (a steam engine, for example) the total amount of energy is constant. The second law stated that heat cannot pass from a colder to a hotter body on its own accord; for this to happen some external cause must come into operation. This last law is often expressed by saying that entropy always increases, entropy (Greek *trope*, transformation) being the measure of the unavailability of energy.

What Clausius's work did was to show that Carnot's 'caloric' was Joule's 'heat', and Kelvin now had the clue he needed. The quantity of heat taken from a source and the quantity of heat absorbed by the receiver or 'heat-sink' depend on the difference in temperature between source and sink. This is irrespective of where both lie on the temperature scale. What the heat sink does not absorb is converted into mechanical work. What is more, if the sink is at 'zero temperature' and remains so, it is taking no heat from the source. All energy is therefore available for conversion to mechanical work. The entropy is zero because no energy is unavailable. For convenience Kelvin took 0° on his 'absolute' scale as the freezing point of water and 100° for the boiling point of water; this gave him his 'absolute zero' at −273.1°C (−459.7°F).

For Kelvin's intervention in the debate on Darwin, see page 427; for his work on magnetism and electricity, see pages 466–467.

This absolute zero point on Kelvin's scale is not in fact attainable due to the way atoms are constructed, though Kelvin did not know this. But that is not the point. What did matter was that the work of Clausius and Kelvin made it clear that heat was no mysterious weightless fluid but was instead a form of energy. So, too, was mechanical work. It also became clear that neither form of energy could be destroyed, though one could be converted into the other. This led to the principle that became known as the conservation of energy, a principle that had already been stated in 1847 by the German physicist Hermann von Helmholtz but to which Clausius and Kelvin had now given a deeper meaning.

Electricity

If heat was conceived of as an imponderable fluid at the beginning of the nineteenth century, it was not the only physical entity to be looked at in this way. Electricity was another; in fact in 1800 the main question seemed to be whether it was one fluid or two. Yet here again the picture was to change due to more extensive and penetrating research. Some clues were given by Davy and others on their work on the chemical reactions occurring inside batteries, and further evidence was extracted from studies on the passage of electricity along wires. Henry Cavendish had experimented on conduction though he had not published his work, but Davy did make his investigations known and showed that when using a metal wire (not silk or thread as in the case of electrostatic experiments the century before), the conducting properties depended upon its diameter and the kind of metal used. This was in 1821 but not for four years did an exact relationship emerge between the strength and quantity of electricity flowing through a conductor. Then a German school-teacher, Georg Ohm, carried out experiments using wires of the same thickness but different lengths and made his measurements using a Coulomb torsion-balance. He found that the resistance of a wire was independent of the quantity of electricity flowing through it (i.e. of the current), but he did more than this. Ohm continued his research and in 1826 and 1827 formulated a theory to explain his results. He suggested that electricity moved through a wire passing from particle to particle (the heat fluid caloric had been thought of as moving rather in this way) and calculated that such a movement must be caused by an electrical tension or potential, analogous to the way a difference in temperature caused a flow of heat. This last concept came to be known as the electromotive force and brought us our unit the volt, named after Volta. Ohm's name became attached to the unit of resistance.

At the same time Ohm was at work two other physicists were carrying out important electrical research, the Danish physicist Hans Christian Oersted, then a 'professor extraordinary' at Copenhagen University, and the French physicist, mathematician and chemist André Marie Ampère. Oersted, who for philosophical reasons believed there must be a connection between electricity and magnetism, was able to prove experimentally that when an electric current passed along a wire there was a magnetic field associated with it. This was in 1820 and soon after, between 1821 and 1825, Ampère traced the whole relationship between currents and their effects on magnets, as well as the opposite effect, the action of magnets on electric currents. Such research then led him to suggest that a magnet was composed of magnetic 'molecules' in each of which a current was forever circulating, a view that correlated the various experimental results then known and was of immense importance.

Illustration page 451

The stage was now set for a new series of developments in electromagnetism and it is fortunate that just at this time the Royal Institution had a physicist with great experimental flair who was able to take the whole matter further. The man concerned was Humphry Davy's assistant Michael Faraday, whose acquaintance we made briefly when discussing electrochemistry. Born in 1791 in Newington (now part of London), Faraday received only a rudimentary education. At the age of thirteen he had to contribute to the family income and he went to work for a bookseller, a Mr Riebau, delivering newspapers and running errands. A year later he was apprenticed to Riebau to learn bookbinding and for the next seven years developed a dexterity that was to serve him well later as an experimental scientist. It was while at the bookseller's that Faraday came across scientific books; these fascinated him and, as luck would have it, this interest in science was noticed by one of Riebau's customers who gave him tickets to attend a series of lectures by Davy at the Royal Institution.

Left William Perkin (1838–1907) holding a skein of material dyed with his new aniline dyes. National Portrait Gallery, London.

Below left A miner working by the light of a Davy safety lamp, invented in 1815. Sir Humphry Davy (1778–1829) combined pure chemical research with work on practical problems such as the protection of ships from corrosion.

Below A condenser designed by Justus von Liebig (1803–73), useful in the analysis of organic compounds. Science Museum, London.

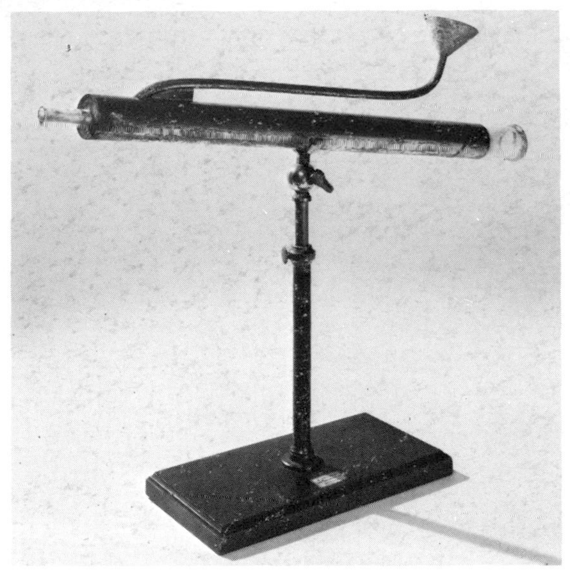

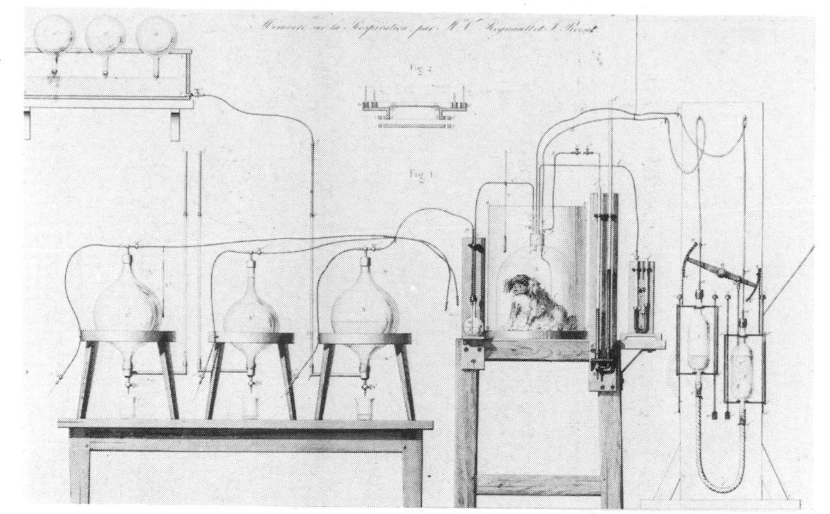

Right The first apparatus in which respiration could be monitored and analyzed accurately for changes in temperature, pressure and composition. From V. Regnault and Jules Reiset, *Annales de Chimie et de Physique*, 1849.

Right James Glaisher (1809–1903) and Henry Coxwell (1819–1900) making a balloon ascent to about 11,000 m (35,000 ft) to make meteorological measurements. In the mid-19th century meteorological data collection was organized on an international basis. From Glaisher's *Voyages Aeriennes*, 1870.

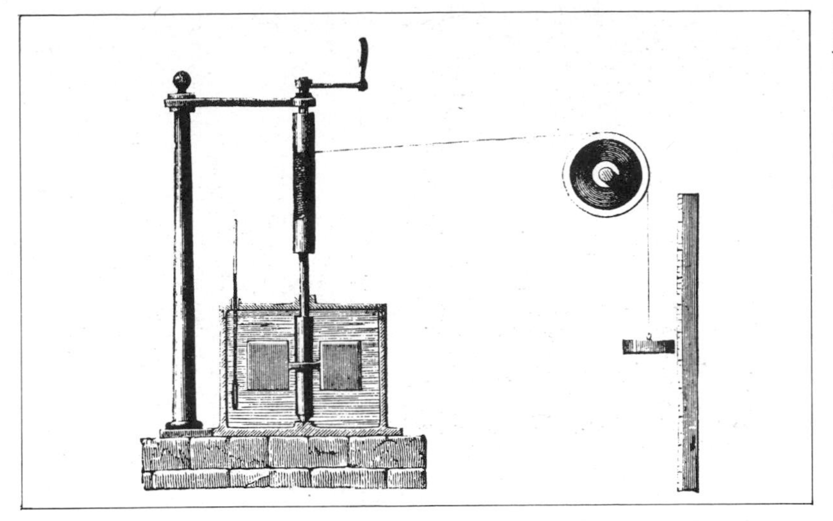

Left The apparatus devised by James Joule in 1847 to measure the mechanical equivalent of heat, by using a regulated weight to drive vanes in a liquid, the temperature of which could be measured by thermometer, thus monitoring changes caused by a given amount of work.

Left Hans Christian Oersted (1771–1851) discovering in 1820 the effects of an electrical current on a magnetic compass.

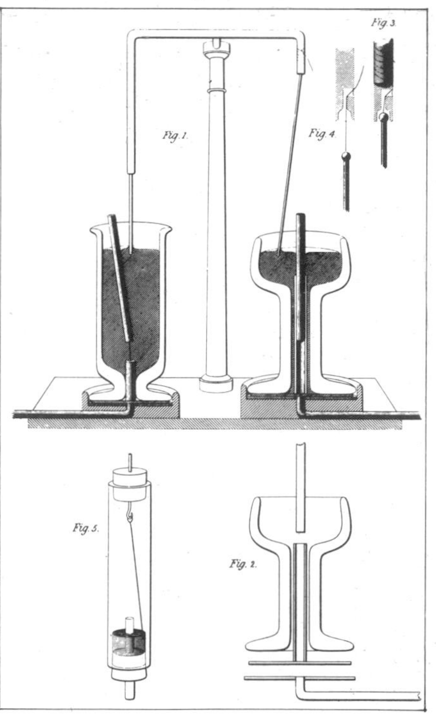
Fig. 3
Fig. 1.
Fig. 4
Fig. 5.
Fig. 2.

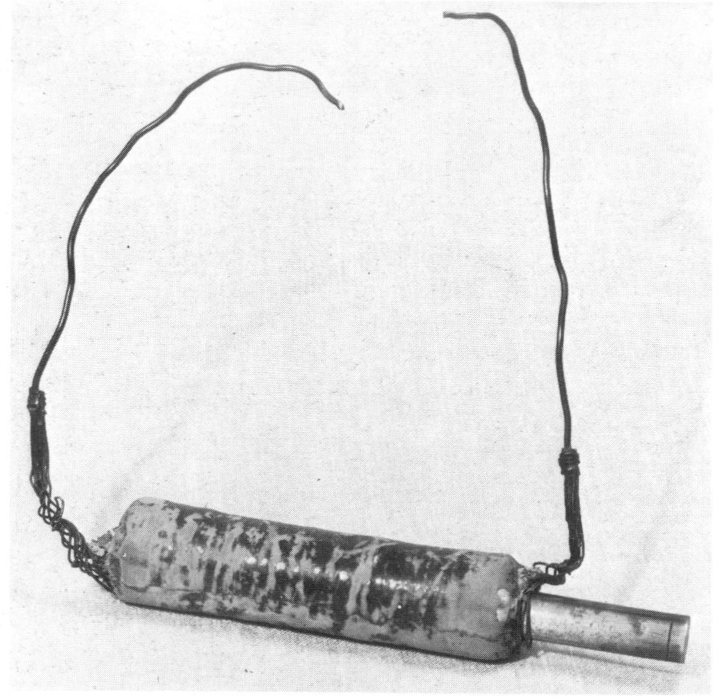

Left James Clerk Maxwell (1831–79) as a young man. His important contributions to science began with his work on elastic solids, done when he was 18.

Opposite above Michael Faraday (1791–1867) in his laboratory. Royal Institution, London.

Opposite below left Faraday's apparatus for investigating electromagnetic rotation, a study that led directly to the development of the electric motor and the dynamo.

Opposite below right A solenoid (coil of wire) and sliding magnet devised by Faraday to test the relationship between the movement of the magnet through the solenoid, and the induction of an electrical current in the coil. Royal Institution, London.

Above Jean Léon Foucault (1819–68), the French physicist whose work included the measurement of the velocity of light, the demonstration of the diurnal motion of the Earth and the invention of the gyroscope, as well as work on astronomy.

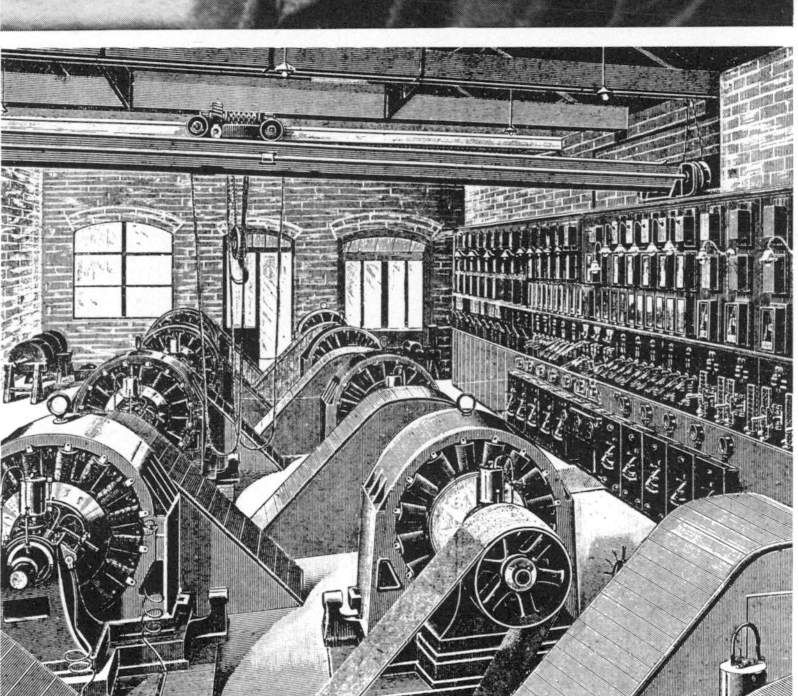

Left Faraday's work on electricity and magnetism in the early 1830's made it possible to generate electricity to provide power for industrial or domestic use; but this development did not occur for more than 50 years. This illustration shows one of the early high-voltage power stations, opened in London in 1896.

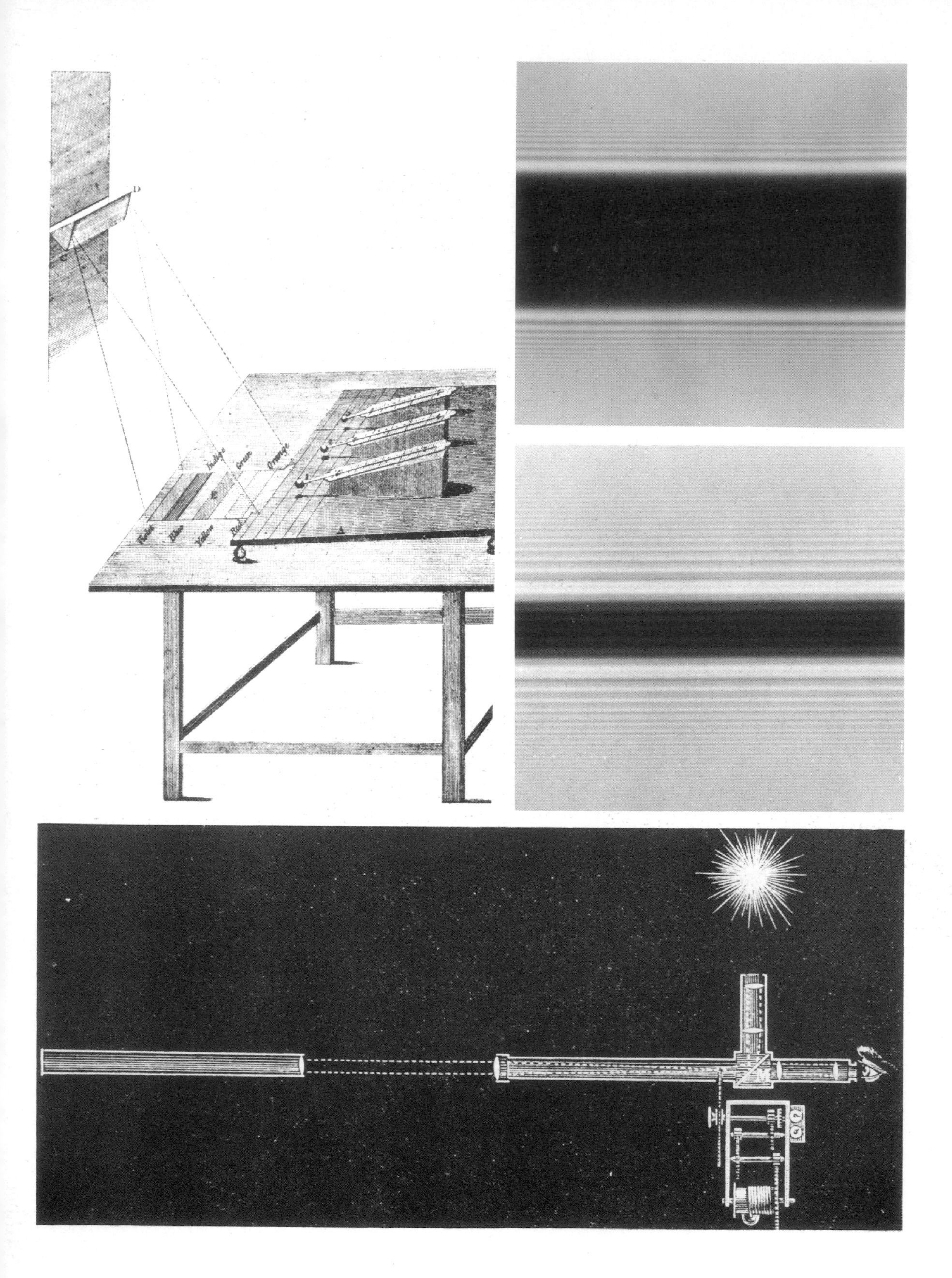
Indigo
Green
Orange
Blue
Yellow
Red

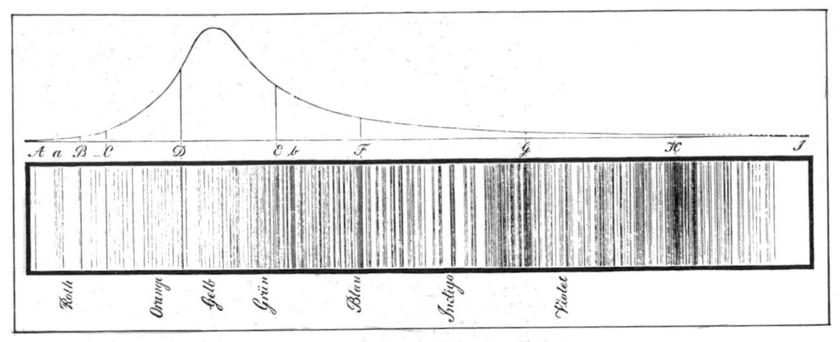

Left Joseph von Fraunhofer's drawing, made in 1814, of the solar spectrum, showing many thin black lines. Above is a curve indicating the intensity of the light at different parts of the spectrum. The investigation of these lines gave rise to the very fruitful field of spectroscopy.

Centre A photograph of the 1.8 m (72 in) aperture reflector built in 1845 by the third Earl of Rosse at Birr Castle, Abbeyleix, Ireland.

Opposite above left Demonstration by William Herschel in 1800 of the heating effects of infra-red radiation. Light was passed through a prism, and thermometers set up on the table beyond the red end of the spectrum.

Opposite above and centre right Two photographs of diffraction showing the patterns, discovered by Augustin Fresnel in 1815, of the interference fringes caused by objects of varying sizes. The smaller the object the more intense are the fringes. The discovery of these patterns helped to support the wave theory of light.

Opposite below Experiment devised by Hippolyte Fizeau (1819–96) and published in 1849, to calculate the speed of light. A light beam is reflected by a thinly-silvered mirror (M) to a distant mirror (left), passing through the teeth of a wheel rotating at high speed. The speed of the wheel is adjusted until the return beam is stopped by the tooth following the gap through which the outward beam passed. The speed of rotation allows the speed of light to be calculated.

Left Photograph published by the Lick Observatory in 1908 of the spiral galaxy in the constellation of Canes Venatici, an object whose spiral nature had been detected by Rosse.

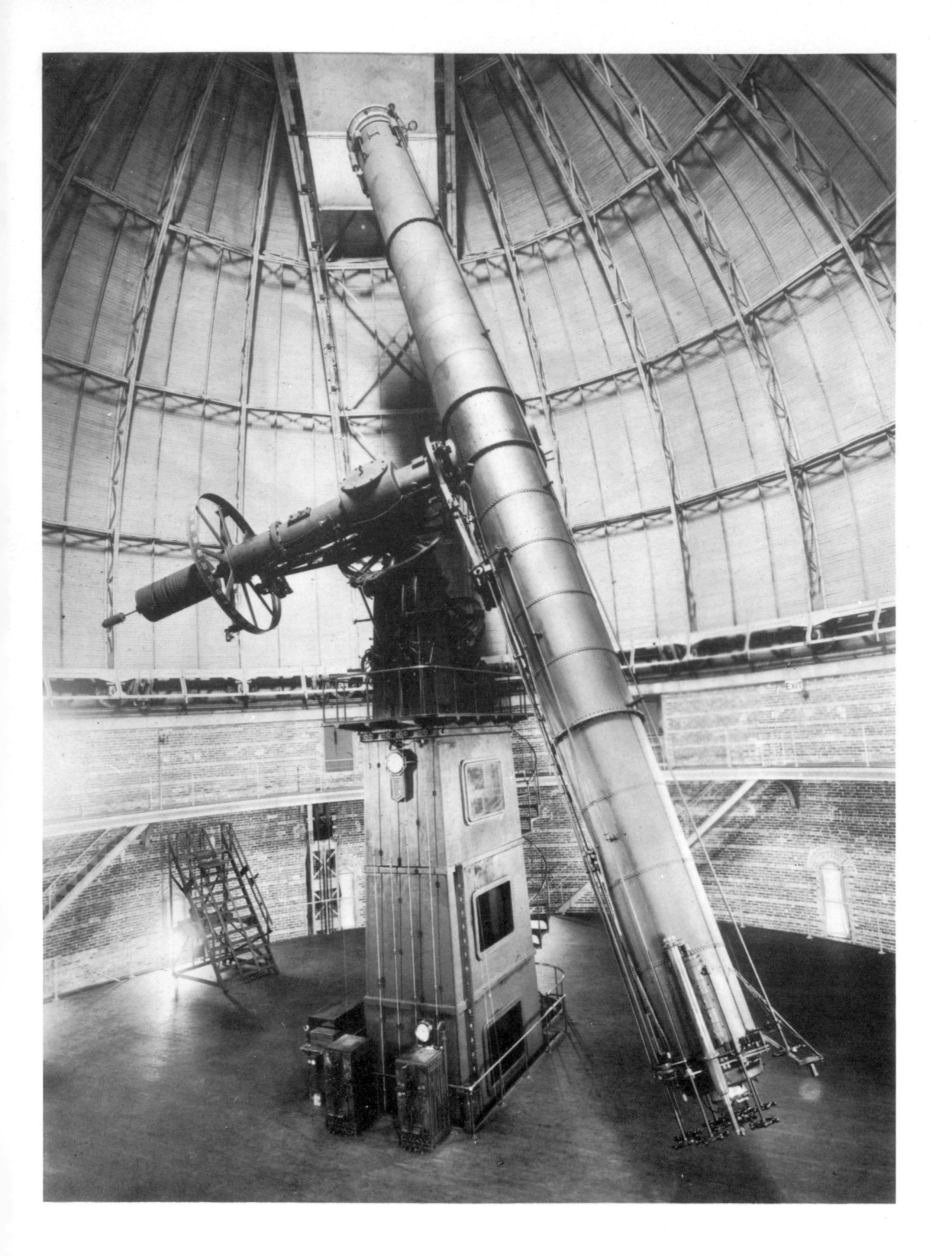
EXIT

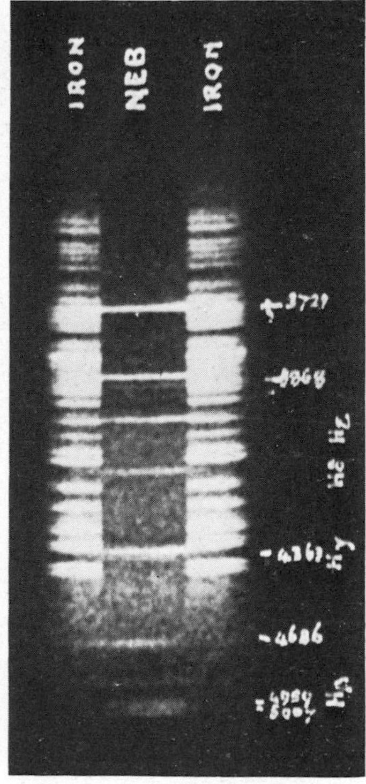

Above Spectrum of the 'Dumb-Bell' nebula photographed by Max Wolf about 1906. The spectrum of the nebula lies in between the companion spectra of iron. Observations such as these sorted out gaseous nebulae like the Dumb-bell, which had bright line spectra, from those which did not. These latter proved to be galaxies.

Above left The importance of photography in astronomy is epitomized by the detail recorded for future study in this picture of 1877 by Andrew Common of the Moon, just after full.

Opposite The 1 m (40 in) aperture refractor at Yerkes Observatory near Chicago built in the 1880s. It is a fine example of the marriage of the advanced optical skills and the heavy engineering of the 19th century.

Left A spectroscope, with a train of six prisms to give a wide dispersion, being used on the Sun, about 1870. Such visual observations were soon to be replaced by photographic ones.

Right The Jacquard loom. Invented in 1801, this loom could produce complex patterns, controlled by the punched-cards seen at the top of the machine. The principle of instructing machines by means of punched cards was developed later in the century, and formed the basis on which computer technology was to be founded after World War II.

Opposite The analysis of US census results by means of punched cards and counting machines in 1890 allowed a more powerful and flexible analysis of statistics than previously possible.

Below right The use of hypnosis for medical purposes was common throughout the 19th century, even though its nature was not properly explored until the work of James Braid in the middle of the century. Here the French doctor Jean-Martin Charcot (1825–93) uses a large tuning-fork to induce hypnotism.

Below Part of the 'analytical engine' devised in England by Charles Babbage (1792–1871) in 1834. This was the true forerunner of the 20th-century computer. Science Museum, London.

[Entered at the Post Office of New York, N. Y., as Second Class Matter. Copyrighted, 1890, by Munn & Co.]

A WEEKLY JOURNAL OF PRACTICAL INFORMATION, ART, SCIENCE, MECHANICS, CHEMISTRY, AND MANUFACTURES.

Vol. LXIII.—No. 9. ESTABLISHED 1845. NEW YORK, AUGUST 30, 1890. [$3.00 A YEAR. WEEKLY.

THE NEW CENSUS OF THE UNITED STATES—THE ELECTRICAL ENUMERATING MECHANISM.—[See page 132.]

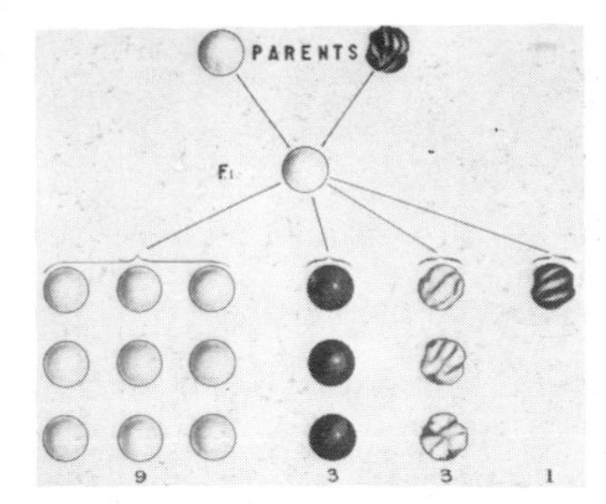

Above Thomas H. Morgan's illustration of dominant and recessive genes from *A Critique of the Theory of Evolution*, 1919. A cross between yellow-round and green-wrinkled peas gives a ratio of nine with both dominant traits, three showing one dominant trait and the other recessive, three showing the opposite and only one displaying both recessive traits.

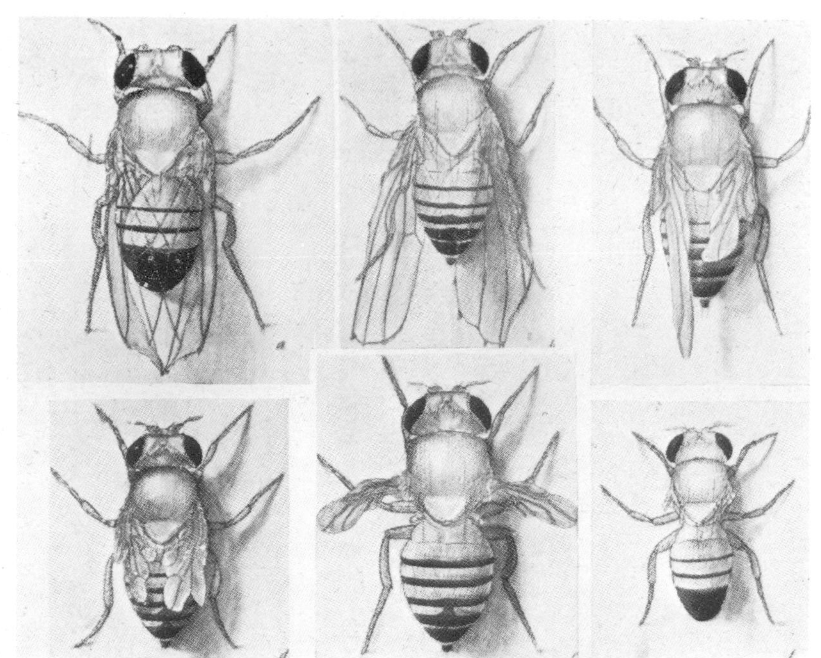

Right Some mutations of wings in the fruit fly (Drosophila), from *A Critique of the Theory of Evolution*.

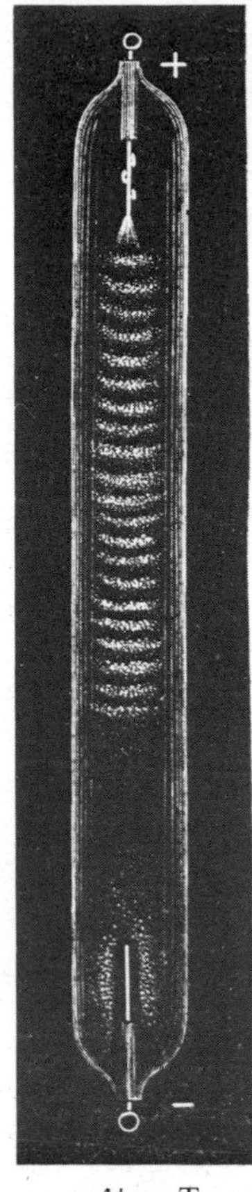

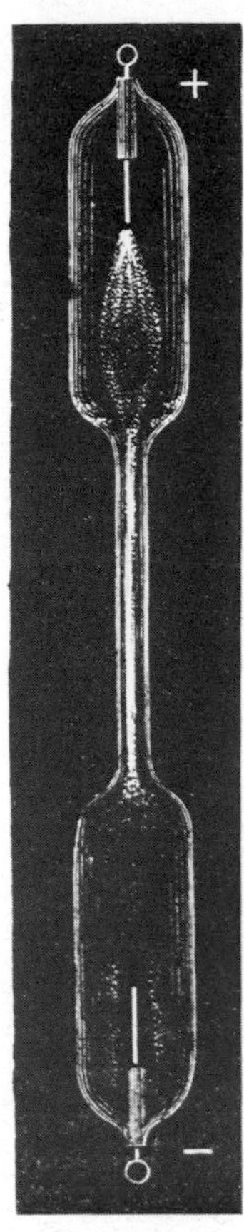

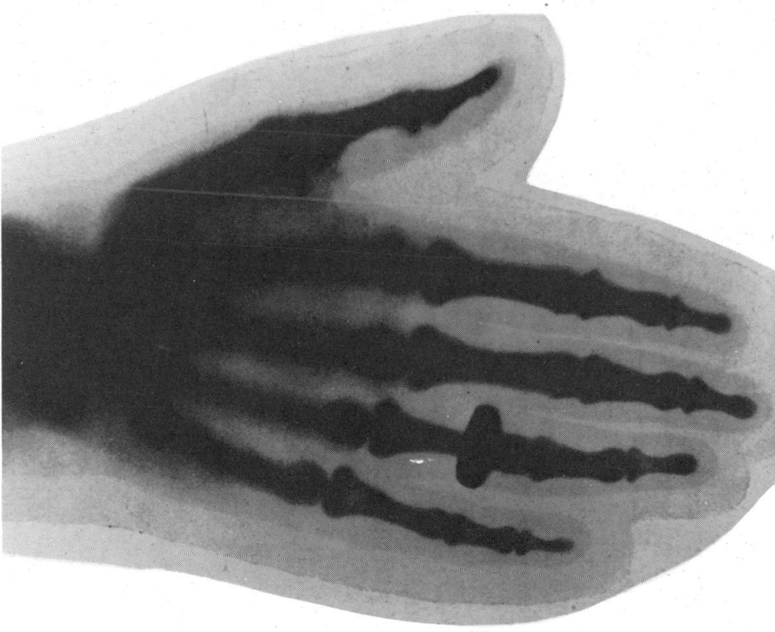

Above Two Geissler tubes from the 1890s, showing discharges in gases. The tube with the narrow middle section was much used for the analysis of gases by spectroscopy.

Above left The production of electric light bulbs using a mercury vacuum pump in the late 19th century. The pump was modified from that developed by Johann Geissler to permit the study of electrical discharges in gases at reduced pressures. This device was adapted by Röntgen in his discovery of X-rays.

Left A very early X-ray photograph of a hand. In the decade after Röntgen's discovery of X-rays in 1895, many different hypotheses were put forward as to their nature.

Opposite below Mutations of maythorpe barley, showing differences in head emergence and maturity. The seeds had been irradiated with gamma rays which affected their chromosomes.

Right Ernest Rutherford (1871–1937) and Hans Geiger (1882–1945) in their laboratory in Manchester where they studied the nature of alpha- and beta-ray particles from 1907.

Below The 'Cloud Chamber' devised by Charles Wilson between 1896 and 1912. The chamber contains air strongly saturated with water vapour. Moving a piston at the bottom of the chamber causes the space inside the chamber to expand suddenly, with the result that the air cools and the water vapour condenses as droplets on any electrified air particles (ions) caused by the passage through the chamber of nuclear particles. This renders the tracks of the nuclear particles visible. Science Museum, London.

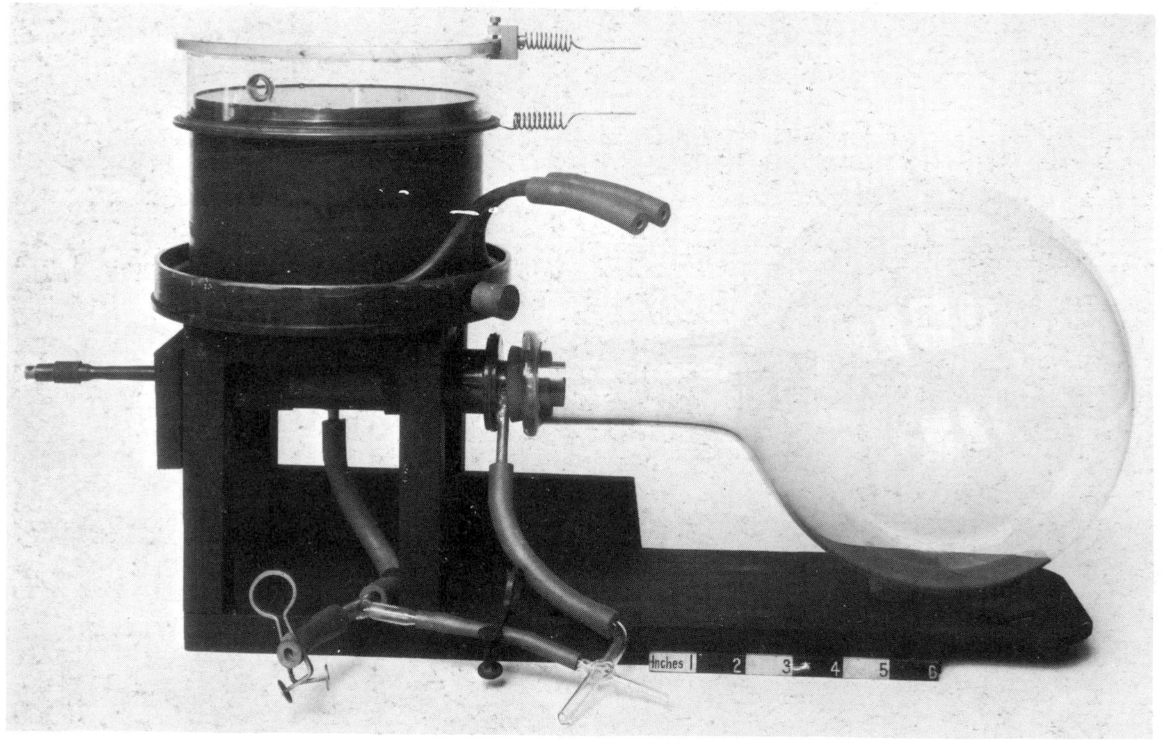

Left Pierre (1859–1906) and Marie (1867–1934) Curie in their laboratory in Paris in 1903, the year in which they shared the Nobel Prize with Henri Becquerel.

Below J. J. Thomson (1856–1940), at the Cavendish Laboratory, Cambridge University.

Right The Russian chemist Dmitri Mendeleyev (1834–1907), whose formulation of the Periodic Table of chemical elements in 1869 (*below*) and its later refinement led to the discovery of new elements. It also underpinned the development of quantum physics and study of the structure of the atom from 1913.

Tabelle I.

			K = 39	Rb = 85	Cs = 133	—	—
			Ca = 40	Sr = 87	Ba = 137	—	—
			—	?Yt = 88?	?Di = 138?	Er = 178?	—
			Ti = 48?	Zr = 90	Ce = 140?	?La = 180?	Th = 231
			V = 51	Nb = 94	—	Ta = 182	—
			Cr = 52	Mo = 96	—	W = 184	U = 240
			Mn = 55	—	—	—	—
			Fe = 56	Ru = 104	—	Os = 195?	—
Typische Elemente			Co = 59	Rh = 104	—	Ir = 197	—
			Ni = 59	Pd = 106	—	Pt = 198?	—
H = 1	Li = 7	Na = 23	Cu = 63	Ag = 108	—	Au = 199?	—
	Be = 9,4	Mg = 24	Zn = 65	Cd = 112	—	Hg = 200	—
	B = 11	Al = 27,3	—	In = 113	—	Tl = 204	—
	C = 12	Si = 28	—	Sn = 118	—	Pb = 207	—
	N = 14	P = 31	As = 75	Sb = 122	—	Bi = 208	—
	O = 16	S = 32	Se = 78	Te = 125?	—	—	—
	F = 19	Cl = 35,5	Br = 80	J = 127	—	—	—

These Faraday enjoyed immensely, making copious notes and then writing them up with the greatest care. In October that year Davy was temporarily blinded by an explosion in the laboratory and Faraday was recommended as an amanuensis. Later, when Davy had his sight again, Faraday sent him his set of lecture notes with the result that in February the following year, when an assistant in the laboratory was sacked for brawling, Davy offered the post to Faraday. The two obviously got on well and when he went on his continental tour with his wife in the late summer, Davy took Faraday with him. Two years later Faraday was appointed superintendent of apparatus, in 1825 director of the laboratory and in 1833 professor of chemistry.

Illustrations page 452

It was while he was director of the laboratory that Faraday first turned his whole attention to the question of electromagnetism and started to produce those results that were to have such far-reaching effects on industry as well as science. To begin with he argued that if electricity flowing in a wire gave magnetic effects, as Ampère had shown, then the converse should be true; a magnetic effect should give rise to an electric current. To test this experimentally he took an iron ring and wound round it two separate coils of wire. One coil went to a battery, the other to a 'galvanometer' (a sensitive detector of an electric current), and Faraday found that when he connected and disconnected the battery, an electric current temporarily flowed in the other wire. Clearly this was generated by magnetic effects of the first current. A second experiment using a coil of wire wound on an iron rod and two bar-shaped magnets demonstrated clearly that magnets alone could induce a current, a direct 'conversion of magnetism to electricity' as Faraday put it. His theoretical imagination had been proved correct. Other experiments followed and from them he found that a coil of wire would induce an electric current in itself at those moments when a current is turned on or off – the phenomenon of 'self-induction'.

These experiments led to all kinds of practical results – the development of the electric motor and the electric generator and thus to electric trains, tramways and a public electricity supply, as well as to the electric telegraph and, in the hands of an inventor like Alexander Graham Bell, to the telephone. They also raised a theoretical problem that was not new, though in the light of them had become a serious challenge. This was the question of how electricity and magnetism could affect each other across empty space, the question of action at a distance. Faraday himself used the useful and productive idea of a field. He imagined that lines of magnetic force existed, being closer together the stronger the magnetic field. These lines, he also thought, tended to shorten when they could and to repel one another. Such explanations accounted for the results of his experiments.

In 1837 Faraday introduced the parallel concept of lines of electric force and the next year was in a position to propose a theory of electricity. Particles of matter were composed of forces arranged in complex patterns; these patterns gave them their characteristics.

However, the patterns distorted under strain such as that imposed by electrical forces. Faraday then used the idea to explain the phenomenon of lightning, and electrostatics and electrochemistry. It was not a theory that commended itself particularly to the scientific community and Faraday himself proposed it with great diffidence, but coupled with his superb experimental work, it did enable him to bring together in one grand unity all the previously disparate elements of electrical study. Voltaic currents, electricity from friction machines, from lightning, electricity by induction, electromagnetic effects, animal electricity (such as that displayed by a torpedo-fish, for instance) and even thermoelectricity (the electricity produced by heating two dissimilar metals in contact) had all been shown to be the same sort of electricity. As Faraday himself put it, 'Electricity, whatever it may be, is identical in its nature'.

In discussing electricity and the concept of lines of force, Faraday had suggested that space might be filled with such lines, and that perhaps light and radiant heat were vibrations travelling along them. But this was an idea that needed thorough mathematical analysis to give it precision if it was to become anything other than an interesting suggestion. The man who took up the challenge was a Scot, James Clerk Maxwell (1831–79). At the age of twenty-four he became professor of natural philosophy at Marischal College, Aberdeen and then very soon after at King's College, London. In 1865, at the early age of thirty-four, he retired from regular academic life to his estate in south-west Scotland and it was here that he wrote his celebrated *Treatise on Electricity and Magnetism*. This important text came out in 1873, by which time Maxwell had moved back into the academic world, having been appointed in 1871 to be the first professor of experimental physics at Cambridge, where he planned and developed the famous Cavendish Laboratory.

Illustration page 453

In his brief life he made remarkable achievements in mathematics proving, amongst other things, that the rings of the planet Saturn could not be solid but must be composed of myriads of minute particles, in optics (he invented the 'fish-eye' lens) and in colour vision, and he also helped advance the theoretical understanding of rarified gases. It was, however, his contribution to electricity and magnetism that was the most significant.

Maxwell's interest in the subject arose both from meetings and correspondence with Faraday and from some work which Kelvin had carried out in 1842 while an undergraduate at Cambridge. Kelvin had compared the charge in a body generated by an electrical machine with the way heat spreads out in a hot body large enough to have no boundaries (since these would complicate the study too much). Kelvin had used this comparison because the appropriate mathematical technique was already to hand. Surprisingly his results showed that the mathematical answers to the electrical problem were similar. As Maxwell put it, Kelvin's work 'introduced into mathematical science the idea of electrical action carried on by means of a continuous medium';

it was an idea proposed by Faraday but never before worked out in mathematical detail. In 1846 Kelvin wrote again on the subject, this time taking things a little further and making use of the then prevalent idea that all space was permeated by an aether which, although it could neither be weighed nor measured, acted as the carrier for light beams. Kelvin compared the electrical effects in an aether that carried electrical and magnetic effects with the strains to be found in a solid body that was undergoing tension. This was a useful exercise in studying how such an aether could transmit effects from one place or another. In the hands of Maxwell it was developed with great imagination and the most able mathematical skill.

Maxwell began his analysis in 1855, and for a start tried to find a mathematically correct explanation of the lines of force surrounding a magnet, that is of Faraday's magnetic field. A year later he was ready to publish a paper in which he attempted to connect together all Faraday's electromagnetic experimental results using methods similar to those with which Kelvin had been so successful. But this was only a first step, more remained to be done and not until five years later did Maxwell attain his goal. At last, in 1861, he was in a position to fit electric currents, electric charges and magnetism into one comprehensive scheme, postulating an aether to explain how electric currents and their assorted magnetic fields were always interacting with one another. Published with full mathematical details in 1864, it marked an immense stride forward in the understanding of electricity and electromagnetic effects. Yet it was more than this, for the mathematical results were astonishing in their implications. The equations that Maxwell had derived for expressing the behaviour of an electric current and its associated magnetic field were similar in every way to those already derived for expressing the behaviour of light waves (a wave theory of light had become accepted by this time). Thus what Maxwell showed was that light must be an electromagnetic wave of some kind, and, conversely, that electromagnetic waves should undergo reflection, refraction and all the other effects which light waves experience. Moreover, his results showed that radiation of shorter and greater wavelengths than light should exist. In 1888, nine years after Maxwell's death, long electromagnetic waves were indeed discovered; then Heinrich Hertz, professor of physics at Karlsruhe, generated such waves which, though they could not be seen visually, were detectable electrically, and could also be transmitted and reflected. Hertz had discovered radio waves; yet this, as it turned out, was to be only one of the consequences accruing from Maxwell's work.

Light

Newton's theory of light as a stream of corpuscles held the field until the beginning of the nineteenth century when a new approach was taken by Thomas Young. Born at Milverton in Somerset in 1773, the son of a Quaker banker and mercer, Young had an interest in science from an early age and also a flair for languages, a flair which

For this aspect of Newton's work, see page 377.

later in life was to enable him to be one of the first to translate Egyptian hieroglyphics. In the 1790s, Young went to Edinburgh, London and Götttingen to study medicine. After gaining his MD degree he went up to Cambridge for still further education, but in 1800 he settled in London, having been left a fine house and a small fortune by a maternal uncle.

Young now tried to set up in medical practice but with little success and when Count Rumford began to look for a professor of natural philosophy for the Royal Institution, Young with his broad interests seemed an ideal candidate, both socially and scientifically. He was appointed in 1801 but his lectures, unlike the clear and popular ones of his colleague Humphry Davy, were too obscure and technical for the kind of audience at which the Royal Institution aimed, and during 1803 he was forced to resign. From this time on Young spent his time in writing, in research and in unpaid public appointments; he also acted as foreign secretary of the Royal Society, another honorary post.

Young's work on the nature of light arose out of his interest in the problem of vision and from 1793 onwards he wrote a number of papers on what has become known as physiological optics; he discovered how the eye focuses images and discussed colour vision. His ideas about the nature of light itself appeared from 1800 onwards in the *Philosophical Transactions*. To begin with Young showed how, if light were a wave disturbance, it would be possible to overcome some of the difficulties that arose from the corpuscular theory. For instance, if light is due to corpuscles shot off from a body why, Young asked, should they always travel at the same speed whether they come from a spark generated by a flint or from the intense rays of the Sun. Surely, the speed should depend on the conditions that gave rise to the light? Again, if light is a stream of corpuscles why are only some refracted through a lens and some reflected? Certainly Newton had proposed his theory of 'easy fits' of reflection and refraction depending on how the corpuscles were vibrating when they were emitted, but as Young pointed out this came very close to a theory of light as vibrations.

Young did not, of course, confine himself to criticism of the corpuscles theory; he had a very definite positive contribution to make. Assuming that space was filled with a luminous aether and that light was a wave disturbance in such an aether, Young found it possible to explain all the usual effects of reflection and refraction. But his theory did more than this. It gave a simple explanation of the phenomenon of 'Newton's rings' (the coloured rings seen when a convex lens is placed in contact with a really flat surface such as a piece of glass) by supposing that every colour is due to waves of a particular wavelength. Taking the analogy of sea waves (though the theory used longitudinal or compression waves), Young said that when the waves meet 'in such a manner that the elevations of one series coincide with those of the other, they must together produce a series of greater elevations; but if the elevations of one series are so situated as to correspond to the depressions of the other, they must exactly fill those depressions'. In the first instance we get light reinforcing light, in the second light cancelling out light of a particular wavelength. The colours were due to the wavelengths of light which were not cancelled in this way. But the theory could go further for Young went on to say, 'I maintain that similar effects take place when any two portions of light are thus mixed; and this I call the general law of the interference of light'. Thus when two beams of light from the same source meet they interfere with each other, producing alternate bands of lightness and darkness. Young proved the point by a simple experiment and thus provided evidence that was not readily explicable, if explicable at all, by the old corpuscular theory.

The problem of diffraction, where light bends round a solid body or spreads out when passing through a minute aperture, was more difficult to account for. The corpuscular theory considered a refraction effect was caused by the gravitational attraction of the corpuscles as they passed very close to a body, but the more experiments that were done, the less satisfactory this explanation appeared to be. By 1803 Young claimed diffraction was caused by interference at the edges of a body – a reason with which physicists still agree – but his explanation of the precise way in which this happened was not convincing. Moreover, diffraction was not the only stumbling block for Young's theory; another was the double-refraction of light that occurred in certain crystals. In 1808 Laplace claimed that this splitting of light into two rays could be explained on the corpuscular theory by assuming that the corpuscles were split into two rays each with different velocities. Young attacked the idea though with no great success, and in the end the Académie des Sciences, of which Young was a foreign associate, offered a prize which was to be given two years later, in 1810, to anyone who could come up with a mathematical theory to explain the phenomenon.

Étienne-Louis Malus, who had served in Egypt with Napoleon and was then in his thirties and a colonel in the engineers, decided to answer the challenge. In December 1808, at the beginning of his research, he came across a new and surprising fact; light reflected at certain angles from a polished surface was unusual in some respects, and similar to the rays of light that emerged from a crystal after double refraction. Malus called the effect 'polarization' and found himself able to explain it on the corpuscular theory though it was inexplicable on Young's theory using longitudinal waves. Malus won the prize and the supporters of the corpuscular theory were triumphant. They could, they now believed, demolish Young's theory once and for all and, at their suggestion, the Académie offered another prize, this time for an explanation of diffraction.

The diffraction prize was won by Augustin Fresnel (1788–1827). This young engineer – he was still in his twenties – produced results which, to the amazement of the promoters, were to lead to the

Illustrations page 454

corpuscular theory's overthrow. In 1815 Fresnel had carried out some research similar to Young's and also some diffraction experiments which showed that diffraction was the same at the edge of an object whether it was sharp or blunt. On either Young's hypothesis or the corpuscular theory this should not happen, and Fresnel plumped for interference as the answer, as Young had done, but used a different explanation for the details. Considering, like Huygens before him, that each light wave gives rise to secondary wavelets, Fresnel than suggested that most of such wavelets are absorbed at the edge of a body and only a few remain to give rise to others. It is the interference of these that we observe as diffraction. The explanation also fitted the observed behaviour of light passing through small apertures.

Fresnel's work won him the support of François Arago, an influential member of the Académie and later its permanent secretary. Arago put Fresnel in touch with Young and the two began to work together. In the years that followed they brought forward new experiments that made the corpuscular theory untenable, though polarization was still a problem. Then, back in France, Arago and Fresnel came upon the strange experimental fact that two differently polarized beams did not interfere with one another as one would expect, and Arago visited Young to discuss the matter with him. This was in 1816 and the next year Young wrote two letters to Arago that have since become famous. For Young had solved the problem and his explanation was simplicity itself; he imagined light waves in the aether really were transverse, like the sea waves he had used in his early explanations. Every experimental result could now be explained, including the anomalous behaviour of beams of polarized light; from thenceforth longitudinal waves were out and the corpuscular theory discredited as well.

Another question about light that arose in the nineteenth century was the true nature of the spectrum. Newton had shown how sunlight was composed of light of all colours and the subject was opened up again by William Wollaston, a chemist and physiologist with very broad interests. In 1802 he decided to investigate the nature of the spectrum in the hope that he might be able to separate the colours from each other. To this end he built an instrument – a spectroscope – in which sunlight was only allowed to enter after passing through a narrow slit. Yet contrary to his expectations no separation of colours was shown by this technique; instead all he saw was that the spectrum was crossed by a number of black lines. After this Wollaston seems to have lost interest in the question and nothing more appears to have been done until twelve years later, when the Bavarian instrument-maker Joseph von Fraunhofer, then in his twenties and directing a glass-making plant, decided to study the solar spectrum. Fraunhofer's aim was to obtain deeper insight into the dispersive power of the flint and crown glass he was making for achromatic refracting telescopes, but the results he obtained using a Wollaston-type spectroscope intrigued him so much that he decided to map the 576 thin black lines

Illustration page 455

he could see. Then, using the Young-Fresnel wave theory, he measured the wavelengths of the darkest ones and presented his complete results to the Munich Academy.

The presence of the dark lines were to remain a mystery for some time, but meanwhile other spectroscopic investigations were taking place. Investigations by the astronomer John Herschel and his friend W. H. Fox Talbot, remembered now for his pioneering work in the development of photography, showed that when certain chemical substances were heated and their flames examined in a spectroscope, each element gave its own characteristic bright lines. In 1826 Fox Talbot published a paper in which he wrote 'a glance at the prismatic spectrum of a flame may show it to contain substances which it would otherwise require a laborious chemical analysis to detect'. He carried this further in 1834 when, after using some lithium which Faraday had given him, and some strontium, both of which burn with a red light, he was able to write 'the prism betrays between them the most marked distrinction which can be imagined'. Others followed up this work, notably William Allen Miller, professor of chemistry at King's College, London, who investigated the spectra of many materials and took photographs of some, a technique that allowed a more detailed analysis to be made. Miller's experiments of 1833 also showed him that if sunlight was passed through various gases in the laboratory, some additional dark lines appeared, and it became generally realized that the Fraunhofer lines might well be due to gases on the Sun. This seemed confirmed in 1855 when the French physicist Jean Foucault identified a pair of close lines in the Fraunhofer map with a pair of bright lines obtained by heating sodium in the laboratory. Indeed Kelvin expressed the general opinion very ably when he said that other 'vapours' besides that of sodium might be found in the Sun and stars by searching in the laboratory for bright lines equivalent to the dark ones in solar and stellar spectra.

Illustration page 453

The whole subject of spectroscopy, a technique which had the most important consequences for later research, was finally put on a more precise footing by two physicists, Balfour Stewart at Manchester University and Gustav Kirchhoff in Heidelberg. In 1858 Stewart showed that if a body emitted radiation at specific wavelengths then it also absorbed radiation best at those wavelengths. This discovery was also made by Kirchhoff, who was then working with the chemist Robert Bunsen, inventor of the bunsen burner so useful for giving a clear hot flame for laboratory use, and who had been instrumental in getting Kirchhoff the chair of physics at Heidelberg. Between them they carried out a series of detailed spectroscopic investigations, and in the end were able to draw up a set of three apparently simple laws governing the various types of spectra observed in the laboratory. Bright lines on a dark background were due to glowing gas and the positions of the lines characteristic of the chemical substance radiating them. A continuous spectrum (like that from the Sun) indicated the source as an incandescent solid or a very dense gaseous body such as the Sun. The dark lines superimposed on such a spectrum were due to cooler gas lying between the source of the continuous spectrum and the observer, the positions of the lines again being characteristic of the particular gases concerned. By 1860, then, the position had become clear, the mystery of the Fraunhofer lines had been solved. The implications of this in chemical analysis were profound, for the spectroscopic technique was able to show up the presence of only the minutest traces of a substance. But probably the most astounding implications of spectrum analysis were in the field of astronomy.

Astronomy

During the nineteenth century the whole face of astronomy changed; from being a science concerned mainly with the planets and their motions, a science of dynamics, it turned into a science concerned primarily with the stellar universe and with the physics of the bodies in it. That is not to say that no dynamical astronomy was done, nor that discoveries failed to be made in the planetary field, but the emphasis changed. This was due partly to developments in the manufacture of telescopes and partly to the influence of William Herschel, a man who seems to have been the most indefatigable observer on record. As we have seen, Herschel's dedicated observing led him to study nebulae but brought him to no decision about whether they were clouds or collections of stars in space.

For Herschel, see pages 351–352.

Some twenty years after Herschel's death the challenge was taken up by William Parsons, the third earl of Rosse, who in 1845 completed the construction of a gargantuan reflector with an aperture of 1.8 metres (72 inches) diameter on his estate in Eire. Yet even this could not answer the question completely; Rosse did observe that certain of the nebulae possessed a bright central core and a spiral structure but their precise nature still eluded him and his contemporaries.

Illustrations page 455

Rosse's telescope was the last large instrument to be constructed with a mirror of speculum metal or to be mounted without the benefits of heavy engineering techniques. In the 1850s and 1860s the English Astronomer Royal, George Airy, and the manufacturer and inventor of the steam-hammer, James Nasmyth began to use methods developed for building steamships and railway engines for mounting telescopes and by the 1880s supporting systems for telescopes had been transformed. In 1888 a vast refractor of 0.9 metres (36 inches) diameter was constructed at Mount Hamilton in California at what became known as the Lick Observatory, and one of 1 metre (40 inches) diameter at the Yerkes Observatory near Chicago. Without the use of these new techniques telescopes of this kind would have been unmanageable and no attempt made to construct them. The other great fillip to the construction of large telescopes was the development of the silver-on-glass mirror. The problem that the instrument-maker had to face was that, to avoid spurious images, the astronomer required his telescope mirrors to be silvered on the surface and no method was known for doing this until the 1850s when Leibig

Illustration page 456

invented such a technique for quite other purposes. In 1856 Karl von Steinheil in Germany and Léon Foucault in France constructed reflecting telescopes with silvered glass mirrors; their efficiency in reflecting light was far superior to mirrors of speculum metal and some large efficient reflectors were built in the latter half of the nineteenth century. These finally demonstrated the superiority of the reflector for telescopes of really large aperture, though the full results of this change were not to become fully evident until the twentieth century.

However, the large telescopes built towards the end of the nineteenth century at Lick and Yerkes in the United States were refractors, and were chosen because of optical developments comparatively early in the century in telescope optics. Techniques for making really suitable glass for achromatic lenses were devised by the Swiss Pierre Guinand and taken up in Germany by Fraunhofer and his colleagues who produced some excellent instruments. In some of these the object glass (front lens) was split across the centre to form a heliometer, a device for more conveniently measuring the apparent diameter of the Sun with precision and it was just such an instrument that was delivered to the Königsberg Observatory in 1820. The director of the observatory was Friedrich Wilhelm Bessel, who had been born at Minden in Germany in 1784 and, after a time in commerce, turned to astronomy, first making a name for himself by calculating cometary paths. Using the new Fraunhofer refractor, he found the heliometer excellent for measurement of star positions and began to try to make measures of stellar parallax. He chose the star 61 Cygni because this seemed to have a comparatively large motion against its background stars, and this he took to mean that it was near. In 1838 he was able to announce that he had successfully measured the star's distance; it was 600,000 times the distance Earth to Sun. The same year another German astronomer, Friedrich Struve, announced a distance measurement for the bright star Vega, though later research has shown that Struve's measures were mistaken. The next year Thomas Henderson, who had recently been making observations in South Africa, gave a distance for the star Alpha Centauri, only excessive caution having prevented him presenting his results in 1833. But now at last this crucial measurement had been made and as the nineteenth century wore on so the number of stars whose distances could be measured directly in this way increased. A glimpse of the truly vast nature of the universe had at last been obtained. From the 1850s onwards photography came to be used in astronomical observation and by the end of the century it had become well established as a means of recording fields of stars for precise measurement, the photographic plates being examined with a specially mounted microscope. Thus the nineteenth century saw the advent of a new degree of precision in astronomical measurement.

Illustrations pages 457, 503

Herschel's discovery of Uranus emphasized the possibility that there might be other bodies in the Sun's planetary system – the solar system – that awaited discovery. This view had already been given some weight in 1771 with a peculiar numerical sequence discovered by Johann Titius and made known by Johann Bode. This sequence – the so-called Titius-Bode law – indicated an unfilled gap between those numbers representing the distances from the Sun of Mars and Jupiter; when Uranus was discovered to fit into the sequence, the gap aroused more interest and a search was made for a planet between the orbits of Mars and Jupiter. Success came in 1801 when an Italian astronomer, Giuseppe Piazzi, observed just such a body, though later research uncovered many others, and it became clear that the orbital gap was filled with myriads of planetoids – 'asteroids' as Herschel called them – a kind of planetary debris whose origin is still uncertain.

Study of the planets and of planetary detail continued throughout the century but as early as the 1840s astronomers became aware that Uranus was not behaving as it should do if Newtonian gravitation theory were correct. By this time Newton's theory seemed so well established that the one satisfactory explanation of the discrepancies in the behaviour of Uranus seemed to be that another planet existed still further from the Sun and was affecting Uranus gravitationally. The problem of calculating the existence of such a hypothetical planet seemed insoluble to most astronomers, though it was tackled by two men, John Couch Adams, a young Cambridge mathematician who had just obtained his first degree, and Urbain Le Verrier, a French mathematical astronomer, eight years Adams's senior. Adams tackled the problem in 1842 and by 1845 had calculated the position of the hypothetical planet; he then sought the assistance of the Astronomer Royal, George Airy, in locating it, but Airy seems to have been doubtful about the solution, and such steps as were taken at Greenwich and Cambridge to observe the supposed planet were lacking in any urgency. As for Le Verrier, he also tackled the question methodically and came up with his solution in 1846, some ten months after Adams. His solution, too, met with some disbelief in France, and it was Johann Galle, the director of the Berlin Observatory, whom he finally persuaded to initiate a search. As luck would have it the observatory had just recently completed a chart of the particular area of the sky where Le Verrier had calculated the planet would be; his calculations proved to be correct. Since those of Adams also gave the same area of sky, there was for a time some argument over priority, though now it is clear both men shared equally in this triumph of Newtonian gravitation. The new planet was named Neptune by international agreement.

The other great stride forward that took place in nineteenth-century astronomy concerned the studies that were made of the spectra of stars following on the publication of Kirchhoff and Bunsen's discovery of the nature of the Fraunhofer lines. Much of the pioneering astronomical work on this subject was done by an amateur astronomer William Huggins. Born in London in 1824, he had to run his father's business until in 1854 he was able to devote himself to astronomy, which had become his predominant interest. In 1875 he

Illustration page 457

married and his wife, Margaret, helped much in his scientific work; her name was associated with his in many of Huggins's publications. In 1900 Huggins became president of the Royal Society, a signal honour due especially to his spectroscopic work, which laid the foundations for what was to become called 'astrophysics'.

Huggins had become friendly with William Miller, whom we have already met in connection with spectroscopy, and was intrigued with the possibility of observing and photographing the spectra of celestial bodies. Between them they devised a spectroscope that would not only display celestial spectra but would also allow those spectra to be directly compared with a laboratory-generated spectrum, a very necessary observational procedure if the lines in celestial spectra were to be correctly identified. Huggins was also determined to photograph celestial spectra in order to obtain permanent records, and in spite of Miller's scepticism, finally managed to do so. But these techniques were only the means to an end, the examination of spectra, and here Huggins had two notable successes.

The first was announced in 1864 in a joint paper with Miller. Huggins had observed a 'planetary nebula' (a disc-shaped nebulous patch) but had seen no such spectrum as he expected; instead of dark lines on a coloured background, he saw a bright line spectrum. As they wrote in their paper, 'The riddle of the nebulas was solved. The answer which had come to us in the light itself, read: Not an aggregation of stars, but a luminous gas'. So clearly there were clouds of gas in space; Halley and Herschel were both proved right. However, observations of Rosse's 'spiral nebulae' with the spectroscope showed them to be some kind of collection of stars, but it was not until well into the twentieth century that the true nature of the spiral objects was to be revealed.

Illustration page 457

Huggins's second success arose out of a discovery made in 1842 by the Austrian physicist Christian Doppler. Doppler realized that for both sound waves (sound had been proved to be a wave disturbance by Laplace in the eighteenth century) and light waves, the wavelength we observe if the source is moving should be different from the wavelength observed if the source is stationary. For instance, if the source is moving towards us then the waves will be closed up – they will arrive, in fact, at a greater frequency – and their wavelength appears shorter. On the other hand if the source is moving away the wavelength will appear longer. Such an effect is now a common occurrence in the sound of a siren emitted by a moving vehicle. As far as light is concerned, Doppler conjectured that stars moving towards us should appear bluer and those moving away, redder. Six years later Hippolyte Fizeau, a French experimental physicist who, with Foucault, devised accurate laboratory methods for measuring the velocity of light, asserted that no such change would be observed. What would happen, however, is that the spectral lines would change position; if the source were moving towards the observer there would be a shift towards the blue end of the spectrum, if moving away a

Illustration page 454

red shift would result. This Doppler shift – or more correctly Doppler-Fizeau shift – was difficult to observe, but this Huggins set out to do and in 1868, with improved apparatus, he was able to measure a red shift of Sirius, the brightest star in northern skies. He found its velocity to be about 48 km per second (30 miles per second) though later observations showed this to be an overestimate, and 32 km per second (20 miles per second) was nearer the mark. This opened up a new technique of observation, for previously velocities of stars towards or away from the observer – radial velocities – could not be detected, let alone measured, since the stars are so far off that they only appear as points of light even in a telescope. Yet the technique did more than this, for it later gave results that were to help lead to a whole new outlook on the universe in the twentieth century.

Mathematics

During the nineteenth century mathematics continued to develop in every field, in geometry, in algebraic analysis and the use of the calculus, and to open out also into the new field of statistics. In doing so its techniques and discoveries became increasingly complex and specialized. Nevertheless, technical though these matters are, some mention must be made of the development in two fields – statistics and geometry.

Illustrations pages 458, 459

Statistics may be said to have begun in the seventeenth century with a series of bills of mortality drawn up in London in 1662 by John Gaunt with the help of William Petty. Their work made it clear that such tables could be of use to a government in its administrative role, and various European states began to collect the appropriate information, extending it to births, marriages, migration, etc. Mathematical statistics were developed by the Belgian Lambert Quételet who was born in Ghent in 1796, studied at the university there and became its first professor of elementary mathematics. Later he became astronomer at the Brussels observatory and in 1834 permanent secretary of the Brussels Academy. In 1825 the philosopher Auguste Comte was teaching what he called *sociologie* and because of this Quételet became interested in 'social physics'. He tried to find the determing influences for observed trends, and discussed natural causes and the perturbations caused by man. He considered great numbers of cases and took averages of various kinds; he also drew up a definition of the 'average man'. All this appeared in 1835 in his *On man and the development of his faculties, an essay on social physics*. In more theoretical work carried out a decade later Quételet mentioned what is now known as a 'normal' or 'Gaussian' distribution (the latter after the German mathematician Carl Friedrich Gauss); it was an important result which shows mathematically and in graphical form the distribution of normal variations (the different heights of an average collection of people, for instance, or the way random errors are distributed among a series of observations).

Some other early work in statistics was done by the French mathematical physicist Siméon Poisson, who was born in 1781 at Pithiviers. Poisson turned his hand to various investigations; he made contributions to the mathematics of perturbations in the solar system by the planets on one another, worked on a mathematical analysis of the theory of heat and, in 1837, three years before his death, tackled questions of probability. It is to him we owe the term 'law of large numbers' and a special curve to show the probability of the occurrence of events of unequal likelihood.

The work of Quételet and Poisson was to bear fruit in many branches of science where large numbers of observations were being handled, as for example in astronomy with an analysis of a great number of stellar motions, in medicine where great numbers of patients are being treated, and so on, while the study of statistics was also to play a considerable part in genetics, the study of inherited characteristics. The man who had much to do with the latter was Francis Galton, a grandson of Erasmus Darwin and therefore Charles Darwin's cousin. Born in 1822 Galton studied medicine for a time but gave up his formal education when his father left him a fortune. He has been called the last of the now-extinct breed of gentleman scientist, for Galton, like Huggins, never held an academic or other official post. His interests covered a wide range; he introduced the use of fingerprints for identification, he did a little research in meteorology and, incidentally, introduced the term 'anticyclone', but his main contributions were in genetics and psychology. These were based on his conviction that everything was quantifiable, a view he held so strongly that at one stage he attempted to analyse numerically the efficacy of prayer!

In psychology, a subject which aroused increasing interest in the nineteenth century – the first psychological laboratory was set up at Leipzig University in 1879 – Galton sowed the seeds of the idea of mental testing and the measurement of the acuteness of the senses, and carried out some experimental psychology himself. He realized that here as in his main interest, the subject of genetics and inheritance, statistical methods were of signal importance. Although Galton's own mathematical abilities were not of the kind that would allow him to develop mathematical statistics as an academic discipline, he did introduce the concepts of correlation (how much one variable depends on another) and regression (correlation connected with the deviation of variables), and made much use of the Quételet-Gaussian distribution curve for errors.

Galton also became much concerned with the inheritance of talent, and claimed that an analysis of the histories of notable families showed that this was no myth. His views met with some opposition, those favouring 'nurture-not-nature' claiming that the children of talented families owed their gifts to their environment, but Galton stood his ground; this led him in the end to advocate a programme of 'eugenics' to 'foster talent and health and suppress sickness and stupidity'; he believed this necessary for any society that wished to maintain, let alone promote, its standing in the world. He carried on a long correspondence with Charles Darwin on inherited characteristics, but perhaps the most significant long-term result of his work was that, just after the turn of the century, a eugenics laboratory was established. Connected with Galton's 'biometric' laboratory for collecting human and other statistics which he had first set up twenty years earlier, the new laboratory had close ties with University College at the recently founded University of London. When he died in 1911 Galton left a large sum for the foundation of a chair of eugenics, expressing a strong wish that Karl Pearson should be the first professor. Pearson, who was born in London in 1857, was a Cambridge-trained mathematician as well as being a qualified (but non-practising) lawyer, and had been appointed Goldsmid professor of mathematics at University College in 1884. In 1911 he accepted the professorship Galton founded, and held it with distinction for the next 22 years, laying down firm mathematical foundations for the whole of twentieth-century statistical science, but especially as applied to biological questions.

The second field of mathematics in which nineteenth-century developments were to have a strong effect on later scientific work was geometry, for during the century two quite new geometrical systems were proposed. These arose out of the failure of two theorems in Euclidean geometry. The general scheme of Euclid was masterly and a monument to human logic; for over two thousand years it had ruled supreme and by the nineteenth century it had become synonymous with one aspect of the meaning of 'truth'. Yet there were two theorems for which no logical proof could be adduced; one stated that the segment of a line can be extended in either direction just as far as one pleases, the other that two parallel lines will never meet each other, however far they are extended in either direction. Both seem obvious enough, yet formal proof eluded Euclid and has evaded every other geometer since. Euclid himself was cautious over both theorems and made sure he never used the parallel-lines postulate if he could possibly avoid it.

For Euclid, see pages 110–111.

In the eighteenth century Giorolamo Saccheri, the Jesuit professor of mathematics at Pavia, tried to prove the theorems. His technique was to use the method of *reductio ad absurdum*. Saccheri's work did not convince other mathematicians, but the interesting thing is that, in reaching his conclusions, he came close to what we now call non-Euclidean geometry but rejected this out of hand on the ground that geometries of this kind were repugnant. The first man to take up the matter in the nineteenth century was that great mathematician and physicist Carl Friedrich Gauss. He tried to replace the theorem by simpler ones, but failed, and following in Saccheri's footsteps also came upon non-Euclidean geometrical ideas. But Gauss's attitude of mind was different, and he concluded that non-Euclidean geometries could exist, though he did not publish his opinion at the time.

The first steps in developing such a geometry were taken in the 1820s and early 1830s by two mathematicians, Nikolai Lobachevsky in Russia and Janos Bolyai in Hungary. Lobachevsky was born in Nizhni Novgorod (now Gorki) in 1792, and studied at Kazan University where he later became a professor. Like others before him, Lobachevsky also tried to prove the parallel-lines theorem, but unlike previous mathematicians, when he found this impossible he took the next logical step. He assumed that since it could not be proved that one line and one line only could pass through a point and be parallel to a second line, one was free to postulate that an infinite number of lines could be drawn. Again, Bolyai, an army officer and a brilliant mathematician, came to a similar result, though when he learned that Gauss too had followed the same steps, Bolyai merely added his work as an appendix to a book of his father's on mathematics. Not until the late 1860s was Bolyai's work made known elsewhere, but by then he was dead.

An alternative reaction to the parallel-lines postulate was taken by Bernhard Riemann, who was the son of a Protestant minister in Germany. Born in 1826, Riemann attended Göttingen University to study theology and philology, but got his father's permission to switch to mathematics. Later he moved to the University of Berlin and then back to Göttingen where he carried out work on an important form of mathematical analysis concerned with cyclic changes that had been derived by the French mathematician Jean Fourier in 1822. It was only in 1866, the last year of his short life, that Riemann entered the field of non-Euclidan geometry. Riemann took an opposite approach to Gauss, Bolyai and Lobachevsky, by assuming that if the parallel postulate could not be proved, then no lines could be drawn parallel to other lines, and he then developed a geometry taking this into account. His results were immensely productive because the geometry he devised could be developed to deal with three or even more dimensions. However, like the geometries of Lobachevsky and Bolyai, it was looked upon as nothing more than abstract mathematical theories. Indeed Riemann's geometry in which space is curved, though not at a constant value, seemed far removed from reality, and was even attacked on philosophical grounds. Yet abstract and esoteric though Riemannian geometry appeared at the time, in the early twentieth century it was to be seen to provide what seems to be a truer expression of reality than Euclidean geometry, so radically were concepts of the universe to change.

For the value of his work in general relativity, see page 519.

Chapter Ten

Twentieth-Century Science

Science, which had begun to move forward with such speed during the nineteenth century, has advanced even more quickly during the twentieth. And not only has scientific discovery accelerated. With more scientists than ever before at work, using increasingly powerful and sophisticated equipment, the results they have achieved have often been astonishing, and certainly would have astounded the most imaginative minds of only a few generations ago. So much work has, of course, provided a vast amount of new detailed evidence and led to some complex and specialized concepts about the natural world. But since the twentieth century is still with us, it is in a sense premature to try to consider its science from an historical viewpoint. Much is still going on, so that a great deal of research is too recent to allow us to stand back and see it in its historical perspective. Even so, it is possible to select some aspects of twentieth-century science and trace their growth, and here we shall concentrate on three that have proved to be of prime importance in understanding the world around us. The vast new universe which twentieth-century astronomy has uncovered is one result of these great strides in our comprehension; another, closely allied to it, is the revolution in the physical sciences brought about by the theory of relativity and by the quantum theory, without which modern nuclear science would never have been conceived. Thirdly there are the prodigious developments that have taken place in biology, developments covering human and animal physiology, heredity and evolution, and that have led, also, to the new discipline of molecular biology, a field where physics, chemistry and genetic theory have come together in a way that is clearly of the utmost significance.

Twentieth-century science has also been transformed by the remarkable growth in its technology, which has facilitated research into many new fields. Although it is not possible here to give a thorough account of all the different technologies that have been involved in this process, or the new branches of science that have developed as a result, we must at least mention that of electronics, and computer technology, which since the 1960s have revolutionized the collection and processing of data of all sorts. Similarly the advent of space travel has stimulated much scientific work – including the development of electronics – and has provided a new dimension for research in many

Illustrations pages 510, 511

fields, particularly astronomy and medicine. But to keep this chapter within reasonable bounds, such advances of technological rather than of genuinely theoretical importance are dealt with only incidentally to the main account.

Biology in the Twentieth Century

Consequences of Darwinism

Darwin's theory of evolution deeply affected biology. On the one hand, it stimulated so immense an amount of interest in the development of animals and plants that during the second half of the nineteenth century this occupied the greatest part of biological research; on the other hand, the whole theory seemed to present a classic example of scientific method, an ideal model of scientific induction; Darwin had gathered immense quantities of detailed biological information that he had then assimilated into his powerful theory of emergent evolution. On this count it was to be admired and used as an example of the power of science. Yet in spite of the theory's standing there were still problems associated with it. First, the theory assumed that the small variations which arose could be inherited and would persist within a biological population; but this had to be proved. Second, there was the difficulty that there seemed to be no way of testing experimentally whether natural selection did occur in such a population. Clearly, further research was necessary.

From the 1860s to the 1880s great efforts were made in morphology – the study of the forms of living things – to try to seek evolutionary relationships between species and members within species. Morphologists wanted to find a basic unity behind common forms, to discover common ancestors for two or more groups of organisms, and to construct family trees which would indicate the history of the development of a particular animal. Their leader was the Darwinian champion, Ernst Haeckel, who had been born in Potsdam, Prussia, in 1834 and was a young man of twenty-five when *The Origin of Species* appeared. A Christian, and at the same time a believer in a mechanistic approach to the whole living process, Haeckel became deeply interested in the comparative anatomy of men and animals. This led him to construct a tree or pedigree for man, and to suggest that, during its development, an embryo goes through the major adult stages of its ancestors in this evolutionary pedigree – this became known as Haeckel's 'biogenetic law'.

Illustration page 413

But because of lack of progress, towards the close of the nineteenth century the younger biologists revolted against this approach and set out to try to obtain answers to a new set of questions. This had two effects; it paved the way for the growth in the twentieth century of experimental embryology, and it led also to experiments on plant breeding, especially those of Hugo de Vries. Born at Haarlem in the Netherlands in 1848, de Vries came from a family with an academic tradition which he combined with a great keenness for botany. It was in the 1890s, near Hilversum, that de Vries came across a plant that was to help make his name; this was the evening primrose (*Onothera lamarckiana*). What he noted about it was that there were what appeared to be two quite distinct strains growing side by side in the same field. When self-fertilized he found they bred true, but when crossed three different types emerged which were different enough in leaf shape and other characteristics to represent new species. As a result de Vries believed he had found an ideal organism to demonstrate that small Darwinian changes did not occur, but that the changes came in large jumps. He called these jumps 'mutations', and in 1901 and 1903 published his two-volume book *The Mutation Theory*. This answered some of the objections to Darwinian theory, for instance that small variations would be swamped by the much larger number of normal and unvarying members of a population. De Vries suggested that because of mutations new variants were already separate species and that the new mutations could then be acted on by natural selection, the action of which was to remove those unfitted for survival. It also went some way to answering Lord Kelvin's criticism that the age of the Earth was too short for Darwinian evolution, because the de Vries mutations gave a shorter time scale.

Illustration page 460

Just as mutation theory brought a new and more experimental look to evolution, so did the experimental approach to embryology. The growing dissatisfaction with the morphologists' approach and the biogenetic law came to a head in 1888 when Wilhelm Roux, who had been born in Jena, east Germany in 1850, published results of his experiments on frog embryos, and his analysis of the mechanism by which they changed from eggs to embryos (tadpoles) and then frogs. Roux proposed a new theory – the 'mosaic theory' – in which hereditary particles inside the cell were divided unevenly during cell divisions within the egg. At each division the two new cells would therefore possess different hereditary particles and so different hereditary possibilities, while subsequent divisions would restrict the possibilities further; in the end, a cell would arrive with only one hereditary trait concerned with one particular type of tissue. This was a theory which could be tested by experiment and this Roux proceeded to do. If the theory were correct, then destroying one of the cells developed early on in the divisions within the egg should lead to an abnormal embryo; if the theory were wrong, then the destruction should have no such effect. Roux obtained a result that supported his theory. However Hans Dreisch, a younger biologist working with eggs of the sea urchin performed the experiment with a slight variation; he did not destroy one of the early cells but separated it from its companion. He found both produced normal, though smaller, embryos, not at all what the mosaic theory would lead one to expect. Later research showed that the contradiction between the two experiments was due partly to different experimental techniques and partly to the different organisms involved, thus underlining the many factors involved in this new experimental approach to embryology, and the care needed in interpreting results.

Mendelism

As the new century dawned, another factor appeared that was to exert an influence on later biology; this was the rediscovery of Mendel's work. Gregor Mendel (1822–84), was a Bohemian monk, who, for a decade in the 1850s and 60s carried out in Brno a series of experiments on plant breeding. His results were curious. Crossing a tall pea plant with another tall pea plant gave tall offspring; crossing a short with a short gave short offspring; but crossing tall with short gave a series of unexpected consequences. The first offspring were all tall, but crossing these gave a ratio in the second generation of three tall to one short. The short bred short when crossed with other short plants. Thus the shortness characteristic, though masked in the first generation by the tallness characteristic, reappeared unchanged in the next generation. To explain these results Mendel proposed that each generation contained two factors for every inherited characteristic, one from the male parent, one from the female. Since certain factors masked the appearance of others, he referred to some as 'dominant' and others as 'recessive'. He then enunciated two general laws: the Law of Segregation stated that in the formation of germ cells the two factors for any characteristic (e.g. height) are always separated from each other and end up in a different egg or sperm; the Law of Independent Assortment stated that the maternal and paternal factors for any set of characteristics segregate independently from those of any other set, so that each germ cell gets a random collection of factors derived from each parent.

For the not dissimilar Chinese concept of Yin and Yang, see page 145.

Mendel published his results, complete with a detailed mathematical analysis, in the journal of the Brno Natural Science Society and here they languished for over thirty-five years; certainly Mendel sent a reprint to Darwin but he never read it. The paper was rediscovered in 1900 by de Vries and independently by two other biologists concerned with plant hybridization, yet even then it did not meet with any immediate success; not until 1910-15 did it come into its own with the establishment of the chromosome theory. As early as 1902, however, Walter Stutton, a young research worker in the United States, had pointed out a similarity between Mendel's theory and the separation of chromosomes during a division of the nuclei of cells, and detailed studies a year later had shown that the Mendelian factors might be chromosomes. Again, by 1906, William Bateson of Cambridge University, a staunch Mendelian, had detected evidence that seemed to support this chromosome hypothesis. But the most important work in establishing the theory was done in the United States after 1910; the leader was Thomas Hunt Morgan, once a critic of Mendelism.

Born at Lexington, Kentucky, in 1866, the son of a diplomat and nephew of a Confederate general, Morgan studied biology at the State College of Kentucky and then at the graduate school at Johns Hopkins University. With a keen interest in marine biology he was a regular visitor to the famous marine biological laboratories at Woods

Hole, Massachusetts until the very end of his life. Morgan's change from critic to supporter of Mendelism was a result of his research at Columbia University. In 1908 he had begun to breed a fruit fly (*Drosophila melanogaster*) in an effort to see whether the kind of mutation observed by de Vries in plants also occurred in animals. The fruit fly was chosen because it produced a new generation every ten to fourteen days, and in consequence made the study of genetic changes practical in a relatively short period. Morgan found no astonishing mutations at the level of species with *Drosophila*, but one surprising result did appear in 1910; a white-eyed male fly was found in one of the breeding bottles. Morgan went so far as to call this a mutation though it did not constitute a new species, for he bred normal (red-eyed) flies from it. However, when he crossed some members of this new generation with each other, the white-eyed characteristic appeared again, and almost entirely among the males. But when one of these males was mated with females of the first generation, half the males and half the females were white-eyed. Mendelism would explain the result and Morgan therefore adopted it, assuming that the factor concerned with eyes was linked to that which determined sex. This assumption of sex-linked inheritance paved the way for relating Mendelism and chromosome theory with the whole subject of heredity, for in 1911 Morgan proposed the novel idea that the Mendelian factors were arranged in a line along the string-like chromosomes of Waldeyer-Hartz.

Illustrations page 460

For Waldeyer-Hartz, see page 430.

Morgan had a good research group with him: Alfred Sturtevant, who was particularly adept at the mathematical analysis of results of breeding and the mapping of genetic factors on chromosomes; Hermann Muller, an imaginative theorizer who also had a flair for designing ingenious experiments, and Calvin Bridges, whose forte was the study of cells. The group worked as a unit, each carrying out his own experiments but well aware of what the others were doing, and there was a free exchange of results and ideas. Each made his own discoveries, but together they developed the idea that Mendel's factors were actual physical units, located at definite positions along a chromosome; they adopted the name 'gene' for these factors, a name which had originally been derived in 1909 by the Danish biologist Wilhelm Johannsen (though in a slightly different context) from a term (pangene) used by de Vries. Further and later research, culminating in the 1960s with the isolation of a single gene, proved the Columbia group's far-sightedness as well as the physical reality of the gene itself.

General Physiology

In the early twentieth century a strong belief grew up that all the phenomena of life could be reduced to the basic laws of chemistry and physics. Jacques Loeb, who was born in Prussia in 1859 but had moved to the United States in 1891 after marrying an American, was a leader of this mechanistic school, which really stemmed from a

German movement known as *Entwicklungsmechanik* (developmental mechanics) which Roux had promoted in the 1880s. Loeb emphatically announced his views on 'The Mechanistic Conception of Life' at an international congress in 1911, a conception which was widely accepted during the 1920s. A decade later, however, Loeb's view ran into opposition from a new attitude of mind among some biologists who sought to find relationships in the behaviour and organization of the different parts of an organism; to them it was not sufficient merely to study parts of the body in isolation. Living systems were, they believed, no mere collections of molecules but systems with a high degree of organized behaviour. A somewhat similar view had been taken earlier by Claude Bernard in the 1870s but now it was to be given more precision by developments in experimental physiology, and most particularly in studies of nerve conduction and the organization of the nervous system.

These two views – the mechanistic and the integrated or holistic outlook – were both to be found in studies of the nervous system. Those who took a materialist 'reductivist' approach – typified by the school of teaching headed by the physicist and physiologist Hermann von Helmholtz – were concerned to reduce everything to basic physical relationships. They studied the conduction of impulses and other details of nerves isolated from their surroundings. The other approach, taken by the English and French researchers, looked at the nervous system as a whole; this led to the appreciation that there was a difference between those nerves leading to the spinal column and those leading away from it. The holistic school also made a particular study of the reflex system (the system which gives such results as the involuntary jerking of the knee when tapped below the kneecap). By 1880 reflexes were acknowledged to be a function of the spinal cord, which Pierre Flourens, permanent secretary of the Académie des Sciences in succession to Cuvier, had demonstrated to be a very intricate structure. But in spite of Flourens's work, there was still confusion about how the entire nervous system functioned.

Notable attacks on the problem were made by the Russian Ivan Pavlov and the Englishman Charles Sherrington. Pavlov worked most of his life in and around Leningrad, where he died in 1936. A pupil of Ivan Sechenov, who had studied with Helmholtz and with Bernard, and who had come to the belief that all behaviour was due to a balance between incoming and outgoing stimulation of the nerves, Pavlov first took a very mechanistic, reductivist approach. Later he modified his attitude slightly, though he was always essentially a mechanist at heart. Famous now for his studies of reflex action, Pavlov first came to be interested in these after he noticed that when the customary routines for feeding laboratory dogs were carried out, the animals began to secrete saliva, and this before any food was actually present. He knew already that the dogs would salivate as soon as food reached their mouths, but how could other actions incidentally connected with food cause this reflex? Careful research and experiment eventually led Pavlov to conclude not only that a learning process was at work but also that this learning was done by a build-up of reflexes: the repeated use of a particular stimulus would give rise to a response – a conditioned reflex – that the experiments could trigger at will. He conceived of temporary nerve connections being formed in the cortex of the brain, though it was hard, if not impossible, to demonstrate this. Moreover, there was the problem posed by the fact that some conditioned reflexes seemed more permanent than others; this was still unsolved. Nevertheless Pavlov's work opened up the important relationship between behaviour and the physiology of the nervous system.

Charles Sherrington, born in London in 1857, was an anti-reductionist. Not for him the simple mechanistic view; he thought the mind was a non-physical entity, whereas the body itself was a physical organism. Sherrington's research was concerned with the transmission of nerve impulses, being based on some able studies by the Spanish tissue expert, Ramon y Cajal who, in the 1880s, through careful microscopic work, had discovered that the nerves of all animals are made up of individual units or 'neurons', each distinct from the next and separated by a gap (later to be called the 'synapse' by Sherrington). This replaced the older idea that nerves had a fine thread-like construction. Sherrington took this discovery by Cajal and showed, first, how important the synapses were if impulses down the nerves were not to be transmitted to a random assortment of outgoing paths. He pictured the spinal cord as an input-output system, it is true, but his studies led him to see that it was more than this because he realized it not only received various impulses but integrated them in a common pathway to the appropriate muscle to cause the correct response. He also appreciated that input at one level could modify input at another level, and developed this idea to show that very fine levels of control were possible. This was a more holistic approach, one that considered the whole organism not just its separate units, and in 1906 he published his *Integrative Action of the Nervous System*, a book of the utmost importance that was to exert wide influence for a long time, its fifth edition coming out in 1947, forty years after its original appearance. It incorporated his considered opinion that there were three levels at which one should study animal behaviour; the physical-chemical level, concerned with actions within individual neurons, and giving the 'machine' aspect of an animal, the level of the 'psyche', in which neurological processes came together to form a percipient thinking creature and thirdly, a level of mind-body relationship.

Sherrington gave what was essentially an integrated picture of the nervous system of an organism right down to the neuron level. More comprehensive still was the later work of Lawrence Henderson of the Harvard Medical School in the United States, a physician, physiological chemist and philosopher who later became a sociologist. His studies on the way the body regulates its chemical balance, so that its fluids are neither too acid nor too alkaline, were fundamental. So too was his work on the blood; it made clear that this and the other tissue fluids could also help the body to adjust its acid-alkali balance. Indeed, Henderson's research clearly showed that the chemical reactions in a living body were those that worked best for an organism to maintain a constant internal *status quo*; another piece of evidence to support the holistic approach of Sherrington. The response of a body to shock, a subject studied during World War I by Walter Cannon, a colleague of Henderson's at Harvard, also confirmed this view.

Biochemistry

World War I was a human holocaust, and its aftermath brought into sharper focus changes that had been simmering in politics and economics during the late nineteenth century and the early years of the twentieth. One of these was the breakdown of the old social order and the growth of a more egalitarian outlook, a slow welding together of hitherto disparate strands of society. This process had already been at work in the scientific community, and in biology were underlined by a gradual convergence of what had previously been separate disciplines, so what had once been the province of the embryologist alone, or of the geneticist, was welded together with the holistic approach of the physiologists. This was particularly true of the 1930s, though biochemistry, one fruit of this convergence, had begun to draw together some of these strands at the beginning of the century. Essentially biochemistry is concerned with the chemical reactions involved in living processes such as respiration or the metabolism of proteins (i.e. the way they are broken down within the body).

Notable among the early successes of this new subject were the results obtained by Gowland Hopkins, the founder of British biochemistry and a biochemist with international influence. Appointed as the first professor of biochemistry at Cambridge University in 1914, Hopkins discovered many substances important in animal metabolism and found that some essential protein fragments known as amino-acids could not be manufactured in the body but had to be obtained from outside in the food. These vital 'accessory substances' were what we now call 'vitamins'.

Illustration page 498

Another study in biochemistry which has had highly successful results is that concerned with the way in which living cells break down the molecules of fats and carbohydrates to produce energy for the organism, together with various waste products – carbon dioxide, water, etc. – which are then disposed of. The technical name for this breakdown is 'respiration' (though it has nothing to do with breathing air) and two types are known; one is 'aerobic respiration', which requires oxygen and occurs in all the cells of higher animals, in protozoa (microscopic single-celled animals), fungi and many bacteria; the other is 'anaerobic respiration' which, at this time was known as fermentation, and occurs in yeast, in some bacteria and some muscle cells. Another aspect has been the growth in knowledge about the nature of proteins and, in particular, that class of proteins now known as 'enzymes', though originally called 'ferments' (the word enzyme – Greek *zymosis*, leaven – was coined in 1878). Enzymes act as catalysts, i.e. their presence makes certain chemical reactions occur more vigorously, or inhibits others.

These two aspects of biochemical research can conveniently be divided into two periods, one from 1890 to 1925, when the nature of enzymes and the part they play in respiration were investigated, and second from 1925 to 1960 during which the structure of proteins was worked out and details of the breakdown of fuel or energy-giving molecules such as sugar were discovered. During the first, research by Liebig and Wöhler led to the claim that plants are the only organisms which can transform inorganic into organic material, but this turned out to be wrong: Claude Bernard later showed animals could also do this. Next, in 1897, the German chemist Eduard Buchner discovered what he called 'zymase' in crushed yeast cells, and found that it was a substance capable of causing sugar to ferment. This was a crucial step; it moved biochemistry away from any dependence on physiology and established it as a separate study, since it was now evident that enzymes acted as catalysts and cell function could be studied as a chemical process without a need for any physiological theory about the nature of cells themselves.

By the beginning of the twentieth century, then, it became clear that when proteins were broken down by acids or by certain enzymes a number of chemically separate amino-acids were produced, but the relationship of these to proteins themselves was not clear. At the time there were two theories about the nature of proteins; one, supported strongly by the German chemist Wilhelm Ostwald, thought of them merely as large collections of smaller groupings of molecules and maintained they had no definite chemical composition. The other, dominated by another German chemist, Emil Fischer, a student of Kekulé, believed them to be like molecules of other substances, and so composed of definite numbers of specific atoms. Fischer's work, which set new standards in chemical research, finally made it clear that amino-acids were the building blocks from which protein molecules were constructed. Indeed, by 1907 he was able to synthesize a unit of eighteen amino-acids which, due to its size, he called a 'polypeptide' (a peptide is the type of chemical bond that exists between amino-acids). Fischer therefore concluded that proteins were large, though he put a limit of 5000 on their atomic weights; with the atomic weight of hydrogen as 1.008, of carbon 12.01 and oxygen 16.00, this still gave very big molecules. However in 1917 the Danish chemist Sören Sörensen performed experiments which led him to think that many were much larger than this; he obtained 35,000 for egg-white, a value seven times greater than Fischer's maximum (though still too small by some 20%). Then in 1925 Theodor Svedberg in Sweden developed the ultra-centrifuge, which spun solutions at great speeds with the result that the larger (and therefore denser) a molecule, the faster it settled out at a given speed of rotation. This revolutionized the determination of molecular weights, and with it Svedberg was able to show that molecules of proteins like haemoglobin (which is found in the red blood cells of vertebrates) have weights of the order of 65,000. Later still, in the 1940s, the technique of chromatography began to be applied to protein chemistry. The technique itself had been discovered in 1906 by the Russian chemist Michael Tswett, who had separated dyes by taking advantage of the fact that molecules of different substances each have their own tendency to stay in solution. Tswett realized that this meant different substances would move up a strip of absorbent paper, or a column packed with resin or starch, by different amounts and would, therefore, lie at a different position on the paper or in the column. The technique is very sensitive and accurate, and is of great help in determining the composition of complex chemical substances.

In the mid-1940s, as a result of these developments, a group of British biochemists at Cambridge under the leadership of Frederick Sanger were able to study the amino-acid arrangements in the protein insulin, and this showed conclusively that proteins were in fact long-chain molecules of amino-acids linked together by peptide bonds. This was, of course, a result of great significance and one which fitted in well with the research done during the mid-1930s by Linus Pauling and his colleague Robert Corey at the California Institute of Technology, who discovered that the long protein molecules were in fact coiled back on themselves in the shape of a helix.

While the structure of proteins was being studied, another quite different line of biochemical investigation was also bearing fruit; this concerned respiration, the breakdown of protein molecules. In the Berlin laboratory of Otto Warburg studies suggested that an enzyme containing iron was responsible for respiration in the sea urchin, in yeast and in other cells, though the final chemical proof was still lacking in the late 1920s. One problem was that the particular iron-enzyme existed in such small quantities, but using a spectroscope and examining the intensities of various lines – the technique of spectrophotometry, which the English biochemist David Keilin had used in 1925 to study substances in insect muscles – Warburg was at last able in the 1930s to identify his iron-enzyme with respiration. This proved that a specific enzyme was concerned with a metabolic process, and Warburg was even able to discern which part of the molecule acted as catalyst. His work also made clear that here were chemical reactions that could occur separately in the laboratory as well as in living creatures. Yet, as research during the next thirty years showed, when they do occur in an organism, such biochemical reactions change the conditions existing around them, and thus the future course of the reaction is charged; in other words the organism is an inter-reacting entity. The old simple mechanistic outlook had to give rise to a holistic materialism, a view that treats the organism as a whole, a view which, incidentally, echoes the old Chinese picture of the universe itself as a self-dependent organism.

For the holistic Chinese approach, see page 133.

Molecular Biology

After World War II it was the victorious Western countries of Britain, France and the United States that were first able to return to research in pure science and, in particular, were the only countries that could afford expensive research projects. Moreover, it was in just these countries that native scientific talent had been enriched from the 1930s by immigrants who had fled from Nazi persecution. As far as the USSR was concerned, although it had been on the side of the victors, and was a country where great store was set by scientific research, destruction during the war made a vast economic readjustment necessary which could not but detract somewhat from research in pure science. Moreover, Russian biology was suffering acutely in the 1940s from the Lysenko theory, an affair which, interestingly, had certain parallels with the action three centuries earlier of the Church against Copernicanism, though here political not religious ideology was at work.

The central question involved was heredity, for Lysenko's adherents rejected the whole of Mendelian genetics. Trofim Lysenko was an agricultural scientist, who had been born 1898 in the Ukraine, studied at the Uman School of Agriculture and then obtained a doctorate in agricultural science from the Kiev Agricultural Institute. Lysenko, whose interests were selection and genetics, first rose to prominence by his work on vernalization, a process whereby seeds shed in the autumn were immersed in water and then frozen. This resulted in quicker germination, though the process was not Lysenko's own; it was known in the nineteenth century. When Lysenko tried it on winter wheat he found it successful, so that even seeds planted in the spring matured before the killing autumn frosts. He then tried it on spring wheat and by 1929 was claiming that the changes induced by vernalization were inherited by subsequent generations of plants so that there was no need to repeat the vernalization treatment each year. This essentially Lamarckian view fitted in well with the Marxist belief that environment, not heredity, is crucial, and Lysenko's view became official Soviet policy, even though there was no increase of production as Lysenko forecast.

Lysenko next claimed that chromosome theory was 'idealistic', and that new plants could be caused to evolve purely by a change of nutrients. Mendelian genetics were now extirpated from Soviet biology and by 1937 Lysenko's power was such that to disagree with his theories was dangerous; indeed the world-famous Russian geneticist and agricultural scientist Nikolai Vavilov, who supported gene theory, was finally arrested for his views in 1941 and committed to prison, where he died two years later. Not until 1952 was it possible for Soviet biologists to repudiate Lysenko.

Meanwhile, nucleic acids and molecular genetics had begun to be studied in detail in the West, where the technique of X-ray analysis was at last being applied with great effect in biochemistry. X-ray analysis was invented by William Bragg and his son Lawrence, British physicists who in 1912 showed that if a beam of X-rays (discovered, as we shall see, by Röntgen in 1895) is passed through a crystal of a substance, it is scattered by the atoms and molecules in the crystal. Because such atoms and molecules are arranged in a latticework pattern – that is why they form a crystal – the scattering gives a clue to the arrangements. In the hands of Max Perutz and John Kendrew at Lawrence Bragg's laboratory (the Cavendish Laboratory) at Cambridge, the technique came into its own in the early 1960s not only to show the immense complexity of protein molecules such as haemoglobin, but also to demonstrate how a knowledge of the three-dimensional structure of the atom gives clues to the way it functions. Indeed their work was to be very important when it came to studying nucleic acids which, although discovered in 1869 by the German chemist Friedrich Miescher, had been neglected and were only to rise to prominence in genetic studies in the late 1960s.

Illustration page 499

Another early discovery that was only to be appreciated later was that of Archibald Garrod, who in 1909 explained that Mendelian genes functioned by blocking some steps in the metabolic process. Clearly this was a crucial discovery – the fact that genes have an effect on what goes on in cells – but although others realized this nothing was done for thirty years, perhaps because it was then so difficult to distinguish one enzyme from another and so to discover whether perhaps genes were themselves enzymes. However, in the 1930s the American geneticist George Beadle met the microbiologist Edward Tatum, and together at Stanford University they began to study the problem. Using the bacterium *Neurospora*, whose metabolic characteristics were comparatively easy to study, they showed not only that metabolic blocks were connected with segregation of genes, but also that each gene controlled the synthesis of a particular enzyme. As it turned out this was only part of the story, but it was a vital step in an important direction. Next, in a series of researches on the human genetic disorder known as sickle-cell anaemia between 1949 and 1957, culminating with the work of V. M. Ingram, it became evident that genes determined the sequence in which amino-acids were to be found.

The picture of heredity was now beginning to build up, though there was still the problem of what the hereditary material really was. A step had been taken unwittingly in 1938 by Max Delbrück, an atomic physicist who had turned to biology and was working at the California Institute of Technology, when he introduced a new research 'animal', the 'bacteriophage', a virus which attacks bacteria. Easy to culture, with a time between generations similar to that of bacteria (half-an-hour or less), viruses were later found to be useful because they consist only of two types of molecule, a protein and a nucleic acid. Early research on hereditary substances concentrated on the proteins concerned, but this changed after it was found that if living, benign types of bacteria together with dead but virulent types are injected into a living creature, some of the benign types also

Illustrations page 498

become virulent. The change comes because the activating agent from the virulent bacteria was found to be a particular nucleic acid, DNA (deoxyribonucleicacid). What therefore became clear in the 1940s was that DNA seemed to be able to convert one genetic type of bacterium into another, and the matter came to a climax in 1952 when radioactive tracers were used by Alfred Hershey and Martha Chase to follow precisely how this occurred. They found that only DNA and not the protein was involved, and that DNA could make copies of itself as well as inducing virus-type proteins to be synthesized so that viruses (bacteriophages) could grow. This was a firm step in the right direction, though there still remained the question of how precisely this happened.

The final stages in the research were taken by the American biochemist James Watson and the English biophysicist Francis Crick. Crick worked for a time at the Cavendish with Max Perutz on the use of X-ray crystallography to determine the structure of haemoglobin, but with his training in pure physics he brought a new look to problems of this kind; it was this attitude of mind that attracted Watson, twelve years Crick's junior, to go to work with him in 1951. They decided to examine DNA as a hereditary material, and started work more than a year before the Hershey-Chase results were known. In this both Crick and Watson were helped by Maurice Wilkins, a biophysicist from New Zealand, who had been doing X-ray crystallography on DNA at King's College in London, and whose research showed that DNA seemed to consist of layers which had a general spiral shape that kept repeating itself. It was also evident from Wilkins's work that molecules of DNA were long chains of atoms and, surprisingly, that the molecule was constant in width throughout its entire length. What Crick and Watson set themselves to find was how the component atoms were arranged so that they would give a regular structure to the molecule, allow it to be chemically stable and permit it to copy itself faithfully. This meant that they had to know what forces could hold the atoms and the molecules together. In the spring of 1952 Crick realized that certain atomic groups were paired within the molecule and this prompted a return to the spiral molecular structure originally discovered by Linus Pauling and modified by Wilkins. Crick and Watson knew that Pauling was himself engaged in working out a model of a DNA molecule. There was much to be done; they had to discover how the backbone of outer atoms was arranged in the spiral. By April 1953 they had divined the answer; the DNA molecule consisted of two helixes wound round each other; indeed the whole structure is rather like a spiral staircase with 'steps' composed of paired chemical groups of atoms.

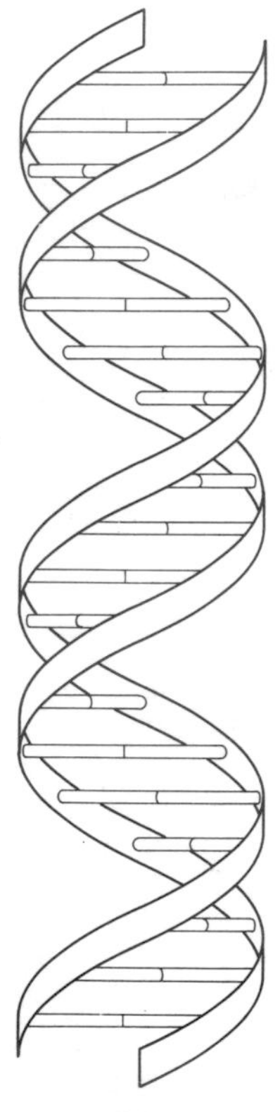

The structure of the DNA molecule, consisting of two strings of alternate sugar and phosphate molecules, bound in spirals and held together by combinations of paired groups of bases. See also illustration page 499.

The new model of DNA made it possible to imagine how it could operate to guide the construction of other molecules, for each strand of DNA could act as a framework to carry other nucleic acids such as RNA (ribonucleic acid), and between 1953 and 1963 research provided the answer, so that it became possible to build a complete scheme to give a biochemical explanation for the gene protein, and to see how genetic 'coding' can give the necessary 'instructions' for the formation of the correct types of molecules. Thus it was that the basic principle of how one species can reproduce its own kind was at last discovered. This work has all kinds of social as well as scientific implications, for it led to the development of the new technique of 'genetic engineering' whereby specific hereditary characteristics can be emphasized or suppressed in future offspring.

Atomic Physics and the Quantum Theory

The next stage in a deeper understanding of the elements, of the nature of atoms themselves, arose out of the practice of passing electricity through rarified gases. The technique was developed so that gases could be forced to glow and thus be examined spectroscopically. What, however, physicists and chemists also began to notice was that the kind of discharge that occurred depended on the degree to which the tube was exhausted of air. In particular, they were interested by a glow on the glass walls of the tube that seemed to be caused by something emanating from one of the metal pins, or electrodes, at the end of the tube. It was the negative electrode or 'cathode' that was the one concerned, and the physicist William Crookes in London tried to explain these 'cathode rays' by suggesting that they were due to the few gas molecules still remaining in the tube becoming electrified and then being repelled from the cathode. However, in 1895 an accidental discovery led not only to the downfall of Crookes's explanation but also to the beginnings of a complete revolution in ideas about the atom.

Illustrations page 461

The discovery occurred in Würzburg in southern Germany where Wilhelm Röntgen was using a gas-evacuated tube for various experiments in his laboratory. The tube was covered in black cardboard and the laboratory was darkened, when Röntgen noticed to his surprise that a sheet of paper coated with platinocyanide which happened to be lying on the bench started to glow, but ceased to do so as soon as the evacuated tube was switched off. Clearly some penetrating rays were passing from the tube through the air and on to the platinocyanide. They could not be particles, because they were not deflected by an electric or a magnetic field, though if they were rays there must be something curious about them because they were not refracted by a lens. Röntgen concluded that they might be very short wavelength rays, but as they were still puzzling, he named them 'X-rays', though many preferred to call them 'Röntgen rays'.

At the Cavendish Laboratory at Cambridge University the professor of physics, John Joseph Thomson, began to study the new rays with a research graduate, Ernest Rutherford. They noticed that when the rays passed through a gas it was able to conduct electricity, a result that led Thomson to speak of the gas as 'ionized' because of its similarity to the effect of a liquid becoming electrically conducting when 'ions' or electrified molecules are present. Since the X-rays

Illustration page 463

were generated by cathode rays, Thomson then left Rutherford to investigate them and turned his attention to the cathode rays themselves.

The German physicist Philipp Lenard had already shown in 1894 that cathode rays could penetrate a piece of metal foil, so they could not be gas molecules as Crookes supposed. Lenard believed them to be some form of electromagnetic radiation. Thomson, on the other hand, measured their velocity as some 1600 times slower than light, so he took the view that they were particles; by 1897 after a series of experiments he concluded that they must be 'small compared with the dimensions of ordinary atoms or molecules'. To settle the matter it was necessary to know the electric charge on the particles and their mass, but for a time Thomson was only able to devise experiments to find the ratio of the two quantities, not the quantities separately. Nevertheless, his results were the same for all gases and this, indeed, made it seem certain he was dealing with something smaller than an atom. However, by 1898 he had managed to measure the electric charge alone, though other, later, experiments gave more precise values, Robert Millikan at the University of Chicago obtaining very accurate results in 1909. It became clear, though, from Thomson's experiments alone, that the mass of the particles was very small (the mass could be determined separately once the charge was known) and was some 1,800 times less than that of the lightest atom. Thus, by the turn of the century, the existence of the 'electron', as it had come to be called, was established. The significance of the discovery was that it proved atoms were not the smallest particles (though they were the smallest that took part in chemical reactions).

In 1896, at the time Thomson was working on cathode rays, the French physicist Henri Becquerel discovered that the heavy element uranium was continuously emitting rays which, like Röntgen's X-rays, enabled a gas to conduct electricity. The question then arose whether uranium was unique or whether there were other chemical elements that behaved like this. Marie Curie, a Polish chemist who was married to the French physicist Pierre Curie, decided to try to find the answer, and together with her husband, working in a wretched laboratory at the School of Physics and Chemistry in Paris, in 1898 discovered radiation from the mineral pitchblende. After a long and careful chemical analysis, the Curies isolated the active ingredient, which turned out to be composed of two very active elements, one of which they named 'polonium', in honour of Marie's homeland, Poland, and the other 'radium'. The isolation of these elements demanded an elaborate series of chemical reactions, yet since these did nothing to impair their 'radioactivity' (a term coined by the Curies), it became clear that this must be due to some property of the atoms themselves.

Illustration page 463

Rutherford, who had left Cambridge in 1898 to move to McGill University in Canada as professor of physics, immediately set himself the task of studying radioactivity in detail. Soon he discovered that two sets of rays – alpha and beta rays, he called them – were being emitted, while further research with a colleague, the English physicist Frederick Soddy, led to the realization that these were not in fact rays at all but particles, and by 1903 brought them to the conclusion that the atoms of radioactive substances were spontaneously breaking up. In 1907 Rutherford moved back to England to Manchester University and here, with the assistance of the young German physicist Hans Geiger, known today as the inventor of the Geiger counter, he discovered that from whatever radioactive element they came, the alpha particles gave the same spectrum as atoms of the gas helium and they also had the same mass. Clearly, the alpha-ray particles were helium atoms. Further investigation showed that the beta-ray particles were high-speed electrons. It was now clearer what was happening when the atoms of radioactive substances broke down; they emitted helium atoms and high-speed electrons. But they did more than this for, in 1900, the French physicist Paul Villard had discovered that during radioactivity radiation was emitted as well. His investigations showed him that these rays were more penetrating than X-rays but behaved in a similar way. He called them 'gamma-rays', but since the alpha and beta 'rays' had turned out to be particles, not rays, Villard's conclusions were doubted.

Illustration page 462

The matter was finally cleared up in 1910 by William Bragg, then working at the University of Leeds, who found that when the rays struck atoms in a rarified gas, they caused these atoms to emit high-speed electrons which, in their turn, 'knocked off' electrons from other atoms. The gas thus became electrically conducting or ionized. It was a two-stage process, and it happened with both X-rays and gamma-rays; Villard was correct. But Bragg's research did more than this, for it uncovered a strange fact. When the X-rays caused ionization they behaved just as if they were particles, not electromagnetic waves, and he had to conclude that they seemed to be both. The significance of this we shall see shortly.

Evidence had accumulated to show that Thomson was right, and that the electron was a sub-atomic particle. It had a negative electric charge, however, which was equivalent (though opposite in polarity) to the positive electric charge displayed by the helium (alpha) particles. Since atoms were electrically neutral under normal conditions, in 1911 Rutherford proposed a theory of the atom which would fit in with all the results and incorporate the experimental evidence of Bragg on ionized gases. His suggestion was that every atom has a central core or nucleus; this has a positive electrical charge and contains most of the mass of the atom. The positive electric charge of this nucleus is balanced by one or more electrons which lie outside it. This 'plum pudding' atom was a bold step forward, even though it was soon to be replaced.

In 1912, the year after Rutherford's atomic model had been proposed, Niels Bohr, a young Danish physicist, went to work at Manchester, and it was while he was there that he began to lay the foundations for his own immensely powerful theory of the atom which he was able to complete the next year. Basically, Bohr's model, which replaced Rutherford's, consisted of a nucleus with a positive charge, as in Rutherford's model, but with the electrons orbiting around it. However it was more subtle than this simple description might lead one to suppose. The orbiting electrons were able to move only in certain specific orbits, rather as the planets only move round the Sun in particular orbits. When the Bohr atom receives energy, by heating, perhaps, or by receiving electromagnetic radiation, the energy is distributed among one or more of its electrons, causing them to jump instantaneously into other fixed orbits further out from the nucleus. After a tiny fraction of a second such disturbed electrons will drop back and, in doing so, cause the atom to emit electromagnetic radiation. The wavelength of this energy, Bohr claimed, depends on two factors. One is the number of orbits over which the electrons jump, the other the closeness of these orbits to the nucleus. Thus the wavelength may lie anywhere within the entire range of electromagnetic radiation, from gamma-rays and X-rays at the very short wavelength end, through the visible spectrum and right on to infrared or heat rays (discovered in 1800 by the astronomer William Herschel) and radio waves at the very long wavelength end. And since the orbits are fixed in definite positions, so each atom will have its own characteristic wavelengths which, in the spectroscope, are observed as specific lines.

A simple model of the structure of the atom, with electrons orbiting around the nucleus.

Bohr's atomic model was supported by other research that had been carried out at the close of the nineteenth century and early in the twentieth. It explained why the spectral lines of the element hydrogen appeared at a series of specific wavelengths which fitted in with a simple equivalent mathematical series, discovered in 1885 by the Swiss physicist and schoolmaster Johann Balmer. But even more significantly, it corresponded with some work by the German physicists Max Planck and Albert Einstein. In 1900 Planck, the professor of physics at Berlin, had put forward the suggestion that radiation does not appear in a continuous stream but in discrete packets or 'quanta' of energy, the number of quanta, and thus the total amount of energy, being greater the shorter the wavelength of the radiation. This had been confirmed in 1905 by Planck's colleague Einstein, who analyzed the 'photoelectric' effect in which short-wave ultraviolet radiation falling on a metal surface causes electrons to be emitted. The problem Einstein had to face was that the energies of the electrons were independent of the intensity of the ultraviolet light, but did depend on its wavelength, and he found that while it was inexplicable on the wave theory of light, it was just the result to be expected if light arrived in quanta, in discrete energy packets. Thus here was an observed fact, and a theoretical analysis, both of which made it clear that light and all electromagnetic radiation should be considered as a both wave and a particle; Bragg's results of 1910 were a confirmation of this.

Opposite A modern atom smasher used for examining the interactions of nuclear particles. Able to accelerate particles to velocities which are a sizeable fraction of the speed of light, this synchroton at the international research unit CERN (*Conseil Européen pour la Recherche Nucleaire*) was established at Geneva in 1952. The area shown (*above*) is an intersection point where particles are drawn off and their reactions after collisions are examined. *Below* Apparatus at the intersection of two storage rings for high-speed nuclear particles.

DO NOT LEAN

EXIT

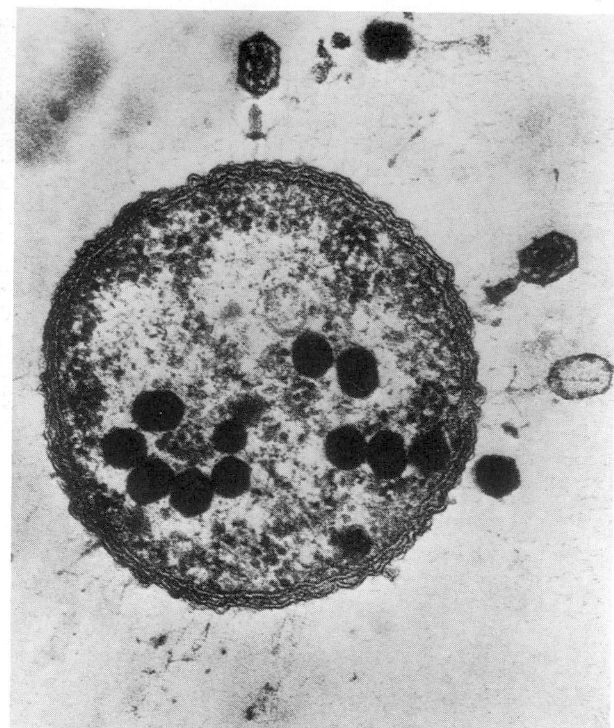

Above Gowland Hopkins (1861–1947), in a portrait by Meredith Frampton. Hopkins made important contributions to the study of amino-acids and vitamins, and the oxidation of living cells. Royal Society, London.

Above right Electron-microscope photograph of bacteriophages infecting a cell of the colon bacterium *Escherichia coli*.

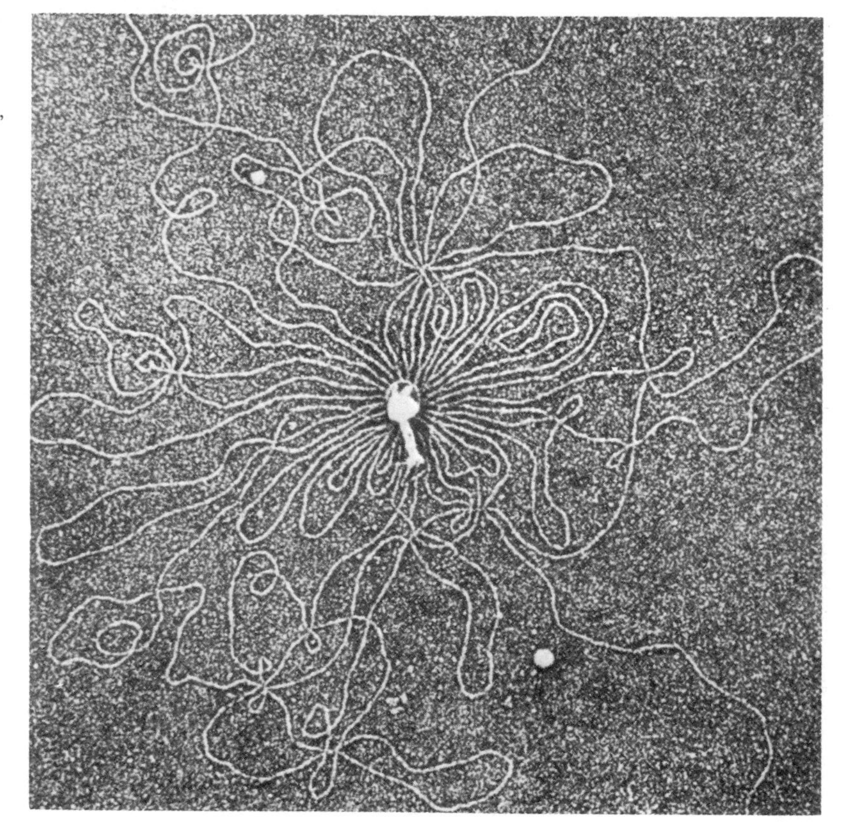

Right Photograph of a bacteriophage, taken at Frankfurt University through an electron microscope at a magnification of 8000 times. The threads are DNA molecules, and the head of the bacteriophage is about 0.75 microns (75 hundred thousandths of a millimetre).

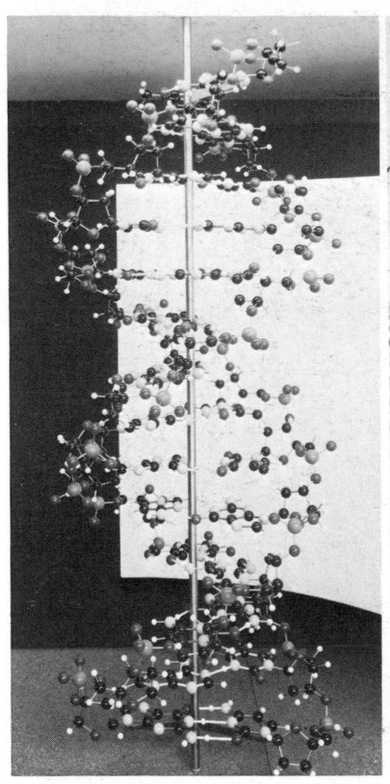

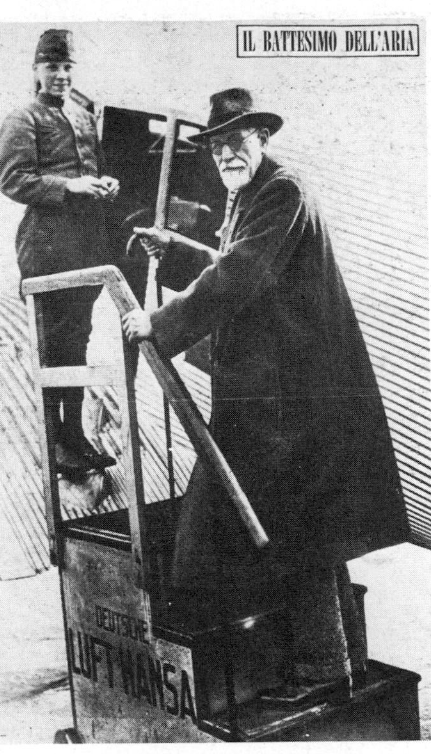

Left Sigmund Freud (1856–1939), taking his first flight at the age of 70. Freud's studies of the unconscious mind, based on his clinical psychiatric experience, and his theory of the importance of infantile sexual experience founded a school of psychology of great importance in the 20th century, complementing the more experimental methods of Pavlov's followers.

Far left Model of the DNA molecule, showing its double-helix structure. Science Museum, London.

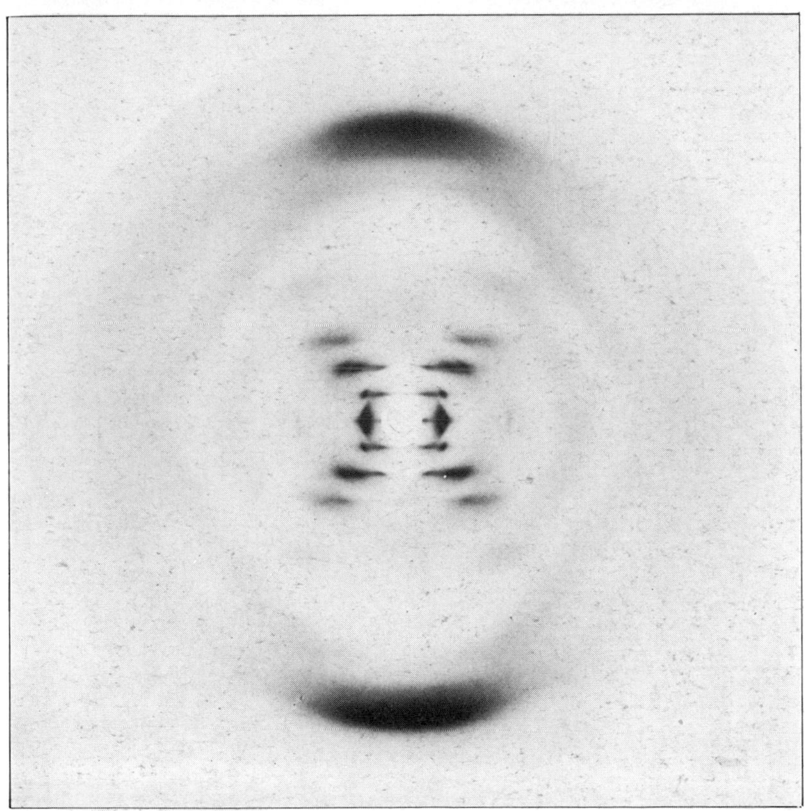

Left An X-ray diffraction pattern of B-DNA.

Above The participants of the fifth international physics conference at Brussels, arranged by the Solvay Institute, set up by the industrial chemist and social reformer Ernest Solvay (1838–1922). Those present included Pauli and Heisenberg (back row, fourth and third from right), Niels Bohr (centre row, far right), Max Planck, Marie Curie and Albert Einstein (front row, second, third and fifth from left).

Right Henri Poincaré (1854–1912), whose mathematical work developed the ideas of Lobachevsky and Riemann in a way that opened up many new applications for physics.

Right Three notable astronomers of the 1930s: Walter Adams (left) director of the Mount Wilson Observatory, California, the Briton James Jeans (centre), and the American Edwin Hubble (right), examining a model of the 100–inch reflector at Mount Wilson, then the world's largest telescope.

Above A detector for gravity waves built in the mid-1970s and here being adjusted by the designer Joseph Weber. The existence of such waves is predicted by relativity theory, and, although direct detection of them has yet to be made, there seems astronomical evidence that they exist. A pair of stars in orbit around each other in one of the host of such binary systems have been found to show a specific change in orbit, which is to be expected if such a system is emitting gravity waves as the theory predicts.

Above left Albert Einstein (left), with Hendrik Lorentz and Arthur Eddington (right).

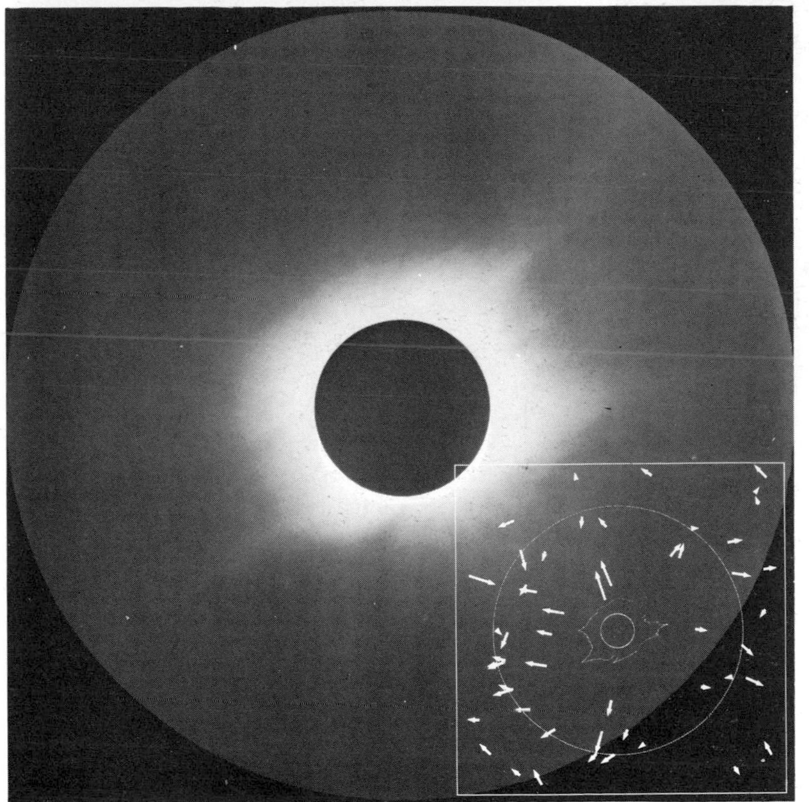

Left Photograph of the total solar eclipse of 1922, one of the occasions when observations were made to measure the deflection of starlight as predicted by relativity theory. The inset diagram shows the stellar displacements observed.

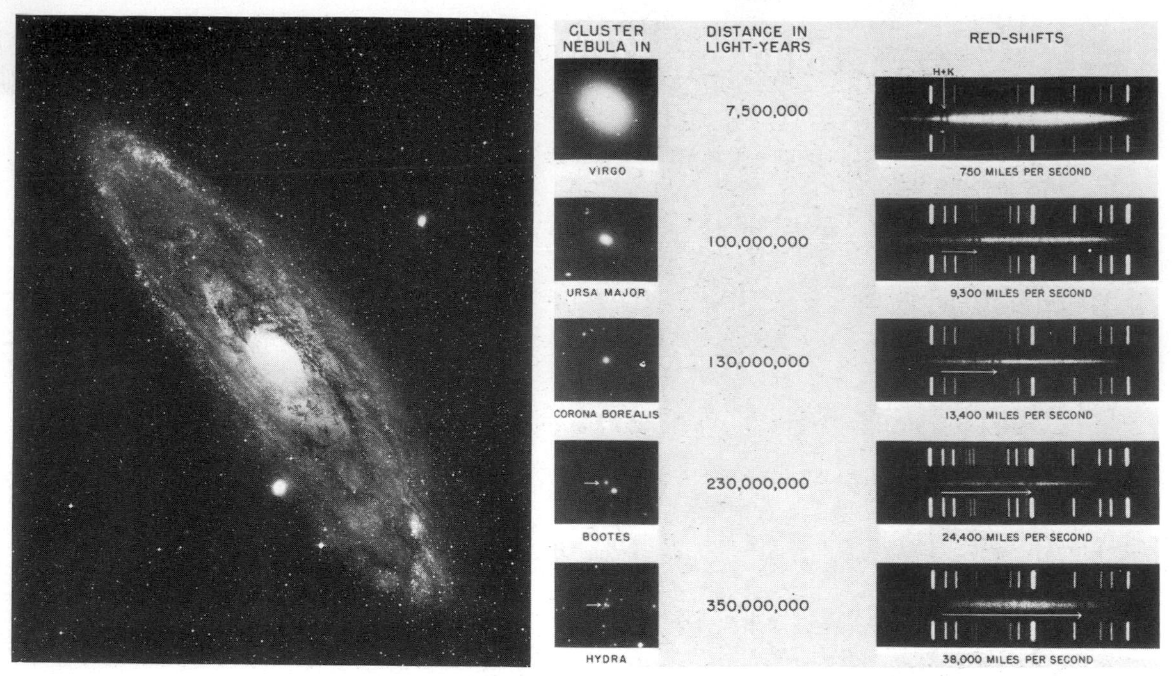
CLUSTER NEBULA IN
DISTANCE IN LIGHT-YEARS
RED-SHIFTS
H+K
7,500,000
VIRGO
750 MILES PER SECOND
100,000,000
URSA MAJOR
9,300 MILES PER SECOND
130,000,000
CORONA BOREALIS
13,400 MILES PER SECOND
230,000,000
BOOTES
24,400 MILES PER SECOND
350,000,000
HYDRA
38,000 MILES PER SECOND

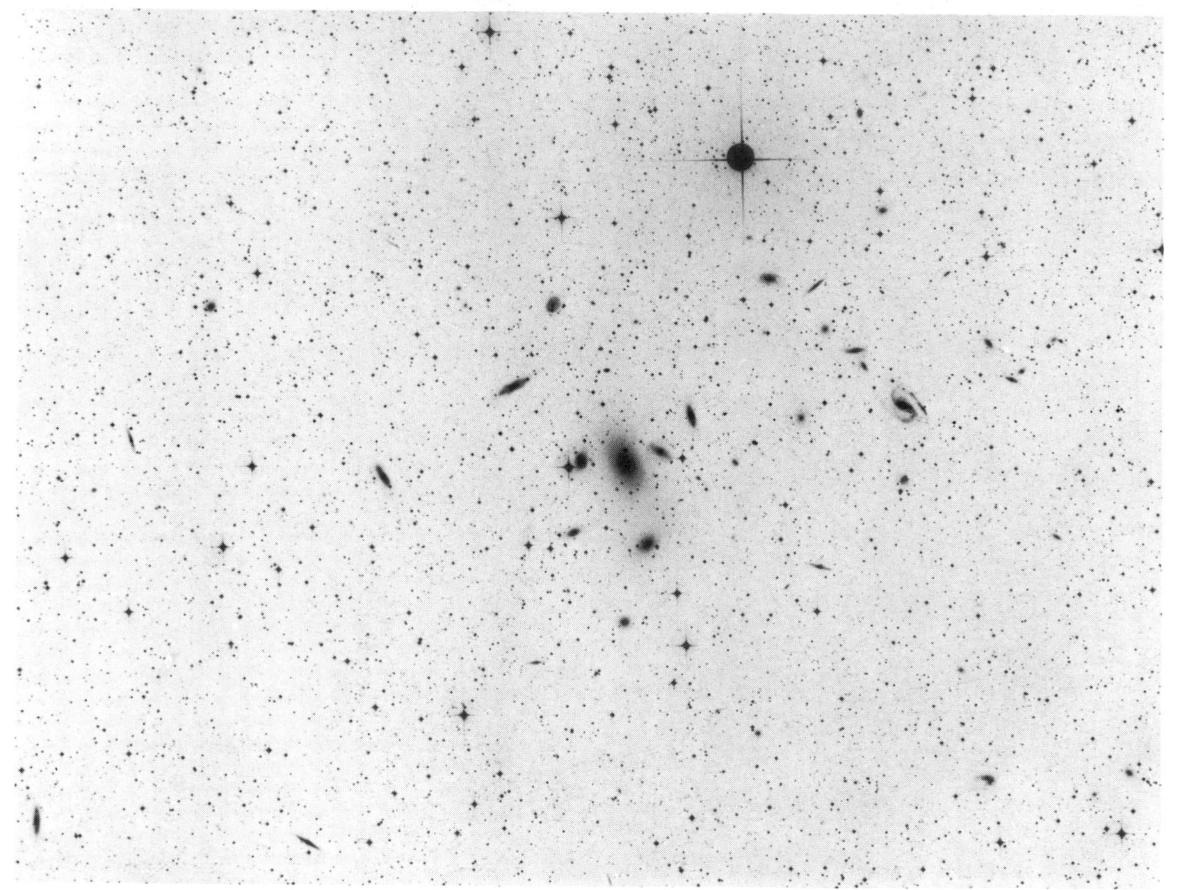

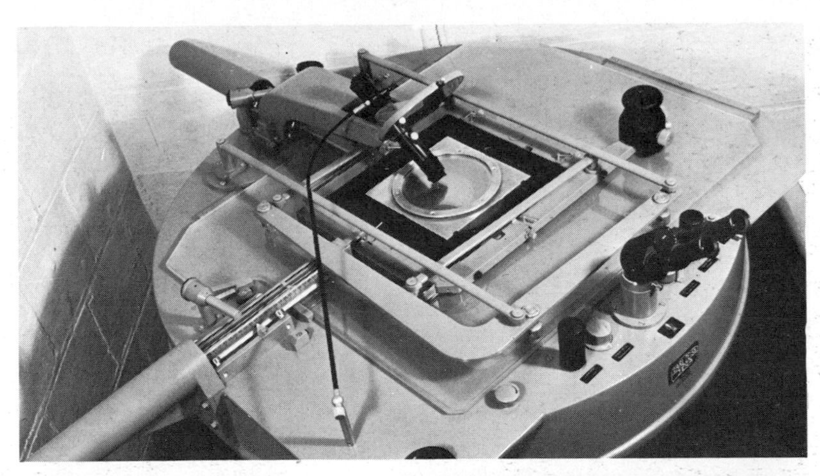

Left A manually operated measuring microscope for determining the positions of celestial objects on photographs.

Centre left A modern plate-measuring machine for determining positions of celestial objects. This works some 40,000 times faster than the manual machine shown above.

Opposite above left The great spiral galaxy in Andromeda, light from which takes more than two million years to reach us. This was the first galaxy to be resolved optically into separate stars, though this could not be achieved until the 5 m (200-inch) telescope was built at Palomar, California, after World War II.

Opposite above right The relationship between distance of a galaxy and the red shift of the lines in its spectrum. The spectra (on the right) of the galaxies are each sandwiched between companion spectra and the shift is indicated by the arrow. The distance in light years (1 light year = 9.6 million km) were worked out by Edwin Hubble and need revising in the light of recent research; they should read 78,000,000 light-years; 1,000 million light-years; 1,400 million light years; 2,500 million light-years, and 4,000 million – an increase of some 10 times. The velocities remain unaltered.

Opposite below A photographic negative – the type of photograph now used by astronomers for automatic analysis – of a cluster of galaxies in the southern hemisphere constellation Paro (the Peacock), photographed with the Anglo-Australian 3.9 m telescope

Left One of a new generation of optical telescopes; the Multiple Mirror Telescope at Mount Hopkins in Arizona. Each mirror has a diameter of 1.8 m but the whole instrument is equivalent to a telescope with an aperture of 6 m.

Right Using an optical interferometer and the computer analysis of a large series of photographs, the distortions caused by the Earth's atmosphere can be minimized to give this unique photograph of the red supergiant star Betelgeuse.

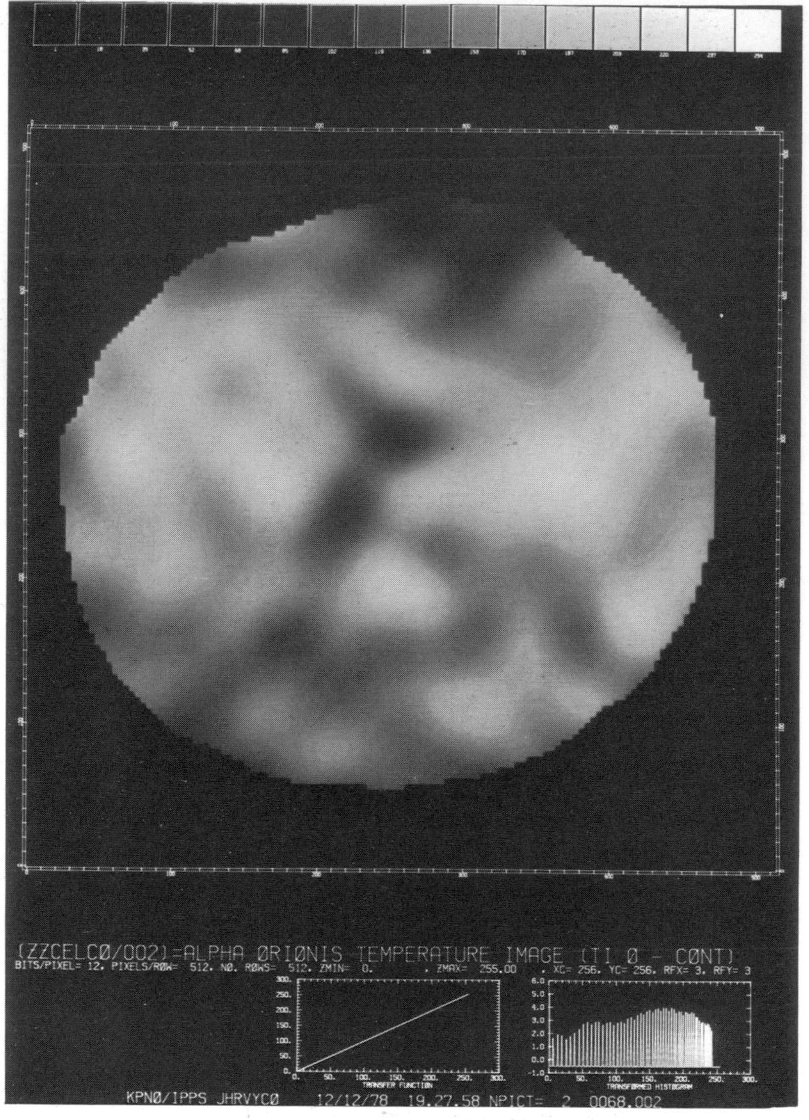

Right The precursor of radio telescopes: the steerable antenna used from 1931 onwards by Karl Jansky (1905–50) at Holmdell, New Jersey, for investigating short-wave radio reception.

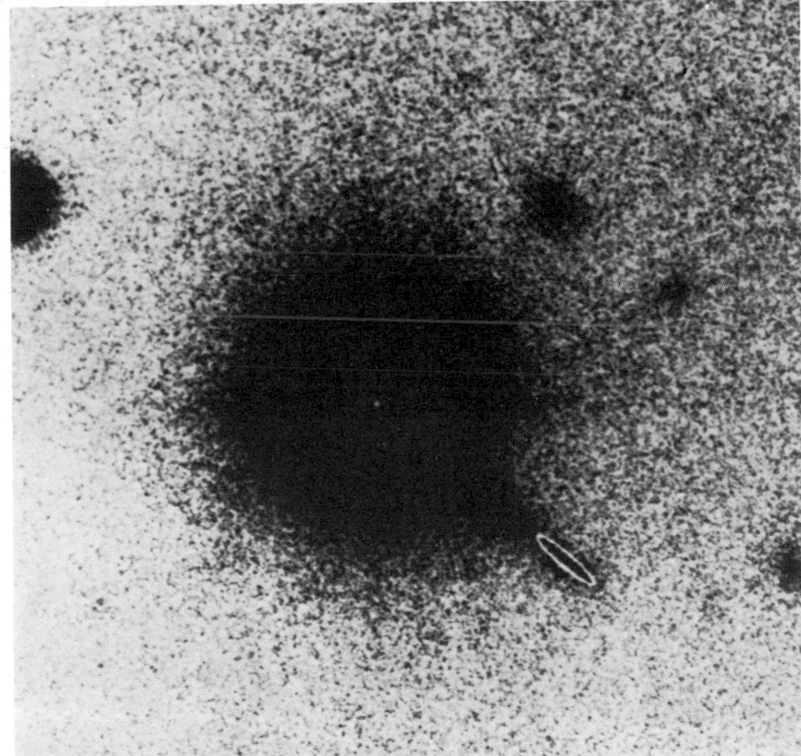

Above left The first very large dish-type radio-telescope which could be pointed in any direction was the 76 m diameter radio-telescope at Jodrell Bank near Macclesfield, England, built in 1956.

Above Aerial view of the 5 km aperture synthesis interferometer at the Mullard Radio Astronomy Observatory at Cambridge University. It contains eight 13m reflectors in all.

Left The quasar 3C273. Still something of a mystery, quasars, which were first detected by radio astronomers in the early 1960s, seem likely to be the cores of very distant galaxies. This one is ejecting a jet of material (the small white oval).

Right The original atomic clock, whose 'pendulum' is a collection of vibrating caesium atoms. This was built at the National Physical Laboratory at Teddington, England, in 1955, and led directly to the development of atomic time standards, now adopted throughout the world. This clock is also the precursor of the modern maser timekeepers, that are correct to 1 second in 100,000 years and the advent of which has made possible the linking of observations made by radio-telescopes on different continents.

Right Robert Goddard (1882–1945), photographed in 1935 with a liquid fuel rocket of his own design. Despite his pioneering work, rocketry was not fully developed in the United States until after World War II.

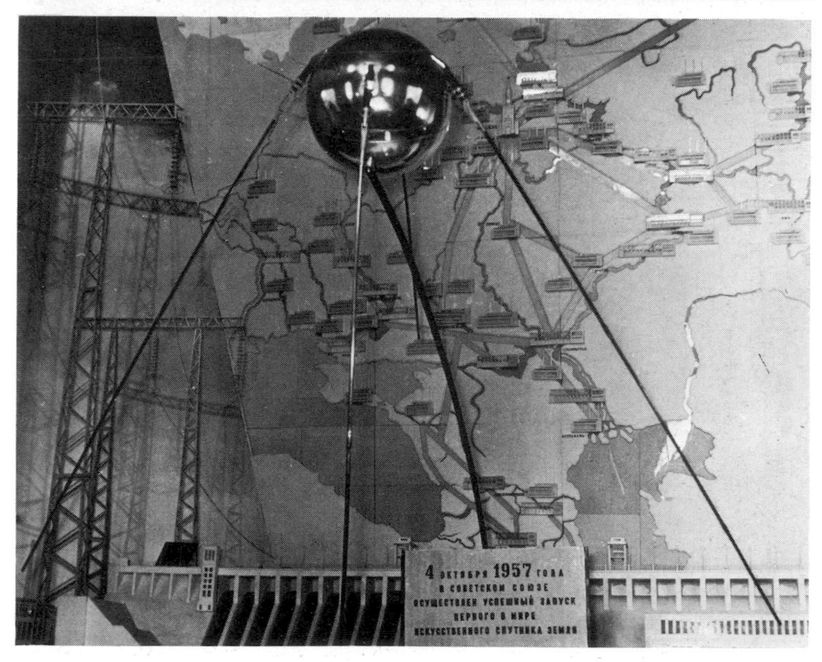

Right The first artificial satellite to be successfully launched, the Russian Sputnik of August 1957.

Left Man in space – the astronaut Edward H. White outside the Gemini 4 orbiting spacecraft in June, 1965. As well as exploration, the human presence in space has had medical and astronomical importance.

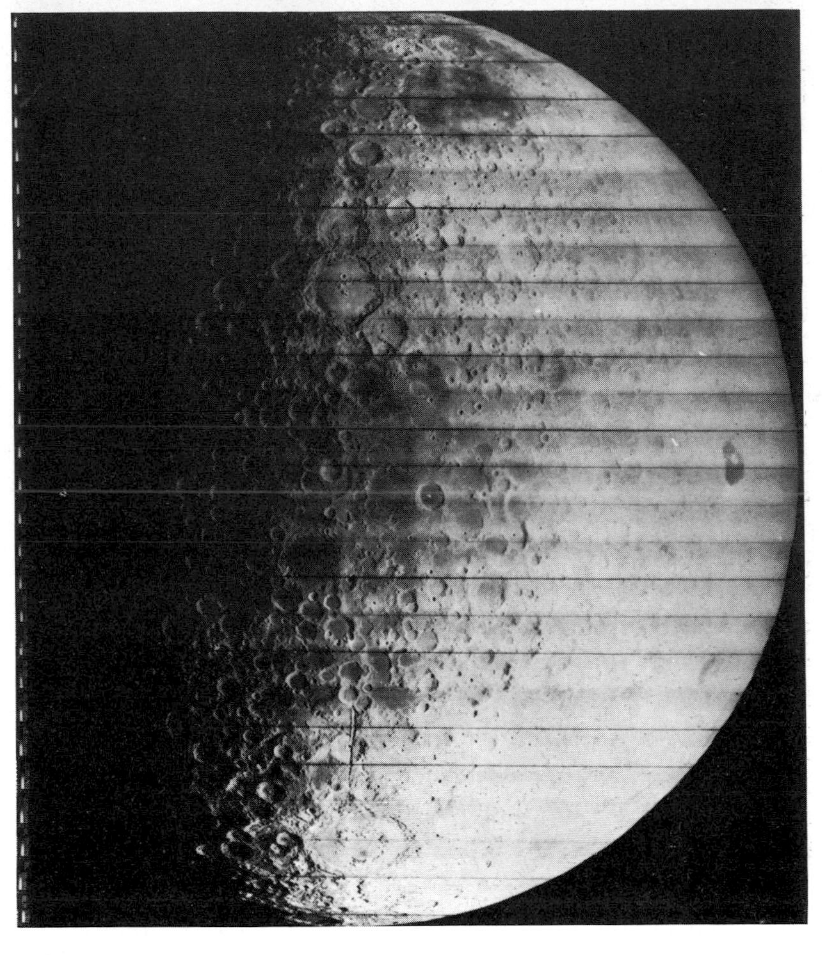

Left The far side of the Moon, which is always turned away from the Earth, but here photographed by the American Moon-orbiting spacecraft Lunar Orbiter IV in 1967.

Left The American Skylab in orbit around the Earth, photographed from the command and service module, 1973.

Opposite above A thin section of lunar rock viewed through a microscope by polarized light, allowing the various minerals to be clearly differentiated. Sample obtained by astronauts on Apollo 17, in 1972.

Opposite below Two spacecraft on the Moon. The Apollo 12 manned spacecraft is on the horizon and in the foreground is the unmanned soft-landing craft Surveyor III. Photograph taken in November, 1969.

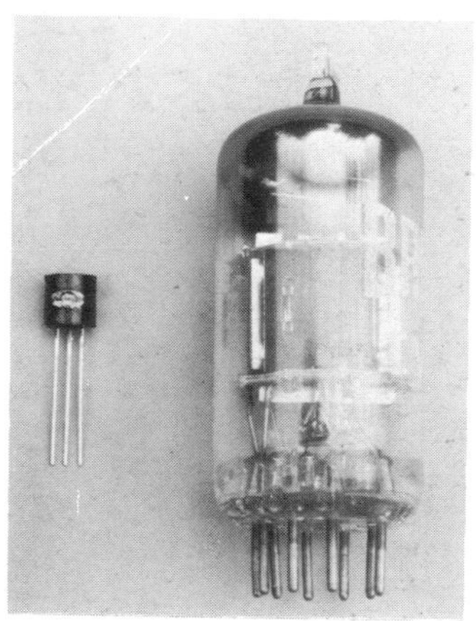

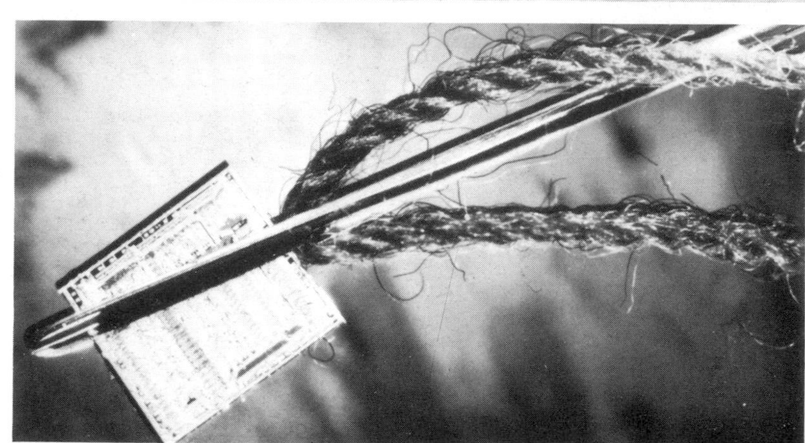

Left An electronic computer of the 1950s. Although it used valves (vacuum tubes) and occupied a large space, it could perform very many more mathematical operations than the differential analyzer.

Opposite above left The first thermionic valve, or vacuum tube (diode), used by John Ambrose Fleming in 1904 at University College, London. This was the forerunner of all later thermionic valves, particularly those developed in the United States by Lee de Forest. Science Museum, London.

Opposite above right An early radio transmission from the Marconi Works at Chelmsford, England, in 1920.

Opposite centre left On the right is a miniature thermionic valve (vacuum tube), some 80 to 100 times less bulky than the first examples, compared with a transistor (left). The other advantage of the transistor, which was developed in 1948, is the low voltage it uses – 9 volts compared with 200–300 for a vacuum tube.

Opposite centre right The micro-miniature circuit is etched on to silicon and contains the equivalent of many transistors and other electronic components. Such circuits, which were developed in the 1960s, are very reliable. The one shown here fitting into the eye of a needle would now be considered large.

Opposite below The mechanical computer, or 'differential analyzer', which could perform the mathematical operation of integration, was designed and built at Manchester University, England, in 1935. It was only capable of a comparatively few mathematical operations. It was used, among other things, for computing railway timetables.

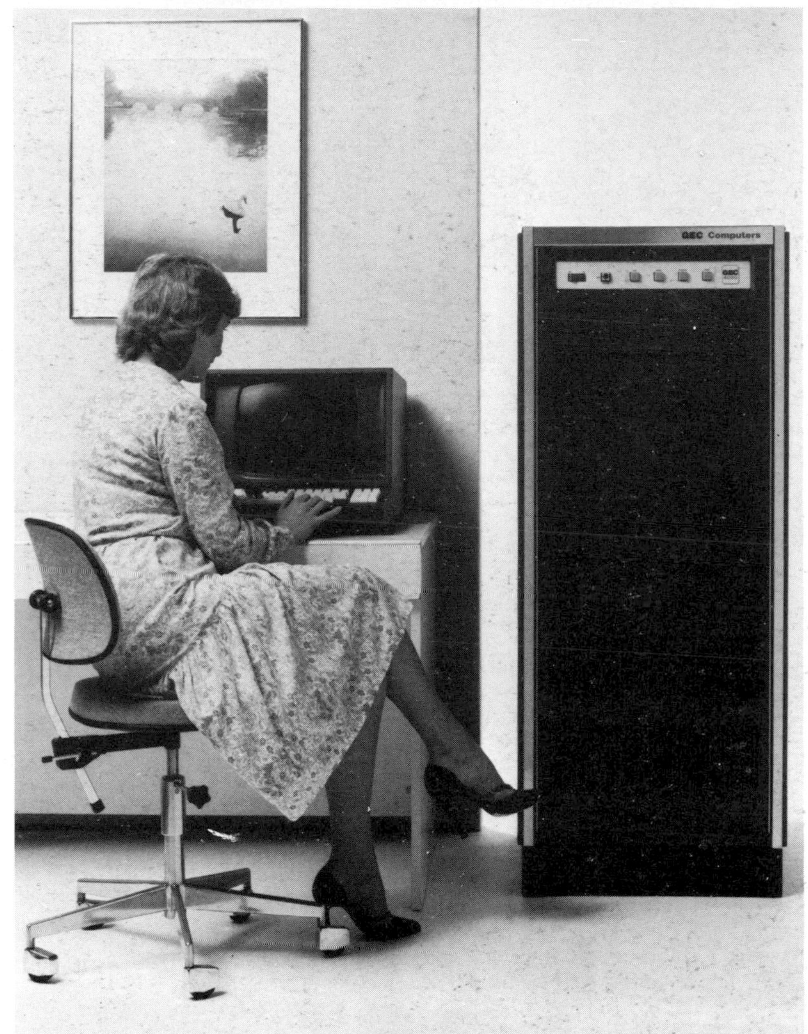

Left An astonishingly compact 1980s computer and display unit, made possible by a new generation of silicon chips. Such compact units enable the computers to be widely used in every branch of scientific research.

Right During the International Geophysical Year (IGY) held in 1957–58, international scientific co-operation took place on an unprecedented scale, crossing political barriers. One of the positive scientific results was the discovery that the aurora at one magnetic pole was accompanied by an aurora at the other magnetic pole; this was an important factor in coming to a complete explanation of the phenomenon. An aurora (shown here) is caused by the influx of electrically charged atomic particles from space, some which are captured by the magnetic field of the Earth and are then ejected into the atmosphere near the Earth's magnetic poles. The particles electrify (ionize) the air molecules which glow, giving rise to the aurora.

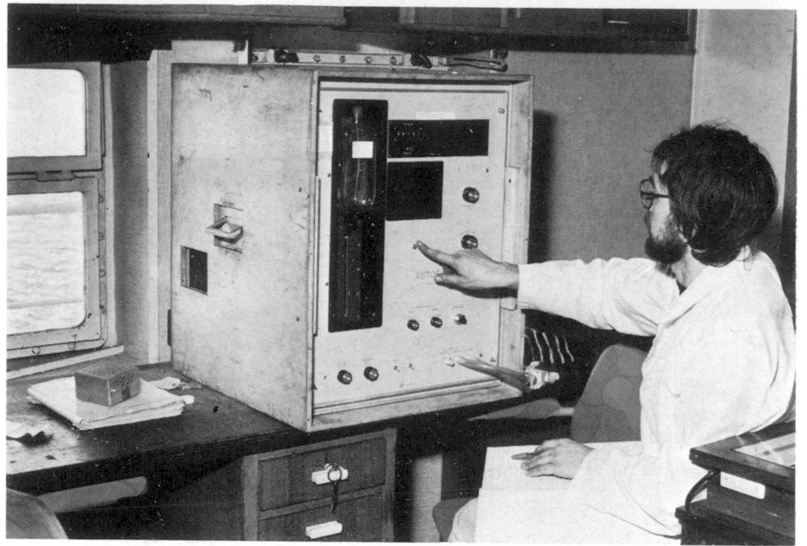

Centre Geophysics – the study of the Earth as a physical body in space – is one of the new scientific disciplines to emerge in the 20th century. The taking of soundings in the oceans to chart the physical structure of the ocean beds and the chemical composition of the oceans themselves. As a result of these studies, the revolutionary theory of plate tectonics was developed in 1965, to explain continental drift, the distribution of volcanic zones and of magnetized rock.

Right Robert Wilson (left) and Arno Penzias (right) standing in front of the large antenna which they designed for work in radio wave propagation at the Bell Telephone Laboratories. They discovered the radio microwave radiation (wavelength 7 cm), visible all over the sky. Believed to be due to hot gas which has cooled over the aeons of time since the big-bang creation of the universe, its discovery has had the most profound effect on late 20th-century concepts of the universe.

The quantum theory, as this theory of discrete wave-packets of energy came to be called, was developed in rigorous detail by a number of mathematical physicists, most notably Louis de Broglie, Erwin Schrödinger, Paul Dirac and Werner Heisenberg. The full theory is extremely mathematical and cannot be discussed here, though one important consequence must be mentioned, and that is Heisenberg's principle of indeterminacy. Heisenberg, who was born in 1901 in Würzburg, Germany and studied for a time under Bohr, proposed his indeterminacy principle in 1927. He showed that the very act of assuming that energy moves in discrete quanta means that certain pairs of variables that constantly affect one another, such as time and energy, cannot be determined with complete accuracy. The more precisely one is defined, the less precisely can one define the other. There is always a specific, though very small, amount of indeterminacy present. Not all physicists were ready to accept this principle, though it certainly seems a consequence arising from the quantization of energy, and it led Einstein to reject the quantum theory; 'God does not play dice', he often used to say. Clearly, it is a principle with the most profound philosophical implications.

Illustration page 500

Bohr's theory of the atom was also able to provide a theoretical background to the Periodic Table of chemical elements, drawn up first in 1869 by the Russian chemist Dmitri Mendeleyev (sometimes spelled Mendeleéff). The table displays the chemical elements arranged according to atomic weight, and lays out in vertical columns those with similar chemical properties. What the Bohr model explained was why they should be arranged like this; chemical properties were due to the number of orbiting electrons and these, in their turn, depended on the electric charge on the nucleus. This electric charge itself depended on the mass of the nucleus and so on the atomic weight.

Illustrations page 464

The explanation of the Mendeleyev Table drew attention to the nucleus, which as Rutherford had already demonstrated was very small, some ten thousand times smaller than the atom itself. What was it really like and of what was it composed? The first really clear work towards answering these questions was carried out by Francis Aston, who spent most of his life at the Cavendish Laboratory as Thomson's assistant. By 1919 Aston had discovered that the gas neon could exist in different forms, having different atomic weights though similar chemical properties, and later research showed this to be true of other elements, which also possessed such 'isotopes'. Clearly, then, something could change within the nucleus without affecting the number of orbiting electrons and thus the electric charge on the nucleus.

In the same year, 1919, Rutherford bombarded nitrogen atoms with alpha particles and disintegrated the nitrogen nucleus. The products of this disintegration were positively charged particles, soon to be called 'protons'; clearly these were components of the nucleus, but were they the only ones? The answer to this had been hinted at by

Aston's work, but had not been fully explained; much more experimental work was needed before this could be done. Nevertheless much had been achieved by the time 1920 dawned, for it had been shown that the atom certainly consisted of still smaller components, the electron and the proton. Moreover, Rutherford's bombardment of the nitrogen nucleus had initiated a new and powerful technique, the forerunner of all the later artificial atomic-disintegration methods to be used by John Cockroft and Ernest Walton in 1932 to break down the nuclei of lithium atoms, and the giant accelerators and other atom-smashers which were to be so productive for research after World War II. Incidentally, Rutherford's 1919 experiment was also the culmination of the alchemist's dream, for by removing a proton from the nitrogen nucleus Rutherford had transmuted nitrogen into another chemical element (an isotope of carbon). This process occurs naturally, of course, during radioactivity because radium and other radioactive elements become transmuted as they disintegrate.

Illustrations page 497

The next step in the discovery of the nature of the nucleus did not occur until 1932 when James Chadwick, another member of Rutherford's research team, discovered the neutron, a nuclear particle with the same large mass as the proton but having no electric charge. This immediately cleared up the anomaly of oxygen in the Periodic Table, where it appeared with a chemical combining-power of eight, which meant it had eight electrons, though its atomic weight was 16, which is twice too much. But with the discovery of the neutron this could be explained; the oxygen nucleus possessed eight protons (to give it a positive charge of eight to balance the eight electrons) and eight neutrons to account for the extra mass. Chadwick's discovery also, of course, explained the nature of Aston's isotopes. Here were nuclei from which neutrons had escaped or which had gained an extra neutron or two.

Meanwhile more studies had been made of radioactive decay, and particularly of what precisely happened when a radioactive nucleus emits beta particles. By 1928 Paul Dirac had worked out that, from a theoretical point of view, the emission of beta particles could be accounted for if a new sub-atomic particle were created at the moment the beta particles left the nucleus. Such a particle should have a positive electric charge and a mass equal to that of the electron. In 1932 Carl Anderson at the California Institute of Technology observed just such a particle; it was named the 'positron'. Yet beta-particle production still presented problems and not until 1934 were these all solved, when Enrico Fermi, an Italian physicist who was later to emigrate to the United States to avoid Italian Fascism, gave a complete explanation using a particle known as the neutrino, invented for theoretical reasons by the Swiss physicist Wolfgang Pauli the year before. The neutrino had no electric charge but had no mass either; its one property was that it seemed to spin and thus contribute the necessary momentum (angular momentum) to the beta-particle production, though not until 1956 did a nuclear-reactor team in the

United States provide experimental evidence for its actual existence.

Fermi's theory of beta-particle production was highly productive of new ideas, but it too raised new questions, in particular how protons and neutrons were held together in the nucleus. The forces involved seemed far too great until in 1935, the year after Fermi proposed his theory, the Japanese physicist Hideki Yukawa hit on an explanation using yet another particle, the meson, which acted as a kind of intra-nuclear glue and so gave the necessary force. The meson, however, turned out to be a particle with a lifetime not greater than one ten-thousand-millionth of a second, and one that could take a number of forms. Nevertheless, in spite of its very transitory nature, it has been observed, first in 1947 and then, in other forms, in 1948 and 1962.

With the discovery of neutrinos, mesons and even more esoteric particles like those with 'strangeness' (a factor devised to explain the peculiar observed results of very high-speed sub-atomic particle collisions), the theory of the atom and its constituent parts has, since the mid-1950's, entered a new and complex phase. New nuclear particles proliferate, with the inevitable consequence that nuclear physicists have sought still more fundamental underlying units. Indeed, three new particles (and their accompanying anti-particles) have been proposed – they are called 'quarks' after a passage in James Joyce's *Finnegan's Wake*, though the German word 'quark' also has the colloquial meaning of 'nothing'; at the time of writing (1982) no quarks have yet been observed. Yet quarks or no, there is evidence from the work of Murray Gell-Mann and others, especially in the United States, that a more tightly-knit theory of the atomic nucleus may be evolved, forging some links with more recent ideas about gravitation, and so give a 'unified field theory' incorporating into one all-embracing scheme, the micro-world of the nucleus and the macro-world of the universe taken on the largest scale.

For Plato's doctrine of the correlation between microcosm and macrocosm, see page 97.

Another scientific development arising from the atomic and quantum theories has been that of solid state physics, that new investigation into the behaviour, properties and structure of solids. This has led to new investigations of magnetism, of substances that conduct electricity well and those that do not, and to an examination of materials half-way between, the semi-conductors. The last has had the most profound practical as well as theoretical results; it lies behind the invention in the United States in 1948 of the transistor by John Bardeen, Walter Brittain and William Shockley, and the more recent development of microminiature circuits – the ubiquitous 'chips' – at the heart of the new microcomputer, and thus of the important new field of artificial intelligence. Another consequence of great practical use in scientific research as well as in general applied technology and medicine, is the laser, and its microwave counterpart, the maser. In these devices light and very short wavelength radio waves (microwaves) are generated using techniques derived from atomic physics. These microwaves appear in an unusual form because the entire radiation output has all its waves in step with one another, thus giving narow beams of very high energy. Among the applications of these is the development of highly accurate measuring equipment. Thus twentieth-century physics has forged new and powerful weapons for both theoretical reseach and also for application in every walk of life.

Relativity

The twentieth century has been a time of more fundamental changes in scientific outlook than any other age. As we have just seen, research by Rutherford and others led to the once-solid atom, the basic unit of every chemical element, being found to comprise even smaller units; Planck, Einstein and others demonstrated that the apparently continuous flow of energy from Sun, stars and other sources comes in discrete units, and that all radiation behaves like separate packets of energy each of which has the properties of a wave. The classical physics of the nineteenth century had gone, never to return; in its stead was a world of transitory particles and an element of indeterminacy when it comes to determining events at the most fundamental level. Yet quantum theory and the nuclear atom were not the only changes that the new century brought. During its first two decades the entire foundations of everyday physics were overturned by the theory of relativity.

Like other revolutionary theories before it, relativity had a venerable ancestry. As early as 1632 in his *Dialogue concerning the Two Chief World Systems*, Galileo had written about relative motion, pointing out that physics experiments on moving bodies made on board ship in a cabin below decks, would not tell the observer whether the ship were stationary or sailing along at an even speed. Galileo's argument was aimed at answering his critics about the laws of physics and the moving Earth, but no matter; the principle was there. Descartes appreciated the point, too, and Newton also claimed that the relative motion of two bodies in a given space would be the same whether that space is at rest or moves uniformly in a straight line though, unfortunately, he clouded the issue by his conceptions of absolute motion and absolute time. In a sense, of course, it does seem to us that there are absolute standards; we think we can tell whether something is stationary or moving just by observing it. Yet in fact we cannot make an absolute determination; if a body is stationary with respect to the Earth, it is certainly not so with respect to the Sun, since the Earth moves continually in orbit round the Sun. Yet we cannot take the Sun as absolutely fixed, either, for, as William Herschel proved, the Sun is moving in space with respect to the stars, and the stars themselves are all in motion. There seems, therefore, to be no place which we can regard as absolutely at rest, a conclusion confirmed in 1889 after a very careful study by the French mathematician Henri Poincaré.

For the Dialogue*, see page 343.*

Illustration page 500

Poincaré's analysis had profound implications, for in his day the laws of physics were all based on observations from Earth which, for convenience, was tacitly considered at rest. Yet if nowhere were at rest these laws needed re-examining; were they still valid in a universe where everything is in relative motion? The significance of this question was underlined by the strange results of an experiment carried out first in 1881 by the American physicist Albert Michelson. At the time there was considerable discussion about how the motion of the Earth in space should affect light waves, and various experiments had been tried but all had given equivocal results. What Michelson realized was that since light waves travel in the aether (for this was what was thought at the time) and the Earth is moving relative to this aether, the velocity of the light waves ought to appear different if one measures them in the direction of the Earth's motion in space, and then compares the result in a direction at right-angles to this. In the first case one ought to obtain a higher velocity (speed of light + the velocity of the Earth) than, in the second case (the speed of light alone). For making his observations Michelson used a special interferometer of his own design to cause the two light beams to interfere with one another; this was a very sensitive technique, yet the experiment seemed to be a failure, for it showed no difference at all between the two velocities. However, Michelson put his failure down to lack of refinement in the experimental arrangements; after all, the speed of light is almost ten thousand times faster than the velocity of the Earth in its orbit, so the difference he was trying to detect was very small indeed. With the aid of Edward Morley, Michelson redesigned the experiment which, in its improved form, was carried out in 1887 at the Chase School of Applied Mechanics at Cleveland, Ohio. Yet to their surprise, they still obtained a nil result; there appeared to be no difference whatsoever between the two velocities. The speed of light was the same in both directions.

In spite of the fact that Michelson was mistaken in thinking that the velocity at right-angles to the Earth's motion should be exactly the speed of light, as the Dutch physicist Hendrik Lorentz was quick to point out, the nil result should not have occurred; that it did showed that something very peculiar was happening. Indeed, what the experiment seemed to demonstrate was that if any other velocity were added to the speed of light, one still ended up with the speed of light. Clearly, this went against all common sense and some explanation had to be sought. Two years later, in 1889, the Irish physicist George FitzGerald suggested that as the apparatus was carried through the aether with the Earth, the arm of the interferometer lying in the direction of that motion contracted a little, so the light had less far to travel than Michelson had thought; it would certainly lead to the result he obtained. Others, most notably the British physicist Joseph Larmor and the Dutchman Hendrik Lorentz, independently studied the situation posed by an aether, carrying electromagnetic waves, and a moving Earth. In essence what they were seeking was a way to transform Maxwell's equations expressing the behaviour of electromagnetic waves so that they remained valid in such a situation. This led, finally, to a set of equations by Lorentz covering length,

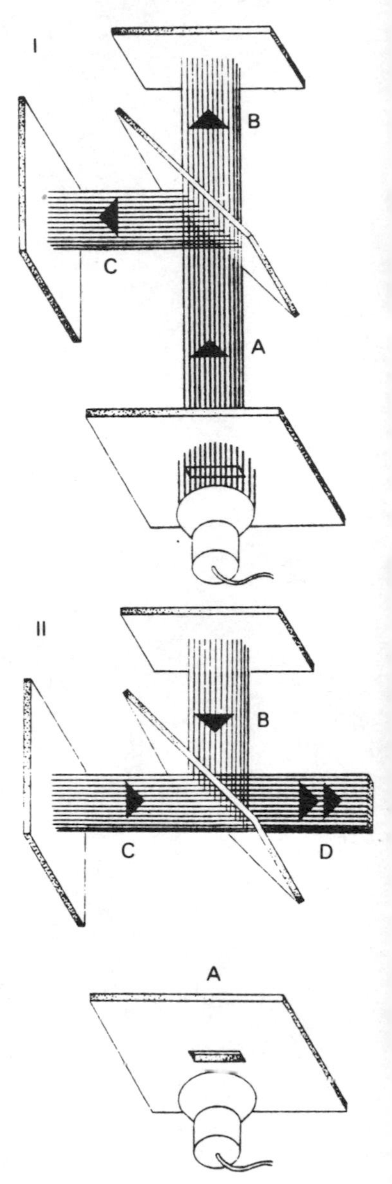

The Michelson-Morley experiment of 1881. Light from source A was partly diverted at right angles to mirror C, while other light proceeded directly to mirror B. The two beams were reflected to point D. It was expected that they should show a difference in interference fringes on account of the movement of the Earth on the axis AB, but none was found.

mass and time, where the velocities concerned were all less than the speed of light. The stage was now set for the theoretical advances of Albert Einstein.

Einstein, born at Ulm in Germany in 1879 and a trained physicist, was working as an examiner in the Patent Office at Berne, Switzerland at the time he published his first scientific paper on relativity. The paper appeared in 1905 in the *Annals of Physics* under the title 'On the Electrodynamics of Moving Bodies' and was a model of clarity, showing that Einstein had taken a long hard look at the problem and rethought the basic physics involved in the most fundamental way. This forced him to reject any absolute stationary space and the existence of an aether; it also led him to formulate new equations, from which the Michelson-Morley result turned out to be an expected consequence; Einstein also concluded that the velocity of light was the greatest velocity in Nature; nothing could travel faster than this. Known later as the 'Special Theory of Relativity', it restricted itself to bodies moving relative to each other at uniform (i.e. not accelerated) velocities, but it was immediately recognized as of immense importance, even though some physicists failed to accept it. It brought Einstein an academic position in Berne, and then a professorship of physics at Zürich; in 1910 he moved to the chair of physics at Prague.

Einstein did not rest content with his 'special' theory; he set about dealing with the far more difficult situation in which bodies are moving relative to one another in accelerated motion, that is in the general conditions we find in the natural world. Moreover, almost immediately, and still in 1905, he published another paper in the *Annals of Physics* 'Does the Inertia of a Body Depend on Its Energy Content?' It was here that he proposed his now famous equation $E = mc^2$, a formula that expresses the relationship between energy (E) and mass (m). The letter c represents the speed of light, so the formula shows that if one could annihilate a mass then the energy emitted would be enormous (because it is multiplied by $c \times c$, and c is a very large number). This energy equation lies at the basis of nuclear-power generation and, of course, of the atomic bomb; it also has important implications in astronomy, as we shall see in a moment.

The development of a general theory of relativity which would incorporate accelerated motion took time; it required a special mathematics – the tensor calculus – and it was not until 1915 that Einstein was in a position to publish it. When it did arrive it was clear that here was another great step forward in man's understanding of Nature. Because the theory dealt with accelerated motion, and since gravity causes bodies to fall at an accelerated pace, general relativity was also a theory of gravity. It showed that while gravity is associated with the mass of the body, this is because space is distorted by the presence of a large mass. Indeed, one can say that it is the distortion of space that gives us what we call gravity. Newton's theory stated that the force of gravity between bodies depends on the distance between them, but in general relativity this distance is affected by the presence of matter. The distance is not the simple straight line we usually think of, but a curve – for in relativity the equations reach their most elegant, and certainly their simplest, form when space is considered to be not Euclidean (flat) but curved in the way suggested in the late 1850s by Riemann. The difference is negligibly small over terrestrial distances, but is significant when it comes to astronomical distances as, for example, when one is computing planetary orbits. Indeed, one of the early proofs of the validity of general relativity was the fact that it could account with great precision for the orbit of the planet Mercury, for which Newtonian gravitation gave a value that was far too small.

For Riemann, see page 479.

General relativity had many other consequences that have been checked over the years and found to be confirmed by observation, strange though some may seem. Thus, it has been found that time is not absolute; it goes more quickly to an observer who is watching time passing on a body which is travelling very fast relative to him. This has been seen in the observed lifetime of mesons, and in other ways using modern very precise atomic clocks. Again, bodies moving very fast relative to an observer appear to increase in mass, as has been confirmed by measurements made in atom-smashing machines where nuclear particles are accelerated to velocities that are a sizeable fraction of the speed of light. For here, as elsewhere, these relativity effects are only noticeable when the relative velocities concerned are very fast indeed.

Perhaps the most spectacular, and certainly the most significant of the 'predictions' of general relativity, was the bending of starlight. We think of light travelling in straight lines, but if space is curved the path of light will be a curve (a 'geodesic'); more particularly, if light passes close to a massive body like the Sun, then space will be more curved and light will travel in a path deviating still further from a straight line. A total eclipse of the Sun is an ideal time to observe such an effect, since when the Moon passes in front of the Sun and obscures it completely, stars near in the sky to the Sun can be observed. According to general relativity, such beams of starlight should be deflected by the presence of the Sun, and in 1918 a total eclipse occurred at which this aspect of the theory could be tested. Arthur Eddington, then at the Royal Observatory at Greenwich and the first British physicist to understand thoroughly the full significance of Einstein's general theory, went to Principe Island in the Gulf of Guinea to observe the eclipse and try to detect this 'Einstein shift'. His expedition was successful and this unusual result – for it was not a consequence expected on the basis of Newtonian gravitation – did much to encourage the acceptance of relativity theory. Subsequent research has confirmed it in many ways and shown it to be more accurate than Newtonian gravitation, to which, however, it reduces in far less precise applications. The question now is not whether relativity is correct – it clearly marks a vast step forward in our

Illustrations page 501

understanding – but what its successor, a still closer approximation to the facts, will be.

Astronomy in the Twentieth Century

Astronomy has made great strides during this century, most significantly in the field of stellar astronomy, though more recently spacecraft have brought a new look to studies of the solar system, while a truly scientific approach to cosmology – the study of the universe as a whole, its beginning and possible end – has at last become possible. Some of these advances have been due to totally new equipment and techniques, as will become clear in a moment.

As a start, it will be most convenient to consider the main advances in stellar astronomy, which were particularly associated in the first half of the century with the Danish astronomer Ejnar Hertzsprung, the American Henry Norris Russell, and the Englishman Arthur Eddington. Their work was based on the foundations laid by Huggins and other astronomical spectroscopists, especially the Italian Jesuit Angelo Secchi and the American Edward Pickering. There were two problems: one to confirm that the spectral lines really did represent chemical elements familiar on Earth; second, and more important, to classify stellar spectra in a way that would give meaning to the great variety of stars which are observed and which differ in intrinsic brightness and in colour. The first problem was solved, at least for one chemical element, by William Ramsay, a Scots chemist who held the chair of chemistry at University College, London University. In 1895 he isolated in the laboratory a gas whose presence was indicated in the solar spectrum but was hitherto unknown on Earth; the gas was named helium (Greek *helios,* Sun) and provided useful confirmation of the interpretation of the presence of chemical elements in celestial objects.

The second problem began to be tackled by Secchi and Pickering, who were both concerned with the classification of stellar spectra based on the dark lines and bands which could be observed crossing the continuous coloured background. Many false starts were made, but in the late 1860s Secchi devised what seemed a workable classification, dividing the stars into five classes. It later turned out that the system devised by Edward Pickering's research team at Harvard University observatory, which numbered Annie Cannon among its members, was the more satisfactory. Based on the work started by Henry Draper, it became known as the Draper classification. Though modified by Annie Cannon between 1918 and 1924, it was nevertheless useful from the start, as the work of Hertzsprung and Russell was soon to make clear.

Hertzsprung's main interest was measuring stellar distances, and he believed the Draper classification could help him, for it showed that blue stars were intrinsically brighter than red stars (though there were some exceptions), and he thought that by plotting the intrinsic brightness of stars against their spectra he would have a diagram which would show him the true brightness of other stars, once he knew their spectra. Once he had the intrinsic brightness, he could compare this with their apparent brightness in the sky and compute their distances, since the greater the distance, the dimmer a star of a given intrinsic brightness would appear. Russell's aim was to gain an insight into the stars themselves and he appreciated that the Draper classification could help him, too, because it was clearly a classification by temperature. Details of the spectra made it evident that the blue stars were the hottest and the red stars the coolest. Quite independently, and with another purpose in mind, Russell also plotted intrinsic brightness against spectral class. A link between temperatures and spectral classes, using a modified form of Secchi's classification, had been plotted in 1888 by Norman Lockyer in England, but the Hertzsprung-Russell approach was based on vastly improved evidence and the graph they produced, known ever after as the Hertzsprung-Russell or H-R diagram, was productive of all kinds of useful astrophysical results. (Spectroscopic study of the stars had, since 1890, been called 'astrophysics'.) In Russell's hands the H-R diagram gave a clue to the life cycle or 'evolution' of stars – Darwinian ideas were still making a powerful impact in the early twentieth century – for the diagram showed that most, though not all, stars, lay in a line ranging from bright blue stars to dim red ones. Russell believed these 'main sequence' stars showed that during its lifetime a star evolved from a hot blue body to a cool red one as it expended its energy.

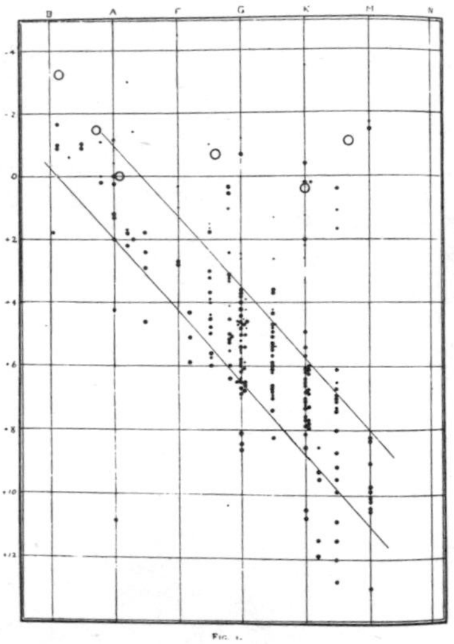

The original H–R diagram published in 1914. The letters along the top indicate various classes of stars from hot blue B-type stars on the left to cool N-type stars on the right. The vertical scale shows true brightness ranging from 'absolute magnitude' 14 (dimmest) at the bottom to -4 (brightest) near the top, a total brightness difference of nearly 16 million times. The relationships indicated on this diagram have proved a fruitful source of research ever since its original publication.

Coupled with the question of stellar evolution was the problem of the source of stellar energy, and between 1916 and 1924 this was hotly debated. In the late nineteenth century Kelvin had suggested that the Sun's energy came from heat generated by its contraction, and during the period in question this view was still strongly advocated by the English astronomer James Jeans. Certainly it had long been recognized that no chemical reaction could generate the energy required, but once Einstein had published his $E = mc^2$ formula, it was clear that here was a process that could produce the energy needed for as long as required, even for the immensely long periods demanded by Darwinian evolution. In 1919 Russell made some suggestions about energy locked up in stellar atoms, but the man whose advocacy turned the scales was Eddington, who in 1920 was strongly supporting atomic energy as a source of power, not only because of Einstein's work but also because of Rutherford's artificial disintegration of nitrogen atoms in the Cavendish Laboratory. As he remarked, 'What is possible in the Cavendish Laboratory may not be too difficult in the Sun'. Eddington also made another and no less pregnant remark, 'If, indeed, sub-atomic energy in the stars is being freely used to maintain their great furnaces, it seems to bring a little nearer to fulfilment our dream of controlling this latent power for the well-being of the human race – or for its suicide'; he died some eight months before the first atom bomb was dropped on Hiroshima.

Later research has shown that nuclear energy is indeed the power source of the Sun and all other stars, and much work on the details has been done by Robert Atkinson, Fritz Houtermans and especially Carl von Weizsäcker, Hans Bethe and George Gamow. Moreover, the whole process of the transport of energy within the stars and the realization that their main constituent is the gas hydrogen has come out of this pioneering work of the 1920s and 30s.

Astronomical research has always been limited by the available means of observation, but since the Scientific Revolution this dependence has assumed far greater importance than it ever had in ancient times. The most significant change was the introduction of the telescope; it brought a new dimension into astronomy, a new precision in measurement and the ability to map the lunar surface and study the planets in detail. As William Herschel and Lord Rosse were to demonstrate so ably, large-aperture reflectors made it possible to probe deeply into space. The twentieth century has seen vast improvements in telescope power, and until well after the end of the World War II, the lead in this side of observational astronomy lay with the United States, due mainly to the efforts of one man, George Ellery Hale. Born in Chicago in 1868, son of a wealthy businessman, Hale studied at the Massachussets Institute of Technology; he soon developed a great interest in solar physics and design and constructed special solar telescopes. But Hale's greatest contributions lay in his conception of grand plans for large space-penetrating telescopes and his extraordinary flair for persuading rich men, and the charitable foundations they established, to provide funds for astronomical equipment in days when astronomy was still a science with comparatively few research workers. His first notable success was the establishment in 1897 of the Yerkes Observatory with what is still the world's largest refractor of 40-inches (1 metre) aperture as its main instrument, but no sooner was this in being than he turned his mind to an even larger aperture. In 1908 a 60-inch (1.5 metre) reflector was set up at Mount Wilson in California, to be followed in 1917 by the Mount Wilson 100-inch (2.5 metre) reflector – the telescope which so moved Alfred Noyes – and then in 1948 by the 200-inch (5 metre) reflector at Palomar, although Hale himself was by then dead. Yet the conception was his and though the 100-inch and 200-inch instruments are no longer the largest telescopes in the world, the work done with them has been the backbone of much twentieth-century astronomy. When it is realized that doubling the aperture of a telescope quadruples its space-penetrating power, the significance of Hale's advocacy of bigger and better instruments becomes clear.

Illustrations pages 361, 455

For Noyes, see page 7.

The value of Hale's large 100-inch reflector emerged in the 1920s with the discovery of the existence of galaxies. At the beginning of the twentieth century it seemed that the general layout of the universe was established, along the broad lines discerned by William Herschel a century before; stars and nebulae were thought to be members of a vast disc-like island in space. But there were some detailed differ-

For Herschel's concept of the lay-out of the universe, see page 369.

ences, for photography had enabled astronomers to observe some objects too dim to be seen by the eye, even in very large instruments, and it was now possible to measure stellar distances in a number of ways besides the method adopted by Bessel in 1839. Indeed, in 1912 Henrietta Leavitt, working at Harvard University observatory, had devised a particularly powerful method of distance measurement for regions of space too remote for the 'direct' Bessel method when she discovered Cepheid variables. (These are stars which vary their brightness regularly over a matter of days, and are named after their prototype in the constellation of Cepheus.) What Leavitt found was that the period of variation depended on the star's intrinsic brightness, so by determining this period such a star's distance could be calculated from its intrinsic and its apparent brightness. This was soon to prove a crucial observation.

During 1919 and 1918 Harlow Shapley, an American astronomer working at Mount Wilson on the shape of the star 'island', and making use of Leavitt's method, found previous estimates wrong and discovered that it was very much larger than previously supposed. On the other hand he classed the spiral nebulae, discovered originally by Rosse and now found to be very numerous, as part of our own disc-like system. Heber Curtis, at Lick Observatory, as well as many other astronomers, believed the spirals to be independent star islands not dissimilar to our own and so envisaged a much vaster universe even than Shapley's. In 1920 a debate between Curtis and Shapley was organized for a meeting of the National Academy of Sciences, but this did not resolve the question; it was not until 1922 and 1924 when, using the 100-inch telescope, Cepheid variables were found in a number of spiral nebulae, that it was proved beyond all doubt that they lay beyond the confines of our own Milky Way system or galaxy. It also became evident that Shapley's size for our galaxy was an overestimate, though his figure was a great improvement over previous assessments.

The leader in obtaining the final proof that the universe was composed of myriads of star islands or galaxies similar to our own, containing either stars, dust and gas or stars alone, was Edwin Hubble. Born in Marshfield, Missouri in 1889, Hubble studied at the University of Chicago where he came under the influence of Hale, who fired him with a love for astronomy. A period at Oxford University reading law followed, but in 1914 all thoughts of a legal career went when he joined the staff of Yerkes Observatory; in 1919, after a period in the US Army during World War I, Hubble accepted Hale's invitation to join the staff at Mount Wilson Observatory. It was here that he not only proved the existence of a universe of galaxies but made a second vital discovery. Analysing the spectra of these distant galaxies taken with the powerful equipment at Mount Wilson, Hubble showed that almost without exception they displayed red shifts; in other words they were all found to be moving away from us and, what is more, with velocities that became greater the

Illustration page 502

further off they were. In fact, the universe of galaxies was seen to be an expanding universe. Such a universe was also shown to be one of the consequences of general relativity, and since all subsequent research has confirmed it, it now lies at the basis of astronomy.

Hale's wisdom in pursuing the ideal of mountain-top observatories which lie above the denser layers of the atmosphere, and constructing large aperture telescopes, was proved beyond doubt by Hubble's epoch-making results, with the result that his philosophy has influenced the siting of all subsequent large telescopes. However, due to developments in optical manufacturing techniques, big apertures are not the rarities they were when the 100-inch and 200-inch instruments were built, and in the 1970s telescopes of 120 inches (3 metres) aperture came into general use.

Yet optical observations with big telescopes are not the only way of examining the universe; one of the most significant advances in astronomy has been appreciation of this fact, and the development of new means for detecting the radiation emitted from celestial sources. The first successful steps in this new direction arose from a chance discovery in 1932-33 by the radio engineer Karl Jansky, working in the United States. Analyzing radio static (the crackling noise heard on short-wave radio transmissions), Jansky noticed that there appeared to be some sources of this static lying in the direction of the Milky Way. Certainly, after Hertz's discovery of radio waves in 1888, attempts had been made to detect radio waves from the Sun, but all had failed for various technical reasons, and Jansky's detection of extra-terrestrial radio emission was the first success. Yet, surprisingly, it did not set the astronomical world aflame; the only man to follow it up was the American amateur astronomer Grote Reber who also detected radiation in 1938.

Illustration page 504

Not until after World War II was the modern technique of radio astronomy finally developed, and then this took place primarily in Australia and England, possibly because these were countries which had no large optical telescopes to occupy their observational astronomers. However it soon became clear that radio astronomy had a serious problem to face, and this was the power of a radio telescope to pick out detail. Optical telescopes could readily separate objects and see detail no larger than a few thousandths of a degree, whereas even a large radio telescope would have a resolution fifty or more times less, merely due to the fact that radio waves are some ten thousand times longer than light waves. To try to ameliorate this to some extent a number of large radio telescopes have been built for special purposes, such as the construction in 1956 of Bernard Lovell's 250-foot (76 metre) diameter steerable 'dish' at Jodrell Bank in England, or the giant 1000-foot (305 metre) 'bowl' reflector at Arecibo, Puerto Rico, built in 1963 and administered by Cornell University. But the question of the resolution of fine detail has only really been overcome by the use of two or more radio telescopes at once, because greater detail is discernible by mixing the radio waves each

Illustration page 505

receives. Pioneers in developing this interferometer technique have been Joseph Pawsey and Bernard Mills in Australia, who achieved much success as early as 1948 and 1953, and Martin Ryle and colleagues at Cambridge University, who, from 1955 onwards, developed special 'aperture synthesis' interferometers which are the equivalent of 'dish' telescopes one mile (1.6 km) and more in diameter. Moreover, the linking in the 1970s of radio telescopes situated on different continents has now made it possible for the radio astronomer to achieve superior resolution to that obtainable with optical telescopes.

Illustration page 505

Radio astronomy has had many notable successes, ranging from the detection of gas in space not previously observable to the discovery of objects formerly unknown to optical astronomers, such as quasars, detected in 1963 by the combined efforts of radio and optical astronomers in England, Australia and the United States, and pulsars, found by Jocelyn Bell and Anthony Hewish of Cambridge University in 1967. Pulsars have been identified as special super-dense stars (neutron stars), but quasars are still something of a mystery, being apparently some kind of galaxy in a primitive state of development.

Probably the most spectacular advances in astronomy during the twentieth century have been those achieved by the actual exploration of space. Travel to the Moon and planets was always a dream of mankind, and the adaption of World War II rocket technology to peaceful use by both Americans and Russians has made the dream a reality. This new era of observation opened in October 1957 with the launch of the Russian 'sputnik', a small body put into orbit round the Earth, and followed within three months by an American artificial satellite. Space technology has developed apace, and has brought a vast amount of detailed information about the solar system available in no other way. In 1959, for the first time in human history, man saw the rear face of the Moon (which always remains turned away from the Earth), and he obtained an actual foothold on the lunar surface in July 1969. Moreover, it has been possible to view the Earth from space and add more information to our knowledge of its properties as a planet and as a physical environment (the subject of geophysics has been a notable development in twentieth-century science) as well as to study closely the planets in ways never before possible because of their distance from us. And, as might be expected, the technology developed to achieve all this, and to transmit colour pictures back over distances of more than 800 million miles (1,300 million kilometres) has found all kinds of applications in everyday life as well as in other branches of scientific study. In astronomy it has at last made possible the examination of the universe in the light of very short wavelengths – short ultraviolet, X-rays and gamma rays – that never reach the Earth's surface. Already in the late 1940s and early 1950s cosmic rays (sub-atomic particles from space) had been detected and these, together with the results obtained by radio astronomers, made clear the importance of this kind of observation; space

Illustrations pages 506, 507, 508, 509

technology was necessary for very short wavelength observations. At last in the 1950s X-rays from the Sun were detected by high-flying rockets, and in 1956 Herbert Friedman of the US Naval Observatory reported evidence of what seemed to be X-ray sources elsewhere in space. Since that time short-wave astronomy has been enthusiastically pursued, providing evidence of all kinds of very energetic explosive sources in space, including information that makes those intense collapsed bodies known as 'black holes' (since they emit no radiation yet intensely distort space) appear to have a real existence.

Twentieth-century astronomy has, then, brought knowledge of a vastly larger, more energetic and more complex universe than ever conceived in earlier centuries. It has also given a hitherto unavailable degree of precision to a subject that has drawn men from the earliest times, the question of the beginning and end of the universe. This subject of cosmology had previously been little more than pure speculation; in 1775, when astronomy was still primarily concerned with the solar system, the German philosopher Immanuel Kant suggested that the Sun and planets condensed out of a disc-like nebula, but this was little more than an attractive idea. Admittedly it was taken up in more detail by Pierre Laplace and is often known as 'Laplace's nebular hypothesis', but still it was very speculative. In the nineteenth century the theory became unpopular – it seemed to be physically unsound – and two other views were suggested. In 1887 Norman Lockyer in London put forward the idea that the solar system (or even perhaps the whole universe) had condensed from meteoric material (i.e. from pieces of rock and metal, samples of which are found to orbit the solar system), while in 1900 two American astronomers, Thomas Chamberlin and Forest Moulton of the University of Chicago, suggested another star had come close to the Sun and drawn off material that formed the planets, an idea taken up in more detail in 1916 by James Jeans. This theory held the field until 1945 when, in the light of new evidence, Carl von Weizsäcker revived the nebular hypothesis in an improved form, and its consequences have been worked on ever since, for in the light of more recent observational evidence, it is now thought to be the most likely possibility. But what of the universe itself?

The expanding universe observed by Hubble, and since confirmed by other evidence, leads one to suppose that the universe began a long time ago – a figure of some twenty thousand million years is calculated at the moment – and in a very concentrated state. The consequence of this can be worked out using general relativity and quantum theory; in 1922 the Russian mathematician Alexsandr Friedman studied the matter and a decade later, in 1933 and 1934, the Belgian astronomer Georges Lemaître put forward a similar idea; the universe began as a concentrated lump, a superatom of material which disintegrated (as all very large atoms do) and then expanded outwards. On the other hand, in 1939, George Gamow, Ralph Alpher and Hans Bethe in the United States considered a 'hot' nuclear explosion of highly concentrated material to have started the expansion. This 'hot big bang' theory is now the favoured view, especially since the announcement in 1965 by Arno Penzias and Robert Wilson of the detection of a background of radio radiation throughout space, the temperature of which is just that to be expected if the hot big bang theory is correct.

None of these theories explains the origin of the superatom from which the universe began. One way out of this difficulty was the 'steady state' theory, proposed by Herman Bondi and Thomas Gold in England in 1948 and followed up in detail by Fred Hoyle. Their suggestion was that matter is continually being created in the universe to replace that which moves away due to expansion; thus to an observer the universe always presents the same general appearance. It is perpetually in a steady state. The view has much to commend it and for a while many cosmologists gave it support, but the discovery by Penzias and Wilson of the background radiation seems inexplicable in terms of any steady state-model. We are left, at the moment, with the hot big bang theory holding the field. But this does not mean it will do so for ever. One of the obvious lessons to be learned from the history of scientific achievement is that no theory survives for ever, and that often when things seem most settled new observations and fresh ideas replace them with new concepts. But, then, this is part of the adventure that is science, part of the slow conquest of the puzzle that is the natural world, part of what Alfred Noyes so elegantly termed the 'long battle for the light' in which man has engaged since the first days of his earliest civilization.

Illustration page 512

Bibliography

Chapter One – **The Origins of Science**
There is little generally available on Egyptian and Mesopotamian science. The most readable account, though a little dated, is in volume 1 of *A History of Science* by George Sarton, Harvard 1952 & Oxford, 1953. Another readable book, though covering more than Egypt and Mesopotamia and also a longer period of time than the present chapter is *Number Words and Number Symbols* by Karl Meninnger, Cambridge, Mass., 1969.

On some specialized fields – mathematics and astronomy – some further reading is available, though much of it is at a high academic level. One not too specialized book is *The Exact Sciences in Antiquity* by Otto Neugebauer, Princeton & Oxford, 1952. There are also four notable articles on the subject in volume XV of the *Dictionary of Scientific Biography*, New York, 1978 (hereinafter referred to as the DSB). These are: 'The Mathematics of Ancient Egypt' by R. J. Gillings; 'Egyptian Astronomy, Astrology and Calendrical Reckoning' by Richard Parker; 'Mathematics and Astronomy in Mesopotamia' by B. L. van der Waerden; and also 'Man and Nature in Mesopotamian Civilization' by A. Leo Oppenheim. The volume further contains an article 'Maya Numeration, Computation and Calendrical Astronomy' by Floyd Lansbury.

Chapter Two – **Greek Science**
A readable account of Greek science is to be found in volumes 1 & 2 of *A History of Science* by George Sarton, Harvard, 1952 & Oxford, 1953 (vol.1), and Harvard & Oxford, 1959 (vol.2).

Rather more specialized and technical is *Early Physics and Astronomy* by O. Pedersen & M. Pihl, London & New York, 1974. Still more specialized is *A Manual of Greek Mathematics* by Thomas L. Heath, New York & London, 1974 (reprint).

Biographical articles will, of course, be found in the DSB.

Chapter Three – **Chinese Science**
The most complete account of Chinese science is given in *Science and Civilisation in China* by Joseph Needham, Cambridge, 1954 (vol.1) onwards; the work is still in progress. This is a large and specialized work, and for more general readers there is *The Shorter Science and Civilisation in China* by Colin A. Ronan. This is an abridgement of the larger work, re-written in simpler terms. Volume 1 appeared in 1978 and vol.2 in 1980, both published at Cambridge. This work is also in progress.

Some illuminating short essays on the subject will also be found in *Clerks and Craftsmen in China and the West* by Joseph Needham, Cambridge, 1970, while the articles on Li Shih-Chen, Shen Kua and Wang Hsi-Shan by Nathan Sivin in the DSB are worth reading also.

Chapter Four – **Hindu and Indian Science**
The one standard work on Indian science is *A Concise History of Science in India* edited by D. M. Bose, S. N. Sen & B. V. Subbarayappa, New Delhi, 1971. However, there is also an informative article on the 'History of Mathematical Astronomy in India' by David Pingree in volume XV of the DSB, though this is rather specialized.

Chapter Five – **Arabian Science**
The most popular account of Islamic science is *Islamic Science – An Illustrated Study* by Seyyed H. Nasr, London, 1976. Again, there are some biographical articles in the DSB but no biographies in English of any Islamic scientists.

Chapter Six – **Roman and Medieval Science**
For a general picture of medieval science in more detail than can be given here, see *Augustine to Galileo* by A. C. Crombie, London, 1952. More specialized are *Robert Grosseteste and the Origins of Experimental Science* by A. C. Crombie, Oxford, 1952 and also *Physical Science in the Middle Ages* by Edward Grant, New York & London, 1971.

Chapter Seven – **From Renaissance to Scientific Revolution**
A general book on some Renaissance scientists is *Six Wings* by George Sarton, Indiana, 1957 & London, 1958. Among biographies are: *Copernicus, The Founder of Modern Astronomy* by Angus Armitage, London, 1938 and, rather more advanced, *Andreas Vesalius of Brussels, 1514-1564*, by C. D. O'Malley, Berkeley, 1964 and *Paracelsus* by Walter Pagel, London, 1958.

Chapter Eight – **The Seventeenth and Eighteenth Centuries**

Biographies are very numerous for this period, and only a selection is given below.

The Life and Times of Tycho Brahe, by J. A. Gade, Princeton, 1947.
Tycho Brahe by J. L. E. Dreyer, New York, 1963 (reprint).
Galileo by Colin A. Ronan, New York & London, 1974.
Galileo At Work, His Scientific Biography, by Stillman Drake, Chicago & London, 1978.
Edmond Halley, Genius in Eclipse, by Colin A. Ronan, New York, 1969 & London, 1970.
The Life of William Harvey, by Geoffrey Keynes, Oxford, 1966.
William Harvey, the Man, the Physician, the Scientist by Kenneth D. Keele, London, 1965.
The Herschel Chronicle by Constance A. Lubbock, Cambridge, 1933.
William Herschel, by Angus Armitage, New York & London, 1962.
William Herschel, Pioneer of Sidereal Astronomy, by Michael W. Hoskin, New York & London, 1959.
Robert Hooke by Margaret Espinasse, London, 1956.
Johannes Kepler by Angus Armitage, London, 1966.
Kepler by Max Caspar, trans. Doris Hellman, New York & London, 1959.
Measuring the Invisible World, The Life and Works of Antoni van Leeuwenhoek, by A. Schierbeek, New York & London, 1959.
The Compleat Naturalist. A Life of Linnaeus, by Wilfrid Blunt, London, 1971.
Never at Rest. A Biography of Isaac Newton, by Richard S. Westfall, Cambridge & New York, 1980.
Joseph Priestley: Adventurer in Science and Champion of Truth, by F. W. Gibbs, London, 1965.

Chapter Nine – **Science in the Nineteenth Century**

Jac. Berzelius, His Life and Work, by J. Erik Jorpes, Stockholm, 1966.
Georges Cuvier Zoologist, by W. Coleman, Cambridge, Mass., 1964.
John Dalton and the Atom, by Frank Greenaway, London, 1966.
John Dalton and the Atomic Theory, E. Patterson, New York, 1960.
Charles Darwin. A Scientific Biography, by Gavin de Beer, New York, 1965.
Humphry Davy, by Harold Hartley, London, 1967.
Michael Faraday, by L. Pearce Williams, New York & London, 1965.
Explorer of the Universe, A Biography of George Ellery Hale, by Helen Wright, New York, 1966.
T. H. Huxley: Scientist, Humanist and Educator, by Cyril Bibby, London, 1959.
Lord Kelvin, Physicist, Mathematician, Engineer, by A. P. Young, London, 1948.
Clerk Maxwell and Modern Science, by C. Domb, London, 1963.
Louis Pasteur: A Great Life in Brief, by René Vallery-Radot, trans. Alfred Joseph, New York, 1958.
William Henry Fox Talbot, by A. J. P. Arnold, London, 1977.
Rudolf Virchow, Doctor, Statesman and Anthropologist, by Erwin H. Ackerknecht, Madison, Wiss., 1953.

Chapter Ten – **Twentieth-Century Science**

Niels Bohr, His Life and Work as Seen by His Friends and Colleagues, ed. S. Rozental, Amsterdam, 1967.
Madame Curie, by Eve Curie, trans. Vincent Sheean, New York & London, 1939.
Arthur Stanley Eddington, by A. Vibert Douglas, New York & London, 1956.
Einstein. A Centenary Volume, ed. A. P. French, London, 1979.
Gowland Hopkins, by Ernest Baldwin, London, 1961.
H. A. Lorentz. Impressions of His Life and Work, ed. G. L. de Haas-Lorentz, Amsterdam, 1957.
Life of Mendel, by H. Iltis, New York, 1966.
Dmitry Ivanovich Medeleev, by N. A. Figurovsky, Moscow, 1961.
The Master of Light (A biography of A. A. Michelson), by D. Michelson Livingston, New York, 1973.
Ivan Petrovitch Pavlov, Work, by E. A. Asratyan, Moscow, 1974.
Rutherford and the Nature of the Atom, by E. N. da C. Andrade, London, 1964.
Charles Scott Sherrington, by E. G. T. Liddell, London, 1966.
J. J. Thomson and the Cavendish Laboratory in His Day, by G. P. Thomson, New York, 1965.

Acknowledgments

The illustrations at the bottom left and bottom right of page 297 are reproduced by gracious permission of Her Majesty the Queen.

The illustration at the top of page 243 is reproduced by courtesy of the Dean and Chapter of Hereford Cathedral, that at the bottom of page 292 of the Governing Body, Christ Church, Oxford, that at the bottom of page 244 of the Master and Fellows of Gonville and Caius College Cambridge and that at the top of page 244 of the Master and Fellows of Magdalene College, Cambridge.

The author would like to thank Mr Storm Dunlop for compiling the index and express his warmest appreciation for the help and co-operation of his patient editor, Mr Peter Furtado.

Photographs

Aberdeen City Library 416 top; Aerofilms, Boreham Wood 26 bottom, 48 top; Graphische Sammlung Albertina, Vienna 296; Alinari, Florence 293 top right; American Institute of Physics, Niels Bohr Library 501 top left; American Museum of Natural History, New York 58 bottom; Archaeological Survey of India, Delhi 200 top; Archives Photographiques, Paris 25 bottom right; Ashmolean Museum, Oxford 289 bottom right; B.B.C. Hulton Picture Library, London 364 bottom, 412 top; Bell Laboratories 512 bottom; Biblioteca Ambrosiana, Milan 232 top; Biblioteca Estense, Modena 85 top centre, 85 top right; Biblioteca Medicea Laurenziana, Florence 261 top; Biblioteca Nazionale Marciana, Venice 262 bottom, 289 top left; Bibliothèque Nationale, Paris 262 top; Bodleian Library, Oxford 221 top left; J. Bottin, Paris 48 bottom right; British Library, London 85 top left, 87 top right, 87 bottom, 88 top, 138 centre, 138 bottom, 209 bottom right, 221 top right, 243 bottom, 263 top left, 263 bottom, 264 bottom, 291 top, 295 bottom right, 298 top right, 298 bottom right, 310 top, 304, 357, 363 bottom, 401 top, 406 top, 450 top; British Museum, London 28 top, 28 bottom, 45 top, 45 bottom, 46 top left, 46 bottom, 48 bottom left, 289 top right, 361 bottom; Cavendish Laboratory, University of Cambridge 463 bottom; CERN-Science Photo Library 497 top, 497 bottom; Cambridge University Press 157 top, 158 top, 159 top, 178 bottom left, 178 bottom right; Central Office of Information, London 505 top left; John Combridge, Ilford 140 top; Danske Kunstindustrimuseum, Copenhagen 160 top right; Department of Archaeology, Pakistan 197 top, 198 top 3 rows; Deutsche Fotothek Dresden 363 top right, 463 top; Edinburgh University Library 242 centre; Arpad Elfer 197 bottom; Mary Evans Picture Library – Sigmund Freud 499 top right; Ferranti – Science Photo Library 510 centre right; Fogg Art Museum, Harvard University, Cambridge, Massachusetts 179 bottom; G.E.C. Computers, Boreham Wood 511 top, 511 bottom; Germanisches Nationalmuseum, Nuremberg 290; Photographie Giraudon, Paris 75 top, 242 top left, 242 top right, 261 bottom, 291 bottom, 415 bottom; Hamlyn Group Picture Library 25 left, 25 top right, 27 top left, 57 bottom, 75 bottom, 76 top left, 85 bottom right, 87 top left, 140 bottom, 198 4th row, 222 bottom, 292 top, 293 bottom, 295 top, 354 top left, 355 top left, 362 top, 412 bottom right, 413 top, 413 bottom, 415 top, 460 top left, 460 top right; Hamlyn Group – S. M. Carr 454 top right, 454 centre right; Hamlyn Group – Graham Portlock 294 top left, 294 top right, 294 right centre, 294 bottom, 295 bottom left, 301 bottom, 302, 353 bottom left, 353 bottom right, 355 top right, 358 bottom left, 367 top, 368 top, 402 top, 403 top, 404 top, 404 centre left, 404 bottom right, 406 bottom, 407, 408 top left, 408 top right, 408 bottom, 409 top, 409 bottom, 411 bottom left, 414 top left, 414 top right, 414 bottom; Robert Harding Picture Library, London 160 top left; Hereford Cathedral Library 243 top; Hirmer Fotoarchiv, Munich 27 top right, 27 bottom, 46 top right; Michael Holford, Loughton 76 bottom; Indian Museum, Calcutta 199 top; Istanbul Üniversitesi Kütüphanesi 219 bottom, 232 bottom; Institut International de Physique et de Chemique Solvay, Brussels 500 top; Institute of Oceanographic Sciences, Godalming 512 centre; V. K. Jain 200 bottom; Kitt Peak National Observatory, Tucson, Arizona 504 top; Professor A. K. Kleinschmidt 498 bottom; Librairie Hachette, Paris 303 top; Lick Observatory, University of California, Pasadena 502 top left; Mansell Collection, London 461 bottom; Mansell – Alinari 76 top right, 293 top left; Bildarchiv Foto Marburg 179 top; Marconi, Chelmsford 510 top right; Metropolitan Museum of Art, New York 180 top, 219 top right, 220; Miln Marsters Group, Chester 460 bottom; Mullard Radio Astronomy Observatory, Cavendish Laboratory, University of Cambridge 505 top right; Musée des Antiquités

Nationales, Saint Germain en Laye 26 top left; Musées Nationaux, Paris 47, 356 bottom left; Museo Nazionale della Scienza e della Tecnica, Milan 401 bottom; Museum of Fine Arts, Boston, Massachusetts 158 bottom, 219 top left; Museum of the History of Science, Oxford 209 top; NASA, Washington DC 506 centre, 507 top, 508 top, 508 bottom, 509; National Gallery, London 300 top; National Maritime Museum, London 360 top, 360 bottom; National Museum of Taiwan 180 bottom; National Physical Laboratory, Teddington 506 top; National Portrait Gallery, London 356 bottom right, 358 top right, 361 top, 410 top, 412 bottom left, 449 top; Novosti Press Agency, London 210 right, 506 bottom; Josephine Powell, Rome 86 top, 198 bottom row, 209 bottom left; Professor Derek de Solla Price, Yale University, New Haven, Connecticut 88 bottom; Rijksmuseum, Amsterdam 178 top; Pedro Rojas 57 top; Ann Ronan Picture Library, Taunton 26 top right, 86 bottom left, 86 bottom right, 264 top, 297 top, 298 top left, 298 bottom left, 299, 300 bottom right, 303 bottom, 353 top, 354 top right, 354 centre right, 356 top, 356 centre, 358 top left, 364 centre, 365, 366 top, 366 bottom, 367 bottom, 368 bottom, 402 bottom, 403 bottom, 404 bottom left, 405 top right, 405 centre, 405 bottom, 410 bottom, 411 top, 411 bottom right, 449 bottom left, 450 bottom, 451 top, 451 bottom, 453 bottom left, 454 top left, 454 bottom, 455 top, 455 bottom, 457 top left, 457 top right, 457 bottom, 458 top, 458 bottom right, 461 top left, 461 top right, 503 top, 504 bottom; Royal Astronomical Society, London 501 bottom, 502 top right, 505 bottom, 507 bottom; Royal Institution, London 452 top, 452 bottom left, 452 bottom right; Royal Observatory, Edinburgh 502 bottom, 503 centre; Royal Ontario Museum, Toronto 137 top; Royal Society, London 364 top, 498 top left; Sächsisches Landesbibliothek, Dresden 58 top; Science Museum, London 138 top, 159 bottom, 177 top, 177 bottom, 242 bottom, 263 top right, 289 bottom left, 300 bottom left, 354 bottom right, 355 bottom, 358 bottom right, 359 top, 359 bottom, 363 top left, 416 bottom left, 416 bottom right, 449 bottom right, 453 top left, 453 right, 455 centre, 458 bottom centre, 459, 462 bottom, 464 top, 499 top left, 510 top left, 510 bottom; Seattle Art Museum, Washington 157 bottom, Dr. Lee D. Simon Science Photo Library 498 top right; Süleymaniye Kütüphanesi, Istanbul 221 bottom, 222 top; Topkapi Sarayi Müzesi, Istanbul 231 top, 231 bottom; K. P. and S. B. Tritton, Edinburgh 139 top, 139 centre, 139 bottom; U.S.I.S. London 512 top; Universitäts Bibliothek, Basel, 85 bottom left, 262 centre; University of Arizona and the Smithsonian Institution 503 bottom; University of Hong Kong 160 bottom; University of Manchester, Physics Department, Schuster Laboratory 462 top; University of Maryland, College Park, Maryland 501 top right; University of Teheran Central Library and Documentation Centre 210 left; Roger-Viollet, Paris 137 bottom, 199 bottom, 362 bottom, 405 top left, 500 centre; Vorderasiatisches Museum, Berlin 45 centre; David Wade, London 510 centre left; Wellcome Institute for History of Medicine, London 241; Professor M. F. Wilkins, Biophysics Department, King's College, London 499 bottom; Wide World 500 bottom; Yerkes Observatory, Williams Bay, Wisconsin 456.

The photographs taken by Graham Portlock for the Hamlyn Group are all of illustrations from books in the Science Museum Library and are reproduced by courtesy of the Keeper of the Library.

Index

References to the captions to the illustrations are indicated in *italic* type.